AF251966

DISCOVERIES IN PLANT BIOLOGY

VOLUME III

DISCOVERIES IN PLANT BIOLOGY

VOLUME III

Editors

Shain-Dow Kung

Hong Kong University of Science and Technology

Shang-Fa Yang

Academia Sinica, Taipei

World Scientific
Singapore • New Jersey • London • Hong Kong

Published by

World Scientific Publishing Co. Pte. Ltd.

P O Box 128, Farrer Road, Singapore 912805

USA office: Suite 1B, 1060 Main Street, River Edge, NJ 07661

UK office: 57 Shelton Street, Covent Garden, London WC2H 9HE

British Library Cataloguing-in-Publication Data
A catalogue record for this book is available from the British Library.

Acknowledgements

Figure 1 on p. 7 is reprinted from *Lectures on the Physiology of Plants*, J. von Sachs © 1887, with permission of Clarendon Press, Oxford.

Figure 2 on p. 7 is reprinted from *Mineral Nutrition of Plants: Principles and Perspectives*, E. Epstein © 1972, with permission of John Wiley & Sons, New York.

Figure 6 on p. 112 is reprinted from "Structural studies on porphobilinogen deaminase", in *The Biosynthesis of the Tetrapyrrole Pigments*, Ciba Foundation Symposium 180, eds. D. J. Chadwick and K. Ackrill © 1994, with permission of John Wiley & Sons, Chichester.

pp. 327–345 is reprinted from *Photosynthesis Research* **44**, 3–22 © 1995, with permission of Kluwer Academic Publishers, The Netherlands.

Figure 3 on p. 417 is reprinted from *Mineral Nutrition of Plants: Principles and Perspectives*, E. Epstein © 1972, with permission of John Wiley & Sons, New York.

ISBN 981-02-3882-7

Printed in Singapore.

PREFACE

It is our pleasure to edit the series, *Discoveries in Plant Biology*, a collection of comprehensive and informative chapters contributed by renowned plant biologists many of whom were the researchers who made the breakthrough discoveries themselves.

In recent decades, the advances in molecular biology have led to rapid development in plant biology research. We noted that while there are many books that describe recent advances in plant biology, volumes that specifically record the history of landmark discoveries are scarce. As scientific progress hinges on the continual discovery and extension of previous discoveries, we feel that there ought to be volumes of the latter kind. When we approached scientists who made the original discoveries in plant biology or who are knowledgeable about such discoveries for contribution in different specialized fields, we feel so privileged to have the excellent response. Also given the wide span of plant biology, we decided to have a series of five volumes. The first two volumes were published in 1998 and this third volume contains 24 chapters. Volume 4 is in the editing stage and the final one is in preparation.

The main aim of this series is to summarize the history of how classic discoveries in plant biology were made and how they have served as the foundation for many subsequent new discoveries. This will enhance our ability to advance our knowledge based on the advancements made previously by others.

The series covers most subjects in plant biology, from Physiology to Pathology, from Cell Biology to Genetics, and from Biochemistry to Molecular Biology. Each of its chapters presents an overview in a specific research area. This helps readers to understand better the background, development, current status and future direction of each of these specific areas. It is therefore a helpful reference for researchers, teachers as well as students. An underlying purpose of this collection is to encourage readers to ask questions such as: *How do scientists make their discoveries? What factors* vi

are involved in the scientific discoveries? Hopefully, this will inspire young researchers and students in designing and conducting their experiments. Moreover, it can also serve as a text for postgraduate and advanced undergraduate courses in the discipline of plant biology.

This series may also provide authors, as well as those readers who are closely associated with the discoveries, with the pleasure to relive the early years of hardwork, frustrations and achievements that he went through with his research team. We feel encouraged that many colleagues informed us about how much they enjoyed writing and/or reading the chapters.

We would like to express our gratitude to all those contributors who have, through the chapters in this series, shared their experiences and knowledge with us and with our readers. We feel particularly touched by the fact that many of the dear contributors undertook to write the chapters despite their other heavy commitments, and some, who are retired, managed to complete the great task even with no or very limited secretarial and technical support.

We also wish to thank the publisher, World Scientific Publishing Co. Ltd., for their support. Last but not the least, we express our appreciation for the hardwork of Betty Law and Mecell Lee, who helped with all the behind-the-scene preparations.

Shain-Dow Kung, Hong Kong University of Science and Technology
Shang-Fa Yang, Academia Sinica, Taipei

CONTRIBUTORS

Adams, Douglas O.
Department of Viticulture and Enology
University of California, Davis, USA
Email: doadams@ucdavis.edu

Bloom, Arnold J.
Professor and Chair, Department of Vegetable Crops
University of California, Davis, USA
Fax: +1- 530-752-9659
Email: ajbloom@ucdavis.edu

Chang, Pi-Fang Linda
Assistant Professor, Department of Plant Pathology
National Chung Hsing University, Taichung
Fax: +886-4-286-0442
Email: pfchang@nchu.edu.tw

Delmer, Deborah P.
Professor and Chair, Section of Plant Biology
University of California, Davis, USA
Fax: +1- 530-752-5410
Email: dpdelmer@ucdavis.edu

Epstein, Emanuel
Research Professor
Department of Land, Air and Water Resources – Soils and Biogeochemistry
University of California, Davis, USA
Email: eqepstein@ucdavis.edu

French, Alfred D.
Research Chemist and Research Leader, Cotton Fiber Quality Research Unit
Southern Regional Research Center, New Orleans, USA
Email: afrench@nola.srrc.usda.gov

Gray, John C.
Professorof Plant Molecular Biology, Department of Plant Sciences
University of Cambridge, UK
Fax: +44-1223-333953
Tel: + 44-1223-333925
Email: jcg2@mole.bio.cam.ac.uk

Guo, Xiangrong
Ph.D., Associate Professor of Plant Molecular Genetics, Institute of Genetics
Chinese Academy of Sciences, Beijing
Email: guoxr@public.c.s.hn.cn

Hodges, Thomas K.
J. C. Arthur Distinguished Professor
Department of Botany and Plant Pathology, Purdue University, USA
Email: hodges@btny.purdue.edu

Hu, Han
Ph.D., Professor of Plant Somatic Genetics, Institute of Genetics
Chinese Academy of Sciences, Beijing
Fax: +86-10-64873482
Email: huhan@igtp.ac.cn

Huang, Anthony H. C.
Professor, Department of Botany and Plant Sciences
University of California, Riverside, USA
Fax: +1-909-787-4437
Email: Anthony.huang@ucr.edu

Kamisaka, Seiichiro
Professor of Plant Physiology, Department of Biology
Osaka City University, Japan

Koncz, Csaba
Max-Planck-Institut für Züchtungsforschung, Germany
Fax: +49-221-5062-213
Email: koncz@mpiz-koeln.mpg.de

Kung, Shain-Dow
Professor, Department of Biology, Hong Kong University of Science and Technology
Fax: +852-2358-1559
Email: aakung@ust.hk

Leonard, Robert T.
Professor and Head, Department of Plant Sciences
The University of Arizona, USA
Email: plshead@ag.arizona.edu

Lichtenthaler, Hartmut K. (Professor Dr. Dr. h.c. mult.)
Botanical Institute (Plant Physiology and Biochemistry)
University of Karlsruhe, Germany
Fax: +49-721-608-4874
Tel: +49-721-608-3833
Email: hartmut.lichtenthaler @bio-geo.uni-karlsruhe.de

Lin, Chu-Yung
Professor, Department of Plant Pathology
National Taiwan University, Taipei
Fax: +886-2-2363-8598
Email: chuyung@ccms.ntu.edu.tw

Masuda, Yoshio
Emeritus Professor, Department of Biology
Osaka City University, Japan
Correspondence: 2-37 Matsugamoto-cho, Ibaraki-shi, Osaka 567-0033, Japan

Ogawa, Masahiro
Doctor of Agriculture, Professor of Department of Life Science
Yamaguchi Prefectural University, Japan
Fax: +81-839-28-2251
Email: ogawa@wsl.yamaguchi-pu.ac.jp

Okita, Thomas W.
Institute of Biological Chemistry
Washington State University, USA

Preiss, Jack
Professor, Department of Biochemistry
Michigan State University, USA
Fax: +1- 517-353-9334
Tel: +1-517-353-3137
Email: preiss@pilot.msu.edu

Saltveit, Mikal E.
Professor, Mann Laboratory, Department of Vegetable Crops
University of California, Davis, USA
Fax: +1- 530-752-4554
Tel: +1-530-752-1815
Email: mesaltveit@ucdavis.edu

Schell, Jeff
Max-Planck-Institut für Züchtungsforschung, Germany
Fax: +49-331-5062-213
Email: schell@mpiz-koeln.mpg.de

Shen, San Chiun
Professor of Molecular Genetics
Shanghai Institute of Plant Physiology, Academia Sinica
Fax: +86-21-64042385
Email: shen@iris.sipp.ac.cn

Sivak, Mirta N.
Research Professor, Department of Biochemistry
Michigan State University, USA
Fax: +1-517-353-9334
Tel: +1-517-353-3247
Email: sivak@pilot.msu.edu

Su, Jong-Ching
Professor of Biochemistry, Emeritus, Department of Agricultural Chemistry
National Taiwan University, Taipei
Fax: +886-2-27320643(home)
Email: jcs@ccms.ntu.edu.tw

Taylor, Alison R.
Leverhulme Research Fellow, Marine Biological Association
The Laboratory, Citadel Hill, Plymouth, UK
Fax: +44 (0) 1752-633102
Tel: +44 (0) 1752-633256
Email: arta@pml.ac.uk

Thornber, J. Philip (gone but not forgotten)
Department of Biology
University of California, Los Angeles, USA

Vierstra, Richard D.
Professor, Department of Horticulture
and theCellular and Molecular Biology Program
University of Wisconsin-Madison, USA
Fax: +1-608-262-4743
Email: vierstra@facstaff.wisc.edu

von Wettstein, Diter
R. A. Nilan Distinguished Professor
Departments of Crop and Soil Sciences & Genetics and Cell Biology
Washington State University, USA
Email: diter@wsu.edu

Watson, John C.
Department of Biology
Indiana University – Purdue University at Indianapolis, USA
Email: jcwatso@iupui.edu

Youle, Richard J.
Chief, Department of Surgical Neurology
National Institute of Health, USA
Email: youle@helix.nih.gov

CONTENTS

Chapter 1

The Discovery of the Essential Elements

Emanuel Epstein
Department of Land, Air and Water Resources -
 Soils and Biogeochemistry
University of California, Davis, CA 95616-8627, USA

ABSTRACT

An account of the development of knowledge of the elements essential for the growth and development of higher plants is given. It was only since the end of the 18th century that the very concept of chemical elements and chemical reactions was understood. It took a further half-century before plant scientists became convinced that chemical elements go into the make-up of plants, and that most of these are mineral nutrients absorbed by roots. By now, in addition to carbon, hydrogen and oxygen, the number of elements known to be essential for higher green plants is 14; three more are essential for at least some plants, or under certain conditions. The discovery of the essentiality of elements depended upon refinements of the solution culture technique, the only method of withholding an element required by plants in very small amounts to an extent where its deficiency becomes apparent.

Introduction

The discovery of the elements essential for plants depended upon the recognition that plants are composed of chemical elements. That in turn could not occur until there was a basic understanding of what chemical elements and chemical reactions are. A number of elements such as carbon, gold, iron, and zinc had been known since antiquity. But until the end of the 18th century, there existed no rationale that would unify, and lead to an understanding of, the considerable body of empirical knowledge that had accumulated over the ages. That conceptualization of what is now understood to constitute chemistry was accomplished by a Frenchman, Antoine Laurent Lavoisier (1743–1794). He disproved such ancient notions as that water could be transmitted into "earth" (solid matter) by repeated distillations. He did away with a will-ò-the-wisp called "phlogiston", an imaginary substance that was invoked whenever it was convenient to explain chemical processes that could not be understood in terms of the then prevailing knowledge of chemistry.

It was Lavoisier who swept away, once and for all, a whole host of invalid ideas and established the basic tenets of chemistry as we now know it. The first comprehensive exposition of this revolutionary view of chemistry was Lavoisier's "Traité Élémentaire de Chimie, Présenté dans un Ordre Nouveau et d'après les Découvertes Modernes; avec Figures": "Treatise on the Elements of Chemistry, Presented in a Novel Order and according to the Modern Discoveries; with Illustrations", published in Paris in 1789. Eventually the "Oeuvres de Lavoisier" were published in six volumes in 1862–1893 (Lavoisier, 1862–1893). Kerr (1790) published an English translation of Lavoisier's "Traité Élémentaire" which, with translations into several other languages, caused his new chemistry to be widely accepted by the end of the 18th century.

The following telling sentences from Lavoisier's treatise are taken from Browne (1944). "We may lay it down as an incontestible axiom, that, in all the operations of art and nature, nothing is created; an equal quantity of matter exists both before and after the experiment; the quality and quantity of the elements remain precisely the same; and nothing takes place beyond changes and modifications in the combinations of these elements. Upon this principle, the whole art of performing chemical experiments depends: we must always suppose an exact equality between the elements of the body examined, and those of the products of its analysis."

Lavoisier's chemical revolution took place in a country, France, and at a time, toward the end of the 18th century, that were revolutionary in politics as well. He made his living as a member of the "establishment" and in the turmoil of the French revolution he was condemned to the guillotine and killed by it on May 8, 1794. "They needed only a moment to take off his head and more than a hundred years perhaps will elapse before another like it will be produced" (a contemporary, quoted in Browne, 1944). An excellent account of Lavoisier's life and magnificent accomplishments has been given by Poirier (1996).

Discovery of Oxygen, Hydrogen, Carbon and Nitrogen

The four elements oxygen, hydrogen, carbon and nitrogen are distinguished from all other elements essential for plants in several ways: they make up the bulk of plant matter, are the elements common to all proteins, and in their entry into plants, their exit from them, or both, are gaseous, or in the case of nitrogen, the ultimate reservoir of it is gaseous. The discovery and characterization of these elements therefore caused exceptional difficulties for early chemists, and it was not till the end of the 18th century that these elements were recognized as such, as we shall see.

Oxygen was discovered by Joseph Priestley (1733–1804). He was an English clergyman, subsequently a librarian to Lord Shelbourne, and a self-taught chemist. He prepared oxygen by heating mercuric oxide. At that time, he did not understand that he had discovered a chemical element, but concluded that it was an "air" in which a candle burned "with a remarkably vigorous flame", and that a mouse lived three times as long in this "air" as in ordinary air (Browne, 1944; Magner, 1994). Priestley conducted experiments with plants as well, but remained an adherent of the "phlogiston" theory, which greatly hampered him in soundly interpreting his results. It was Lavoisier who correctly identified oxygen as an element, and named it.

Priestley, like Lavoisier, was a victim of persecution. He was the pastor of a Unitarian church, which dissented from the Church of England. That, and his liberal views in politics as well, in which he was aided and abetted by Benjamin Franklin, led to the incitement of a mob which burned his church, his house, and his laboratory. He emigrated to America in 1794 and built a new home and laboratory in Pennsylvania (Browne, 1944). Thus, Priestley became the first of a great number of European scientists who moved to America because they were persecuted in their native lands.

Hydrogen was discovered by another Englishman, Henry Cavendish (1731–1810), "a man said to be as wealthy as he was eccentric" (Magner, 1994). He dissolved zinc, iron, or tin in "vitriolic" (sulfuric) acid. The gas so generated he called "inflammable air." Cavendish published his findings in 1766 in the Philosophical Transactions of the Royal Society. Like Priestley, however, Cavendish was unable to free himself from the intellectual fetters of the "phlogiston" fallacy and was therefore limited in understanding the meaning of his discoveries. That understanding was provided by Lavoisier, who also coined the name for this element, hydrogen.

Carbon was known from time immemorial. As for science, as early as the beginning of the 17th century, a Belgian physician and chemist, Jan Baptista van Helmont (1577–1644), had discovered that charcoal upon burning gives off a gas (a term he coined, on the basis of the Greek "chaos"), which he called "gas sylvestre," later to be known as "fixed air." It was left to Lavoisier, however, to understand this gas to be a compound of carbon and oxygen, our carbon dioxide.

The element nitrogen was discovered in 1772 by Rutherford, then a medical student in Edinburgh, Scotland. He called it "mephitic air." Virtually simultaneous discoveries of this element were made by Joseph Priestley and Henry Cavendish, and the Swedish chemist, Carl Wilhelm Scheele (Weeks, 1956; Partington, 1960).

Rutherford showed that a small volume of this "air" breathed by a small animal would not support life, although the carbon dioxide (not yet known by that name) was absorbed by alkali. Thus, he reasoned, there was another gas incapable of supporting life. Lavoisier clearly distinguished between nitrogen and carbon dioxide, and realized that nitrogen played no role in respiration (Browne, 1944). It was he who understood nitrogen to be a chemical element, which he called azote, still the word for it in French.

Because of the fact that oxygen, nitrogen and hydrogen are gases when not parts of solid or liquid compounds with other elements, their discovery was exceptionally difficult at a time when there was no valid rationale of chemistry. But once Lavoisier had provided such a rationale he was able, in his Traité élémentaire, to list no less than 31 elements, including nearly all those now known to be essential for plants.

Discovery of the Essentiality of Oxygen, Hydrogen, Carbon and Nitrogen

The first to apply Lavoisier's new chemistry to the study of plant nutrition was a Swiss, Théodore de Saussure (1767–1845). He published a landmark book in 1804: "Recherches chimiques sur la Végétation." This is the first report of numerous quantitative experiments and measurements concerning the science of plant nutrition. Many of his experiments and reports strike the reader as essentially modern in approach and presentation.

De Saussure established that the plant takes up the elements of water concurrently with carbon, and thus may be credited with the discovery of the essentiality of oxygen, hydrogen and carbon. He understood, further, that the source of the carbon is not the soil, but the carbon dioxide of the atmosphere. De Saussure also found that plants failed to grow unless their roots absorbed nitrogen, thus establishing its essentiality. (There was, in his time, a widely held view that plants absorbed nitrogen from the air. For a discussion of the discovery of dinitrogen fixation, see the chapter by R. H. Burris in the present work.) In experiments with an early version of solution culture, de Saussure established that plants absorb elements selectively, and that the amounts taken up depend on both the abundance of the element in the substrate and the plant species.

It is necessary here to draw attention to the work of several investigators of that period who, while not discoverers of the essentiality of any element, were very important contributors to the developing science of plant nutrition. The Frenchman J. B. Boussingault (1802–1887) conducted quantitative experiments on the absorption of nitrogen by plants and firmly rejected the idea that plants absorb it from the air. He also affirmed that humus is unnecessary for the growth of plants: the roots of plants are engaged in inorganic plant nutrition, to use modern terminology. For an account of this extraordinary, versatile man, see McCosh (1984).

The German scientist, Justus von Liebig (1803–1873) published in 1840 a book, *Organic Chemistry in Its Applications to Agriculture and Physiology* (Liebig, 1840). Liebig was an assertive, belligerent (and by no means fair-minded) man. His book, which went through many editions and was translated into several languages, did much to spread the word on the "mineral theory of fertilizers." For an account of his accomplishments and failings see Moulton (1942).

Finally, J. B. Lawes and J. H. Gilbert initiated field experiments at the now famous Rothamsted Experiment Station in England which they founded in 1843. The experiments, which are being continued to the present day, produced convincing evidence that mineral nutrients added to the field plots, i.e., chemical fertilizers, were needed to sustain the fertility of soil cropped year after year (Reed, 1942). These experiments were done in the same quantitative manner introduced by Boussingault and de Saussure.

Sachs (1906), himself a major contributor to plant biology later on, has remarked upon the fact that the extraordinary advances made by de Saussure were followed by several decades in the early 1800s during which little progress was made in this science. As late as 1800, the Berlin Academy of Sciences announced a prize for the best answer to the following question concerning the inorganic components of plants (Schneidewind, 1917): Do they enter [plants] as they occur, or are they generated through the activity of plant organs? The prize winning answer to this question was: the ash components of plants are generated by the vital process. Against this background, de Saussure's 1804 work can be recognized as the breakthrough it was.

The lag phase that Sachs had commented upon ended after the early decades of the 19th century and a series of discoveries added to the list of essential elements. It should be interjected here that a formal definition of essentiality was not proposed until Arnon and Stout (1939a) published one in a paper devoted to micronutrients, especially copper. We shall come to it later, but it is worth keeping in mind that all the earlier investigators simply referred to an element as "necessary" or "indispensable," without precisely spelling out the meaning of the terms.

Another reservation is in order. "Discovery" refers to a particular, distinct event, with a date and a place and a name, or names, attached to it. But "discoveries" may evolve over a period of time, when initial but inconclusive findings and interpretations are succeeded by subsequent more definite ones. In such cases, and they are not infrequent, the assignment of a discovery to some certain person or team becomes a matter of judgment, and an account of the actual complexity of the "discovery" is preferable to *ex cathedra* pronouncements. At the same time, undue credit may be given to persons who made tentative, inconclusive findings. Such minor contributors may not appear in this review, which, in keeping with its title, puts emphasis on discoveries and those who made them.

A good example of progressive discoveries is one particularly relevant to the present topic: the discovery that plants can grow without soil, provided the essential mineral nutrients are dissolved in water bathing the roots. As early as 1692, John Woodward (1665–1728), in London, experimented with mint, *Mentha* sp., which he grew for 56 days in distilled water, Hyde Park water, and Hyde Park water shaken with a small amount of garden soil. The plants grown in Hyde Park water gained three times as much in weight as did the distilled water "controls" (not his terminology), and the plants in water shaken with soil gained seven times as much (Browne, 1944). Woodward concluded that mineral matter contributes to the nutrition of plants. This was a very early version of what eventually became, in the hands of Sachs and Knop, the technique of solution culture in essentially its modern form, and indispensable for the discovery of most of the essential elements.

One intermediate step in the evolution of solution culture and the understanding that it provided was taken by Wiegmann and Polstorff in 1842 (Schneidewind, 1917; Reed, 1942; Browne, 1944). The University of Göttingen, Germany, in 1838 had announced a prize for the best reply to a question about the significance of inorganic substances for the life of plants, not unlike the one posed by the Berlin Academy of Sciences in 1800 already referred to. Wiegmann and Polstorff grew plants of several species in sand irrigated either with distilled water, or with water to which they had added mineral salts of the kind they knew to be present in fertile soil. The plants in the distilled water treatment failed, but those supplied with mineral salts grew and absorbed minerals from the solution bathing their roots. The method of growing plants in sand or other solid medium considered chemically inert, and supplying the nutrients in solution, is a variant of solution culture. The method of sand culture does, however, have the disadvantage that the sand itself may contribute mineral salts to the solution. Indeed, this objection was raised in connection with the Wiegmann and Polstorff experiment. To settle the point once and for all, Wiegmann modified the experiment as follows (Browne, 1944). He used a platinum crucible as container, and a mass of fine platinum wire as a porous, inert medium, kept moist with distilled water. The experimental plants, garden cress, *Lepidium sativum*, germinated but died in less than a month. Their ash weighed exactly as much as the ash of the same number of seeds used in the experiment. Wiegmann and Polstorff won the prize in 1842, by thus establishing what no knowledgeable person should have doubted on the basis of de Saussure's work of four decades earlier.

The development of the solution culture technique in essentially its present form was simultaneously accomplished by two German investigators in 1860, W. Knop (1817–1891) and J. von Sachs (1832–1897) (Hewitt, 1963; Baumeister and Ernst, 1978). Sachs, the more assertive, prolific and versatile scientist, is usually given the main credit. Figure 1 is taken from the 1887 English translation of his 1882 "Lectures on the Physiology of Plants." It will be seen from Fig. 2 (Epstein, 1972) that a contemporary solution culture set-up does not differ much from the one of Sachs. Forced aeration of the nutrient solution is the most obvious difference. More subtle ones are the use of polyethylene or polypropylene containers, the much better purified (distilled-demineralized) water, and the greater purity of the salts used to make up the nutrient solutions. For studies of some micronutrients, even the amount furnished by the seeds may be adequate to support the plants. It is then necessary to use the seeds from plants grown in solutions to which the nutrient in question had not been added. Those seeds, very low in their content of the nutrient, are then used in solutions made up without deliberate addition of the nutrient, and deficiency disorders may then show up.

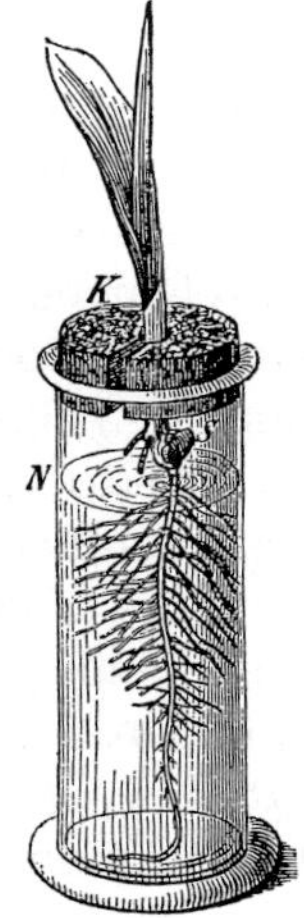

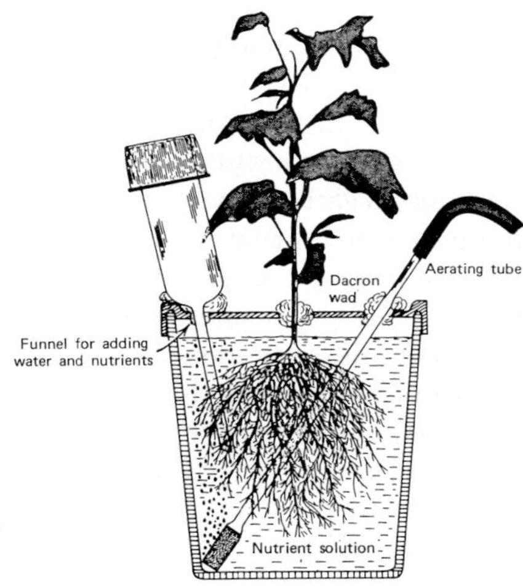

Fig. 1. Solution culture. From Sachs (1887), Clarendon Press.

Fig. 2. Solution culture. From Epstein (1972), *Mineral Nutrition of Plants: Principles and Perspectives.* John Wiley & Sons, New York. Reprinted by permission of John Wiley & Sons, Inc.

It is fair to say that with the advent of the method of solution culture, the "heroic" period of the discovery of plant nutrients ended. With this method in place, the discovery of the essentiality of inorganic nutrients was in a sense inevitable: there remained the application of the method, with the use of ever more highly purified nutrient salts and refined methodology, to determine the elements which, when omitted, would prevent plants from growing.

Discovery of the Essentiality Of Phosphorus, Sulfur, Potassium, Calcium, Magnesium and Iron

Georges Ville of France showed in 1861 by means of sand culture experiments that phosphorus is essential for the growth of plants (Reed, 1942). Salm-Horstmar showed the essentiality of calcium in 1856 (Miller, 1938), and Stohmann observed in 1862 that root tips died in solutions lacking this element, according to Reed (1942).

The credit for the discovery of the essentiality of potassium goes to Birner and Lucanus (1866). They experimented with oats grown in solution culture and showed in 1866 that potassium, but no other alkaline metal, was indispensable for the growth of plants. The finding was soon confirmed and extended by other investigators (Reed, 1942).

Through these and his own investigations, and those of Knop, Sachs (1865) could state categorically that in addition to those elements already named, sulfur and magnesium are essential elements for plants. Thus at that time, in addition to carbon, hydrogen and oxygen, the following elements were known to be essential nutrients, all derived from soil and absorbed by roots: nitrogen, phosphorus, sulfur, potassium, calcium and magnesium, i.e., all the elements now known as macronutrients. In addition, Sachs included iron among the indispensable elements, the only element in his list subsequently classified as a micronutrient. The inclusion of iron was due to the discovery of its essentiality by Eusèbe Gris, a professor of chemistry in Châtillon-sur-Seine. He found in the 1840s that the application of soluble iron salts would cure chlorotic plants, make them green up, and resume growth (Sachs, 1865). Gris in addition showed that when a dilute iron solution was applied to a small area of a chlorotic leaf, the treated spot would green up — the first use of the method of foliar application in the scientific study of plant nutrition.

Thus, by the mid-1860s, the essentiality of all the macronutrients now recognized as such was known, and that of one micronutrient, iron. Furthermore, the technique of solution culture, indispensable for further progress in this type of research, was available. Nevertheless, there ensued one of those inexplicable lapses in progress, of the kind that Sachs had remarked upon in regard to the first decades of the 19th century, and it was not until the early 1920s, over half a century after Sachs could list ten elements as essential for plants, that the next elements, manganese and boron, were added to that list.

Discovery of the Essentiality of the Micronutrients Other Than Iron

It was the virtually simultaneous research of two scientists, one American and one English, that put an end to this long sterile spell. They worked independently and were at first unaware of each other's discoveries.

McHargue (1878–1960) of Kentucky, using carefully prepared solution and sand cultures, showed manganese to be indispensable for the growth of several species (McHargue, 1922, 1923). Chlorosis being the initial and obvious symptom of manganese deficiency, he concluded that the element "has a function to fulfill in the formation of chlorophyll and consequently in carbon assimilation and possibly in the synthesis of protein." He amply acknowledged earlier work on manganese in plants. McHargue later became the head of the Department of Chemistry at the Kentucky Experiment Station, in which capacity he served from 1927 till his retirement in 1948.

The other pioneer in the modern era of plant nutrition was Katherine Warington. Born in Harpenden, she was a staff member at the famous Rothamsted Experiment Station for 36 years, and died at nearby St. Albans in 1993. She was a modest person not given to self-advertising, but her discovery of the essentiality of boron brought her worldwide fame among plant scientists. She grew bean plants in solution culture, using meticulous methods to keep contamination of the plant cultures to a minimum. To give but one example of the precautions she took, all visitors to the greenhouse had to wash their hands first, even if they were not going to handle the plants. The result of her labors was a classical paper, Warington (1923), documenting unequivocally the essentially of boron for the broad bean. Warington was meticulous not only in her experimentation but as well in fully acknowledging the work of prior scientists who had done research on the significance of boron for plants. None of those, however, had achieved clear-cut evidence of its essentiality. One question left unanswered in her research was the essentiality of boron for non-leguminous plants, but that issue was soon to be settled, as will be seen.

McHargue's and Warington's discoveries in the 1920s showed that through painstaking, meticulous solution culture work, elements could be shown to be indispensable for the growth of plants, although the amounts needed were small. Every one of the subsequent identifications of essential elements owed its success to the twin features that formed the foundation of the discoveries made by these pioneers. First, the insight that the major nutrient salts necessary to make up culture solutions contained impurities, including elements possibly essential, as did the ambient air, plant containers, etc., and second, the ability to apply the necessary experimental finesse to prepare nutrient solutions as free from the suspected essential element as then available methods permitted, and minimize the input of impurities from the experimental environment.   That understood, progress was rapid, and continues unabated to this day.

Soon after the publication of the papers by McHargue and Warington, Sommer and Lipman at the University of California, Berkeley, published the results of their research on the essentiality of zinc and boron for a number of plant species (Sommer and Lipman, 1926). Inasmuch as they included non-legumes in their studies, they showed the need for boron to be general for plants, not just for legumes. Lipman, born in Moscow in 1883, was brought to the U.S. at age 6 and died in Berkeley in 1944, having distinguished himself in both research and in university service, as Dean of the Graduate Division for many years. He was a person of high integrity and gave unstinting credit to the Frenchman, P. Mazé, who in the previous decade had claimed to have shown the essentiality of boron, aluminum, fluorine and iodine for corn, *Zea mays*. None of these claims, except that for boron, has stood the test of time. Mazé thus deserves credit for having drawn attention to the possibility that elements in addition to those then known to be essential would be shown to be so, but his experimental work was not equal to the task of determining their identity.

A few years after the publication of Sommer and Lipman's paper on the essentiality of zinc and boron for several species, there appeared two papers in which the essentiality of copper was reported. They were by Sommer (1931) and Lipman and MacKinney (1931), published a few months apart in the same journal, Plant Physiology. Sommer, formerly Lipman's graduate student and research assistant, did her work on copper at the University of Minnesota but had moved to the Alabama Experiment Station at Auburn when her paper appeared. She failed to cite her previous work with her mentor, Lipman. Lipman and MacKinney only heard of her work after they submitted their own paper to the same journal, and in a final footnote drew attention to Sommer's work. Evidently, there was a breakdown in communication between Lipman and Sommer after the latter left Berkeley. Be that as it may, the discovery of the essentiality of copper must be credited to both Sommer and to Lipman and MacKinney.

The next discoveries of essential elements came about through the influence of Lipman at the University of California at Berkeley, and that of Dennis R. Hoagland, at the same university. Hoagland (1884–1949) was the principal architect of plant nutrition research in this century. His influence was particularly pronounced in the matter of ion transport in plants. It was he, too, who formulated what is still called Hoagland solution, a formulation of a nutrient solution (two solutions, actually, one with and one without ammonium ions) which, with modifications, is still widely used. His book, *Lectures on the Inorganic Nutrition of Plants* (Hoagland, 1944) is a classic.

Hoagland was a keen observer of plants. He suspected that elements other than those then known to be essential may be so. He prepared a number of "A to Z" solutions containing, in addition to the then known nutrients, small concentrations of a raft of elements that might be essential (Stout, 1956). He concluded that indeed one or more elements in these fortified solutions benefited the plants. In the event, he was right, as usual.

Hoagland's associate, Perry R. Stout (1909–1975) carried the system of purifying nutrient solutions to unprecedented lengths, and, with D. I. Arnon, described these methods in detail (Stout and Arnon, 1939). Their application led to the discovery of the essentiality of molybdenum for the tomato (Arnon and Stout, 1939b), and subsequently, for plants generally (Stout and Johnson, 1956). "For a while it appeared that information on the molybdenum nutrition of plants might remain in the realm of merely interesting knowledge, but of no practical significance in the business of growing field crops. However, it was not too long before the discovery of molybdenum deficiencies in the field" (Stout, 1956). It was soon found, first in Australia and subsequently in many parts of the world, that applications of molybdenum on the order of 75 gram per hectare could make the difference between poor pastures and excellent, thrifty ones. So much for "merely interesting knowledge." "Scientific research is by nature an uncertain undertaking. Like any exploratory process, it is not possible to predict what one will find or what its eventual utility might be" (Myhrvold, 1998).

Next, a team, led again by P. R. Stout, discovered chlorine to be an essential element (Broyer *et al.*, 1954). The paper describing their experiments with the tomato plant makes fascinating reading, for it describes the successive steps whereby an investigation having as its initial object the test of cobalt as a possibly essential element led instead to clear-cut evidence of the essentiality of chlorine. Subsequent experiments established the essentiality of chlorine for other species (Johnson *et al.*, 1957).

Nickel is the most recent addition to the roster of elements known to be generally essential for plants. The initial, conclusive evidence for its essentiality was obtained in experiments with soybean plants by Eskew, Welch, and Cary (1983). Their work was prompted by, among other things (Welch, 1981), the discovery that jackbean urease is a nickel metalloenzyme (Dixon *et al.*, 1975). Further investigations showed nickel to be essential for cowpeas as well (Eskew *et al.*, 1984) and for barley (Brown *et al.*, 1987). Nickel is therefore firmly established as a micronutrient essential for plants generally.

As was the case in the discovery of all the micronutrients in the 20th century, success in establishing nickel as a micronutrient depended on sophisticated microchemistry to free the nutrient solutions from the element to such an extent that deficiencies of it could be demonstrated.

Cobalt, Sodium and Silicon

Some elements in addition to those already discussed need to be considered (Asher, 1991; Loneragan, 1997). Those for which no convincing evidence of essentiality has been produced, such as aluminum, will not, however, be discussed.

Cobalt is not known to be generally essential for higher plants, but is required when legumes and certain other plants acquire atmospheric nitrogen through association with nitrogen-fixing symbionts. This discovery was made simultaneously at Berkeley in work with alfalfa, *Medicago sativa* (Reisenauer, 1960; Delwiche *et al.*, 1961), and at North Carolina Agricultural Experiment Station, Raleigh, with the soybean, *Glycine max* (Shaukat-Ahmed and Evans, 1959, 1960, 1961). Cobalt is essential in all nitrogen-fixing systems.

Sodium, like the other elements discussed in this section, is not known to be generally essential for higher plants. But Brownell and Wood (1957) showed it to be a micronutrient for the Australian bladder salt bush, *Atriplex vesicaria*, and subsequent research showed all Australian *Atriplex* species tested, ten in all, to have this requirement. No other species of various groups, however, including non-Australian *Atriplex* species, could be shown to respond to sodium (Brownell, 1968).

These baffling, inconsistent findings were finally clarified when it was shown that the sodium requirement was consistently found in plants having the C_4 photosynthetic pathway, but not in those with the C_3 pathway (Brownell and Crossland, 1972). CAM plants also have a sodium requirement (Brownell and Crossland, 1974), depending on the conditions of growth. For a definitive discussion of sodium as a nutrient, see Brownell (1979). Plants native to saline habitats may absorb large amounts of sodium and their growth may be enhanced thereby, but they seem to have no absolute requirement for it.

The role of silicon as a plant nutrient is an "anomaly" (Epstein, 1994). Its essentiality for diatoms is well-known, but among higher plants, only the Equisitaceae (scouring rushes or horsetails) have an absolute and quantitatively major requirement for it (Chen and Lewin, 1969). Nevertheless, all soil-grown plants contain the element in amounts comparable to those of such macronutrient elements as calcium and sulfur, and many plants ranging from rice, *Oryza sativa*, to cucumber, *Cucumis sativus*, exhibit marked abnormalities when deprived of it (Epstein, 1994, 1999).

Heretofore in this review, little attention has been paid to the rigid definition of essentiality laid down by Arnon and Stout (1939a). In connection with silicon, however, a brief discussion of this flawed definition is in order. According to it, an element is essential if (i) a deficiency of it keeps the plant from completing its life cycle; (ii) no other element is able to substitute for it and (iii) it is directly involved in the nutrition of the plant and does not merely meliorate an unfavorable chemical or microbiological condition of the substrate. This is not the place for a critical assessment of this definition. We have, however, already encountered situations where its usefulness is limited, such as the fact that even in one genus, *Atriplex*, there may or may not be a sodium requirement, depending on whether it is a species with the C_4 or C_3 photosynthetic pathway.

Nowhere is the usefulness of this definition more clearly in doubt than in the context of the silicon nutrition of plants. The failure of silicon to qualify as essential by that definition is unlikely to be helpful to researchers finding silicon-deprived plants to lodge and generally, to be structurally weak, to be abnormally sensitive to aluminum and heavy metal toxicities, and far more prone to damage by disease and pest organisms, including phytophagous insects, than are silicon replete plants. Silicon must therefore be included in such a chapter as this, whether or not it is given the official stamp of approval, "essential," by the plant physiological establishment. This failure of silicon to be recognized as essential by the Arnon-Stout definition has led to its omission from the formulation of nutrient solutions, and more generally, to its being considered a plant biological nonentity (Epstein, 1994, 1999). If the accepted definition of essentiality were simply the need of an element for normal plant growth and development, silicon would assuredly be accorded essential status for many species. Discovery, as has been pointed out, is not always a simple, straightforward process; thus there is need for ongoing, critical review and assessment.

Acknowledgments

I thank Paul Brooks, Bob B. Buchanan, William M. Roberts and James Vlamis, all at the University of California, Berkeley, for providing information on C. B. Lipman and A. L. Sommer; Brian C. Loughman of the University of Oxford for sending me background material on Katherine Warington; and Frank B. Stanger, Jr. of the University of Kentucky for supplying biographical information on J. S. McHargue.

References

Arnon, D. I. and Stout, P. R. (1939a) The essentiality of certain elements in minute quantity for plants with special reference to copper. *Plant Physiol.* **14**: 371-375.

Arnon, D. I. and Stout, P. R. (1939b) Molybdenum as an essential element for higher plants. *Plant Physiol.* **14**: 599-602.

Asher, C. J. (1991) Beneficial elements, functional nutrients, and possible new essential elements. In: *Micronutrients in Agriculture. Second Edition.* J.J. Mortvedt, F.R. Cox, L.M. Shuman, R.M. Welch, eds., pp. 703-722. Soil Science Society of America, Madison.

Baumeister, W. and Ernst, W. (1978) *Mineralstoffe und Pflanzenwachstum.* 3rd ed., Gustav Fischer Verlag, Stuttgart.

Birner, H. and Lucanus, B. (1866) Wasserkulturversuche mit Hafer in der Agric.-Chem. Versuchsstation zu Regenwalde. *Landw. Versuchsst.* **8**: 128-177.

Brown, P. H., Welch, R. M. and Cary, E. E. (1987) Nickel: a micronutrient essential for higher plants. *Plant Physiol.* **85**: 801-803.

Browne, C. A. (1944) *A Source Book of Agricultural Chemistry.* Chronica Botanica 8: VI-X, 1-290.

Brownell, P. F. (1968) Sodium as an essential micronutrient element for some higher plants. *Plant Soil* **28**: 161-164.

Brownell, P. F. (1979) Sodium as an essential micronutrient element for plants and its possible role in metabolism. *Adv. Bot. Res.* **7**: 117-224.

Brownell, P. F. and Crossland, C. J. (1972) The requirement for sodium as a micronutrient by species having the C_4 dicarboxylic photosynthetic pathway. *Plant Physiol.* **49**: 794-797.

Brownell, P. F. and Crossland, C. J. (1974) Growth responses to sodium by *Bryophyllum tibuflorum* under conditions inducing crassulacean acid metabolism. *Plant Physiol.* **54**: 416-417.

Brownell, P. F. and Wood, J. G. (1957) Sodium as an essential micronutrient element for *Atriplex versicaria*, Heward. *Nature* **179**: 635-636.

Broyer, T. C., Carlton, A. B., Johnson, C. M. and Stout, P. R. (1954) Chlorine - a micronutrient element for higher plants. *Plant Physiol.* **29**: 526-532.

Chen, C.-H. and Lewin, J. (1969) Silicon as a nutrient element for *Equisetum arvense. Can. J. Bot.* **47**: 125-131.

Delwiche, C. C., Johnson, C. M. and Reisenauer, H. M. (1961) Influence of cobalt on nitrogen fixation by Medicago. *Plant Physiol.* **36**: 73-78.

De Saussure, T. (1804) *Recherches chimiques sur la végétation.* Nyon, Paris, as cited by Browne (1944), Reed (1942), and Sachs (1906).

Dixon, N. E., Gazzola, C., Blakeley, R. L. and Zerner, B. (1975) Jack Bean urease (E.C. 3.5.1.5). A metalloenzyme. A simple biological role for nickel? *J. Am. Chem. Soc.* **97**: 4131-4133.

Epstein, E. (1972) *Mineral Nutrition of Plants: Principles and Perspectives.* John Wiley and Sons, New York.

Epstein, E. (1994) The anomaly of silicon in plant biology. *Proc. Natl. Acad. Sci. USA.* **91**: 11-17.

Epstein, E. (1999) Silicon. *Annu. Rev. Plant Physiol. Plant Molec. Biol.* **50**: 641-664.

Eskew, D. L., Welch, R. M. and Cary, E. E. (1983) Nickel: an essential micronutrient for legumes and possibly all higher plants. *Science* **222**: 621-623.

Eskew, D. L., Welch, R. M. and Norvell, W. A. (1984) Nickel in higher plants. Further evidence for an essential role. *Plant Physiol.* **76**: 691-693.

Hewitt, E. J. (1963) Mineral nutrition of plants in culture media. In *Plant Physiology. A Treatise.* Steward, F. C., ed., III: 97-133. Academic Press, New York.

Hoagland, D. R. (1944) *Lectures on the Inorganic Nutrition of Plants.* Chronica Botanica Company, Waltham.

Johnson, C. M., Stout, P. R., Broyer, T. C. and Carlton, A. B. (1957) Comparative chlorine requirements of different plant species. *Plant Soil* **8**: 337-353.

Kerr, R. (1790) *Elements of Chemistry in a New Systematic Order, Containing all the Modern Discoveries by Mr. Lavoisier. English Translation of Lavoisier's "Traité élémentaire."* 2nd ed., 1793; 3rd ed., 1796.

Lavoisier, A. L. (1789) *Traité élémentaire de Chimie Présenté dans un Ordre Nouveau et d'après les Découvertes Modernes.* Paris, 2 volumes.

Lavoisier, A. L. (1862-1893) *Oeuvres de Lavoisier publiees par les Soins de son Excellence le Ministre de l'Instruction Publique et des Cultes.* Paris, 6 volumes.

Liebig, J. (1840) *Organic Chemistry in Its Applications to Agriculture and Physiology,* edited from the manuscript of the author by Lyon Playfair. Taylor and Walton, London.

Lipman, C. B. and MacKinney, G. (1931) Proof of the essential nature of copper for higher green plants. *Plant Physiol.* **6**: 593-599.

Loneragan, J. F. (1997) Plant nutrition in the 20th and perspectives for the 21st century. In *Plant Nutrition for Sustainable Food Production and Environment.* Ando, T., Fujita, K., Mae, T., Matsumoto, H., Mori, S. and Sekiya, J., eds., pp. 3-14. Kluwer Academic Publishers, Dordrecht.

Magner, L. N. (1994) *A History of the Life Sciences.* 2nd ed., Marcel Dekker, New York.

McCosh, F. W. J. (1984) *Boussingault: Chemist and Agriculturist.* D. Reidel Publishing Company, Dordrecht.

McHargue, J. S. (1922) The role of manganese in plants. *J. Am. Chem. Soc.* **44**: 1592-1598.

McHargue, J. S. (1923) Effect of different concentrations of manganese sulphate on the growth of plants in acid and neutral soils and the necessity of manganese as a plant nutrient. *J. Agric. Res.* **24**: 781-794.

Miller, E. C. (1938) *Plant Physiology.* 2nd ed., MacGraw Hill Book Company, New York.

Moulton, F. R., ed. (1942) *Liebig and after Liebig. A Century of Progress in Agricultural Chemistry.* American Association for the Advancement of Science, Washington.

Myhrvold, N. (1998) Supporting science. *Science* **282**: 621-622.

Partington, J. R. (1960) *A Short History of Chemistry.* 3rd ed., MacMillan & Co., London.

Poirier, J.-P. (1996) *Lavoisier: Chemist, Biologist, Economist.* Translated from the French by Rebecca Balinski. University of Pennsylvania Press, Philadelphia.

Reed, H. S. (1942) *A Short History of the Plant Sciences.* The Ronald Press Company, New York.

Reisenauer, H. M. (1960) Cobalt in nitrogen fixation by a legume. *Nature* **186**: 375-376.

Sachs, J. von (1865) *Handbuch der Experimental-Physiologie der Pflanzen.* Verlag von Wilhelm Engelmann, Leipzig.

Sachs, J. von (1887) *Lectures on the Physiology of Plants.* Oxford, at the Clarendon Press.

Sachs, J. von (1906) *History of Botany (1530-1860).* 2nd Impression. Oxford, at the Clarendon Press.

Schneidewind, W. (1917) *Die Ernährung der landwirtschaftl. Kulturpflanzen.* 2nd ed., Verlagsbuchhandlung. Paul Parey, Berlin.

Shaukat-Ahmed and Evans, H. J. (1959) Effect of cobalt on the growth of soybeans in the absence of supplied nitrogen. *Biochem. Biophys. Res. Commun.* **1**: 271-275.

Shaukat-Ahmed and Evans, H. J. (1960) Cobalt: a micronutrient element for the growth of soybean plants under symbiotic conditions. *Soil Sci.* **90**: 205-210.

Shaukat-Ahmed and Evans, H. J. (1961) The essentiality of cobalt for soybean plants grown under symbiotic conditions. *Proc. Natl. Acad. Sci. USA.* **47**: 24-36.

Sommer, A. L. (1931) Copper as an essential for plant growth. *Plant Physiol.* **6**: 339-345.

Sommer, A. L. and Lipman, C. B. (1926) Evidence on the indispensable nature of zinc and boron for higher green plants. *Plant Physiol.* **1**: 231-249.

Stout, P. R. (1956) Micronutrients in crop vigor. *Agric. Food Chem.* **4**: 1000-1006.

Stout, P. R. and Arnon, D. I. (1939) Experimental methods for the study of the role of copper, manganese, and zinc in the nutrition of higher plants. *Am. J. Bot.* **26**: 144-149.

Stout, P. R. and Johnson, C. M. (1956) Molybdenum deficiency in horticultural and field crops. *Soil Sci.* **81**: 183-258.

Warington, K. (1923) The effect of boric acid and borax on the broad bean and certain other plants. *Ann. Bot.* **27**: 629-672.

Weeks, M. E. (1956) *Discovery of the Elements*. Journal of Chemical Education, Eastron, Pennsylvania.
Welch, R. M. (1981) The biological significance of nickel. *J. Plant Nutr.* **3**: 345-356.

Chapter 2

The Discovery of 1-Aminocyclopropane-1-Carboxylic Acid as the Immediate Precursor of Ethylene

Douglas O. Adams
Department of Viticulture and Enology
University of California, Davis, CA 95616-8740, USA

ABSTRACT

This chapter provides an account of how 1-aminocyclopropane-1-carboxylic acid (ACC) came to be recognized as the immediate precursor of the plant hormone ethylene. This discovery relied heavily on earlier observations and technological advances in the field, such as the introduction of gas chromatography with flame ionization detection as a rapid and sensitive analytical method for the gas. Likewise, the discovery that methionine is a close precursor was an important pre-requisite for complete elucidation of ethylene's biosynthetic pathway. The first step in establishing the pathway from methionine to ethylene was obtaining evidence that S-adenosylmethionine was an intermediate. This was done by showing that 5′-methylthioadenosine and methylthioribose were products of methionine metabolism in ethylene-producing tissues. A brief explanation is given of the role in the discovery played by a class of inhibitors, 3,4-unsaturated amino acids, the most important member of the class being aminoethoxyvinylglycine (AVG). The way that these inhibitors were used to establish the order of intermediates in the pathway is summarized. This account describes how feeding studies using radioactive methionine under anaerobic conditions led to direct observation of a labeled metabolite which fit the criteria for a hypothetical intermediate that had been proposed by several previous investigators. The experiments undertaken to identify the unknown metabolite (ACC) are described; the result being that this unusual amino acid, which had been isolated from apples and pears twenty years prior, was finally recognized as the immediate precursor of ethylene.

Introduction

The discovery that 1-aminocyclopropane-1-carboxylic acid is the immediate precursor of the plant hormone ethylene represents a landmark in the research on this truly remarkable but simple plant hormone. It was the culmination of many years of inquiry concerning the biochemical reactions that produce the gas, and set the field on a search for how the process is controlled, which continues today. The history of how ethylene itself was recognized as a plant growth substance has been described in detail by Saltveit, Yang and Kim in this series (Saltveit, Yang and Kim, 1998). The account given here is actually a story within that larger one and is not intended to be a general review of the field. Instead, selected features of the early history of ethylene in plant biology are recounted so as to highlight the influence of ideas that led to the current discovery. This background is important if we are to recognize factors that guided the conceptual schemes and experimental approaches used in discovering this new information.

Ethylene was first recognized only as an exogenous agent that could affect the growth behavior of plants. Early workers tended to think of ethylene as an air pollutant that was responsible for defoliating greenhouse-grown plants, or in the more positive role of promoting degreening of lemons shipped in kerosene-heated boxcars from California to markets in the eastern United States.

That ethylene is a natural product produced by plants was probably suspected by several early workers, but was not demonstrated conclusively until the work of Gane who collected the volatiles produced by 27 kg of apples by passing air over them and then through a solution of bromine water at $-65°C$ for four weeks. The resulting dibromoethane was identified by reaction with aniline and confirmed by melting point depression studies with synthetic N,N'-diphenyl ethylene diamine (R. Gane, 1934).

It is clear from Gane's report that he recognized the significance of his work in establishing the critical link between the well-known effects of ethylene on plants and similar effects caused by some plant tissues themselves, especially ripening fruit.

Since ethylene is produced by most plant tissues in very minute quantities, development of sensitive analytical systems was absolutely essential in characterizing the ethylene production patterns of plants. Gane's method for trapping ethylene as ethylene dibromide was the basis of the first analytical procedures, but it was not until the development of a more convenient manometric method developed by Young *et al.* (1952) that researchers could assay ethylene production by plant tissues in a flow system with the sensitivity required for accurate rate measurements.

The manometric method was based on formation of an ethylene-mercury complex by passing the air from a flow system through a solution of mercuric perchlorate. The principle of this assay later became the basis of trapping and measuring radiolabeled ethylene produced from labeled precursors. Trapping labeled ethylene as the mercury complex, either in mercury perchlorate or mercuric acetate in methanol, was an important technique in experiments using radioactive precursors to study the biosynthetic pathway of ethylene. The method is still widely used even though the researchers may not recognize its origin in a manometric method that was soon replaced by an even more sensitive technique, that of gas chromatography.

The combination of gas chromatography with flame ionization detection made all earlier chemical methods for ethylene determination obsolete. The first report using this technology was in 1959 but it was quickly adopted because it was much more simple, accurate and sensitive than chemically based manometric methods (Burg and Stolwijk, 1959). The importance of gas chromatography to the study of ethylene in plant biology cannot be overstated. It enabled workers to make many more determinations with much more sensitivity and accuracy. It transformed what had been a laborious assay into one of the easiest in the field of plant hormone research. and the results obtained by this technique laid the groundwork for understanding the biochemical steps that give rise to the gas. Nevertheless, there were numerous observations in the pre-gas chromatography era that played vital roles in the research story.

Early Observations Critical for Understanding the Pathway

As will be seen, inhibition of ethylene production by anaerobic conditions was an important tool for understanding its biosynthesis. It was earlier recognized that ethylene production by plant tissue is an aerobic process and that it is inhibited at high temperature (greater than 40°C).

Gane (1934) showed that apples deprived of oxygen produced no ethylene, and Hansen (1942) found a similar effect with pears held in nitrogen, hydrogen or helium. Hansen added the important observation that pears transferred back to air after a period in nitrogen produced ethylene at a higher rate than pears held in air continuously (Hansen, 1942).

In one of the earliest studies employing gas chromatography as an analytical method for ethylene analysis, Burg and Thiman (1959) showed that the effect of reduced oxygen levels had a parallel effect on oxygen consumption and ethylene production. This result clearly indicated that ethylene evolution by plant tissue was dependent on aerobic respiration. In this study, Burg and Thiman also showed conclusively that apple tissue held under nitrogen for four hours exhibited an accelerated rate of ethylene production when returned to air. Their interpretation of this result was that nitrogen treatment led to the accumulation of some metabolite that could then be converted to ethylene very rapidly when the tissue was returned to oxygen (aerobic conditions). Thus, the suggestion that nitrogen atmospheres led to accumulation of a metabolite that was readily converted to ethylene in the presence of oxygen appeared in the earlier literature, thereby influencing the thinking and experimental approaches of subsequent workers.

Methionine as a Precursor of Ethylene

In early work to establish the pathway for ethylene biosynthesis, several compounds were proposed as precursors. These included linolenic acid, propanal, acrylic acid, ethanol, ethane, and even another recognized plant hormone, indoleacetic acid. Each of these proposed precursors were either disproved or abandoned for lack of evidence (Yang, 1974), and it was not until Lieberman and his co-workers in the mid 1960s proposed the amino acid methionine as a physiological precursor, that work on the pathway progressed (Lieberman *et al.*, 1965). The proposal that methionine is an ethylene precursor was based on the observation that methionine was converted to ethylene in a model system containing cuprous ion and ascorbate. The use of methionine labeled with ^{14}C at specific carbons in this model system showed that ethylene was derived from carbons 3 and 4, and that the carboxyl carbon was lost as carbon dioxide. Carbon 2 gave rise to a small amount of carbon dioxide, but ethylene was not formed from carbon 2 or the methyl group.

Abeles and Rubenstein (1964) discovered that crude extracts of pea seedlings could form ethylene in the presence of flavin mononucleotide (FMN) and Yang showed that this system was light-dependent and that the substrate in the crude pea extract was methionine (Yang *et al.*, 1966). Labeling studies on this FMN/light system showed a pattern similar to the copper/ascorbate system with the additional observation that carbon 2 was converted to formic acid and that the sulfur and methyl groups appeared as dimethyl disulfide (Yang *et al.*, 1967).

The most important result to come out of work prompted by interest in the model systems was that methionine labeled in carbons 3 and 4 could also give rise to labeled ethylene when fed to apple tissue. This was reported by Lieberman *et al.* in 1966. Confirmation quickly followed Lieberman's report, and Burg and Clagett added the observations that methionine labeled with ^{35}S gave few volatile counts when fed to apple tissue. They also reported that the methyl group accompanied the sulfur into a "neutral compound" in apple tissue (Burg and Clagett, 1967).

Further work on both of the model systems showed that they were probably unrelated to the mechanism whereby plants produce ethylene from methionine *in vivo*. Nevertheless, from these studies came the key observation that methionine could act as a physiological precursor. Various schemes were proposed, whereby ethylene could be produced directly from methionine, and the scheme that emerged from the model systems and *in vivo* studies with labeled methionine guided research in the field for over ten years. During this time, application of appropriately labeled methionine demonstrated that the oxygen requirement observed long before occurred at a step between methionine and ethylene, as was the block imposed by respiratory inhibitors (Baur *et al.*, 1971; Murr and Yang, 1975). These observations along with the fact that conversion of methionine to ethylene was inhibited by arsenate led Burg to the very perceptive and fruitful suggestion that S-adenosylmethionine (AdoMet) might be an intermediate in the conversion of methionine to ethylene (Burg, 1973).

Rhizobitoxine and AVG as Key Inhibitors in the Study of the Ethylene Pathway

In many fields of research, we can point to a particular tool and remark on its importance in helping to establish a new piece of information. The present account would be incomplete if we did not mention the role that a family of inhibitors played in helping to establish the sequence of intermediates between methionine and ethylene. This account also illustrates how misapprehension can sometimes be a helpful element in the ultimate progress that occurs in a field of research.

The inhibitors in question are α-amino acids having a double bond between carbons 3 and 4. The first of these compounds was recognized as a metabolite produced by certain strains of *Rhizobium japonicum* that was shown to be responsible for "rhizobial induced chlorosis" of soybean plants (Owens and Wright, 1965). It was given the trivial name rhizobitoxine (I). The finding that some strains of the bacteria could produce the toxin in culture facilitated the isolation of milligram quantities for chemical structure determination and physiological studies (Owens and Wright, 1965a).

Initial structural work suggested that the rhizobitoxine molecule contained a sulfur atom and although this was later disproved, it led Owens and co-workers to suspect that its toxic effect might be due to interference with methionine biosynthesis or utilization. They were first able to show that growth inhibition of *Salmonella typhimurium* by rhizobitoxine could be overcome by methionine, and later that the toxin was a very powerful inhibitor of β-cystathionase, a key enzyme in methionine biosynthesis (Owens *et al.*, 1968).

I

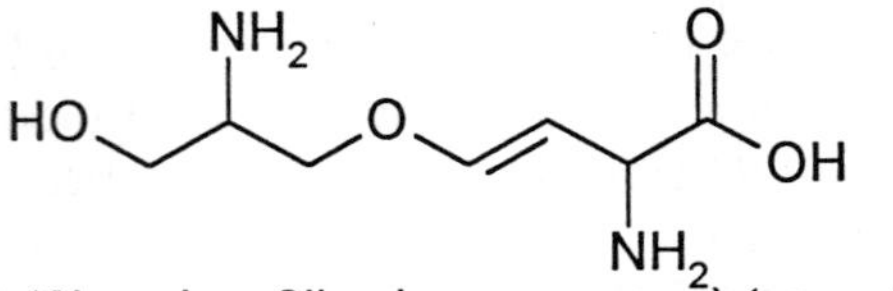

L-2-amino-4-(2'-amino-3'hydroxypropoxy)-trans-3-butenoic acid
(Rhizobitoxine)

II

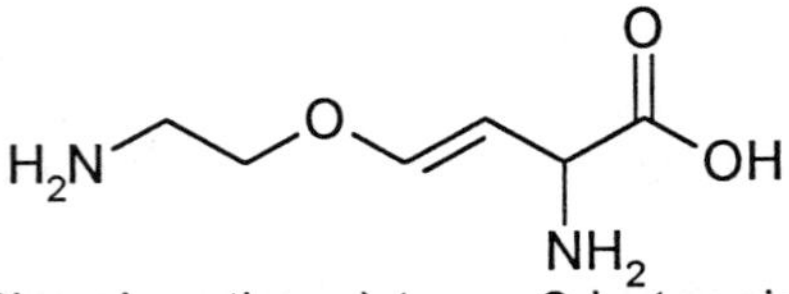

L-2-amino-4-(2'-aminoethoxy)-trans-3-butenoic acid
(Aminoethoxyvinylglycine, AVG)

III

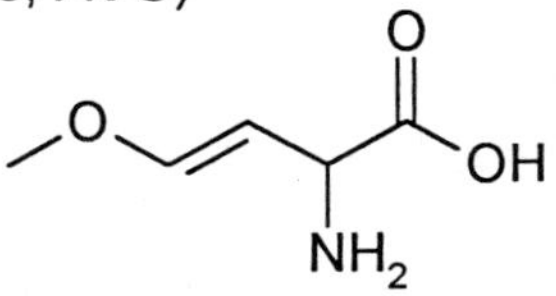

L-2-amino-4-methoxy-trans-3-butenoic acid

IV

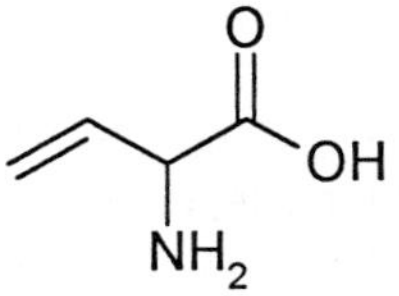

L-2-amino-3-butenoic acid   (Vinylglycine)

Lowell Owens and Morris Lieberman (who was first to show that methionine was the physiological precursor of ethylene) worked at the U.S. Department of Agriculture's Beltsville Agricultural Research Center. Discussions between them led to the testing of effects of rhizobitoxine on ethylene production, based on the finding that it blocked methionine biosynthesis in the *Salmonella typhimurium* bacteria. In a very original and formative paper, Owens, Lieberman and Kunishi (1971) reported that rhizobitoxine was indeed a very good inhibitor of ethylene production in apple tissue. Since methionine could overcome growth inhibition in the bacteria, they expected that inhibition of ethylene production by rhizobitoxine in apple tissue would likewise be alleviated. Instead, they found that methionine could not overcome the inhibition, and that rhizobitoxine strongly inhibited conversion of labeled methionine to ethylene. This result pointed to the very important conclusion that rhizobitoxine blocked ethylene production by inhibiting the conversion of methionine to ethylene.

It was not until the following year that the completed chemical structure of rhizobitoxine was reported, and it did not contain a sulfur atom as was first thought (Owens *et al.*, 1972). Thus, the mistaken belief that it contained sulfur led to the discovery that it interfered with methionine biosynthesis in bacteria, which in turn led to the observation that it could very effectively inhibit conversion of methionine to ethylene in plant tissue. This serendipitous sequence of results proved very fortunate for the field of ethylene biosynthesis, but it turned out to be a related inhibitor and not rhizobitoxine itself that went on to play such an important role in many studies on ethylene biosynthesis and physiology.

The wide use of aminoethoxyvinylglycine (AVG, II) as opposed to rhizobitoxine occurred entirely as a result of its availability. It was generously provided to researchers by J. P. Scannell of Hoffman-La Roche, Inc., Nutley, New Jersey. The availability of this inhibitor made possible numerous experiments to confirm the role of ethylene in a large variety of physiological responses, and without it progress in this field would have certainly been much slower. It was first isolated from a fermentation broth of an unspecified species of *Streptomyces*, and showed anti-microbial activity against several bacteria that was alleviated by a variety of amino acids (Pruess *et al.*, 1974). In the first paper reporting the isolation and identification of AVG, the authors state as a personal communication from Morris Lieberman that AVG inhibited ethylene production by plant tissue. Lieberman later published work showing that AVG and the closely related methoxy analog (III) were effective inhibitors of ethylene production, and like rhizobitoxine, they blocked by inhibiting conversion of methionine to ethylene (Lieberman *et al.*, 1975).

Rando studied the mechanism of inhibition of specific enzymes by this group of 3,4–unsaturated amino acids, and found that it involved irreversible inhibition of enzymes that use pyridoxal phosphate as cofactor (Rando 1974). In the paper describing the effects of rhizobitoxine on ethylene production, Owens *et al.* (1971) first suggested that pyridoxal phosphate was involved in the conversion of methionine to ethylene. Yang and Baur (1972) subsequently proposed a reaction mechanism whereby methionine itself could be converted directly to ethylene using pyridoxal phosphate as cofactor. Later, combining the idea of methionine activation with the earlier suggestion of Burg (that AdoMet might be an intermediate in the pathway from methionine to ethylene), Murr and Yang (1975) proposed a mechanism whereby AdoMet was directly converted to ethylene in a pyridoxal phosphate-catalyzed elimination reaction. Thus, the idea that pyridoxal phosphate was somehow involved in the synthesis of ethylene from methionine was firmly entrenched in the minds of researchers working on this problem at the time. The source of that idea was the mechanism by which this class of inhibitors exerted their toxic effects, and that the target enzymes which they inhibited utilized pyridoxal phosphate as cofactor.

Sulfur Recycling and the Involvement of Adomet in the Ethylene Pathway

Baur and Yang (1972) had pointed out the necessity of recycling methionine sulfur if plant tissues were to produce ethylene continuously from small methionine pools. However, some researchers roundly criticized this notion, saying that ethylene production was small compared to the plant's capacity to reduce sulfate, and therefore the assumption of sulfur recycling was an intellectually attractive but unnecessary part of the ethylene pathway. Nevertheless, the idea of sulfur recycling proved to be a fruitful avenue of research. Murr and Yang (1975a) were able to show that 5'-methylthioadenosine (MTA), the hypothetical product of the elimination reaction involving AdoMet, when labeled in the methyl carbon could be a source of the methyl group for methionine in apple tissue. While this did not speak directly to the question of sulfur atom recycling, it implied that perhaps both the sulfur atom and the methyl group were recycled after degradation of the AdoMet's aminobutryl moiety to ethylene.

Dennis Murr had just recently obtained evidence for the incorporation of the methyl group of MTA into methionine when I joined Shang Fa Yang's research group in the Department of Vegetable Crops at U.C. Davis. The importance of the scheme proposed by Murr and Yang, showing the involvement of AdoMet and the recycling of MTA into methionine, was that it gave testable predictions as to the fate of the methionine methyl group and sulfur atom during ethylene production, and it was these features of the scheme that we set out to test.

I first spent an unproductive year studying ethylene production in a fungal system *Penicillium digitatum*. In this system, glutamic acid was shown to be the closest known precursor rather than methionine (Chou and Yang, 1973), but work in Yang's laboratory had proceeded on both systems even though by this time it was well-recognized that the pathway in the fungal system and higher plants were quite different. When Dennis Murr accepted a faculty position at the University of Guelph, I saw an opportunity to switch from studying ethylene biosynthesis in *Penicillium*, to the system in higher plants. It was clear to me from the beginning that while work on the fungal system was interesting from a biochemical viewpoint, the real significance in the ethylene problem was in understanding its biosynthesis and mode of action in higher plants. This was nowhere more evident at the time than in the Mann Laboratory where many of the postharvest biology research projects involved ethylene in one way or another. In our conversation about changing systems, Yang warned me that the problem was not an easy one. He explained that a Ph.D. thesis must show results and the risk would be that it could take substantial time to get enough data for a satisfactory thesis. Nevertheless, I was eager to get my new project underway. I felt that I had moved into a more important role because I was now working on ethylene production in higher plants. I realized that my experiments and results would attract more attention and criticism, and this is not something graduate students take on without a certain amount of anxiety.

Choice of an Experimental System

In choosing what plant tissue to use for our experiments, we considered two different experimental systems being used in Yang's laboratory at the time to study ethylene production. Oi-Lim Lau was using an etiolated mung bean hypocotyl system to study the synergistic induction of ethylene production by auxin and cytokinins. This system had at least two advantages; tissue could be produced from seed whenever needed, and tissue not treated with auxin and cytokinins produced very little ethylene and could therefore provide an inducible system or serve as a control in some kinds of experiments. The apple tissue system that Dennis Murr used had both advantages and disadvantages. One disadvantage was that ethylene production in stored or store-bought fruit was constitutive. This meant that inhibitors could be used to study the pathway, but looking for critically inducible steps was not an option except in preclimacteric fruit which could only be obtained once a year prior to the onset of ripening. Another disadvantage of the apple system was the variability of experimental material during the year. The nearby supermarket could provide apples all year, but they were always from different sources and each variety had a particular season when they were available. This was problematic because some varieties produce very little ethylene compared to others. The Golden Delicious variety was preferred because it had less tendency to brown when tissue samples were cut from the fruit and ethylene production was among the highest.

The advantage of the apple system over the mung bean system was in the amount of labeled methionine converted to ethylene. In mung bean, it was a small percentage whereas in apple tissue the majority of the methionine fed to the tissue was converted to ethylene. I was swayed toward the apple system because of this feature and a remark made by Paul Stumpf from whom I had taken a plant biochemistry class. He remarked several times during his lectures that we should study a biochemical pathway in a tissue that produced large quantities of the product. So while our decision about the choice of an experimental system carefully weighed the advantages and disadvantages of apple versus etiolated mung bean hypocotyls, it was ultimately Stumpf's point and our awareness of the long history and demonstrated utility of studying ethylene production in apple fruit that convinced us that the work should continue with apple. The source and availability problems were addressed by travelling to a nearby apple producing area at harvest time, collecting Golden Delicious apples from the same side of several adjacent trees and storing them in the Mann Lab at 0°C until required for the experiments. This turned out to be a very pleasant chore that occurred just before classes began in the fall quarter.

Implication of Adomet in the Ethylene Pathway

The first experiments we undertook were aimed at testing the scheme proposed by Murr and Yang showing AdoMet as an intermediate in the conversion of methionine to ethylene. Dennis Murr's experiments showing recycling of MTA into methionine had been conducted using methyl-labeled methionine, but the scheme proposed for conversion of AdoMet to ethylene suggested that the sulfur would accompany the methyl into MTA during ethylene production. It was this feature of the scheme that we first set out to test by feeding ^{35}S-methionine to apple plugs and looking for labeled MTA. The first experiments showed that even when apple tissue was incubated for considerable time with ^{35}S-methionine, incorporation of sulfur into the spot on paper chromatograms that corresponded to MTA was disappointingly small. Yang's suggestion was to feed unlabeled MTA to the apple tissue along with the ^{35}S-methionine so as to trap the labeled MTA by isotope dilution in the event that MTA was quickly being metabolized and thus not accumulating to any extent. When this was done, we observed much greater incorporation into the regions of the paper chromatograms where MTA migrated, and we next attempted to verify that the labeled material was indeed MTA by co-electrophoresis of the labeled material with authentic MTA.

The electrophoresis results were unambiguous and disappointing. The labeled metabolite that migrated with MTA on paper chromatography did not move with MTA on paper electrophoresis at pHs where MTA had a positive charge due to the presence of the adenine. In fact, the metabolite did not move at any pH on paper electrophoresis. The results were discouraging for two reasons; first, because the metabolite was clearly not MTA, and secondly, because the metabolite appeared to be a neutral molecule which seemed to suggest that electrophoresis would be of little value in establishing its identity. Having a neutral radioactive metabolite limited the amount of information available to establish its identity. Repeating the experiment with methyl-labeled methionine showed that the neutral metabolite retained the methyl group as well as the sulfur atom. This was the extent of the structural information with which we began the task of identifying the metabolite that accumulated when sulfur or methyl-labeled methionine was fed to apple tissue in the presence of unlabeled MTA. Noting that the positive charge on MTA molecule was contributed by the adenine moiety, Professor Yang suggested that the neutral metabolite might well be the de-adenylated form of MTA.

It was at this time that I began my acquaintance with searching the various indexes of chemical abstracts "by hand". Of course the term "by hand" was not used at the time because "by computer" was not yet available. I was prompted to develop this skill on the advice of Harlan Pratt given at Mann Laboratory noon seminars. The advice was something to the effect that when you think you have discovered something new, go to the library and find out who discovered it first. This skill, and that advice, would later prove valuable in unraveling the pathway.

My library excursion turned up work on the fate of MTA, showing that the first product of its metabolism was methylthioribose (Shapiro and Mather, 1958; Pegg and Williams-Ashman, 1969). The first step was to make some MTR from MTA and determine if it co-chromatographed with the labeled metabolite, which it did. Yang's suggestion to perform electrophoresis in borate buffer turned the handicap of the metabolite's neutrality into a distinct advantage. If the metabolite was indeed MTR then complexation of the vicinal hydroxyl groups with borate should make it mobile during electrophoresis at high pH. Co-electrophoresis of the metabolite with authentic MTR in borate buffer helped confirm that the metabolite was MTR, and added evidence for the involvement of AdoMet in the pathway from methionine to ethylene. Additional experiments showed that when labeled MTA was fed, it was readily converted to MTR, suggesting that the primary product was MTA, which was further metabolized into MTR. Our data were in agreement with the idea that during conversion of methionine to ethylene, methionine was first metabolized into AdoMet which was then converted into MTA and ethylene. This idea was further supported by experiments using AVG.

We conducted experiments with AVG showing that it inhibited incorporation of label from methionine into MTR, which was expected because of its known ability to inhibit production of labeled ethylene from methionine. Pre-climacteric apple tissue produced little labeled ethylene from L-[U-^{14}C] methionine and gave very little labeled MTR, even when unlabeled MTA was fed along with it. Thus, incorporation of sulfur- or methyl-labeled methionine into MTR correlated with incorporation of L-[U-^{14}C] methionine into ethylene, both with the AVG inhibitor and in the physiological pre-climacteric stage where ethylene production was minimal.

Further work with ^{35}S-MTA showed good incorporation into methionine, and reinforced the idea of sulfur recycling as an important process in sustaining ethylene production from small methionine pools. The results implicating AdoMet as an intermediate in the conversion of methionine to ethylene were published in 1977 (Adams and Yang, 1977). The pathway proposed in that paper showed AdoMet as the precursor. This scheme incorporated what was known at the time, and in that paper, we suggested that MTR was the "neutral compound" that Burg and Claggett (1967) had observed in apple tissue when feeding methyl- or sulfur-labeled methionine ten years before.

Our work, immediately following publication of those results, was aimed at obtaining additional evidence for the involvement of AdoMet, and determining how other well-known inhibitors affected the pathway. We confirmed that apple tissue incorporated methionine into AdoMet, which we identified by characterizing its hydrolysis products as MTA and homoserine lactone by paper chromatography and paper electrophoresis. This work also demonstrated that S-methylmethionine was formed from labeled methionine, but this was shown to be unrelated to ethylene production and thus not an intermediate.

Our inhibitor study clearly showed incorporation of methionine into AdoMet but the behavior of the pathway in the presence of the inhibitors was confusing. AVG, which was thought to directly block ethylene production by inhibiting AdoMet conversion to ethylene, was found to cause an increase in labeled AdoMet from labeled methionine, suggesting that AVG inhibits methionine to ethylene conversion not by blocking the synthesis of AdoMet from methionine but at a step between AdoMet and ethylene. Worse yet, the uncoupler DNP which would have been thought to act by preventing formation of AdoMet by decreasing the amount of ATP available for its synthesis, actually caused an increase in labeled AdoMet from methionine. A second uncoupler CCCP had no effect on the amount of labeled AdoMet even though both uncouplers inhibited conversion of methionine to ethylene.

The inhibitor studies were inconclusive and it appeared that the basis for showing that AdoMet was the immediate precursor of ethylene would eventually come down to a kinetic argument. I sensed that Professor Yang was becoming uneasy with my inconclusive inhibitor studies, but I felt we needed to have a reversible inhibitor if a kinetic argument was to be convincing. We needed one where we could first block ethylene production from labeled methionine, then relieve the inhibition and follow the decrease in labeled AdoMet along with the evolution of ethylene. High temperature was considered as a candidate, but we decided to first try anaerobic conditions because of the long-standing use in ethylene studies, and because both Burg and Thiman (1960) and Baur *et al.* (1971) had hypothesized that an intermediate accumulated under nitrogen which was converted to ethylene upon re-exposure to oxygen. We thought AdoMet might be that intermediate.

Several configurations were tried before we found a system good enough for maintaining adequately low oxygen levels and for storing apple tissue under nitrogen for the duration of feeding experiments. Simple incubation in flasks sealed with serum caps proved inadequate, because even after repeated flushing, the oxygen levels remained high enough to give inconsistent low levels of ethylene production. The solution turned out to be simple but important for observing the elusive intermediate that had been postulated by several workers for many years.

We needed something that would allow for good flushing so that all of the oxygen could be removed from the incubation atmosphere. It was also important to have an accurate value for the volume of the chamber because total ethylene production by the tissue was calculated based on that volume. We needed to be able to re-introduce air very quickly, and then rapidly reseal the chamber if precise kinetic measurements were to be taken. Finally, we wanted to be able to freeze the tissue while still under nitrogen so that any "intermediates" would be preserved because of their expected rapid conversion to ethylene once the tissue was re-exposed to oxygen (air).

The system we developed was based on a disposable 12 ml syringe that became the incubation chamber. Disposable syringes of various sizes were used routinely for gas chromatography associated with ethylene determinations, and as disposable columns for purification of radiolabeled metabolites from feeding experiments. Thus, it was a very natural extension to use them for incubation chambers.

The key features of the system were two holes drilled between the mouth of the syringe and the first (12 ml) calibration mark. A three-way syringe stopcock sealed with a small serum cap was installed on the needle hub, such that when the plunger was just inserted into the barrel, the holes permitted venting of the barrel chamber, but when the plunger was set to the 12 ml mark, the barrel was sealed (Fig. 1). This system allowed nitrogen to be introduced through the three-way stopcock so that the chamber could be continuously flushed with nitrogen without completely removing the plunger from the syringe. After flushing, the plunger was set to the 12 ml mark sealing the barrel, and the sealed chamber could then be sampled through the serum cap. Tissue could be frozen in the chamber by immersing the syringe barrel in a dry ice-ethanol bath which allowed the freezing of the tissue without exposure to oxygen.

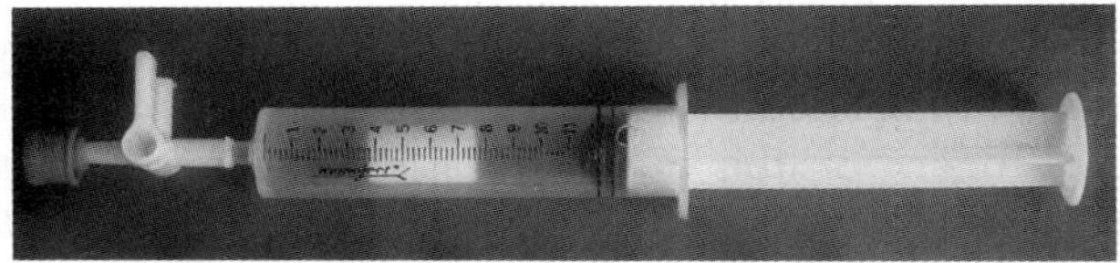

Fig. 1. Syringe chamber used for incubation of apple tissue under anaerobic conditions. The chamber could be flushed by introducing nitrogen through the three-way stopcock allowing exit through two small holes drilled near the opposite end of the syringe barrel. The chamber could be sealed quickly by closing the stopcock and setting the plunger to the 12 ml calibration mark. After incubation the chamber was sampled through the serum stopper on the stopcock.

The first experiments with this simple system were performed using apple plugs alone without labeled substrates. As others had previously reported, we found that ethylene production could be nearly eliminated under nitrogen and that there was a very high ethylene production rate upon re-exposure to air. The elevated ethylene production rates sometimes lasted for up to three hours before returning to pre-anaerobic levels.

Labeling Study Under Nitrogen as a Crucial Experiment

It was a very simple experiment that finally revealed a frequently postulated but elusive intermediate that accumulated under nitrogen and was rapidly converted to ethylene after re-exposure to oxygen. Three plugs cut from a single apple were infiltrated with L-[U-^{14}C] methionine. One plug was incubated continuously in air, a second plug was incubated continuously in nitrogen, and a third plug was incubated in nitrogen for six hours and then in air for six hours. Each hour, the amount of labeled ethylene produced was determined by removing a sample of the atmosphere from the incubation chamber and trapping the labeled ethylene in mercuric perchlorate. The chamber was flushed and resealed so that the amount of labeled ethylene produced during the next hour could be determined, thus giving accurate rate measurements for ethylene production during the experiment. After the twelve-hour time course the plugs were frozen while still in the incubation syringes and then extracted for paper chromatographic analysis of the labeled metabolites remaining in the tissue.   The chromatograms from this experiment showed very unexpected results.

The plug incubated continuously in air showed little methionine remaining in the tissue.   The big surprise was that the plug held continuously in nitrogen contained a metabolite that we had not seen in any of our previous experiments, and this metabolite had disappeared from the plug first held in nitrogen for six hours and then returned to air. During the air incubation, the plug held in nitrogen and then returned to air produced a quantity of labeled ethylene nearly equal to that produced by the plug held continuously in air.   The clear implication was that the unknown metabolite present in the plug kept under nitrogen represented the elusive intermediate that was presumed to accumulate under anaerobic conditions.

One important aspect of this experiment was that the metabolite appearing on paper chromatograms at Rf 0.37 was new. This insight can be attributed to our experience in characterizing many of the radioactive spots that appeared on paper chromatograms in methionine-feeding experiments. We catalogued the Rfs of many of the methionine metabolites, MTA and MTR because of the work implicating AdoMet. From that work, we also recognized that the new 0.37 spot was neither homoserine nor its lactone. The unknown metabolite's mobility on paper chromatograms was similar but not identical to 2-aminobutyric acid.   As methionine is so easily oxidized, we learnt where methionine sulfoxide and methionine sulfone ran; and the product of methionine methylation, S-methyl methionine was also characterized in the AdoMet experiments described earlier. We were surprised by the large spot at 0.37, and after realizing that it fit the criteria for an intermediate, we knew that we must focus on its identity and its relationship to the ethylene pathway.

One of the things that led us to believe that this metabolite was an intermediate was an unpublished plot of the experimental data. The plot shown in Fig. 2 is based on data from the experiment. We believed that a kinetic argument would be needed to implicate AdoMet in the pathway. The labeled ethylene produced in each hour was measured so that the most accurate rates of labeled ethylene evolution could be obtained. The plot of the data that was eventually published is shown in Fig. 3. In Fig. 3, the shapes of the curve from the plug incubated continuously in air, and that from the one kept in nitrogen, and then returned to air appear to be somewhat similar. However, in the plot of the same data shown in Fig. 2, they are seen to be entirely different. In Fig. 2, ethylene produced from methionine by the plug incubated continuously in air is a good example of a product coming from a precursor through intermediate metabolic pools, characterized by a lag, then a rise to a maximum, and finally a decline. But the plot of the data from the plug held in nitrogen and then returned to air shows nearly a first order decline, suggesting that the ethylene was emerging from an immediate precursor. The plot of the data in Fig. 2, which is simply the slope of the lines in Fig. 3, was strong evidence that the unknown labeled metabolite at Rf 0.37 was the intermediate we had been looking for. I presented this explanation at one of the Mann Lab noon seminars before we identified the intermediate, and recall how particularly unconvinced the audience was. It made us realize that kinetic arguments alone would never be convincing without more direct evidence that the metabolite was involved in the pathway. Our research then had two primary objectives: to establish the role of the unknown methionine metabolite in ethylene biosynthesis, and to determine its chemical structure. The first came relatively quickly, but determining its identity took much longer and more unexpected turns than we might have imagined.

Evidence that the metabolite (which by this time we were calling "X"), was directly involved in the ethylene pathway was based on showing that it could be converted to ethylene. We isolated enough of it by paper chromatography to allow a re-feeding experiment, and the results showed nearly quantitative conversion to ethylene when it was fed to apple plugs. The next experiment was to determine if conversion of X to ethylene was blocked by AVG. The results showed very clearly that its conversion to

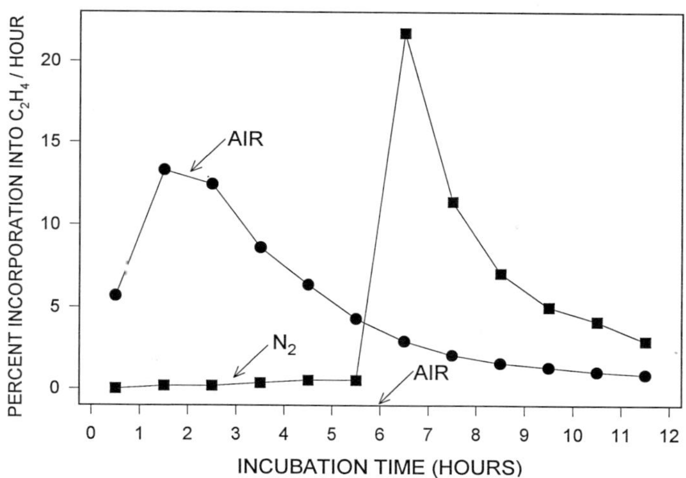

Fig. 2. Incorporation per hour of L-[U-14C] methionine into ethylene by apple plugs. The plugs were infiltrated with 0.5 μCi of L-[U-14C] methionine (100μCi/μmol) and incubated in air (●), or in nitrogen for six hours and then in air (■).

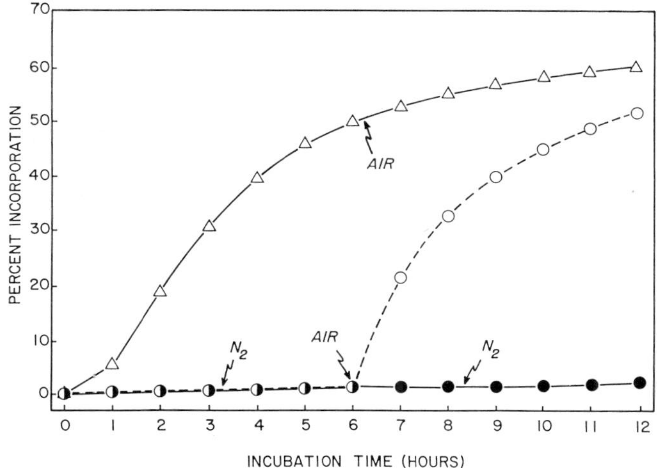

Fig. 3. Total incorporation of L-[U-14C] methionine into ethylene by apple plugs. The plugs were infiltrated with 0 5 μCi of L-[U-14C] methionine (100μCi/μmol) and incubated in air (Δ), in nitrogen (●) or in nitrogen for six hours and then in air (O) .

ethylene was unaffected by AVG. This result established two important points and illustrates the key role that the AVG inhibitor played in establishing the pathway. It showed that X occurred after the well-known block imposed by AVG, and since AVG had not blocked formation of AdoMet from methionine, it meant that the the role of AVG must be to inhibit formation of X from AdoMet. Feeding the metabolite X with and without AVG thus allowed us to propose the biosynthetic sequence of ethylene formation from methionine, (Met → AdoMet → X → C_2H_4) except that the chemical identity of the intermediate X, the proposed immediate precursor of ethylene was still unknown. This single objective became the focus of our research for the next several months.

Identification of "X" as 1-Aminocyclopropane-1-Carboxylic Acid

Feeding studies with appropriately labeled methionine quickly re-affirmed that the unknown metabolite did not contain the sulfur atom or the methyl group from methionine, but that it did retain carbon 1. On electrophoresis at high pH it migrated as an anion, and at low pH as a cation, suggesting the amino group was present and that the molecule was an amino acid. It also behaved as such on ion exchange resins. This work also led us to think that the molecule was relatively stable since chemical methods used to isolate it did not lead to its degradation.

The inference was that it was a four-carbon amino acid derived from carbons 1–4 of methionine. Homoserine and its lactone were quickly eliminated as possibilities as were 2-aminobutyric acid and 4-aminobutyric acid because of their failure to co-chromatograph with the radioactive unknown on paper chromatography. For a while, the imino acid azetidine-2-carboxylic acid was an interesting possibility because of work showing that it was derived from methionine, possibly via AdoMet. However, after obtaining an authentic sample for comparison by co-chromatography, it was also quickly eliminated.

Having excluded the most obvious four-carbon amino acids, we were forced to turn our attention to less obvious ones. We were perhaps influenced in our thinking by the occurrence of the 3,4-unsaturated amino acids such as rhizobitoxine and AVG because they were such important elements in establishing the metabolic sequence. The fact that these compounds were also natural products was encouraging and the corresponding four-carbon member of this family, vinylglycine (IV), was known to be produced by bacteria.

We reasoned that if the unknown was vinylglycine then hydrogenation to reduce the double bond would yield 2-aminobutyric acid (Fig. 4). Yang who once worked as a postdoc in Paul Stumpf's laboratory on campus, had used an apparatus for high pressure hydrogenations. We decided to conduct hydrogenation experiments with the unknown metabolite to see if we obtained 2-aminobutyric acid, which would add support for the notion that the unknown metabolite might be vinylglycine. These experiments were laborious and time-consuming because X could only be obtained by feeding labeled methionine to apple plugs under nitrogen, and then extracting and purifying the metabolite by paper chromatography. Despite this, we conducted several hydrogenation experiments with very encouraging results in favor of the vinylglycine structure. Hydrogenation was found to be very slow and always gave two products in addition to some residual starting material. On the one hand, this gave us the unsettling suspicion that our "purified metabolite" might be a mixture, but the encouraging aspect was that we were able to show that the major one of the two hydrogenation products was 2-aminobutyric acid by co-chromatography and co-electrophoresis. For some time, we believed we had good evidence that the unknown metabolite was in fact vinylglycine.

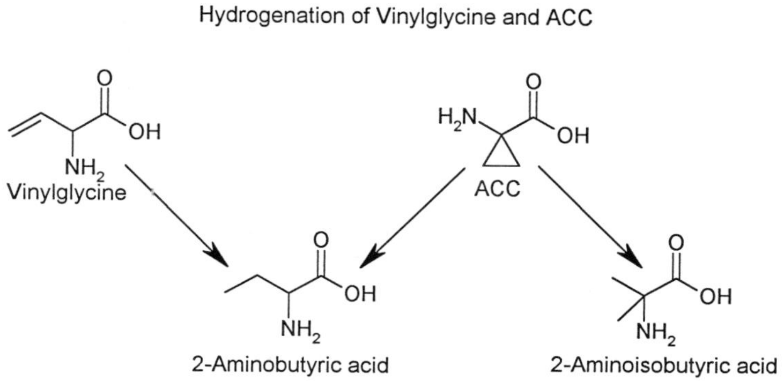

Fig. 4.   Showing how hydrogenation of vinylglycine and ACC can both give rise to 2-aminobutyric acid. The figure also shows the other suspected hydrogenation product of ACC, 2-aminoisobutyric acid.

It was at precisely this time that the deadline for abstracts was approaching for the 1978 meeting of the American Society of Plant Physiologists to be held in Blacksburg Virginia. In the abstract, we meant to say that our preliminary characterization suggested that the compound was vinylglycine. However, because of a typographical error by a secretary in the department, which neither of us caught before it was submitted, the part of the sentence mentioning vinylglycine was left out and our mistaken notion that the unknown was vinylglycine did not appear in the abstract.

It was soon after submitting the abstract that we finally obtained an authentic sample of vinylglycine for comparison with our radioactive metabolite. We felt certain that based on the hydrogenation results, the experiments would confirm its identity as vinylglycine. Chromatography of the radioactive metabolite along with vinylglycine was unequivocal, the metabolite was definitely not vinylglycine. I was horrified and discouraged at the same time; horrified because we had submitted an abstract with the metabolite mis-identified as vinylglycine (we thought), and discouraged because we had seemingly run out of four-carbon amino acids.

My worst moment in graduate school was when I had to walk into Professor Yang's office to tell him that X was not vinylglycine. After looking over the chromatograms carefully, he very calmly said "well, that's a surprise, isn't it?" I was equally surprised by his response as I was by the results since I expected him to be upset. Instead, we ended up discussing the results and going over the compounds, eliminating them one by one. We discussed the possibility that the methionine fragment might be conjugated to another molecule, in which case making determination of its identity very difficult. It would thus require a totally different approach than the one used before.

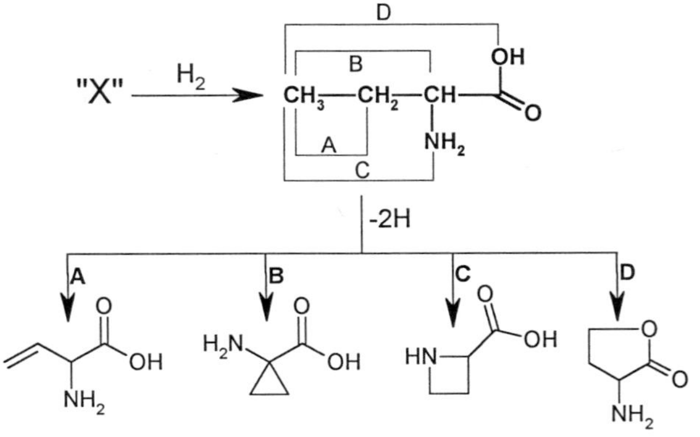

Fig. 5. As hydrogenation of X yielded 2-aminobutyric acid, X must have been a dehydrogenated form of 2-aminobutyric acid. Since dehydrogenation of 2-aminobutyric acid might occur between any of the atom pairs A–D, we reasoned that X had to be one of the compounds shown in the figure. **A**, vinylglycine; **B**, ACC; **C**, azetidine-2-carboxylic acid; **D**, homoserine lactone.

In the ensuing days, we discussed the results repeatedly, tracing our logic and trying to pinpoint where we had gone wrong. Based on the fact that upon hydrogenation, X yielded 2-aminobutyric acid, X must be a dehydrogenated form of 2-aminobutyric acid. By this reasoning, X had to be one of the following compounds (Fig. 5): vinylglycine, azetidine-2-carboxylic acid, 1-aminocyclopropanecarboxylic acid (ACC) or homoserine lactone. By then, we had ruled out the possibility of azetidine-2-carboxylic acid, homoserine lactone and vinyglycine based on their behavior on paper chromatography and paper electrophoresis. The only possibility we had not tested was ACC. We incorrectly reasoned that this cyclopropane compound would be chemically labile whereas we knew X to be quite stable through the steps required to isolate it from paper chromatograms. Although ACC had come up several times during our "paper chemistry" reasoning, this idea was dismissed each time because of the supposed instability of the compound. During our discussions, Professor Yang discovered a particular sentence in J. March's book "Advanced Organic Chemistry". The sentence reads "Cyclopropane rings resemble double bonds in behavior and give analogous products. The direction of ring opening of unsymmetrical cyclopropanes by catalytic hydrogenation is similar to that for attack at double bonds." This description was in full agreement with our results that when X was hydrogenated, it yielded two compounds, the major one being 2-aminobutyric acid. Contrary to our previous assumptions about the stability of X, Yang was now convinced that X must be ACC and that the minor product from the hydrogenation experiments must be 2-aminoisobutyric acid. When I next met Professor Yang, he was very animated and could hardly contain his excitement in relating what he had learnt about the hydrogenation of cyclopropane rings. It was clear that he thought the identity of X might now be solved: X must be ACC, which we had considered earlier but then discounted because of its presumed instability. Although I was not equally enthusiastic about his suggestion, I drew out the proposed structure and set off to the library to practice my "Chemical Abstracts" skills, which I felt by then were quite proficient. I soon found the correct name for the compound, 1-aminocyclopropane-1-carboxylic acid, and found some work on its spectra and chemical synthesis. I was more intrigued to find earlier work showing that the proposed structure was a natural product that had been isolated from cider apples and perry pears by Burroughs in 1957, and from cowberry by Vahatalo and Virtanen also in 1957 (Burroughs, 1957; Vahatalo and Virtanen, 1957). Yang immediately wrote Burroughs a letter requesting a sample of what they abbreviated as ACPC, and which we eventually came to call ACC. I am still amazed that Burroughs was able to retrieve the retrieve the sample from his freezer that he had isolated in 1957, but it was the sample that provided the first evidence that our radioactive metabolite was indeed ACC. I suppose this vindicates those of us who never throw anything away.

When the sample of ACC arrived from Burrough's laboratory, we started through our identification protocols which up till then had easily eliminated all candidates as being the Rf 0.37 radioactive metabolite. This time it was different. Co-chromatography of X with ACC matched perfectly, as did co-electrophoresis. The hydrogenation experiment which had misled us into thinking it was vinylglycine now turned into an advantage by showing that hydrogenation of unlabeled ACC gave the same two products as were observed with the labeled metabolite (Fig. 4).

While these experiments were being completed with the small sample that Burroughs had provided, we learned that ACC was commercially available from CalBiochem. With larger quantities available, we were able to identify the radioactive compound X as ACC by paper chromatography, paper electrophoresis at various pHs and finally by co-crystallization of the radioactive X with authentic ACC to constant specific activity. Yang had done postdoctoral work in Andrew A. Benson's laboratory, where he learnt the powerful technique of confirming the identity of minute amounts of radioactive compounds by co-crystallization to constant specific radioactivity with suspected authentic compounds (Miyano and Benson, 1962). It should be noted that the X compound isolated was highly radioactive but contained very little mass. Hence, co-crystallization experiments were well-suited for this work. A sample containing a known amount of radioactive X, purified by and eluted from paper chromatograms, was added to a sample of authentic ACC. The mixture was then re-crystallized several times and a sample was taken after each re-crystallization for analysis of radioactivity and determination of total ACC. Retention of radioactivity as judged by specific radioactivity indicated that radioactive X was indeed ACC. We also performed a co-crystallization experiment with the major hydrogenation product to firmly establish that it was 2-aminobutyric acid. It was at this point that we considered the structure proven, and the results were communicated to PNAS by Paul Stumpf (Adams and Yang, 1979).

After the identification of ACC and clarification of the pathway, the work in the laboratory proceeded quickly in several directions simultaneously. Yeong-biau Yu who had been working on lipid peroxidation in soybean seeds began working on how auxin induced the ethylene pathway in etiolated mung bean seedlings (Yu *et al.*, 1979; Yu and Yang, 1979). Art Cameron set out to determine the effects of applying ACC to plant tissues, showing that it caused accelerated ethylene production in a wide variety of excised plant tissues (Cameron *et al.*, 1979).

Chemical stability studies started by Neil Hoffman using ninhydrin led to the discovery that ACC could be degraded to ethylene in the presence of hypochlorite. This opened the possibility that a very sensitive assay for ACC could be devised based on the gas chromatographic assay of the ethylene evolved. Connie Lizada conducted experiments to optimize the yield of ethylene in this system, showing that the yield was greatly increased by mercuric ion and further increased if the reaction with hypochlorite was conducted at 0°C. This quickly led to what has become the most widely used assay for ACC (Lizada and Yang, 1979).

Physiological studies clearly showed that the rate-limiting step in the ethylene pathway occurred at the point where AdoMet was converted to ACC. It was also shown by use of the AVG inhibitor that this was the reaction that required pyridoxal phosphate. Together these results suggested that ethylene-producing tissues would contain an as-yet-undescribed enzyme, that could catalyze formation of ACC from AdoMet in the presence of pyridoxal phosphate. Our faithful apple system failed to yield this enzyme activity, as did avocado which was tried because of the large quantity of ethylene it produces during its climacteric. The reason that we were unable to obtain enzyme activity from apple is now clear. In apple tissue, unlike other sources, the enzyme is associated with the pellet fraction in tissue homogenates, and can only be solubilized in an active form by using a detergent such as Triton X-100 in the extraction medium (Yip *et al.*, 1991). Nevertheless, using tomato tissue and the sensitive assay that Connie Lizada had by then developed, we were able to demonstrate and characterize the enzyme now designated ACC synthase E.C. 4.4.1.14 (Y. Yu *et al.*, 1979a).

The story of how ACC came to be recognized as an intermediate in ethylene biosynthesis has a pair of interesting coincidences. First, there were two reports in the same year showing that it occurred as a natural product in fruit (Burroughs, 1957; Vahatalo and Virtanen, 1957). Secondly, there were two reports in the same year implicating it as an intermediate in ethylene biosynthesis (Adams and Yang, 1979; Lürssen *et al.*, 1979). Whereas we identified ACC as the immediate precursor of ethylene by metabolic studies, Lürssen *et al.* came to the same conclusion by screening tests in which they observed that application of ACC enhanced ethylene production in a number of plants and plant organs. The way in which Lürssen and his co-workers came to this conclusion could be an interesting account to add to the current one, and might reveal how a problem that had been the focus of several laboratories reached the same conclusion independently.

As suggested in the introduction to this chapter, this has been a story within a larger account of the role of ethylene in plant biology including both its biosynthesis and its mode of action. For a complete account of the broader picture the reader should consult the chapter in this series on the discovery of ethylene as a plant growth substance (Saltveit, Yang and Kim, 1998).

The discovery of ACC as the immediate precursor of ethylene has stimulated a considerable amount of research in many different areas of plant biology, and has several interesting features which illustrate how research comes about and evolves. The work described here was really the culmination of an effort by several researchers who collaborated with Shang Fa Yang on his long-standing interest in the problem of how plants biosynthesize ethylene.

References

Abeles, F. B and Rubenstein, B. (1964) Cell-free ethylene evolution from etiolated pea seedlings. *Biochim. Biophys. Acta* **93**: 675-677.

Adams, D. O. and Yang, S. F. (1977) Methionine metabolism in apple tissue: Implication of S-adenosylmethionine as an intermediate in the conversion of methionine to ethylene. *Plant Physiol.* **60**: 892-896.

Adams, D. O. and Yang, S. F. (1979) Ethylene biosynthesis: Identification of 1-aminocyclopropane-1-carboxylic acid as an intermediate in the conversion of methionine to ethylene. *Proc. Natl. Acad. Sci. USA.* **76**: 170-174.

Baur, A. H. and Yang, S. F. (1972) Methionine metabolism in apple tissue in relation to ethylene biosynthesis. *Phytochemistry* **11**: 3207-3214.

Baur, A. H., Yang, S. F., Pratt, H. K. and Biale, J. B. (1971) Ethylene biosynthesis in fruit tissues. *Plant Physiol.* **47**: 696-699.

Burg, S. P. (1973) Ethylene in plant growth. *Proc. Natl.Acad. Sci. USA.* **70**: 591-597.

Burg, S. P. and Clagett, C. O. (1967) Conversion of methionine to ethylene in vegetative tissue and fruits. *Biochem. Biophys. Res. Comm.* **27**: 125-130.

Burg, S. P. and Stolwijk, J. A. J. (1959) A highly sensitive kathrometer and its application to the measurement of ethylene and other gasses of biological importance. *J. Biochem. Microbiol. Technol. Eng.* **1**: 245-259.

Burg, S. P. and Thimann, K. V. (1959) The physiology of ethylene formation in apples. *Proc. Natl. Acad. Sci. USA* **45**: 335-344.

Burg, S. P. and Thimann, K. V. (1960) Studies on the ethylene production of apple tissue. *Plant Physiol.* **35**: 24-35.

Burroughs, L. F. (1957) 1-Aminocyclopropane-1-carboxylic acid. A new amino acid in perry pears and cider apples. *Nature* **179**: 360-361.

Cameron, A. C., Fenton, A. C. L., Yu, Y., Adams, D. O. and Yang, S. F. (1979) Increased production of ethylene by plant tissues treated with 1-aminocyclopropane-1-carboxylic acid. *Hort. Science* **14**: 178-180.

Chou, T. W. and Yang, S. F. (1973) The biogenesis of ethylene in *Penicillium digitatum*. *Arch. Biochem. Biophys.* **157**: 73-82.

Gane, R. (1934) Production of ethylene by some ripening fruits. *Nature* **134**: 1008.

Hansen, E. (1942) Quantitative study of ethylene production in relation to respiration of pears. *Bot. Gaz.* **103**: 543-558.

Lieberman, M. A., Kunishi, A., Mapson, L. W. and Wardale, D. A. (1966) Stimulation of ethylene production in apple tissue slices by methionine. *Plant Physiol.* **41**: 376-382.

Lieberman, M. A., Kunishi, A., Mapson, L. W., and Wardale, D. A. (1965) Ethylene production from methionine. *Biochem. J.* **97**: 449-459.

Lieberman, M., Kunishi, A. T., and Owens, L. D. (1975) Specific inhibitors of ethylene production as retardants of the ripening process in fruits. In *Facturs Et Regulation De La Maturation Des Fruits*, Colloques Internationaux Du Centre National De La Recherche Scientifique No. 238. Centre National De La Recherche Scientifique, Paris.

Lizada, M. C. A and Yang, S. F (1979) A simple and sensitive assay for 1-aminocyclopropane-1-carboxylic acid. *Anal. Biochem.* **100**: 140-145.

Lürssen, K., Nauman, K and Schroder, R. (1979) 1-Amino-cyclopropane-1-carboxylic acid - an intermediate of the ethylene biosynthesis in higher plants. *Z. Pflanzenphysiol.* **92**: 285-294.

Miyano, M. and Benson, A. A. (1962) The plant sulfolipid. VI. configuration of the glycerol moiety. *J. Am. Chem. Soc.* **84**: 57-59.

Murr, D. P. and Yang, S. F. (1975) Inhibition of *in vivo* conversion of methionine to ethylene by L-canaline and 2,4-dinitrophenol. *Plant Physiol.* **55**: 79-82.

Murr, D. P. and Yang, S. F. (1975a) Conversion of 5'-methylthioadenosine to methionine by apple tissue. *Phytochemistry* **14**: 1291-1292.

Owens, L. D. and Wright, D. A. (1965) Rhizobial-induced chlorosis in soybeans: isolation, production in nodules, and varietal specificity of the toxin. *Plant Physiol.* **40**: 927-930.

Owens, L. D. and Wright, D. A. (1965a) Production of the soybean chlorosis toxin by Rhizobium japonicum in pure culture. *Plant Physiol.* **40**: 931-933.

Owens, L. D., Guggenheim, S. and Hilton, J. (1968) Rhizobium-synthesized phytotoxin: an inhibitor of β-cystathionase in Salmonella typhimurium. *Biochim Biophys. Acta* **158**: 219-225.

Owens, L. D., Lieberman, M. and Kunishi, A. (1971) Inhibition of ethylene production by rhizobitoxine. *Plant Physiol.* **48**: 1-4.

Owens, L. D., Thompson, J. F., Pitcher, R. G. and Williams, T. (1972) Structure of rhizobitoxine, an antimetabolic enol-ether amino acid from Rhizobium japonicum. *J. Chem. Soc. Chem. Commun.*, p. 714.

Pegg, A. E. and Williams-Ashman, H. G. (1969) Phosphate-stimulated breakdown of 5'-methylthioadenosine by rat ventral prostate. *Biochem. J.* **115**: 241-247.

Pruess, D. L., Scannell, J. P., Kellett, M., Ax, H. A., Janecek, J., Williams, T. H., Stempel, A. and Berger, J. (1974) Antimetabolites produced by microorganisms X, L-2-amino-4-(2-aminoethoxy-trans-3-butenoic acid. *J. Antibiotics* **27**: 229-233.

Rando, R. R. (1974) Chemistry and enzymology of K_{cat} inhibitors. *Science* **185**: 320-324.

Saltveit, M. E., Yang, S. F. and Kim, W. T. (1998) History of the Discovery of Ethylene as a Plant Growth Substance. In *Discoveries in Plant Biology*, Vol. 1., S. D. Kung and S. F. Yang, eds., World Scientific Publishing Co. Pte. Ltd., Singapore, pp. 47-70.

Shapiro, S. K. and Mather, A. N. (1958) The enzymatic decomposition of S-adenosyl-L-methionine. *J. Biol. Chem.* **233**: 631-633.

Vahatalo, M. L. and Virtanen, A. I. (1957) A new cyclic α-aminocarboxylic acid in berries of cowberry. *Acta Chem. Scand.* **11**: 741-743.

Yang, S. F. (1974) The biochemistry of ethylene: biogenesis and metabolism. In *The Chemistry and Biochemistry of Plant Hormones. Recent Advances in Phytochemistry*, Vol. 7., V. C. Runeckles, E. Sondheimer and D. C. Walton., eds., Academic Press, New York, pp. 131-164.

Yang, S. F. and Baur, A. H. (1972) Biosynthesis of ethylene in fruit tissues. In *Plant Growth Substances 1970.*, D. C. Carr, ed., Springer Verlag, Berlin and New York, pp. 510-517.

Yang, S. F., Ku, H. S. and Pratt, H. K. (1966) Ethylene production from methionine as mediated by flavin mononucleotide and light. Biochem. *Biophys. Res. Comm.* **24:** 739-743.

Yang, S. F., Ku, H. S. and Pratt, H. K. (1967) Photochemical production of ethylene from methionine and its analogues in the presence of flavin mononucleotide. *J. Biol. Chem.* **242:** 5274-5280.

Yip, W-K, Dong, J-G. and Yang, S. F. (1991) Purification and Characterization of 1-aminocyclopropane-1-carboxylate synthase from apple fruits. *Plant Physiol.* **95:** 251-257.

Young, R. E., Pratt, H. K. and Biale, J. B. (1952) Manometric determination of low concentrations of ethylene. *Anal. Chem.* **24:** 551-555.

Yu, Y. and Yang, S. F. (1979) Auxin-induced ethylene production and its inhibition by aminoethoxyvinylglycine and cobalt ion. *Plant Physiol.* **64:** 1074-1077.

Yu, Y., Adams, D. O. and Yang, S. F. (1979) Regulation of auxin-induced ethylene production in mung bean hypocotyls. *Plant Physiol.* **63:** 589-590.

Yu, Y-B., Adams, D. O. and Yang, S. F. (1979a) 1-Aminocyclopropane-carboxylate synthase a key enzyme in ethylene biosynthesis. *Arch. Biochem. Biophys.* **198:** 280-286.

Chapter 3

Discovery of Auxin

Yoshio Masuda
2-37 Matsugamoto-cho, Ibaraki-shi,
Osaka 567-0033, Japan

and

Seiichiro Kamisaka
Department of Biology, Faculty of Science, Osaka City University,
Sumiyoshi-ku, Osaka 558-8585, Japan

ABSTRACT

Pioneer studies in the 19th – early 20th century leading to the discovery of auxin are introduced first. In 1928, Dutch botanist Fritz W. Went finally isolated auxin diffused out from the tip of oat coleoptiles in the gelatin block. Following Went's success, auxin, indole-3-acetic acid (IAA) was then isolated first from human urine, then from fungi, and finally from higher plants. Discovery of auxin is thus the result of the work of many botanists and organic chemists from various countries. Biosynthesis, metabolism and physiological actions of auxin are also briefly described.

Early Studies Leading To The Discovery Of Auxin

The discovery of auxin has been thoroughly described in most of the textbooks on phytohormones and plant physiology and hence may not require further description in this article. Nevertheless, it may be worthwhile to review how auxin was discovered from another point of view.

The hormone concept was developed from three discoveries in the 19th century (Went and Thimann, 1937; Hartung, 1984); namely, (1) "organ-forming substances" by Julius Sachs (1882, 1887); (2) "the influence of light" from the study of phototropism by Charles Darwin (1880); and (3) the idea of "growth enzymes" from the study on galls by M. W. Beijerinck (1888, 1897).   The above three discoveries will be described briefly in the following.

(1) Studies by Sachs: when stem segments excised from herb and woody plants were kept vertically or horizontally, shoots and roots were regenerated at the top and bottom end respectively.   This experiment is described in Sachs' textbook "Vorlesungen über Pflanzenphysiologie" (1882), p. 632-, as [Versuchen mit der Reproduktion von Organen an abgeschnittenen Pflanzentheilen].   The whole experiment is explained in detail here.   If we look at the corresponding pages in the English edition of the same book "Lectures on the Physiology of Plants" (1887), p. 520, after describing experimental results, Sachs interpreted them as follows: "It is to be noticed that the "substances" for the formation of organs (apart from germinal stages where they proceed from the reservoirs of reserve material) are produced in the foliage leaves by assimilation, and pass out thence into other parts of the plant.   So long as a vigorous main bud of the shoot is present, mixtures of substances suited chiefly for the formation of shoots pass into it from the leaves, whereas the substances fitted for the formation of roots flow in the opposite direction, into the roots which already exist.   If then a portion of the shoot axis is cut off and kept in a moist warm environment, the substances suitable for the formation of shoots which are already present in it will, as heretofore, move in the acropetal direction, and those which form roots, on the contrary, in the basipetal one."   Sachs also explained his ideas on the "substances", how they were influenced by gravity and light in the following pages.

It is described in "Phytohormones" by Went and Thimann (1937, p. 7) "The phenomenon of correlation was studied in greater detail and Sachs brought forward a complete theory — Sachs' great achievement was that he applied the laws of causality to morphology."   In fact, Sachs assumed the existence of root-forming, flower-forming and other substances which move in different directions through the plant.   The above book also describes Sachs' studies as follows: light and gravity were assumed to affect the distribution of these special substances.   With only two assumptions: 1. the existence of organ-forming substances which, in minute amounts, direct development, and 2. polar distribution of these substances, — a distribution which may be modified by external forces such as light and gravity.   It was not clear however if these organ-forming substances were a single substance or a mixture of different substances. Sachs really had foresight.

(2) The study of phototropism by Charles Darwin using *Avena* and *Phalaris* coleoptiles was the first clue to the isolation of auxin, as indicated some "influence" transported from the tip to the base of coleoptiles (1881). It took thirty years until a Danish botanist, P. Boysen-Jensen, further characterized this "influence" by his most elaborate experiments, showing that when the coleoptile tip was cut off and stuck on again, the coleoptile regained its ability to respond to the light. He also showed that even when the tip was stuck on placing an agar or gelatin block in between (1910). His study clearly indicated that the influence could be transported from the tip to the base not only through the cut wound but also through a foreign object, i.e., agar or gelatin. The fact that this influence was truly the cause for coleoptile curvature was clearly pointed out by a Hungarian botanist, Arpad Paál, who showed that when, in the dark, the tip was cut off and replaced it asymmetrically, the coleoptile showed curvature, similar to the curvature caused by light (1919). He gave a name to this "influence" from the tip to the base as "correlation carrier". In 1928 this "influence" was finally isolated in the agar block by F. W. Went (1928) (Fig. 1). He placed the excised tips of *Avena* coleoptiles on agar or gelatin for a couple of hours and then applied the small agar block asymmetrically to another coleoptile stump. The agar now had received the "influence" and could produce curvature. Thus, the influence was first isolated as a substance into an agar block. This famous experiment has been frequently described in plant physiology text books and other publications (e.g., Gabriel and Fogel, 1955). Taylor (1963) described the work of Went in the following way:

"The crucial experiment which proved that plants also possess 'growth hormones' was carried out in 1923 by a young botany student in Utrecht, named Frits Went. By day, Went was serving in the Dutch army, but at night he returned to his father's plant physiology laboratory at the university to keep up his graduate studies in botany. The test which he thought up was to cut off the tip of an oat seedling, lay it on a small block of gelatin, and leave it for several hours, in the expectation that the supposed growth substance would seep out into the gelatin. Then he would take the block of gelatin and stick it on the side of a second decapitated seedling, in the hope that it would grow faster on the side where the block was stuck, thus causing the stem to bend. This experiment was less easy than it sounds. It had to be done in only the dimmest red light, since roots also tend to curve away from a source of light; and in a very humid atmosphere, to prevent the cut tip drying out. At 3 a.m. on the morning of April 17, 1928, Went returned to the laboratory to see how his experiment was proceeding, and found that the tip had indeed curved as he had foreseen."

This outstanding discovery then led to the isolation and characterization of "auxin" made by Dutch organic chemists, as described later.

Fig. 1. Botanical Laboratory, University of Utrecht. Went did his experiments on isolation of auxin in the gelatin blocks and on the effect of auxin on the curvature of *Avena* coleoptiles in the laboratory in the cellar of this building (taken by Y. Masuda in 1962).

(3) Beijerinck (1888) considered that the development of galls was caused by "a protein", although its action was different from that of ordinary proteins, but resembled that of enzyme. He also found that the gall developed when a bee (*Nematus*) pricked the leaf veins; thus supposed that the growth enzyme existed in the secretion of the insect. Sachs was strongly intrigued by Beijerinck's studies and built up a precise concept of his organ-forming substances (Gimmler, 1984). Later, Beijerinck (1897) extended his view to the development of organisms in general, that form was determined by liquid substances, which move freely through considerable numbers of cells in growing tissues (Went and Thimann, 1937).

Fitting (1910) isolated an extract from pollens which caused the growth of ovary and he was the first to use the name "hormone" in botany. Interestingly, he was also the first to use the term "coleoptile" for what Darwin called "cotyledon" (Fitting, 1907). Yasuda (1934) also found that the water extract of *Petunia* pollen promoted the growth of ovary of egg plants. Laibach (1933) then found that the ether extract not only from pollen but also from human urine and animal tissues promoted the growth of ovary and coleoptiles. This must have been the strong clue why Kögl extracted auxin from human urine. The "pollen hormone" then turned out to be auxin.

It should also be noted here that Haberlandt (1913) studied cell division in potato tuber tissues and found that in small pieces of the parenchymatous tissue, tissue cell division only occurs in the presence of a fragment of a vein containing phloem tissue. He concluded that two substances regulate cell division, one from the wound called "wound hormone" and the other in lesser amount from some other tissues "leptohomone". Although Haberlandt was unable to cultivate the tissue by cell division, he was truly the pioneer of *in vitro* tissue culture. At least one of these hormones proposed by Haberlandt was later found to be auxin.

Isolation and Characterization Of Auxin

Thus, although the three lines of early studies proposing the existence of hormonal substances (with different names depending upon their physiological roles either in organ formation, phototropism and gall formation) contributed to the future isolation of auxin, none of them offered any direct evidence for the existence of such special substances except for the study by Went who isolated the Darwin's influence in the agar block. Apart from the above studies, in 1910 Hans Fitting gave the name "hormone" to an ether extract of pollens which promoted the growth of ovaries. Thus the time, the middle of 1930s, was mature for the isolation and characterization of auxin. The study of Boysen-Jensen was particularly of special significance; namely, his discovery that the "influence" diffused in through the tissue wound to agar blocks and out from the agar blocks to the plant tissue again through the wound. This must have been a strong clue for Went, and possibly for Laibach who further contributed to the chemical isolation of auxin (Wuchsstoff) by the Dutch biochemists.

It is interesting to note that early studies on the isolation of growth substances were carried out by extracting a diversity of biological materials other than plants (Went and Thimann, 1937). The materials were malt, saliva, diastase, pepsin, pathogenic fungi, or even human urine. The first large scale chemical studies involving the isolation of auxin was carried out by Dutch biochemist F. Kögl and his co-workers (1933), who extracted the substances from human urine, probably because they were aware of Laibach's work. They began isolating growth substances from human urine in 1931 (see references), and first isolated and identified "auxins a and b" which caused great confusion later. Taylor (1963) describes:

"After a long and difficult research, Kögl, a chemist, extracted a few drops of the active agent from 100,000 maize seedlings, but there was not enough to analyze. Then came the discovery that human urine was rich in this growth substance: and soon Kögl had analyzed it. In 1934 a substance 50,000 times more potent was discovered. It is now known as auxin a, while that which Went discovered is called heteroauxin. Subsequently, other growth substances, and finally a whole range of hormones controlling fruiting, root development, leaf fall and possibly flowering were found. English children sing a nursery rhyme which asks

How can you or I or anyone know
How oats and beans and barley grow?

It may not be strange that biologists have largely answered this question, but it is a very strange coincidence that it was precisely with work on oats and beans that the growth process was explored, and in barley that the flowering process was chiefly studied."

However, the story of the discovery of "auxins a and b" is a real mystery until the time when a paper from the same Utrecht organic chemistry institute was published in 1966, reporting that sample vials of those auxins and their lactones were found in the laboratory when it was to move to a new building. When the samples were subjected to modern mass spectrometry analysis, it turned out that the substances were not the auxins and lactones but something else. Thus, the authors of this paper concluded that auxins a and b were non-existing (Vliegenthart and Vliegenthart, 1966). In the same year, Nakamura and her co-workers investigated the physiological activity of auxin b lactone which had been synthesized and found that it had no growth promoting activity at all (Nakamura *et al.*, 1966). Went and Thimann wrote in their textbook (1937) "A remarkable property of both auxin a and b is their spontaneous inactivation." Without further investigation into the cause of this false discovery, we conclude that only "heteroauxin" (indole-3-acetic acid), isolated and identified in 1934 (Kögl *et al.*, 1934) and having a simpler structure than auxins a and b, turned out to be the genuine auxin.

The genuine auxin, "heteroauxin" was named upon its isolation and identification. Chemically, indole-3-acetic acid (IAA) was isolated as crystals from a fungus (*Rhizopus*) (Thimann, 1935) or from cornmeal (Haagen-Smit *et al.*, 1942). The latter was the first success in the isolation and identification of auxins in plant materials, but the IAA crystals were finally isolated from the immature seeds by Haagen-Smit and his co-workers (Haagen-Smit *et al.*, 1946).

Upon the deaths of Boysen-Jensen and Kögl, Thimann wrote an obituary for them in 1960 . In the beginning and end of his obituary, he wrote: "The deaths during the past year of both Fritz Kögl and Peter Boysen-Jensen may be said to mark the end of an era in that active field which encompasses plant growth and growth substances. Both men made substantial contributions to knowledge, and both more or less withdrew later on from the field, though for different reasons". Thimann concluded his obituary: "Whether Kögl and Boysen-Jensen ever saw much of one another is not recorded. They were both at the Botanical Congress in Amsterdam in 1935, and Kögl lectured widely in Europe on the work of his laboratory. Boysen-Jensen, perhaps because of indifferent health, was no great traveller or public lecturer. Indeed he had few collaborators and seems to have preferred to work alone, while Kögl did virtually all his work through students or assistants. But by now it is a platitude to say that successful scientists include all types of personality; about the only thing they have in common is a devouring interest in their work. Certainly this characteristic was exemplified in these two men, and botany owes them a great debt."

Occurrence of IAA

After World War II, IAA was found to be widely distributed and synthesized in the tissues of higher plants. With the development of analytical techniques for qualitative and quantitative determination of IAA (e.g., thin-layer chromatography or GC-MS), IAA was thus identified in a variety of higher plants, several gymnosperms and many angiosperms (e.g., Schneider and Wightman, 1974 and other studies). Besides IAA, a new naturally occurring auxin, 4-chloroindole-3-acetic acid (4-Cl-IAA) and its methylester were isolated from immature seeds of peas and other leguminous plants (Marumo *et al.*, 1968; Engvild *et al.*, 1980). There is a possibility that 4-Cl-IAA is widely distributed in plant tissues since this compound was discovered in the seed of *Pinus sylvestris*, a conifer (Ernstsen and Sandberg, 1986). There are also precursors for IAA biosynthesis and related compounds which are found in a wide variety of plants.

The amount of IAA in vegetative tissues of higher plants, as measured by bioassay of extracts, ranges 1-100 micrograms per kg fresh weight, but in some cases the amount determined was found to be more than 300 micrograms per kg, in oat and maize shoots and roots (Rivier and Pilet, 1974).

The amount of IAA varies depending upon the kinds of organs: e.g., in soybean, immature seeds [50-200 ng per g fresh weight], primary leaves

[6-10], trifoliates [6-10], stems [25-48], cotyledon [18] and roots [14], or in tomatoes, axis [3.8 ng per g fresh weight], locule [121], seed [1313], pericarp [11], sepal [46] (Tables in the monograph by Takahashi and Masuda, 1994).

Biosynthesis and Metabolism Of Auxins

It is understood for the last 25-30 years that the precursor for IAA synthesis is an amino acid tryptophan (Try). However, a doubt against it has also been proposed (Wright *et al.*, 1991) and whether L-Try or D-Try is serving as the direct precursor of IAA has been controversial (e.g., Kutacek and Kefeli, 1970; Tsurusaki *et al.*, 1990). The pathway from Try to IAA consists of five routes. In the pathway from L-Try via D-Try, then indolepyruvic acid (IPyA) to indoleacetaldehyde (IAAld) and finally to IAA, L-Try is converted by a racemase to D-Try which in turn converted to IPyA by aminotransferase (e.g., McQueen-Mason and Hamilton, 1989). Two forms of L-Try aminotransferases and one D-Try aminotransferase were reported to be separated from maize coleoptiles (Koshiba *et al.*, 1993).

Secondly, the IPyA pathway starting from L-Try has long been known. The enzymes, amino transferase (L-Try to IPyA), IPyA decarboxylase (IPyA to IAAld), IAAld oxidase or IAAld dehydrogenase (IAAld to IAA) have been known, although the first enzyme has a high Km value. This has been the cause for a high possibility that L-Try could first be converted to D-Try which then produces IPyA. Recently, Koshiba and Matsuyama (1993) claimed that L-Try is converted to IAA in a single step by an enzyme containing IPyA oxidase.

Then comes the tryptamine pathway by which tryptamine (TNH$_2$) is produced from Try by Try decarboxylase, then TNH$_2$ is converted to IAAld by TNH$_2$ oxidase. However, this conversion has been found only in a limited number of plants; therefore this pathway is not considered a universal one in the biosynthesis of IAA in plants. It was reported that the amount of TNH$_2$ in transgenic tobacco plants containing the Try carboxylase gene accumulates to 360 times that of non-transgenic plants, but the amount of IAA remains the same (Songstad *et al.*, 1990).

There are reports stating the presence of a pathway where Try is converted to indole-3-acetaldoxime (IAOX) which is converted to indoleacetonitrile. Also reported, is the pathway of indoleacetamide which is produced from Try and converted to IAA, this pathway being originally found in *Agrobacterium*, but the enzymes involved, Try monooxigenase and indoleacetamide (IAM) hydrolase were found in higher plants too (Kawaguchi *et al.*, 1991, 1993).

The above-mentioned first two pathways from tryptophan to IAA via IPyA or TNH_2 have been widely accepted (Sembdner *et al.*, 1981). These pathways were proposed based upon *in vivo* tracer experiments where radioactive Try is fed for a long period of time (e.g., several hours). If feeding time is one to three hours, the incorporation of the label to the intermediates found in long term experiments was not observed in maize and barley protoplasts (Iino, 1982; Sandberg *et al.*, 1982). The biosynthetic system of one-step conversion from Try to IAA using an *in vitro* coleoptile extracts (Koshiba and Matsuyama, 1993) does not discriminate the L- and D-enantiomers of Try. Interestingly enough, the conversion of Try to IAA was only found in coleoptile tips but not in sub-apical segments (5-7.5 mm from the top), indicating that the synthesis of IAA occurs specifically in the tip region (Koshiba *et al.*, 1995). Koshiba *et al.* (1995) also reported that a pulse irradiation with red light, two hours before excision of tips and the application of radioactive Try, caused a reduction in the amounts of extractable and diffusible IAA by 40-60% without a change in the rate of label incorporation. Their findings seem to be in agreement with the discovery of auxin from the coleoptile tips by F. W. Went.

Libbert and his co-workers studied the pathways of IAA biosynthesis in the homogenates of pea stems and epiphytic bacteria isolated from the stem (1970a, b). They found that the conversion from TNH_2 to IAA occurred rapidly in the stem but the conversion from Try to TNH_2 was barely found; therefore, they concluded that in plants, IAA biosynthesis occurs via IPyA, a non-efficient system.

It was reported that 95% of auxin in plant tissues are in the bound form (van Overbeek, 1941). Auxins are metabolized in plant tissues in two ways: (1) auxins are bound to glucose and aspartic acid; (2) auxins are oxidized by oxidases and peroxidases.

In the plant tissue, it has been established that the active form of auxin is supposed to be diffusible (Scott and Briggs, 1960) and the bound form of auxin has four-fold functions: (1) stored form; (2) protected against degradation; (3) to keep a constant concentration of active auxin, i.e., detoxication of supraoptimal concentration of auxin; (4) related with transport of auxin (Kamisaka, 1983). IAA-aspartate and IAA-glucose were found during the period 1950-60 (Andreae and Good, 1955; Zenk, 1961). If IAA is applied to stems and roots, IAA is found to bind not only to aspartic acid but to other amino acids as well. Another metabolite of IAA, dioxindole-3-acetic acid (DIA) has been found in *Vicia faba* L., and IAA-aspartate is reported to be the precursor of DIA (Tsurumi and Wada, 1980).

Epstein *et al.* (1980) reported that when maize seeds germinate, IAA-inositol present in the endosperm is transported to the coleoptile tip and converted to IAA. Thus, at the germination stage of graminaceous seeds, IAA is not only synthesized in the coleoptile tips (oat), but also formed from IAA-inositol (maize).

IAA in plant tissues is degraded oxidatively by IAA oxidase and peroxidase, to produce 3-hydroximethyloxiindole, 3-methyleneoxindole and 3-methyloxindole, as found in pea seedlings (e.g., Tuli and Moyed, 1967). These enzymes exist as isoenzymes and are functionally divided into four types (see Kamisaka, 1983). They need Mn^{++} and monophenol as cofactors (Wagenkneght and Burris, 1950; Goldacre *et al.*, 1953).

Synthetic Auxins

Many compounds have been synthesized in order to examine the physiological activity of auxin. The representative compounds with auxin activities are chlorinated phenoxy and benzoic acids (Zimmerman and Hitchcock, 1942). The representative is 2,4-dichlorophenoxyacetic acid (2,4-D) which has also been widely used as a herbicide. In addition, naphthaleneacetic acid (NAA), cis-cinnamic acid and phenylacetic acid (PAA) have been found to have auxin activity, PAA was discovered as the naturally occurring auxin in plants (Okamoto *et al.*, 1967; Schneider and Wightman, 1974).

Some substituted synthetic compounds or isomers such as 2,4,6-trichlorophenoxyacetic acid and trans-cinnamic acid (van Overbeek *et al.*, 1951) have been found behaving as anti-auxins which competitively inhibit auxin effects. In addition, 2,3,5-triiodobenzoic acid has also been found as an anti-auxin which inhibits polar transport of auxin (Thimann and Bonner, 1948).

Physiology

Chemists Majima and Hoshino (1925), working at Tohoku Imperial University, synthesized IAA during their studies on indole compounds. This enabled many physiologists to see the effect of applied IAA on a variety of phenomena. Otherwise, natural auxin is very difficult to obtain.

It has long been known that auxin shows a diversity of physiological effects, as have been summarized in many books and reviews (e.g., Went and Thimann, 1937; Letham *et al.*, 1978; Scott, 1984). The main area of study has been the mechanism of auxin action on cell extension, in connection with tropisms. It should be noted that first physiological studies on the mechanism of auxin on cell extension were carried out by Hugo de Vries who was later known as a geneticist, but his early career was as a plant physiologist in the Institute of Julius Sachs in Würzburg, Germany. De Vries (1874), using several techniques, quantitatively measured the elastic (reversible) and plastic (irreversible) deformation of flower stalks of different plant species and found that there was a correlation between the both deformabilities due to applied force and the growth rate. This work was ignored for a long time until Dutch botanist Anton Heyn (1930) and van Overbeek (with Heyn, 1931) (after auxin had been isolated in agar block by F. W. Went) applied auxin-containing agar block to decapitated oat coleoptiles and quantitatively measured the mechanical property of the coleoptile cell wall. They used two techniques, i.e., the stretching method where a stretching force was applied to deform the cell wall specimen and, the bending method where horizontally placed coleoptiles were bent by applying a weight on the tip, and the angle of bent coleoptiles was measured. At around the same time, a German botanist Hans Söding (1931) did a similar study and also found the auxin effect to promote the deformability. The measurement has become extremely elaborated recently and the property is now interpreted in physical terms of rheology e.g., using stress-relaxation phenomenon (Yamamoto *et al.*, 1970). Cell wall studies then expand to biochemical ones, analyzing the change in constituting polysaccharides as the background of the mechanical property, owing to an extensive elucidation of chemical structure of the primary cell wall by Albersheim and his co-workers (summarized 1976). The cell wall studies in connection with auxin effects have been reviewed (Masuda, 1978, 1990). There is plausible evidence that partial degradation of hemicellulosic polysaccharides of the cell wall is involved in auxin-induced changes in the cell wall, using an antibody against polysaccharide fragments such as xyloglucans (Hoson *et al.*, 1991, 1992). In addition, it has been pointed out that biosynthesis of cell wall polysaccharides is also necessary for the continued auxin-induced cell elongation (Inouhe *et al.*, 1987).

Notably, there was a claim that hydrogen ions act as second messenger for auxin action to induce cell extension (Hager *et al.*, 1971). This claim aroused a lot of controversial discussion (Sakurai *et al.*, 1977), thus it requires further scrutiny.

Recently, the mechanism of auxin was studied at the molecular level and it is now known that auxin specifically induces gene expression earlier than the initiation of growth, indicating that the activation of gene expression may be responsible for the growth processes (Theologis, 1986; Koshiba *et al.*, 1995). Although the role of nucleic acid in auxin-induced cell growth was suggested quite some time ago (e.g., Key and Ingle, 1964; Masuda, *et al.*, 1966), it has only recently attracted attention again and in fact, several primary auxin responsive genes have been identified (Kim *et al.*, 1997). It is hoped that the proteins directly involved in auxin action would soon be elucidated.

References

Albersheim, P. (1976) The primary cell wall. In *Plant Biochemistry, 3rd*, J. Bonner and J. E. Varner, eds., pp. 91-114, Academic Press, New York.

Andreae, W. A. and Good, N. E. (1955) The formation of indoleacetylaspartic acid in pea seedlings. *Plant Physiol.* **30**: 380-382.

Beijerinck, M. W. (1888) Ueber das Cecidium von Nematus Capreae auf Salix amgdalina. *Bot. Zeit.* **46**: 17-27.

Beijerinck, M. W. (1897) Sur la cecidiogenese et la generation alternante chez le Cynips calicia. *Verz. Geschr.* **III**: 199-232.

Boysen-Jensen, P. (1910) Über die Leitung des phototropischen Reizes in *Avena*-Koleoptiles. *Ber. Deut. Bot. Ges.* **28**: 28-120.

Darwin, C. (1881) The Power of Movement in Plants. D. Appleton & Co., New York.

De Vries (1974) Über die Dehnbarkeit wachsender Sprosse. Arbeit. d. Bot. Inst. Würzburg **1**: 519-545.

Engvild, K. C., Egsgaard, H. and Larsen, E. (1980) Determination of 4-chloroindole-3-acetic acid methyl ester in *Lathyrus, Vicia* and *Pisum* by gas chromatography-mass spectrometry. *Plant Physiol.* **48**: 499-503.

Epstein, E., Cohen, J. D. and Bandurski, R. S. (1980) Concentration and metabolic turnover of indoles in germinating kernels of *Zea mays* L. *Plant Physiol.* **65**: 415-421.

Ernstsen, A. and Sandberg, G. (1986) Identification of 4-chloroindole-3-acetic acid and indole-3-aldehyde in seeds of *Pinus sylvestris. Physiol. Plant* **68**: 511-518.

Fitting, H. (1907) Die Leitung tropistischer Reize in parallelotropen Pflanzenteilen. *Jahrb. wiss. Bot.* **44**: 177-253.

Fitting, H. (1910) Weitere entwicklungsphysiologische Untersuchungen an Orchideenbluten. *Zeits. f. Bot.* **2**: 225-267.

Gabriel, M. L. and Fogel, S. eds. (1955) Great Experiments in Biology. Englewood Cliffs, N. J. Prentice-Hall, Inc.

Gimmler, H. ed. (1984) Julius Sachs und die Pflanzenphysiologie heute. Verlag der Physik. Med. Gesellschaft.

Goldacre, P. L., Galston, A. W. and Weintraub, R. L. (1953) The effect of substituted phenols on the activity of the indolleacetic acid oxidase of peas. *Arch. Biochem.* **43**: 358-373.

Haagen-Smit, A. J., Leech, W. D. and Bergern, W. R. (1942) The estimation, isolation and identification of auxins in plant materials. *Amer. J. Bot.* **29**: 500-506.

Haagen-Smit, A. J., Dankider, W. B., Wittwer, S. H. and Murneek, A. E. (1946) Isolation of 3-indoleacetic acid from immature corn kernels. *Amer. J. Bot.* **33**: 118-120.

Haberlandt, G. (1913) Zur Physiologie der Zellteilung. Sitz. ber. k. preuss. *Akad. Wiss.* **1913**: 318-345.

Hager, A., Menzel, H. and Kraus, A. (1971) Versuche und Hypothese zur Primärwirkung des Auxins beim Streckungswachstum. *Planta* **100**: 47-75.

Hartung, W. (1984) Der Beitrag von Julius Sachs zur Entdeckung der Phytohormone. In "Julius Sachs und die Pflanzenphysiologie heute", H. Gimmler, ed., Verlag der Physik.-Med. Gesellshaft, pp. 167-180.

Heyn, A. N. J. (1930) On the relation between growth and extensibility of the cell wall. *Proc. Roy. Acad. Amsterdam.* **33**: 1045-1058.

Heyn, A. N. J. and van Overbeek, J. (1931) Weiteres Versuchsmaterial zur plastischen und elastischen Dehnbarkeit der Membran. *Proc. Kon. Akad. Wetens. Amsterdam.* **34**: 1190-1195.

Hoson, T., Masuda, Y., Sone, Y. and Misaki A. (1991) Xyloglucan antibodies inhibit auxin-induced elongation and cell wall loosening of azuki bean epicotyls but not of oat coleoptiles. *Plant Physiol.* **96**: 551-557.

Hoson, T., Masuda, Y. and Nevins, D. J. (1992) Comparison of the outer and inner epidermis. Inhibition of auxin-induced elongation of maize coleoptiles by glucan antibodies. *Plant Physiol.* **98**: 1298-1303.

Iino, M. (1982) Action of red light on indole-3-acetic acid status and growth in coleoptiles of etiolated maize seedlings. *Planta* **156**: 21-32.

Inouhe, M., Yamamoto, R. and Masuda, Y. (1987) Efffects of indoleacetic acid and galactose on the UTP level and UDP-glucose formation in *Avena* coleoptile and *Vigna* epicotyl segments. *Physiol. Plant* **69**: 579-585.

Kamisaka, S. (1983) Biosynthesis and metabolism of indole-3-acetic acid. *Chem. Regul. in Plants* **18**: 113-124 (in Japanese).

Kawaguchi, M., Kobayashi, M., Sakurai, A. and Syono, K. (1991) The presence of an enzyme that converts indole-3-acetamide into IAA in wild and cultivated rice. *Plant Cell Physiol.* **32**: 143-149.

Kawaguchi, M., Fujioka, S., Sakurai, A. and Syono, K. (1993) Presence of a pathway for the biosynthesis of auxin via indole-3-acetamide in trifoliata orange. *Plant Cell Physiol.* **34**: 121-128.

Key, J. L. and Ingle, J. (1964) Requirement for the synthesis of DNA-like RNA for growth of excised plant tissue. *Proc. Nat. Acad. Sci.* **52**: 1382-1388.

Kim, J., Harter, K. and Theologis, A. (1997) Protein-protein interactions among the Aux/IAA proteins. *Proc. Natl. Acad. Sci. USA.* **94**: 11786-11791.

Kögl, F., Erxleben, H. and Haagen-Smit, A. J. (1933) Über ein Phytohormon der Zellstreckung. Zur Chemie des krystallisierten Auxins. V. Mitteilung. *Zeits. Physiol. Chem.* **216**: 31-44.

Kögl, F., Erxleben, H. and Haagen-Smit, A. J. (1934) Über ein neues Auxin ("Heteroauxin") aus Harn. XI. Mitteilung. *Zeits. Physiol. Chem.* **228**: 90-103.

Kögl, F. and Erxleben, H. (1934) Über die Konstitution der Auxine a und b. X. Mitteilung über pflanzliche Wachstumsstoff. *Zeits. Physiol. Chem.* **227**: 51-73.

Kögl, F. and Erxleben, H. (1935) Synthese der "Auxin-glutarusäure" und einiger Isomerer. XV. Mitteilung. *Zeits. Physiol. Chem.* **335**: 181-200.

Koshiba, T. and Matsuyama, H. (1993) An *in vitro* system of indole-3-acetic acid formation from tryptophan in maize (*Zea mays*) coleoptile extracts. *Plant Physiol.* **102**: 1319-1324.

Koshiba, T., Mito, N. and Miyakado, M. (1993) L- and D-Tryptophan aminotransferases from maize coleoptiles. *J. Plant Res.* **106**: 25-29.

Koshiba, T., Kamiya, Y. and Iino, M. (1995) Biosynthesis of indole-3-acetic acid from L-tryptophan in coleoptile tips of maize (Zea mays L.). *Plant Cell Physiol.* **36**: 1503-1510.

Koshiba, T., Ballas, N., Wong, L. M. and Theologis, A. (1995) Transcriptional regulation of PS-IAA4/5 and PS-IAA6 early gene expression by indoleacetic acid and protein synthesis inhibitors in pea (Pisum sativum). *J. Mol. Biol.* **253**: 396-412.

Kutacek, M. and Kefeli, V. (1970) Biogenesis of indole compounds from D- and L-tryptophan in segments of etiolated seedlings of cabbage, maize and pea. *Biol. Plant.* **12**: 145-158.

Laibach, F. (1933) Wuchsstoffversuche mit lebenden Orchideenpollinien. *Ber. d. bot. Ges.* **51**: 336-340.

Letham, D. S., Goodwin, P. B. and Higgins, T. J. V. eds. (1978) Phytohormones and related compounds - A comprehensive treatise. Volumes 1 and 2, Elsevier, Amsterdam.

Libbert, E., Fischer, E., Drawert, A. and Schroder, R. (1970a) Pathways of IAA production from tryptophan by plants and by their epiphytic bacteria: A comparison. II. Establishement of the tryptophan metablites, effects of a native inhinbitor. *Plant Physiol.* **23**:278-286.

Libbert, E., Schroder, R., Drawert, A. and Fischer, E. (1970b) Pathways of IAA production from tryptophan by plants and by their epiphytic bacteria: A comparison. III. Metabolism of tryptamine, indoleacetaldehyde, indoleethanol and indoleacetamide. Effects of a native inhibitor. *Physiol. Plant* **23**: 287-193.

Majima, T. and Hoshino, T. (1925) Synthetische Versuche in der Indol-Gruppe. VI. Eine neue Synthese von β-Indolyl-alkylaminen. *Ber. Detut. Chem. Ges.* **58**: 2042-2046.

Marumo, S., Abe, H., Hattori, H. and Munakata, K. (1968) Isolation of a new auxin, methyl-4-chloroindoleacetate from immature seeds of *Pisum sativum*. *Agric. Biol. Chem.* **32**: 117-118.

McQueen-Mason, S. J. and Hamilton, R. H. (1989) The biosynthesis of indole-3-acetic acid from D-tryptophan in Alaska pea plastids. *Plant Cell Physiol.* **30**: 999-1005.

Masuda, Y. (1978) Auxin-induced cell wall loosening. *Bot. Mag. (Special Issue)* **1**: 103-123.

Masuda, Y. (1990) Auxin-induced cell elongation and cell wall changes. *Bot. Mag.* **103**: 345-370.

Masuda, Y., Setterfield, G. and Bayley, S. T. (1966) Ribonucleic acid metabolism and cell expansion in oat coleoptile. *Plant Cell Physiol.* **7**: 243-262.

Nakamura, T., Takahashi, N., Matui, M. and Hwang, Y. (1966) Activity of synthesized auxin b lactone as the plant growth regulator of auxin type. *Plant Cell Physiol.* **7**: 693-696.

Okamoto, T., Isogai, Y. and Koizumi, T. (1967) Studies on plant growth regulator. II. Isolation of indole-3-acetic acid, phenylacetic acid and several plant growth inhibitors from etiolated seedlings of *Phaseolus*. *Chem. Pharm. Bull.* **15**: 159-163.

Paál, A. (1919) Über phototropische Reizleitungen. *Ber. Deut. Bot. Ges.* **32**: 499-502.

Rivier, L. and Pilet, P. E. (1974) Indolyl-3-acetic acid in cap and apex of maize roots. Identifiation and quantification by mass fragmentography. *Planta* **120**: 107-112.

Sachs, J. (1882) Vorlesungen über Pflanzenphysiologie. Verlag von Wilhelm Engelmann, Leipzig.

Sachs, J. (translated by H. M. Ward) (1887) Lectures on the Physiology of Plants. Clarendon Press, Oxford.

Sakurai, N., Nevins, D. J. and Masuda, Y. (1977) Auxin- and hydrogen ion-induced cell wall loosening and cell extension in *Avena* coleoptile segments. *Plant Cell Physiol.* **18**: 371-380.

Sandberg, G., Jensen, E. and Crozier, A. (1982) Biosynthesis of indole-3-acetic acid in protoplasts, chloroplasts and a cytoplasmic fraction from barley (*Hordeum vulgare* L.). *Planta* **156**: 541-545.

Schneider, E. A. and Wightman, F. (1974) Metabolism of auxin in higher plants. *Ann. Rev. Plant Physiol.* **25**: 487-513.

Scott, T. K. and Briggs, W. R. (1960) Auxin relationships in the Alaska pea. *Amer. J. Bot.* **47**: 492-499.

Scott, T. K. (ed.) (1984) Hormonal regulation of development II. The functions of hormones from the level of the cell to the whole plant. Springer-Verlag, Berlin.

Sembdner, G., Gross, D., Liebisch, H. W. and Schneider, G. (1981) Biosynthesis and metabolism of plant hormones. In *Encyclopedia of Plant Physiology, New Series, Vol. 9*, J. MacMillan, ed., pp. 281-444, Springer-Verlag, Berlin.

Söding, H. (1931) Wachstum und Wanddehnbarkeit bei der Haferkoleoptile. *Jahrb. f. wiss. Bot.* **74**: 127-151.

Songstad, D. D., De Luca, V., Brisson, N., Kurz, W. G. W. and Nessler, C. L. (1990) High levels of tryptamine accumulation in transgenic tobacco expressing tryptophan decarboxylase. *Plant Physiol.* **94**: 1410-1413.

Takahashi, N. and Masuda, Y. (ed.) (1994) Phytohormones, a handbook. Baifukan.

Taylor, G. R. (1963) The Science of Life, A Picture History of Biology. Thames & Hudson, London.

Theologis, A. (1986) Rapid gene regulation by auxin. *Ann. Rev. Plant Physiol.* **37**: 407-438.

Thimann, K. V. (1935) On the plant growth hormone produced by *Rhizopus suinus*. *J. Biol. Chem.* **109**: 279-291.

Thimann, K. V. and Bonner, W. D. (1948) The action of tri-iodobenzoic acid on growth. *Plant Physiol.* **23**: 158-161.

Thimann, K. V. (1960) The passing of two pioneers. *Plant Sci. Bull.* p. 7.

Tsurumi, S. and Wada S. (1980) Metabolism of indole-3-acetic acid and natural occurrence of dioxindole-3-acetic acid derivatives in *Vicia* roots. *Plant Cell Physiol.* **21**: 1515-1625.

Tsurusaki, K., Watanabe, S., Sakurai, N. and Kuraishi, S.(1990) Conversion of D-tryptophan to indole-3-acetic acid in coleoptiles of a normal and a semi-dwarf barley (*Hordeum vulgare*) strain. *Plant Physiol.* **79**: 221-225.

Tuli, V. and Moyed, H. S. (1967) Inhibitory oxidation products of indole-3-acetic acid: 3-hydroxymethyloxindole and 3-methyleneoxindole as plant metabolites. *Plant Physiol.* **42**: 425-430.

van Overbeek, J. (1941) A quantitative study of auxin and its precursor in coleoptiles. *Amer. J. Bot.* **28**: 1-10.

van Overbeek, J., Blondeau, R. and Horne, V. (1951) Trans-cinnamic acid as an anti-auxin. *Amer. J. Bot.* **38**: 589-595.

Vliegenthart, J. A. and Vliegenthart, J. F. G. (1966) Reinvestigation of authentic samples of auxins a and b and related products by mass spectrometry. *Recueil* **85**: 1266-1272.

Wagenkneght, A. C. and Burris, R. H. (1950) IAA inactivating enzymes from bean roots and pea seedlings. *Arch. Biochem.* **25**: 30-53.

Went, F. W. (1928) Wuchsstoff und Wachstum. *Rec. trav. bot. neerl.* **25**: 1-116.

Went, F. W. and Thimann, K. V. (1937) Phytohormones. Experimental Biology Monographs. Macmillan & Co.

Wright, A. D., Sampson, M. B., Neuffer, M. G., Michalczuk, L., Slovin, J. P. and Cohen, J. D. (1991) Indole-3-acetic acid biosynhesis in the mutant maize orange pericarp, a tryptophan auxotoph. *Science* **254**: 998-1000.

Yamamoto, R., Shinozaki, K. and Masuda, Y. (1970) Stress-relaxation properties of plant cell walls with special reference to auxin action. *Plant Cell Physiol.* **11**: 947-956.

Yasuda, S. (1934) The second report on the behavior of the pollen tubes in the production of seedless fruits caused by interspecific pollination. *J. Genet.* **9**: 118-124 (in Japanese with English summary).

Zenk, M. H. (1961) 1-(Indole-3-acetyl)-β-D-glucose, a new compound in the metabolism of indole-3-acetic acic in plants. *Nature* **191**: 493-494.

Zimmerman, P. W. and Hitchcock, A. E. (1942) Substituted phenoxy and benzoic acid growth substances and the relation of structure to physiological activity. Contrib. Boyce Thompson Inst. **12**: 321-343.

Chapter 4

Non-Reducing Saccharides: Floridosides and Sucrose

Jong-Ching Su
Department of Agricultural Chemistry
National Taiwan University, 1 Roosevelt Road Section 4
Taipei

ABSTRACT

The non-reducing sugars floridosides and sucrose, which are abundant in red algae and green plants, respectively, are the initial stable neutral photosynthetic products. Although all three isomers of floridosides, or α-D-galactosylglycerols, are known, their metabolic roles are still under speculation. By radiotracer feeding studies, we demonstrated a cycling of sucrose synthesis cleavage without net gain of sucrose, and the glucose moiety of sucrose was preferentially incorporated into the rice seed starch. Sucrose synthase (SuS) has been implicated as the key enzyme in both systems. The enzyme is common in green plants, and consists of four identical or mixed protomers. Three rice SuS genes were cloned and their structures mapped. One of the isologous genes is expressed in the endosperm of starch-filling rice seed, and the other two are ubiquitously expressed, with one dominating over the other in different tissues. Other sucrose-metabolizing enzymes, invertase and sucrose phosphate synthase, should play distinctly different physiological roles because they catalyze opposite irreversible reactions. Man regard sucrose as the most important natural sweetener of all time. Not many enzymes are directly involved in the metabolic pathways spanning around it. However, further investigations on the exact mechanisms by which sucrose is used as carbon and energy sources and as a metabolic regulatory signal in the maintenance of plant life are still required.

Sucrose — The Mass-Produced Crystalline Food

Sucrose is probably the only organic food material that is massively manufactured and commonly traded in the crystalline form. In Taiwan, the peak production of the "plantation white sugar," or the granulated sucrose directly recovered from the expressed sugar cane juice in one process, surpassed one million metric tons before the onset of the Pacific War. As an agricultural chemistry student in Taiwan at the time (immediately following the conclusion of the War), the chemistry on cane sugar manufacturing was one of the major courses I had to take. I learnt that, the non-reducing sucrose, having both anomeric carbons in α-D-glucopyranose and β-D-fructofuranose linked together, was stable under alkaline conditions but susceptible to acid-catalyzed hydrolysis. So, the expressed cane juice could be treated with slaked lime and heating, an apparently drastic chemical condition, to neutralize the juice acidity and flocculate colloidal impurities in the first step of sugar purification. The primary clarified juice goes through carbonatation or sulfitation process to remove excess lime as well as to adsorb impurities on the precipitating calcium salt. The final step of purification is vacuum evaporation-crystallization followed by removal of molasses by centrifugation to recover sugar crystals. Apart from the lecture course, we had to spend about a month at a couple of sugar-refining factories in the southern Taiwan to learn the whole process on site.

Sucrose — The Primary Neutral Saccharide Produced in Photosynthesis

Undergraduate level biochemistry or physiology of photosynthesis taught me, based on the transient accumulation of starch granules in chloroplasts, that glucose was postulated as the primary stable saccharide produced in the photosynthesis. I was also taught a theory according to which formaldehyde was the intermediate, which when condensed forms a hexose. The synthesis of a product named "methylenitan" from formaldehyde (or known as dioxymethylene) *in vitro* (Butlerow, 1861) was cited as the evidence to support the theory. It was then a great surprise to learn from an article by Melvin Calvin published in the *Journal of Chemical Education* (1949), that sucrose was the first detectable neutral saccharide among many acidic compounds, clearly shown in the two-dimensional autoradiograms of radiolabeled photosynthates from green algae. Based on my knowledge of sucrose chemistry, I fancied that probably the chemical stability conferred by the non-reducing nature of the disaccharide was selected by the nature to serve as the vehicle of transport from the leaf to sink organs, or as the storage material most notably in cane stalks and sugar beets.

Non-Reducing Counterpart of Sucrose in Red Algae — Floridosides

The papers of Melvin Calvin's and Daniel Arnon's work at Berkeley in the fifties greatly stimulated my interest in the role of sucrose in plant physiology, and also the biochemical pathway that would lead to the biosynthesis of sucrose.

By the time I started my job as a teaching assistant in 1950, and was enrolled as a graduate student in the MS program one year later at the department from where I graduated, I had found a research topic that could be done under the prevailing environment. Japanese biochemistry chair professor Suguru Miyake, a carbohydrate chemist who had postdoctoral experience with Haworth of Edinburgh, returned to Japan in 1947. His successor, Professor Hon-Kai Ho, a graduate of our department who worked at the Manchurian National Institute of Sciences as a lipid chemist, returned from Manchuria, escaping from the siege of the Manchurian capital by the communist in 1949. Meanwhile, without significant financial inputs for improving teaching and research from the Chinese authorities for almost ten years since the restoration of Taiwan to China in 1945, the biochemistry laboratory where I began work had only facilities, equipment and chemicals suitable for carbohydrate chemistry left behind by the Japanese. Then, I could only read all of the publications from the laboratory in the past eighteen years (1928–1945), checked out chemicals and equipment available in the laboratory, set up a plan to start the biochemistry teaching laboratory as charged to me by Professor Ho, and find a topic suitable for my own MS dissertation research for Prof. Ho's approval.

The Japanese apparently tried to build a database of carbohydrate resources of Taiwan. Among the materials they studied were rice starches and various plant polysaccharides, especially mucilages and gel-forming matters from local plants for food and industrial use, including those from marine algae. A couple of their publications drew my attention. They reported on the galactans from red algae *Bangia fuscopurpurea* and *Porphyra crispata*. Besides isolating crystalline DL-galactose from the acid hydrolysates of water-soluble polysaccharides, crystalline D-galactose was obtained from the ethanolic extract of *P. crispata*. Crystallization of sugars from the concentrated syrup, characterization of various phenylhydrazone and phenylosazone derivatives and oxidation and reduction products were the main techniques of qualitative analysis employed. They used a dried *Porphyra* specimen harvested at the Pescadores, a group of islands located between Taiwan and mainland China. So, it was possible that D-galactose was derived from a complex saccharide during the course of sample preparation, shipment and treatment rather than occurring as such in the alga. Besides, the most interesting point was that the L-isomer of galactose was present exclusively in the polysaccharide but not in the low molecular weight fraction. It is well-known that, among the naturally occurring aldohexoses, only galactose has D and L forms, and besides, a paper chemistry shows that one form is transformed to the other by simply exchanging their end groups along the carbon chain. This fact was demonstrated by oxidizing galactitol, obtainable by reducing D-galactose, with a 3% hydrogen peroxide solution in the presence of ferric ion to obtain DL-galactose (Tollens and Elsner, 1935). My plan was to find out if free D-galactose was formed in the alga, and derive the biochemical mechanism of L-galactose formation.

With Professor Ho's approval, I obtained some travel funds, set up a portable heating apparatus, obtained some knowledge of seaweed taxonomy by studying dry specimens of local marine algae collected by the Japanese, and ventured out to the northern coastal area for sample collection. We had to laboriously transport all our equipment either on foot, or using buses and trains. After finally arriving at our destination, I began setting up an alcohol burner-water bath to heat ethanol-containing flasks to fix *Porphyra* and some other red algae freshly collected from the sea shore. All of a sudden, I was surrounded by several bayonet-fixed rifles. Soldiers started firing questions in an unfamiliar language. Only upon presentation of identity cards did I prove myself as a student and a university employee. I was then allowed to continue the activity but under their constant surveillance.

Afterwards, I requested the university authority to file an application to the Keelung Garrison Command for a free pass to the coastal area for sample collection. Since the fall of the mainland to the communist in 1949 and the subsequent massive exodus of military as well as civilians from the mainland, the whole Taiwan area had been placed under the martial law. Curfews were severely enforced, and the coastal areas were heavily guarded by armed forces, to fend off seemingly imminent invasion from mainland China. My request was turned down by the university for the reason that they did not want me to undertake the dangerous activity. Thus, the original project that needed continuous sampling of living algae had to be terminated. It was only four years later that I could next obtain a fresh sample of *Porphyra* to work on. By that time, I had learnt the technique of paper chromatography from the paper by Calvin. Then, recognizing the importance of chemical analysis at the sub-milligram level, I dug into the books of microchemistry to learn how to do chemical operations in capillary tubes and spot plates, and the use of a microscope to enhance the sensitivity of chemical detection and melting point determination. I was then able to couple the paper chromatographic separation with the classical chemical detection techniques to further enhance my analytical capability.

I was able to show that the ethanol solubles from all of the red algae I collected in 1951 did not contain reducing sugars with a confidence level of less than one μg per gram fresh weight, but they yielded much D-galactose on acid hydrolysis. Later, I isolated and identified floridoside, or α-D-galactopyranosyl 2-glycerol from *P. crispata*. I could not find other forms of galactoside, and confirmed that DL-galactose could be crystallized from the hydrolysate of its water-soluble polysaccharide (Su, 1956).

By the time I finished this work, Putman and Hassid (1954) published the isolation and structural determination of floridoside from a red alga *Iridophycus flaccidum* (formerly known as *Irideae laminarioides*) belonging to the order Florideophyceae. Professor Hassid further studied the $^{14}CO_2$-photosynthesis of the alga and proposed that floridoside would be the counterpart of sucrose in the marine plant (Bean and Hassid, 1955). I wrote to Professor Hassid to discuss my identification data. This communication brought me to his laboratory as a US-AID trainee in 1958, and also my interest back to sucrose.

However, my Ph.D. dissertation completed at the Hassid laboratory was not on the plant sugar nucleotide biochemistry that was vigorously pursued in there at that time. As I was given by AID only one year of stay in the US, which was later extended to two years, my graduate advisor Professor H.A. Barker advised me to complete all course and residential requirements for a Ph.D. degree but finish the dissertation work in Taiwan. Professor Baker told me that, as long as I was determined to stay in the academic field, a Ph.D. degree would be a necessity. However, as Professor Barker was not aware of the prevailing academic situation in Taiwan, I consulted with Professor Hassid and he agreed that I should continue my line of work on *Porphyra* so that I might be able to finish a dissertation research within my allowed time of stay in the US. Using *P. perforata* collected from the rocky beach beneath the Golden Gate Bridge (with the help of an old friend from National Taiwan University who had just finished a degree on phycology at Berkeley, Dr. Kung-Chu Fan), the work yielded three papers, one short communication reporting two novel nucleotides (Su and Hassid, 1960), and two full-length papers reporting the chemistry of a whole spectrum of nucleotides, two forms of D-galactosylglycerols including floridoside and α-D-galactopyranosyl-1-D-glycerol, two inositols laminitol and scylloinositol, and a DL-galactan sulfate constituting of D- and L-galactoses, 3,6-anhydro-L-galactose, 6-O-methyl-D-galacotse and 6-O-sulfate groups on the unmodified galactose residues in the ratio 1:2:1:1. A possible biosynthetic pathway linking these saccharides together was proposed (Su and Hassid, 1962a,b). After returning to Taiwan, I tried to continue the biochemical aspects of work but was hampered by the fact that the only *Porphyra* species available to us contained a very active adenylate deaminase (Su *et al.*, 1966) which transformed all ATP required in the *in vitro* reactions into ITP.

While mainland China was still in the midst of the cultural revolution turmoil, my friend Dr. Fan, who went back to the mainland soon after helping me with the seaweed collection, died at a young age. Professor Pappenfuss, Dr. Fan's thesis advisor, told me when he visited Taiwan in the late 1960s that he received a postcard from Dr. Fan with a message: "do not send me any scientific periodicals anymore" in Chinese. The news made me very sad indeed.

Biosynthesis of Sucrose: Application of Sucrose Synthetic Enzymes

Although the chemical structure of sucrose was established by methylation studies of Haworth school in 1920s (Avery *et al.*,1927), the feat of its *in vitro* synthesis was first achieved by Hassid, Doudoroff and Barker (Hassid *et al.*, 1944) from α-D-glucopyranose 1-phosphate (G1-P) and D-fructose under the catalysis of sucrose phosphorylase from *Pseudomonas saccharophila*. The chemical synthesis was achieved by Lemieux and Huber (Lemieux and Huber, 1953, 1956) by reacting 3,4,6-tri-O-acetyl-1,2-anhydro-D-glucopyranose and 1,3,4,6-tetra-O-acetyl-D-fructofuranose to yield sucrose octaacetate. Before leaving for the US to join the Hassid laboratory, I tried to find sucrose phosphorylase activity in several plants, but failed. Then the discoveries of the activities of sucrose synthase (SuS) (Cardini *et al.*, 1955) and sucrose phosphate synthase (SPS) (Leloir and Cardini, 1955), both using UDPG as the glucosyl donor, were reported. For the detection of sucrose phosphorylase activity, I could manage to use a G1-P prepared by a potato starch phosphorylase-catalyzed reaction, but the UDPG needed for the detection of SuS and SPS was beyond my reach then. At the biochemistry department of Berkeley, besides sharpening research techniques by doing dissertation work, I had learnt many research methods by taking laboratory courses as well as observing many brilliant postdoctoral fellows, including Elizabeth Neufeld, David Feingold and G.A. Barber, at the Hassid laboratory in action.   The technique that benefited me most was the combined application of chemical and biochemical methods for the synthesis of commercially unavailable radiocarbon-tagged sugar nucleotides from commercially available precursors.

There were no commercial routes for the import of perishable biochemicals available in Taiwan in 1960. Upon my return from the US, I had to prepare many biochemicals, including U-^{14}C-glucose, ATP, NAD, various ^{14}C-sugar nucleotides, in order to continue my biochemical research. Among the synthetic products, a double-labeled sucrose, prepared by applying SuS from asparagus spears, which had straightforward Michaelis-Menten kinetics in the direction of sucrose synthesis, opened up a new avenue of sucrose research (Lee and Su, 1982). The synthesis started with the preparation of G1-P by the phosphorolysis of sucrose by the *P.*

saccharophila enzyme, coupling the G1-P with UMP by cyclohexylcarbodiimide, and the synthesis of sucrose by the SuS catalysis. By choosing appropriate ^{14}C- and ^{3}H-tagged reactants, sucrose preparations having the two monosaccharide residues independently tagged by the radioisotope of choice were obtained. Many SuS's exhibit sigmoidal reaction kinetics and could not utilize the substrates at low concentrations, but the asparagus enzyme could and yielded sucrose of high specific radioactivity in good yield. Our later studies on the rice SuS isozymes revealed that they also had the property of asparagus enzyme, and they are now extensively utilized in the synthesis of sugar nucleotides (see for example, Stein *et al.*, 1998).

Once Professor Hassid told me that their sucrose phosphorylase research attracted the interest of the Coca-Cola Company. Why the sucrose produced by the enzyme-catalyzed reaction attracted the interest of the soft drink company remains a mystery to me. Growing up in Taiwan, I could not imagine why sucrose synthesized from G1-P and fructose could be cheaper than the natural product in abundance. However, the mechanism of the enzyme-catalyzed reaction, thoroughly studied by the three discoverers, has provided an excellent model of enzyme mechanism research, and the basic concept developed by them is still of importance to biochemistry (Doudoroff *et al.*, 1947). Besides, the useful enzyme activity could be so easily obtained from the bacterial cells (Doudoroff, 1955) and the reaction catalyzed by the enzyme is so clean and efficient that one can cleave sucrose, even at a "tracer" concentration level, to obtain easily separable G1-P and fructose in quantitative yield (Abraham and Hassid, 1957).

Futile Cycle of Sucrose in Plants

Among the research strategies that I learned at Berkeley, the mapping of metabolic pathway by the use of radiotracer as the basis of molecular or enzymic studies attracted me most. As mentioned earlier, buying radiocarbon-labeled glucose for my research in Taiwan in 1960 was impossible. However, I could import ^{14}C-barium carbonate, as it remains intact during the year of transportation from the US to Taiwan. So, I, together with my assistant, Mr. Ti-Sheng Lu (who later obtained a Ph.D. degree in plant physiology at the University of California, Davis), used the carbonate as the source of carbon dioxide to obtain starch by the photosynthetic method, devised by Professor Hassid (in Calvin *et al.*, 1949; Abraham and Hassid, 1957). The radioactive starch was recovered from the tobacco leaf that was exposed to ^{14}CO$_2$ under light, hydrolyzed with acid and the recovered glucose was used in a feeding study using bamboo shoot slices.

From the kinetic analysis of radioactivity levels of ethanol-soluble saccharides from the tissue slices, we could show that the radioactivity level of free glucose depleted rapidly but that of sucrose increased in proportion, but free fructose gained practically none of the radioactivity. The radioactive sucrose was hydrolyzed and the recovered hexoses were analyzed. We found that they gained radioactivity in a similar fashion, but with the increase of glucose level leading that of fructose in about 2 to 1. Since the levels of all soluble sugars in the tissue slices remained practically constant in the short duration of tracer feeding, these results suggested that the metabolism of free fructose was segregated from the sucrose synthesis pathway by either not sharing the same intermediate or by compartmentation of metabolic sites. We may further see that, if UDPG is used as the glucosyl donor in the sucrose synthesis, the glucosyl acceptor is generated at the site of UDPG synthesis, or both of them share the same precursor, and a continuous cycling of sucrose synthesis and degradation takes place in the same cellular compartment because the total sucrose remained constant (Su, 1982). Years later, we did the same type of experiment on banana fruit slices and obtained similar results. An interesting finding was that, when the banana fruit was subjected to a chilling injury treatment, the cycle was disrupted, as indicated by the finding that no radio-labeled fructose was incorporated into sucrose (Niu and Su, 1969). We may write such cycling pathway of synthesis and degradation by incorporating known enzyme activities, including SuS, SPS, sucrose phosphate phosphatase (SPase), invertase, hexokinase, phosphohexoisomerase, phosphoglucomutase and UDPG pyrophosphorylase, but these enzyme activities were not affected by the chilling injury. Thus, we had proposed that the injury was incurred on the membrane transport of substrates.

We may see that the overall balance of the cycle constructed from the enzyme activities listed above is "futile" in the sense that pyrophosphate bond energy is lost at no gain of sucrose if the sucrose synthesis in the cycle is catalyzed by SPS and SPase while that of sucrose degradation is catalyzed by either invertase or SuS, especially by invertase, because invertase-catalyzed hydrolysis is as irreversible as the synthetic reaction catalyzed by SPS and SPase. As we know, the SuS-catalyzed reaction is fully reversible and also may utilize nucleoside diphosphates other than UDP, and if only this enzyme is involved in the degradation and synthesis of sucrose, and if the change in direction of reaction is governed by a signal, the response to the signal will be more rapid than the futile cycling, and the sucrose cleavage reaction may be coupled to different requirements of sugar nucleotides. If the cycling is the common phenomenon in plants, then we may see that the one catalyzed by SuS single-handedly would be more versatile than the futile cycle operated by the invertase/SPS-SPase system. However, as we have shown by the radio-labeled glucose feeding done on bamboo shoots and banana fruits, the fructose residue in the sucrose molecule is in equilibrium with the precursor of the glucose residue, not in favor of the cycling catalyzed by SuS only. Then, what is the physiological significance of the cycle, regardless of either "futile" or "non-futile"? This is the question also raised by Pontis (Pontis, 1978) based on the biochemical properties of SuS, SPS and SPase. Unfortunately, besides knowing that the sucrose level will determine the growth, differentiation, etc., of plant tissue culture and plants, clear metabolic pathways have not yet been charted.

What is the Physiological Role Played by Sucrose Synthase?

When Cardini *et al.* discovered sucrose synthase, they undoubtedly considered it as the enzyme for catalyzing sucrose synthesis. Then, on studying the metabolism of UDPG, they further found the presence of SPS, and later sucrose phosphate phosphatase in green plants. The sucrose synthesis catalyzed by SPS is energetically more favorable than that of SuS, probably because fructose-6-phosphate (F6-P) has only the furanose form to readily condense with glucose to form sucrose, while fructose has the more abundant pyranose form in addition. Furthermore, the phosphatase-catalyzed reaction removes sucrose phosphate from the reaction system to render the overall reaction irreversible. With such development, it is natural to look for the physiological significance of SuS-catalyzed reaction in the direction of sucrose degradation. We know that the rate of SuS-catalyzed reversible reaction is rapid in kinetics, and it may be very important in the "sucrose cycle" to regulate the sucrose concentration level to meet the role of sucrose as a metabolic regulator.

Besides being a possible regulator of metabolism, the SuS-catalyzed sucrose degradation reaction has an advantage over the invertase-catalyzed one because it directly gives UDPG, a precursor for complex saccharide synthesis by itself, as well as the starting substrate for a series of saccharide transformation reactions in the formation of uridyl nucleotides of D-galactose, D-glucuronic acid, D-galacturonic acid, D-xylose and L-arabinose, all needed for the biosyntheses of plant cell wall polysaccharides, as elegantly demonstrated by the Hassid school in the 1950s.

Here I have to mention about the substrate specificities of SuS-catalyzed reaction. From our results obtained from nearly ten plant sources (Su, 1982), and also many reported by others, we can say that sucrose is the only disaccharide substrate SuS uses. However, although UDP is invariably the nucleotide substrate with the least Km value, usually ADP or TDP comes to the next best, and other nucleoside diphosphates are also reactive, though at more reduced rates. It is therefore highly possible, with the adenylate nucleotides usually predominating in the soluble nucleotide pool, *in vivo* formation of ADPG, in addition to UDPG, by the SuS-catalyzed reaction is possible (Chen *et al.*, 1982). Besides the multiplicity of sugar nucleotide specificity, we have found that all of the SuS we studied have a quaternary structure constituting four either identical or not identical protomers. Recent studies on the SuS isozymes have revealed that the wide spectrum of sugar nucleotide specificity is not due to the heterotetrameric structures (Yen *et al.*, 1994; Huang and Wang, 1998).

My radiotracer study on bamboo shoot slice indicated the importance of the SuS-catalyzed reaction in the cell wall polysaccharides syntheses. However, trials in solubilizing the enzyme activities identified by the tracer study were a failure, although the catalytic activities were all found in a particulate fraction sedimented at 10^5 x g (Sung *et al.*, 1971). With no means of purifying the enzymes, I had to divert my effort to another class of plant polysaccharide, the starch.

My choice of the system was the rice. By noting that the starch synthesizing activity of the rice seed resided almost exclusively associated with the insoluble starch granules, while the precursor nucleotides synthesizing activity was extractable into a buffer solution, we titrated the two activities in the formation of starch by reconstituting them in different proportions in test tubes. It was found that sucrose plus ADP were not only better substrates than G1-P plus ATP in the starch synthesis, but also used less soluble fraction for a fixed amount of starch granule to achieve a higher level of starch synthesis, indicating that the activity of sucrose synthase to synthesize ADPG was much more adequate than the ADPG pyrophosphorylase activity in the rice seed (Chen *et al.*, 1981). Then we used rice seeds at different maturing stages to feed a double-labeled sucrose for five to sixty minutes. We were excited to find that, when the feeding time was shorter, or the seed growth stage was earlier, the incorporation of the glucosyl residue into starch was higher than that of the fructosyl residue. The segregation ratios of the two hexoses as the precursor of starch were from six to four, the higher when the feeding time was shorter. When the radio-labels were analyzed on ADPG, UDPG, hexose 6-phosphates and G1-P isolated from the seed, the results indicated that glucosyl residues in the two nucleotides were derived mainly from the glucose part of sucrose, while that in all of hexose phosphates were from the fructose part, showing that the pyrophosphorylase- catalyzed reactions were not responsible for the formation of sugar nucleotides if sucrose was fed, and that the SuS-catalyzed reaction directly provided the precursor of starch synthesis in the rice seed (Lee and Su, 1982). Same kind of experiment done on other plants, such as pea seedlings, asparagus spears, bamboo shoot, etc., yielded results not as distinct as that of the rice seed but showed the same trend. However, the result obtained from sweet potato root slices was entirely different; it indicated that G1-P was a more direct substrate than a sugar nucleotide at the initial phase of starch synthesis. These results led us to concentrate on the studies of SuS in the rice grains and the starch phosphorylase in the sweet potato tuberous root (Chang *et al.*, 1987).

Genes Encoding Rice Sucrose Synthase Isozymes

With its richness in genetic backgrounds, the *Sus* mutants and their direct relevance to the carbohydrate biochemistry were first described for the maize system. The maize *shrunken-1* and *sugary* mutants are the best documented examples in which the SuS deficiency is attributed to the poorer biosynthesis of endosperm glucans. Many plants, analogous to maize, are known to have two isoforms of *Sus* genes. From rice, we cloned and established the cDNA and genomic structures of three genes encoding sucrose synthase polypeptides, primary structures of which were deduced (Yu *et al.*, 1992; Wang *et al.*, 1992; Huang *et al.*, 1996). From the homology comparison analysis, we can see that two of the three genes are highly homologous and correspond to the maize *Sus*, while the third one, named as *RSus2*, is closely related to the maize *Sh1*. DNA- and immuno-probes specific to the three genes and polypeptides, respectively, are now available. By applying these identification tools in the mapping of spatial and temporal expression of these genes at the transcription and translation levels (Wang *et al.*, 1999), we can conclude that *RSus2* is a housekeeping gene. Of the two genes similar to maize *Sus*, the one we named *RSus3* is unique in having its expression almost exclusively in the endosperm of milk-ripe stage seeds. The close cousin of *RSus3*, *RSus1*, has its temporal and spatial expression patterns compensating those of *RSus2*. Homotetrameric quaternary structures are prevalent when only one gene is expressed at a time in a confined tissue, while heterotetrameric structures can be found if two or more genes are expressed simultaneously in the same compartment. The gene products of *RSus1*- and *RSus2*-catalyzed reactions show different initial rates in the directions of sucrose synthesis and breakdown, and the rice leaves with earlier and later order of emergence have different activity ratios of sucrose synthesis over breakdown, implicating that the expression patterns of the two genes change as the leaf tissue ages. The temporal and spatial changes, as well as the availability of sugar and oxygen have been reported to affect the expression of *Sus* genes (Yen, 1998). Most plants, with the exception of rice (which contains three), are known to have two functional isologous *Sus* genes. Whether such diversification is due to the adaptation to the semi-anoxic condition by the rice root system merits further studies. In this respect, it is important to note that vascular tissues of rice root and leaf have only one type of *Sus* expressed at the translational level.

The importance of SuS in the polysaccharide biosynthesis has been demonstrated in maize grains. It has been shown that one of the maize genes contributes to the formation of cell wall in the endosperm tissue, while the other enhances the starch synthesis. We specifically inhibited the expression of *RSus3* in the rice grain by the antisense technique and obtained the shrunken phenotype of the grain, thus demonstrating that the conclusion we had drawn from our earlier radiotracer study using double-labeled sucrose was right, and the responsible SuS is encoded by *RSus3*.

Recently, we have demonstrated the presence of an SuS activity in the cell wall fraction of the rice grain. Whether this isozyme is responsible for providing a precursor to the cellulose synthase, as proposed by Delmer *et al.* (Delmer and Amor, 1995), that a SuS constitutes the hypothetical cellulose synthesizing complex on the plasma membrane, and thus have a parallel function of one of the maize SuS isozymes, is a problem worth studying.

Sucrose Metabolism Initiated By Invertases

From the facts that the preferential utilization of glucose over fructose in the double-labeled sucrose feeding is limited to about six to one even though the tracer feeding time was shortened to five minutes, and that the ratio decreased as the feeding time is prolonged, we should take into account the invertase initiated pathway in addition to the SuS pathway. By the enzyme purification and gene cloning-sequencing, at least three invertase genes and isozymes encoded by them could be isolated from the rice grain at the laboratory of Hsien-Yi Sung, one of my colleagues at the same department. The complexity of invertase isozymes is prevalent among plants, and their physiological significance in the apoplastic and symplastic transport of sucrose has been demonstrated. One of the isoforms has been known to be localized in the apoplastic space of many plant species, and its function in the apoplastic sucrose transport is a commonly accepted view. The occurrence of a proteinaceous inhibitor of the apoplastic invertase is also common, and the implication that the inhibitor is a switch to control the apoplastic transport of sucrose is gaining wider acceptance. Two other forms of invertase are known to localize in cytoplasm and vacuole, and their possible roles include symplastic transport and vacuolar deposition of sucrose.

Conclusion

It is interesting to find that the red marine algae produce a series of non-reducing floridoside isomers instead of sucrose as the primary neutral photosynthetic products. Sucrose is produced in all green plants, especially in higher plants as the major transport of photosynthate from the source (leaf) to sink organs through the vascular system. The metabolic links between floridosides and biomacromolecules, including polysaccharides, proteins and others, remain unknown, although they are supposed to play the counterpart role of sucrose in green plants (Bean and Hassid, 1955). On the other hand, the biosynthetic reactions and metabolic fates of sucrose are now well documented. Sucrose is synthesized in the photosynthetic cell, transported to sink organs via vascular system, transferred through cell wall and cytoplasmic membrane and metabolized via invertase or SuS pathways, or further transported into vacuoles as a storage substance. Externally added sucrose has been known as an osmotic agent, and now a sucrose carrier is reported to facilitate its membrane transport. The metabolic pathways of sucrose were established, or deduced, by combining the textbook knowledge with those derived from rational experiments. The unknown part of sucrose metabolism resides mainly in the physiological role it may play as a regulator or a switch of metabolic pathways. We know that sucrose is invariably the best carbon source to support tissue or cell cultures of higher plants. In the normal metabolizing plant cells, sucrose may undergo continuous synthesis and degradation. We may postulate that such metabolic cycling may be a mechanism of regulating the sucrose level *in vivo*. Since the discovery of the Crabtree effect, the available sugar level has been known to play pivotal regulatory roles in carbohydrate catabolism. And with the findings that the sugar level also exerts effects on the expression level of some genes, the sucrose research has thus entered into the era of molecular genetics-physiology-biochemistry that needs concerted efforts of plant scientists from multidisciplinary fields. It is hoped that the physiological significance of the "futile" cycle of sucrose synthesis and degradation will be deciphered soon.

References

Abraham, S. and Hassid, W. Z. (1957) The synthesis and degradation of isotopically labeled carbohydrates and carbohydrate intermediates. *Methods in Enzymol. III.* 489-560.

Avery, J., Haworth, W. N. and Hirst, E. L. (1927) The constitution of the disaccharides. Part XV. Sucrose. *J. Chem. Soc.* 2308-2318.

Bean, P. C. and Hassid, W. Z. (1955) Assimilation of $C^{14}O_2$ by a photosynthesizing red alga *Iridophycus flaccidum. J. Biol. Chem.* **212**: 411-425.

Butlerow, A. (1861) Bildung einer zuckerartigen Substanz durch Synthese. *Ann.* **120**: 295-298.

Calvin, M. (1949) The path of carbon in photosynthesis. *J. Chem. Edu.* **26**: 639-656.

Calvin, M., Heidelberger, L., Reid, J. C., Tolbert, B. M. and Yankwich, P. F. (1949) In "Isotopic Carbon", John Wiley & Sons, New York and Chapman & Hall, London, pp. 263-268.

Cardini, C. E., Leloir, L. F. and Chiriboga, J. (1955) The biosynthesis of sucrose. *J. Biol. Chem.* **214**: 149-155.

Chang, T. C., Lee, S. C. and Su, J. C. (1987) Sweet potato starch phosphorylase - Purification and characterization. *Agric. Biol. Chem.* **51**: 187-195.

Chen, H. J., Sung, H. Y. and Su, J. C. (1981) Role of sucrose synthase in sucrose-starch conversion in rice grains. *Proc. Natl. Sci. Council R.O.C. Part B.* **5**: 165-170.

Delmer, D. P. and Amor, Y. (1995) Cellulose biosynthesis. *Plant Cell* **7**: 987-1000.

Doudoroff, M.(1955) Disaccharide phosphorylases, Methods in Enzymol. I, pp. 225-231.

Doudoroff, M., Barker, H. A. and Hassid, W. Z. (1947) Studies with bacterial sucrose phosphorylase. I. The mechanism of action of sucrose phosphorylase as a glucose transferring enzyme (transglucosidase). *J. Biol. Chem.* **168**: 725-732.

Hassid, W. Z., Doudoroff, M. and Barker, H. A. (1944) Enzymatically synthesized sucrose. *J. Am. Chem. Soc.* **66**: 1416-1419.

Huang, D. Y. and Wang, A. Y. (1998) Purification and characterization of sucrose synthase isozymes from etiolated rice seedling. *Biochem. Mol. Biol. Internatl.* **46**: 107-113.

Huang, J. W., Chen, J. T., Yu, W. P., Shyur, L. F., Wang, A. Y., Sung, H. Y., Lee, P. D. and Su, J. C. (1996) Complete structures of three rice sucrose synthase isogenes and differential regulation of their expressions. *Biosci. Biotech. Biochem.* **60**: 233-239.

Lee, P. D. and Su, J. C. (1982) Sucrose-starch transforming system in rice seeds. A tracer study. *Proc. Natl. Sci. Council R.O.C. Part B.* **6**: 172-180.

Leloir, L. F. and Cardini, C. E. (1955) The biosynthesis of sucrose phosphate. *J. Biol. Chem.* **214**: 157-165.

Lemieux, R. U. and Huber, G. (1953) A chemical synthesis of sucrose. *J. Am. Chem. Soc.* **75**: 4118.

Lemieux, R. U. and Huber, G. (1956) A chemical synthesis of sucrose. A conformational analysis of the reactions of 1,2-anhdro-α-D-glucopyranose triacetate. *J. Am. Chem. Soc.* **78**: 4117-4119.

Niu, Y. H. and Su, J. C. (1969) Effect of chilling injury and storage in nitrogen on the carbohydrate metabolism and respiration of bananas. *J. Chinese Agric. Chem. Soc. Sp. Iss.* 50-63.

Pontis, H. G. (1978) On the scent of riddle of sucrose. *Trends in Biochem. Sci.* **3**: 137-139.

Putman, E. W. and Hassid, W. Z. (1954) Structure of galactosylglycerol from *Irideae laminarioides. J. Am. Chem. Soc.* **76**: 2221-2223.

Stein, A., Kula, M. R. and Elling, L. (1998) Combined preparative enzymatic synthesis of dTDP-6-deoxy-4-keto-D-glucose from dTDP and sucrose. *Glycoconjugate J.* **15**: 139-145.

Su, J. C. (1982) Some aspects of sucrose metabolism in plants (An invited review). *Proc. Natl. Sci. Council R.O.C. Part B.* **6**: 172-180.

Su, J. C. (1956) Studies on saccharides of marine algae. I. Identification of floridoside and a DL-galactan in a red alga, *Porphyra crispata*, Kjellm, *J. Chinese Chem. Soc. Ser. II.* **3**: 65-71.

Su, J. C. and Hassid, W. Z. (1960) Identification of guanosine diphosphate L-galactose, guanosine diphosphate D-mannose and adenosine 3',5'-pyrophsophate from red alga *Porphyra perforata* J. G. Agardh, J. Biol. Chem. **235**: 36-37.

Su, J. C. and Hassid, W. Z. (1962) Carbohydrates and nucleotides in the red alga *Porphyra perforata*. I. Isolation and identification of carbohydrates, *Biochem.* **1**: 468-474.

Su, J. C. and Hassid, W. Z. (1962) Carbohydrates and nucleotides in the red alga *Porphyra perforata*. II. Separation and identification of nucleotides, *Biochem.* **1**: 474-480.

Su, J. C., Li, C. C. and Ting, C. L. (1966) A new adenylate deaminase from red marine alga *Porphyra crispata. Biochem.* **5**: 536-543.

Sung, H. Y., Yuan, H. F. and Su, J. C. (1971) Carbohydrate metabolism in the shoot of bamboo *Leleba oldhami*. IX. Some properties of the pentosan synthesizing system. *J. Chinese Agric. Chem. Soc., Sp. Iss.* 62-73.

Tollens, B. and Elsner, H. (1935) Kurzes Handbuch der Kohlenhydrate. 4th Ed. Johann Ambrosius Barth, Leipzig. 362-363.

Wang, A. Y., Kao, M. H., Yang, W. H., Sayion, Y., Liu, L. F., Lee, P. D. and Su, J. C. (1999) Differentially and developmentally regulated expression of three rice sucrose synthase genes. *Plant Cell Physiol.* **40**: 800-807.

Wang, A. Y., Yu, W. P., Juang, R. H., Huang, J. W., Sung, H. Y. and Su, J. C. (1992) Presence of three rice sucrose synthase genes as revealed by cloning and sequencing of cDNA. *Plant Mol. Biol.* **18**: 1191-1194.

Yang, W. H. (1998) Studies on physiological role of rice sucrose synthase isogenes. Ph.D. Dissertation. Department of Agricultural Chemistry, National Taiwan University, Taipei.

Yen, S. F, Su, J. C. and Sung, H. Y. (1994) Purification and characterization of rice sucrose synthase isozymes. *Biochem. Mol. Biol. Internatl.* **34**: 613-620.

Yu, W. P., Wang, A. Y., Juang, R. H., Sung, H. Y. and Su, J. C. (1992) Isolation and sequences of rice sucrose synthase cDNA and genomic DNA. *Plant Mol. Biol.* **18**: 139-142.

Chapter 5

Chlorophyll Biosynthesis I: From Analysis of Mutants to Genetic Engineering of the Pathway

Diter von Wettstein
*Departments of Crop and Soil Sciences & Genetics and Cell Biology
Washington State University, Pullman, WA 99164-6420, USA*

ABSTRACT

Biochemical, genetic and molecular adventures have resulted in never-ending, exciting discoveries in the elucidation of the chlorophyll biosynthetic pathway from 1945 to 1997. Major developments are traced in this chapter. Special emphasis is placed on the pathway to 5-aminolevulinate by David Shemin, on the information gained from Sam Granick´s mutants in *Chlorella* and from the protochlorophyllide holochrome by James H. C. Smith. Analysis of structural and regulatory mutants in higher plants and photosynthetic bacteria support the generality of the pathway. The bacteriochlorophyll operons of Barry Marrs and John Hearst together with the evolving techniques of nucleotide and protein sequencing, as well as those for crystal structure analyses and site-directed mutations have opened the way to discoveries, how the different enzymes accomplish their task in synthesizing chlorophyll — an indispensable molecule for life on earth. Antisense gene technology teaches us how the plant avoids destruction by light-sensitive intermediates. It has been a privilege to have been able to follow the adventures of many scientists mentioned in this and the following review and to interact with a significant number of them. I am also grateful to the many enthusiastic and skilled students, postdoctoral fellows and colleagues, who have allowed me to participate in their discoveries.

Steps in the Elucidation of Chlorophyll Biosynthesis

The structure of chlorophyll as depicted in Fig. 1 was determined through chemical analysis by Richard Willstätter and Arthur Stoll (1928), Hans Fischer and colleagues (Fischer and Stern, 1940) and Linstead and colleagues (Ficken *et al.*, 1956). Its correctness was further proven by chemical synthesis of the entire molecule from pyrrole by Robert Woodward and colleagues (1960; cf. also Seely, 1966). The biosynthetic route of this cyclic

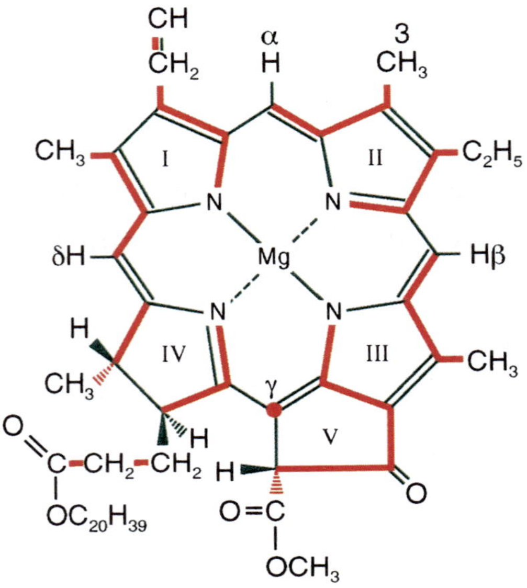

Fig. 1. The chemical structure of chlorophyll *a*. Every chlorophyll molecule is synthesized in the chloroplast from eight molecules of 5-aminolevulinic acid. The location of the atoms derived from these molecules of 5-aminolevulinic acid in the finished chlorophyll molecule are indicated by the heavy lines. Position 3, which is occupied by a methyl group in chlorophyll *a* and a formyl group in chlorophyll *b* is marked.

tetrapyrrole and its closely related iron porphyrin heme has been established by a combination of the following different experiments. (i) Feeding isotope-labeled putative precursor molecules to animals or plants or their cells can detect the appearance of label in heme or chlorophyll and their precursors. (ii) Accumulation of intermediates by mutants of algae, photosynthetic bacteria and higher plants has been successfully exploited to learn about the route of chlorophyll synthesis. (iii) Accumulation of intermediates by application of enzyme inhibitors or conditions halting their synthesis has often been helpful. (iv) Ultimate proof of a biosynthetic step is obtained with a purified enzyme isolated from the organism or produced by expressing its gene in *E. coli*, yeast or another heterologous organism.

The Pathway of David Shemin

In 1945, David Shemin and David Rittenberg reported an experiment in which the first author had consumed over a 3-day period 66g of glycine containing 32.4 atom per cent excess of the heavy isotope ^{15}N in addition to his usual diet. Blood samples were taken over a period of 99 days and the ^{15}N isotope concentration in the hemin of the red blood cells determined. After 18 days, about 20 per cent of the porphyrin was newly synthesized, and it had an average ^{15}N concentration of 4.1 atom per cent excess. The only substance in the diet that could have such a isotope concentration was the labeled glycine and it was therefore concluded that the N atoms in the pyrroles of protoporphyrin of heme originate from glycine. This was confirmed in the following year (Shemin and Rittenberg, 1946) by feeding rats ^{15}N-labeled glycine, ammonium citrate, glutamate, proline and leucine. Only glycine was incorporated at significant concentrations into hemin. Subsequently, David Shemin with his colleagues Wittenberg and Kumin used 2-^{14}C-glycine and 1-^{14}C-succinate to demonstrate that 5-aminolevulinic acid was synthesized in red blood cells and in the photosynthetic bacteria *Rhodobacter spheroides* and *Rhodosprillum rubrum* by condensation of succinyl-CoA and glycine with pyridoxal phosphate as cofactor (Kikuchi *et al.*, 1958; reviewed in Bogorad, 1966). ^{14}C-labeled 5-aminolevulinic acid was synthesized by Shemin and Russel (1953) and shown to be able to replace succinyl-CoA and glycine in the synthesis of porphyrin. 5-^{14}C aminolevulinate was incubated with a protein fraction of duck erythrocytes and the mono-pyrrole porphobilinogen was then isolated, crystallized and chemically analyzed (Schmid and Shemin, 1955).

The molar radioactivity of the porphobilinogen product was twice that of the 5-aminolevulinate substrate, revealing that porphobilinogen is formed by condensation of two molecules of 5-aminolevulinate, and that this precursor pyrrole is the source of all atoms of protoporphyrin and the macrocycles of chlorophyll and heme. This enzyme, now known as 5-aminolevulinate dehydratase or porphobilinogen synthase, was also isolated from aqueous extracts of chicken erythrocytes by Sam Granick (1954).

Sam Granicks Knockout Mutants

Sam Granick was encouraged by the success of George Wells Beadle and Edward Lawrie Tatum´s experiments to determine metabolic pathways in the fungus *Neurospora crassa* with the aid of gene knockout mutations of enzymes. In his Harvey lecture, Granick (1949) presented the X-ray mutants he had isolated in *Chlorella vulgaris,* a favourite alga for photosynthesis experiments, and developed his plan for how he would use these mutants to clarify the intermediates in the pathway leading to chlorophyll and heme. The mutants were pale green and lacked yellow pigments (carotenoids), or brownish red, or variously tinged from orange to almost colorless. This alga was used because the mutants could be grown on glucose-inorganic salt media and, in contrast to e.g., higher plants, they accumulated the precursors for the intermediate blocked by the knockout mutation in substantial amounts. Mutant W_5 developed no chlorophyll but formed huge amounts of brown pigment. Sam Granick isolated, crystallized and analyzed the pigment with the aid of derivatives produced by the methods of Hans Fischer. To his enjoyment, it proved to be protoporphyrin isomer IX, the parent molecule of heme (Granick, 1948a). He thus concluded that protoporphyrin is the metabolic precursor of both heme and chlorophyll and that the biosynthetic steps up to this porphyrin must be identical. In mutant No. 60, he found Mg-protoporphyrin IX (Granick, 1948b) and in mutant No. 31, Mg-monovinyl pheoporphyrin a_5 (Granick, 1950), and he proposed this to be a precursor of protochlorophyll, the phytyl ester of Mg-monovinyl pheoporphyrin a_5. Protochlorophyll was isolated from dark-grown barley seedlings and shown to be a phytyl ester. As a first intermediate between Mg-protoporphyrin and Mg-vinyl pheoporphyrin a_5, he designated the monomethylester of Mg-protoporphyrin accumulated in *Chlorella* mutant No. 60A (Granick, 1961).

As to the enzymatic steps in the common pathway of chlorophyll and heme from porphobilinogen to protoporphyrin IX, broken cell preparations of *Chlorella* formed a number of porphyrins including uroporphyrinogen I (hydroxymethylbilane) and uroporphyrinogen III in addition to protoporphyrin IX (Bogorad and Granick, 1953a). The development of improved separation and identification techniques by Lawrence Bogorad (Granick and Bogorad, 1953) identified in the double mutant W_5B-17, an accumulation of uroporphyrinogens I and III, coproporphyrin, protoporphyrin and hematoporphyrin IX as potential intermediates or conversion products of intermediates (Bogorad and Granick, 1953b; Granick *et al.,* 1953).

Subsequently and in parallel, mutants in *Rhodobacter capsulatus* or *sphaeroides* (formerly *Rhodopseudomonas*) and *Rhodospirillum* accumulating and secreting bacteriochlophyll precursor molecules were identified, and further supported the already rather detailed outline of chlorophyll and bacteriochlorophyll synthesis. A complete listing of these up to 1972 was presented by Simon Gough (1972).

The Protochlorophyllide Holochrome of James H. C. Smith

Seedlings of barley, maize and other angiosperms germinated in the dark lack chlorophyll, but they contain a small amount of protochlorophyllide or its phytylated derivative protochlorophyll. This pigment was first isolated by Pringsheim and Monteverde in the last century and suggested to be a precursor of chlorophyll (cf. Boardman, 1966 for review). In 1929, Noack and Kiessling found protochlorophyll in seed coats of cucurbits, which, however, is not converted to chlorophyll by light and therefore provided a stable source for this pigment. Hans Fischer and August Oestreicher (1940) synthesized protochlorophyll (the phytyl ester of Mg-vinyl pheoporphyrin a_5) and demonstrated that, compared to chlorophyll a, it lacked the two hydrogens at carbons 7 and 8 of ring IV, these two carbons being linked by a double bond. James H. C. Smith at the Department of Plant Biology of the Carnegie Institution of Washington, Stanford, loved to use his then newly developed Beckman spectrophotometer to plot hour after hour, exquisite absorption spectra from highly purified pigment solutions, and he applied this skill to elucidate the photochemical conversion of protochlorophyllide to chlorophyllide a, with a special protein which he named holochrome. Violet Koski and James H. C. Smith purified protochlorophyllide from 5kg of dark-grown barley leaves by aceton extraction and powdered sucrose chromatography, and determined with great precision the specific absorption coefficients at various wavelengths as a basis for quantitative determinations of small amounts of pigments in their coming experiments (Koski and Smith, 1948). Smith (1948) used the coefficients to show direct proportionality between the amount of chlorophyll formed at 0°C during a 2-hour illumination and the amount of protochlorophyllide present at the start of the experiment. Working with maize seedlings, Violet Koski (1950) in her Ph.D. thesis established unequivocally, that it is possible to distinguish between the transformation of the dark-formed protochlorophyllide into chlorophyll a and the formation of chlorophyll a from additional precursor synthesized in the light at 18°C, by making measurements at 10 second intervals. She concluded that it is possible to study the transformation process alone in the seedling leaves of dark-grown barley, and that light is required to convert the protochlorophyllide directly into chlorophyllide a. The carotenoid deficient mutant "white seedling-3" of maize forms normal amounts of protochlorophyllide in its seedling leaves in the dark and converts it into chlorophyllide upon illumination (Koski and Smith, 1951), but the absence of protective carotenoids caused the chlorophyllide or chlorophyll to be bleached by photo-oxidation.

Building on data by Sylvia Frank, Violet Koski, Stacey French and James Smith (1951) exploited the white seedling mutant and its wild type to obtain an action spectrum for the photochemical conversion of protochlorophyllide to chlorophyllide, in order to find out if protochlorophyllide is the light-absorbing pigment for the photochemistry of the reaction. Stacey French loved to build gadgets and in this case, the challenge was to obtain sufficiently strong monochromatic light at intervals of 5 to 10 mμ covering a wavelength spectrum from 680 to 405 mμ for irradiation of the seedlings grown in a dark room. A Huggins capillary mercury lamp was obtained and aimed at a plane reflection grating with 600 lines/mm borrowed from Harold Babcock of the Mount Wilson Observatory. By rotating the grating with a screw drive mechanism, the wavelength of the spectrum could be moved by 10mμ and the desired resolution was obtained. The monochromatic beam was aimed at the leaves in the dark room and the amount of protochlorophyllide conversion determined. Very pretty action spectra were obtained. Maxima were found at 650, 593, 548 and 445 mμ corresponding to the absorption maxima of protochlorophyllide. In wild-type leaves, red light was more efficient than blue light, while the opposite was true for the mutant. The low efficiency of blue light for protochlorophyllide conversion in the wild type is due to the major absorption of this light by carotenoids. Thus, it was discovered that protochlorophyllide is the light absorbing agent for its conversion into chlorophyllide.

If the conversion of protochlorophyllide to chlorophyllide was a strictly photochemical process, illumination should permit conversion at -195°C. This was not the case (Smith and Benitez, 1954). At −70°C, fairly rapid and extensive conversion occurred and increases in the rate continued as the temperature was raised to 40°C. Leaves heated above this temperature lost the capacity for conversion, and freezing and thawing also destroyed the activity completely. These studies led James Smith to the conclusion that the protochlorophyllide or protochlorophyll is bound to a protein, which was thus referred to as the complex protochlorophyll holochrome. He also inferred that the elimination of the double bond is not an intramolecular photochemical transfer of two hydrogens but rather, it involves an enzymatic reaction. James Smith and Donald Kupke (1956) proceeded to isolate with a glycine buffer, a 400,000 molecular weight holochrome complex from dark-grown bean leaves that upon illumination, converted the attached protochlorophyllide to chlorophyllide. This preparation was obtained with the needle valve homogenizer constructed by Stacey French and Harold Milner, subsequently known as the French Press, which until today is thought by most students to be an invention from France.

The opal-glass light scattering technique for recording spectra of intact leaves, permitted one to follow the spectral shifts of the newly formed chlorophyllide or chlorophyll holochrome in a subsequent dark period. The shifts were first observed by A. A. Krasnovskii and L. M. Kosobutskaya (1952), and was subsequently described in detail by Kazuo Shibata (1957), but the molecular events in the holochrome resulting in these spectral changes remain to be clarified. The isolated holochrome was used by James Smith and Stacey French (1958) to determine the quantum yield for the conversion of protochlorophyllide to chlorophyllide and was found to be close to 0.5. They speculated that each quantum transfers a hydrogen atom. This has been extensively debated by other investigators, but may yet turn out to be true in light of newer investigations (see below).

In 1959, James Smith, G. C. McLeod and Janice Coomber came surprisingly close to discovering the hydrogen donors for the reaction (McLeod and Coomber, 1959; Smith and Coomber, 1959). They studied the conversion of protochlorophyllide holochrome by ultraviolet light and found that short wavelengths UV only converted 20 to 30% of the protochlorophyllide, but a subsequent irradiation with light of longer wavelengths effected conversion of the other 70 to 80% protochlorophyllide molecules in the preparation. They concluded that the UV-triggered conversion was due to light absorption by the phenolic group of a tyrosine amino acid side chain, to which part of the protochlorophyllide was non-covalently bound rather than to absorption by the pigment. Extensive analyses of the effects of pH on the rate and extent of inhibition of protochlorophyllide holochrome transformability showed increasing inactivation with increasing alkalinity, total inhibition being observed at pH 10.88. This was taken as evidence for dissociation of the pigment from its coordination with amino acid residues containing phenolic, sulfhydryl or amino groups. As to be reported below, site-directed mutagenesis has recently shown that changing a tyrosine or lysine residue in the pea holochrome leads to inactivation of the reductase activity. It is thus suggested that the proton transferred to the carbon atom carrying the methyl group is donated by the tyrosine (while the other proton is transferred from NADPH).

Mutants In Higher Plants Catch Up

My interest in chlorophyll biosynthesis increased when I took up a position as an assistant to Åke Gustafsson in Stockholm in 1951, and joined the multidisciplinary Swedish group of mutation research created and successfully led by him. One of my tasks was to analyse mutation rates and spectra in the M_2 generation of barley grains treated with various ionizing radiations and chemical mutagens. This was done by a test devised by Gustafsson in the 1930s, and consisted of counting the white, yellow, light-green or tiger-striped lethal seedling mutants emerging from thousands of spikes planted in the greenhouse during the winter season. These tests were done to find the most efficient treatments for inducing mutants suitable in barley breeding programs. Encouraged by the successful efforts of Beadle and Tatum to analyze metabolic pathways by knock-out mutations, it seemed to me that all these hundreds of interesting mutants can be used to dissect the development of chloroplasts and pigment biosynthesis. I thus started collecting representatives of the different types of mutants and conducting crosses to determine allelic relationships by complementation tests. At the time, electron microscopy of thin sections started to reveal the details of the ultrastructure of animal and plants cells. So I took my mutants and travelled every week for a few days to the Biochemistry Institute of Arne Tiselius at Upsala University, where Håkan Leyon had developed microtome and embedding procedures to study chloroplast development, and where I could use the third electron microscope ever built by Siemens (1940). It had been acquired by Svedberg and was installed next to his ultracentrifuges. The mutants turned out to be very useful for characterizing the development of chloroplast structure. The question of finding mutants blocked in chlorophyll synthesis was first addressed during a stay with James Smith and Stacey French at the Carnegie Institution at Stanford in the fall of 1958. Employing absorption, derivative and fluorescence spectrophotometry, the accumulation of protochlorophyllide and chlorophyll was analyzed in a series of barley mutants. In the yellow mutant *xantha-f*[10] (Fig. 2D), no trace of protochlorophyllide or chlorophyll could be detected, but the mutant seemed to synthesize cytochromes. Assuming a common pathway for heme and chlorophyll, the suggestion was made that this mutant may be blocked between protoporphyrin IX and protochlorophyll (von Wettstein, 1959). In contrast to Granick's *Chlorella* mutants or Lascelles' *Rhodobacter* mutants, no intermediates accumulated. It was only 38 years later when we demonstrated that *Xantha-f* codes for one of the three proteins required to insert Mg^{++} into protoporphyrin IX (Jensen *et al.*, 1996).

The usefulness of the mutants in barley and other higher plants for studies of the chlorophyll biosynthetic pathway rests on a discovery made by Sam Granick in (1959). When barley grains are germinated in the dark, the seedlings have yellow leaves because they lack chlorophyll and contain only small amounts of protochlorophyllide (Fig. 2A). If such 7-day-old dark-grown barley shoots are placed into a solution containing

Fig. 2.   Structural and regulatory nuclear gene mutants of barley affecting chlorophyll synthesis.
A. Primary leaves in barley grown for 7 days in darkness at 23°C.
B. Primary leaves of the mutant *tigrina-d* [12] grown for 7 days in darkness. The green color is due to unregulated accumulation of protochlorophyllide, as it also takes place in wild-type leaves fed with 5-aminolevulinate for 24 hours.
C. Mutant *xantha-l* [35] fed for 24 hours with water (left) or 0.01 M 5-aminolevulinate (right). The mutant is unable to carry out an efficient cyclase reaction to form ring V and accumulates red Mg-protoporphyrins IX.
D. Yellow *xantha-f* [10] and *tigrina-b* [19] mutants. *Xantha-f* is the structural gene for the large subunit of Mg-chelatase.
E. Yellow *xantha-h* [57] seedling is unable to insert Mg^{++} into protoporphyrin IX. *Xantha-h* is the structural gene for the 42 kDa subunit of Mg-chelatase. Also seen is the regulatory mutant *tigrina-d* [12] grown in light-dark cycles.

5-aminolevulinate, they will green in a few hours due to the accumulation of protochlorophyllide. The accumulation of large amounts of protochlorophyllide reveals that all enzymes required to convert 5-aminolevulinate into chlorophyll are present in the dark-grown leaves. When the feeding experiment is repeated with shoots from mutants blocked in chlorophyll synthesis, the intermediate that can no longer be converted into protochlorophyllide accumulates.   Thus, the mutant *xantha-l*[35] accumulates red Mg-protoporphyrin IX and its monomethyl ester (Fig. 2C) because it is defective in the cyclase reaction to form ring V of the chlorophyll molecule.   In this way, mutant *xantha-u* is identified to have a block in the conversion of uroporphyrinogen to coproporphyrinogen, and *xantha-f, -g* and *-h* mutants accumulate protoporphyrin, indicating a block in the insertion of Mg^{++} (Gough, 1972; von Wettstein *et al.*, 1971; Henningsen *et al.*, 1993).

The experiments described above have shown that the formation of 5-aminolevulinate is limiting in the etioplasts.   Mutations in four barley genes (*tigrina-b, -d, -n* and *-o*) remove this limitation and accumulate two to ten times the wild-type amount of protochlorophyllide in the dark (Nielsen, 1974).   One of these mutants, *tigrina-d*[12] (Fig. 2B), has no defect other than unregulated 5-aminolevulinate formation, and is therefore involved in excessive protochlorophyllide synthesis in the dark. Homozygous *tigrina-d*[12] mutants are fully viable and fertile, if grown in continuous light but are dark-sensitive (Hansson *et al.*, 1997).   The dark sensitivity accounts for the name of these mutants: In light/dark cycles, the seedlings are green-white banded (Figs. 2D, *tigrina-b*[19] and 2E, *tigrina-d*[12]). The white, often necrotic leaf domains are caused by the accumulation of protochlorophyllide in the dark. Only a normal amount of protochlorophyllide reductase is available to convert the protochlorophyllide into chlorophyllide, and when the light turns on, the excess of protochlorophyllide causes photodynamic damage to the plastids. With short light pulses, the accumulated protochlorophyllide can successively be converted into chlorophyll and the *tigrina-d*[12] mutant can be saved.   This is not so with other regulatory gene mutants, such as *tigrina-b*[19], which accumulate ζ-carotene in addition to protochlorophyllide (Nielsen and Gough, 1974).   Mutations in the *tigrina-n* and *-o* genes deregulate the synthesis of β-carotene as well as protochlorophyllide, thus resulting in the accumulation of lycopenic pigments and the formation of chromoplasts.

Plants homozygous for both the *xantha-l*[35] and *tigrina-d*[12] mutations accumulate Mg-protoporphyrin and its monomethyl ester, just like *xantha-l*[35] seedlings fed with 5-aminolaevulinate (von Wettstein *et al.*, 1974). This was exploited in the following way. Mutant *xantha-l*[35] is leaky because it accumulates   some   protochlorophyllide   in   addition   to   the Mg-protoporphyrins. In order to obtain a completely blocked mutant in this gene, we mutagenized a large number of *tigrina-d*^{12} caryopses. In the M_2 generation, Mg-protoporphyrin accumulaters were then screened out by *in vivo* spectrophotometry and the non-leaky allele *xantha-l*^{81} was obtained (Kahn *et al.*, 1976).

Barry Marrs and John Hearst Dive Into The Bacteriochlorophyll Operons

Barry Marrs (1974, 1978, 1981) discovered that in *Rhodobacter* (*Rhodopseudomonas syn.*) *capsulatus*, a 46kbp chromosome segment contained in operon-type organizations most if not all of the genes for the biosynthesis of carotenoids from farnesylpyrophosphate (Yen and Marrs, 1976), for the synthesis of bacteriochlorophyll from protoporphyrin IX (Taylor *et al.*, 1983), and for the reaction center and light harvesting polypeptides I (Youvan *et al.*, 1984 a,b). This provided novel possibilities of studying structural and regulatory genes of bacteriochlorophyll synthesis. The complete nucleotide sequence of the 45,959 base pairs of Barry Marrs´s photosynthetic gene cluster in his plasmid pRPS404 (Marrs, 1981) was established (Alberti *et al.*, 1995, EMBL DNA sequence accession number Z11165). The function of 31 of the 43 identified open reading frames in the cluster was determined by transposon mutagenesis. From left to right, the cluster contains the following operons (Alberti *et al.*, 1995; Bollivar *et al.*, 1994): 1) the gene for the reaction center H protein, and two genes for the function and assembly of the reaction center, and light harvesting I antenna are transcribed from a common promoter together with six other open reading frames. 2) Adjacent is the operon comprising the structural genes for Mg-protoporphyrin methyl transferase, the iron-containing subunit L of protochlorophyllide reductase, Mg-protoporphyrin IX chelatase subunit H, protochlorophyllide reductase subunits B and N, and 2-vinyl-bacterio-chlorophyllide hydratase. 3) This operon (transcribed in the opposite left to right direction) consists of a regulatory gene for the chlorophyllide reductase and a gene of unknown function. 4) The operon containing the structural genes for Mg-protoporphyrin IX monomethyl ester oxidative cyclase, 4-vinyl reductase, geranylgeranyl bacterio-chlorophyll synthase and reductase, and two genes of unknown function (one possibly being a transacting factor regulating transcription of the light-harvesting II antenna) are transcribed in the same direction. 5) The operon (transcribed in the right to left direction) encodes a gene with unknown function, the genes for the D and I Mg-chelatase subunits and for spheroidine monooxigenase, so far only known to insert a keto group to give the carotenoid spheroidenone. 6) This carotenoid synthesis-specific operon contains the structural genes for hydroxyneurosporene synthase and methoxyneurosporene dehydrogenase. 7) The operon (transcribed in the left to right direction) houses the genes for geranylgeranyl pyrophosphate synthase and hydroxyneurosporene-O-methyltransferase. 8) The operon encoding the genes for 2-α-hydroxyethyl bacterio-chlorophyllide dehydrogenase, for an iron-containing and two other chlorophyllide reductase subunits is transcribed in the same direction. 9) The right most operon encodes the *pufQ* regulator protein for bacteriochlorophyll synthesis, the β and α light harvesting I polypeptides, the reaction center proteins L and M, and an electron transfer protein.

The function of the mentioned genes for bacteriochlorophyll synthesis was deduced by interposon mutagenesis in the following principle way (Yang and Bauer, 1990). Restriction fragments containing the open reading frames to be mutated were subcloned into *E. coli* vectors and a kanamycin-resistant structural gene inserted in the open reading frame to create the desired *orf::Km^r* disruption. The plasmids with the disrupted open reading frame were transformed into the plasmid mobilizing strain of *E. coli* developed by Taylor *et al.* (1983). The transformants were then mated with a special *Rhodobacter capsulatus* strain overproducing gene transfer agent (Young *et al.*, 1989). Cell-free gene transfer agent with the packaged disrupted gene was prepared from the ex-conjugants by filtration and used to transduce the wild type strain of *Rhodobacter capsulatus*. Kanamycin-resistant transductants that arose as a result of homologous recombination with the gene in the *Rhodobacter* chromosome (Scolnik and Haselkorn, 1984) were selected and analyzed for photosynthetic growth capability, as well as for the nature of Mg-tetrapyrroles that accumulated during dark semi-anaerobic growth.

Antisense Technology Elucidates the Effects Of Decreasing Individual Enzymatic Activities

The *tigrina* mutants described above accumulate in the dark, excessive amounts of protochlorophyllide that is not attached to the reductase enzyme and cannot therefore be converted immediately to chlorophyllide and incorporated in chlorophyll binding proteins, when the light is turned on. The protochlorophyllide, like other porphyrins, absorbs radiant energy leading to the formation of singlet oxygen and other highly reactive oxygen radicals that destroy the tissue (Arakane *et al.*, 1996). Cloning of such genes involved in the regulation of chlorophyll synthesis in higher plants is still not possible, but antisense technology has provided a means to determine the effects of deregulated synthesis of intermediates in the chlorophyll pathway. In this procedure, the cDNA sequence of the enzyme is inserted in reverse orientation under the control of a cauliflower mosaic virus 35S promoter or a tissue-specific promoter into an expression vector, and introduced by *Agrobacterium*-mediated transformation into plants. The antisense gene can control the transcript level in the nucleus and the translation efficiency of the target mRNA in the cytoplasm.

Fig. 3. Transformants of tobacco expressing a glutamate 1-semialdehyde aminotransferase antisense gene.

The transformant at left has pale leaves, because of reduced amount of the enzyme. In the transformant at the center, chlorophyll formation is inhibited along the leaf veins. An untransformed plant is at the right. From Höfgen *et al.*, 1994.

Figure 3 provides an example of this approach, in which tobacco plants were transformed with an antisense glutamate 1-semialdehyde aminotransferase gene. This enzyme catalyzes the last step in

5-aminolevulinate synthesis, and antisense transformants show a general or tissue-specific reduction in chlorophyll (Höfgen *et al.*, 1994). Partial or complete suppression of the aminotransferase mimics a wide variety of inheritable chlorophyll variegation patterns elicited by nuclear and organelle gene mutations in different higher plants. Such antisense plants provide a sensitive visible marker for activities of chlorophyll synthesis genes in leaves, stems and flowers. Their cell- or tissue-specific action depends *inter alia* on the position of insertion in the chromosomes of the host, a position we at present cannot control. Relative to the wild type, chlorophyll content varied from 10% to 53% and was accompanied by a glutamate 1-semialdehyde aminotransferase, with activity ranging from 22% to 85%, and a strong reduction in the enzyme protein. Studies of these transgenic plants showed that a reduction of the chlorophyll content to less than 10% of that of the wild type, with a reduction of the enzyme activity to 22% did not lead to accumulation of glutamate 1-semialdehyde. This observation points again to a very tight control of the pathway; in this case to an inhibition of the substrate intermediate even when the enzyme is present in low amounts in contrast to the accumulation of glutamate 1-semialdehyde, when the enzyme is present but inhibited by gabaculine (Kannangara and Schouboe, 1985). In most of the transgenic plants, no endogenous sense mRNA for glutamate 1-semialdehyde aminotransferase could be detected in Northern blots, indicating its rapid turnover elicited by the antisense mRNA. There was no observable effect of the presence of the antisense gene on the steady state levels of the mRNA for the small subunit of the CO_2-fixing enzyme ribulose bisphosphate carboxylase, for a protein of photosystem II, a mitochondrial protein and a light-harvesting chlorophyll binding protein.

The importance of avoiding accumulation of porphyrin intermediates causing photo-oxidative damage is emphatically demonstrated by the tobacco antisense plants with reduced uroporphyrinogen III decarboxylase (Mock *et al.*, 1995; Mock and Grimm, 1997; Mock *et al.*, 1998) and reduced coproporphyrinogen III oxidase (Kruse *et al.*, 1995a,b; Mock *et al.*, 1998). These plants grow almost normally under dim light or with short light periods but then develop extensive necrotic lesions on the leaves under high light intensity. The uroporphyrinogen decarboxylase antisense plants accumulate uroporphyrinogen and uroporphyrin, those with reduced coproporphyrinogen oxidase accumulate coproporphyrinogen and coproporphyrin up to 500-fold of that found in wild type plants. The extent of necrosis correlates with the amount of porphyrins accumulated in the leaves. Deficiency of uroporphyrinogen decarboxylase in the human disease porphyria cutanea tarda results in accumulation of uroporphyrinogen, and its oxidation to the light-absorbing uroporphyrin causes phototoxic skin lesions (Sweeney, 1986).   This illustrates the importance of a tightly regulated tetrapyrrole pathway in all organisms. As studied in the antisense plants with uroporphyrinogen decarboxylase activity and enzyme levels reduced to 45% of wild type, no effect on the developmentally controlled protein steady state levels of tetrapyrrole synthesizing enzymes or the levels of constituents of the photosynthetic apparatus with heme and chlorophyll binding properties was found.   Some decrease in the activity of 5-aminolevulinate dehydratase and porphobilinogen deaminase is attributed to the photodynamic effect of the accumulated porphyrin via oxygen radicals. As the porphyrin-induced necroses resemble induction of cell death encountered in response to pathogenesis and air pollution, the defense responses to tetrapyrrole-induced oxidative stress were investigated with these two types of antisense transgenic plants (Mock *et al.*, 1998). This was facilitated by the knowledge of the mechanism of the diphenyl ether herbicides that inhibit protoporphyrinogen oxidase (Matringe *et al.*, 1989), causing the accumulation of protoporphyrinogen which then diffuses into the cytoplasm of the cell, where it is oxidized to protoporphyrin by peroxidases or non-enzymatically (Jacobs *et al.*, 1996). In the presence of oxygen, protoporphyrin produces toxic singlet oxygen that destroys the cell (Becerril and Duke, 1989). The antisense transformants increased levels of mRNA encoding enzymes detoxifying hydrogen peroxide such as superoxide dismutases, ascorbate peroxidase and catalase, as well as glutathione peroxidase. This resulted in increased activities of the cytosolic Cu/Zn-superoxide dismutase, the mitochondrial Mn-isozyme, and ascorbate peroxidase. Increased activities were also noted for the ascorbate-regenerating enzymes mono- and dihydroascorbate reductases. Thus, the oxidative stress induced in the antisense plants by the accumulation of porphyrins elicited the induction of the enzymatic detoxifying defense system. However, this did not prevent the deterioration of the level of the anti-oxidant compounds, ascorbate and glutathione and development of necrotic leaf lesions.   Conditional overexpression of photosensitizing porphyrins are of interest in the development of herbicides and pathogen-defense strategies and antisense plants are a useful tool in exploring the possibilities.

This is scientific paper 9805-18a from the College of Agriculture and Home Economics Research Center, Washington State University, Pullman, WA, USA-Project 3106
Acknowledgement: Reproduction of Figs. 1 and 2 from D. von Wettstein *et al.* The Plant Cell 7, pp. 1039-1057 is by copyright permission of the American Society of Plant Physiologists.

Recent Reviews:

Alberti, M., Burke, D. H. and Hearst, J. H. (1995) Structure and sequence of the photosynthesis gene cluster. In *Anoxygenic Photosynthetic Bacteria*, Blankenship, R. E., Madigan, M. T. and Bauer, C. E., eds., The Hague: Kluwer, pp. 1083-1106.

Beale, S. I. (1995) Biosynthesis and structures of porphyrins and hemes. In *Anoxygenic Photosynthetic Bacteria*, Blankenship, R. E., Madigan, M. T. and Bauer, C. E., eds., The Hague: Kluwer, pp. 1083-1106.

Porra, J. R. (1997) Recent progress in porphyrin and chlorophyll biosynthesis. *Photochem. Photobiol.* **65:** 492-516.

Reinbothe, S. and Reinbothe, C. (1996) The regulation of enzymes involved in chlorophyll biosynthesis. *Eur. J. Biochem.* **237:** 323-343.

Rüdiger, W. (1997) Chlorophyll metabolism: from outer space down to the molecular level. *Phytochemistry* **6:** 1151-1167.

von Wettstein, D., Gough, S. and Kannangara, C. G. (1995) Chlorophyll biosynthesis. *Plant Cell* **7:** 1039-1057.

This article:

Alberti, M., Burke, D. H. and Hearst, J. H. (1995) Structure and sequence of the photosynthesis gene cluster. In *Anoxygenic Photosynthetic Bacteria*, Blankenship, R. E., Madigan, M. T. and Bauer, C. E., eds., The Hague: Kluwer, pp.1083-1106.

Arakane, K., Ryu, A., Hayashi, C., Masunaga, T., Shinmoto, K., Mashiko, S., Nagao, T. and Hirobe, M. (1996) Singlet oxygen ($^1\Delta_g$) generation from coproporphyrin in *Propionibacterium acnes* on irradiation. *Biochem. Biophys. Res. Commun.* **223:** 578-582.

Becerril, J. M. and Duke, S. O. (1989) Protoporphyrin IX content correlates with activity of photobleaching herbicides. *Plant Physiol.* **90:** 1175-1181.

Boardman, N. K. (1966) Protochlorophyll. In *The Chlorophylls*, Vernon, L. P. and Seely, G. R., eds., New York, Acad. Press, pp. 437-479.

Bogorad, L. (1966) The biosynthesis of chlorophylls. In *The Chlorophylls*, Vernon, L. P. and Seely, G. R., eds., New York, Acad. Press, pp. 481-510.

Bogorad, L. and Granick, S. (1953a) Enzymatic synthesis of porphyrins from porphobilinogen. *Proc. Natl. Acad. Sci. USA.* **39:** 1176-1188.

Bogorad, L. and Granick, S. (1953b) Protoporphyrin precursors produced by a *Chlorella* mutant. *J. Biol. Chem.* **202:** 793-800.

Bollivar, D. W., Suzuki, J. Y., Beatty, J. T., Dobrowolski, J. M. and Bauer, C. E. (1994) Directed mutational analysis of bacteriochlorophyll *a* biosynthesis in *Rhodobacter capsulatus. J. Mol. Biol.* **237:** 622-640.

Ficken, G. E., Johns, R. B. and Linstead, R. P. (1956) Chlorophyll and related compounds IV. The position of the extra hydrogens in chlorophyll. The oxidation of pyropheophorbide-a. *J. Chem. Soc.* **1956:** 2272-2280.

Fischer, H. and Oestreicher, A. (1940) Chlorophyll XCIV. Protochlorophyll and vinyl porphyrins. The oxo reaction. *Z. f. Physiol. Chemie* **262:** 243-269.

Fischer, H. and Stern, A. (1940) Die Chemie des Pyrrols. Vol. II, 2. Leipzig, Akad. Verlagsges.

Gough, S. (1972) Defective synthesis of porphyrins in barley plastids caused by mutation in nuclear genes. *Biochim. Biophys. Acta* **286:** 36-54.

Granick, S. (1948a) Protoporphyrin IX as a precursor of chlorophyll. *J. Biol. Chem.* **172:** 717-727.

Granick, S. (1948b) Magnesium protoporphyrin as a precursor of chlorophyll in *Chlorella. J. Biol. Chem.* **175:** 333-342.

Granick, S. (1949) The structural and functional relationships between heme and chlorophyll. *The Harvey Lectures* **XLIV (1948-1949):** 220-245.

Granick, S. (1950) Magnesium vinyl pheoporphyrin a$_5$, another intermediate in the biological synthesis of chlorophyll. *J. Biol. Chem.* **183:** 713-730.

Granick, S. (1954) Enzymatic conversion of δ-aminolevulinic acid to porphobilinogen. *Science* **120:** 1105-1106.

Granick, S. (1959) Magnesium prophyrins formed by barley seedlings treated with δ-aminolevulinic acid. *Plant Physiol.* **34:** XVIII.

Granick, S. (1961) Magnesium protoporphyrin monoester and protoporphyrin monomethyl ester in chlorophyll biosynthesis. *J. Biol. Chem.* **236:** 1168-1172.

Granick, S. and Bogorad, L. (1953) Separation of porphyrins by counter-current distribution. *J. Biol. Chem.* **202:** 781-792.

Granick, S., Bogorad, L. and Jaffe, H. (1953) Hematoporphyrin IX, a probable precursor of protoporphyrin in the biosynthetic chain of heme and chlorophyll. *J. Biol. Chem.* **202:** 801-813.

Hansson, M., Gough, S., Kannangara, C. G. and von Wettstein, D. (1997) Analysis of RNA and enzymes of potential importance for regulation of 5-aminolevulinic acid synthesis in the protochlorophyllide accumulating barley mutant *tigrina-d*[12]. *Plant Physiol. Biochem.* **35:** 827-836.

Henningsen, K. W., Boynton, J. E. and von Wettstein, D. (1993) Mutants at *xantha* and *albina* loci in relation to chloroplast biogenesis in barley (*Hordeum vulgare* L.). *Royal Dan. Acad. Sci. Lett. Biol. Skrifter* **42:** 1-349.

Höfgen, R., Axelsen, K. B., Kannangara, C. G., Schüttke, J., Pohlenz, H.-D., Willmitzer, L., Grimm, B. and von Wettstein, D. (1994) A visible marker for antisense mRNA expression in plants: Inhibition of chlorophyll synthesis with a glutamate-1-semialdehyde aminotransferase antisense gene. *Proc. Natl. Acad. Sci. USA.* **91:** 1726-1730.

Jacobs, J. M., Jacobs, N. J., Kuhn, C. B., Gorman, N., Dayan, F. E., Duke, S. O., Sinclair, J. F. and Sinclair, P. R. (1996) Oxidation of porphyrinogens by horseradish peroxidase and formation of a green pyrrole pigment. *Biochem. Biophys. Res. Commun.* **227:** 195-199.

Jensen, P. E., Willows, R. D., Larsen Petersen, B., Vothknecht, U. C., Stummann, B. M., Kannagara, C. G., von Wettstein, D. and Henningsen, K. W. (1996) Structural genes for Mg-chelatase subunits in barley: *Xantha-f, -g and -h. Molec. Gen. Genet.* **250:** 383-394.

Kahn, A., Avivi-Bleiser, N. and von Wettstein, D. (1976) Genetic regulation of chlorophyll synthesis analyzed with double mutants in barley. In *Genetics and Biogenesis of Chloroplasts and Mitochondria*, Bücher,T., Neupert, W., Sebald, W. and Werner, S., eds., Amsterdam: North Holland, pp. 119-131.

Kannangara, C. G. and Schouboe, A. (1985) Biosynthesis of δ-aminolevulinate in greening barley leaves VII. Glutamate 1-semialdehyde accumulation in gabaculine treated leaves. *Carlsberg Res. Commun.* **50:** 179-191.

Kikuchi, G., Kumar, A., Talmage, P. and Shemin, D. (1958) The enzymatic synthesis of δ-aminolevulinic acid. *J. Biol. Chem.* **233:** 1214-1219.

Koski, V. M. (1950) Chlorophyll formation in seedlings of *Zea mays. Arch. Biochem.* **29:** 339-343.

Koski, V. M. and Smith, J. H. C. (1948) The isolation and spectral absorption properties of protochlorophyll from barley seedlings. *J. Am. Chem. Soc.* **70:** 3558-3562.

Koski, V. M. and Smith, J. H. C. (1951) Chlorophyll formation in a mutant, white seedling-3. *Arch. Biochem. Biophys.* **34:** 189-195.

Koski, V. M., French, C. S. and Smith, J. H. C. (1951) The action spectrum for the transformation of protochlorophyll to chlorophyll *a* in normal and albino corn seedlings. *Arch. Biochem. Biophys.* **31:** 1-17.

Krasnovskii, A. A. and Kosobutskaya, L. M. (1952) Spectral analysis of the constitution of chlorophyll during its formation in plants and in colloidal solutions of the substance of etiolated leaves. *Doklady Akad. Nauk S.S.S.R.* **85:** 177-180.

Kruse, E., Mock, H.-P. and Grimm, B. (1995a) Reduction of coproporphyrinogen oxidase level by antisense RNA synthesis leads to deregulated gene expression of plastid proteins and affects the oxidative stress defense system. *EMBO J.* **14**: 3712-3720.

Kruse, E., Mock, H.-P. and Grimm, B. (1995b) Coproporphyrinogen III oxidase from barley and tobacco - Sequence analysis and initial expression studies. *Planta* **196**: 796-803.

Marrs, B. L. (1974) Genetic recombination in *Rhodopseudomonas capsulatus*. *Proc. Natl. Acad. Sci. USA.* **71**: 971-973.

Marrs, B. L. (1978) Transduction in *Rhodopseudomanas*. In *The Photosynthetic Bacteria*. Clayton, R. K. and Sistrom W. R., eds., New York, Plenum Press, pp. 873-883.

Marrs, B. L. (1981) Mobilization of the genes for photosynthesis from *Rhodopseudomonas capsulata by a* promiscuous plasmid. *J. Bacteriol.* **146**: 1003-1012.

Matringe, M., Camadro, J. M., Labbe, P. and Scalla, R. (1989) Protoporphyrinogen oxidase as a molecular target for diphenyl ether herbicides. *Biochem. J.* **260**: 231-235.

McLeod, G. C. and Coomber, J. (1959) The transformation of protochlorophyll to chlorophyll *a* in ultraviolett light. *Carnegie Inst. Wash. Year Book* **59**: 324-325.

Mock, H.-P. and Grimm, B. (1997) Reduction of uroporphyrinogen decarboxylase by antisense RNA expression affects activities of other enzymes involved in tetrapyrrole biosynthesis and leads to light dependent necrosis. *Plant Physiol.* **113**: 1101-1112.

Mock, H.-P., Keetman, U., Kruse, E., Rank, B. and Grimm, B. (1998) Defense responses to tetrapyrrole-induced oxidative stress in transgenic plants with reduced uroporphyrinogen decarboxylase or coproporphyrinogen oxidase activity. *Plant Physiol.* **116**: 107-116.

Mock, H.-P., Trainotti, L., Kruse, E. and Grimm, B. (1995) Isolation, sequencing and expression of cDNA sequences encoding uroporphyrinogen decarboxylase from tobacco and barley. *Plant Mol. Biol.* **28**: 245-256.

Nielsen, O. F. (1974) Macromolecular physiology of plastids. XII. *Tigrina* mutants of barley. *Hereditas* **76**: 269-304.

Nielsen, O. F. and Gough, S. (1974) Macromolecular physiology of plastids. XI. Carotenes in etiolated *tigrina* and *xantha* mutants of barley. *Plant Physiol.* **30**: 246-254.

Schmid, R. and Shemin, D. (1955) The enzymatic formation of porphobilinogen from δ-aminolevulinic acid and its conversion to protoporphyrin. *J. Am. Chem. Soc.* **77**: 506-508.

Scolnik, P. A. and Haselkorn, R. (1984) Activation of extra copies of genes coding for nitrogenase. *Nature* **307**: 289-292.

Seely, G. R. (1966) The structure and chemistry of functional groups. In *The chlorophylls*, Vernon, L. P. and Seely, G. R., eds., New York, Acad. Press, pp. 67-109.

Shemin, D. and Rittenberg, D. (1945) The utilization of glycine for the synthesis of a porphyrin. *J. Biol. Chem.* **159**: 567- 568.

Shemin, D. and Rittenberg, D. (1946) The biological utilization of glycine for the synthesis of the protoporphyrin of hemoglobin. *J. Biol. Chem.* **166**: 621-625.

Shemin, D. and Russel, C. S. (1953) δ-aminolevulinic acid, its role in the biosynthesis of porphyrins and purines. *J. Am. Chem. Soc.* **75**: 4873-4874.

Shibata, K. (1957) Spectroscopic studies on chlorophyll formation in intact leaves. *J. Biochem.* **44**: 147-173.

Smith, J. H. C. (1948) Protochlorophyll, precursor of chlorophyll. *Arch. Biochem.* **19**: 449-454.

Smith, J. H. C. and Benitez, A. (1954) The effect of temperature on the conversion of protochlorophyll to chlorophyll *a* in etiolated barley leaves. *Plant Physiol.* **29**: 135-143.

Smith, J. H. C. and Coomber, J. (1959) The effect of pH on the phototransformation of protochlorophyll holochrome. *Carnegie Inst. Wash. Year Book* **59**: 325-330.

Smith, J. H. C. and French, C. S. (1958) Quantum yield for the protochlorophyll-chlorophyll conversion. *Carnegie Inst. Wash. Year Book* **57**: 290-293.

Smith, J. H. C. and Kupke, D. (1956) Some properties of extracted protochlorophyll holochrome. *Nature* **178**: 751-752.

Sweeney, G. D. (1986) Porphyrea cutanea tarda, or the uroporphyrinogen decarboxylase deficiency diseases. *Clin. Biochem.* **19**: 3-15.

Taylor, D. P., Cohen, S. N., Clark, W. G. and Marrs, B. (1983) Alignment of the genetic and restriction maps of the photosynthesis region of the *Rhodopseudomonas capsulata* chromosome by a conjugation-mediated marker rescue technique. *J. Bacteriol.* **154**: 580-590.

von Wettstein, D. (1959) Spectrophotometric studies of chlorophyll mutants in barley. *Carnegie Inst. Wash. Year Book* **59**: 338-339.

von Wettstein, D., Henningsen, K. W., Boynton, J. E., Kannangara, C. G. and Nielsen, O. F. (1971) The genetic control of chloroplast development in barley. In *Autonomy and Biogenesis of Mitochondria and Chloroplasts*, Boardmann, N. K., Linnane, A. W. and Smillie, R. M., eds., Amsterdam: North Holland, pp. 205-223.

von Wettstein, D., Kahn, A., Nielsen, O. F. and Gough, S. P. (1974) Genetic regulation of chlorophyll synthesis analysed with mutants of barley. *Science* **184**: 800-802.

Willstätter, R. and Stoll, A. (1928) Untersuchungen über Chlorophyll. Berlin, Springer Verl., (English edition: Investigations on chlorophyll. Lancaster, Ohio, Science Press).

Woodward, R. B., Ayer, W. A., Beaton, J. M., Bickelhaupt, F., Bonnett, R., Buchschacher, P., Closs, G. L., Dutler, H., Hannah, J., Hauck, F. P., Ito, S., Langemann, A., LeGoff, E., Leimgruber, W., Lwowski, W., Sauer, J., Valenta, Z. and Volz, H. (1960) The total synthesis of chlorophyll. *J. Am. Chem. Soc.* **82**: 3800-3802.

Yang, Z. and Bauer, C. E. (1990) *Rhodobacter capsulatus* genes involved in early steps of the bacteriochlorophyll biosynthesis pathway. *J. Bacteriol.* **172**: 5001-5010.

Yen, H.-C. and Marrs, B. L. (1976) Map of genes for caratenoid and bacteriochlorophyll biosynthesis in *Rhodopseudomonas capsulata. J. Bacteriol.* **126**: 619-629.

Young, D. A., Bauer, C. E., Williams J. C. and Marrs, B. L. (1989) Genetic evidence for superoperonal organization of genes for photosynthetic pigments and pigment-binding proteins in *Rhodobacter capsulatus. Mol. Gen. Genet.* **218**: 1-12.

Youvan, D. C., Alberti, M., Begusch, J., Bylina, E. and Hearst, J. E. (1984a) Reaction center and light-harvesting genes from *Rhodopseudomonas capsulata. Proc. Natl. Acad. Sci., USA.* **81**: 189-192.

Youvan, D. C., Bylina, E. J., Alberti, M., Begusch, H. and Hearst, J. E. (1984b) Nucleotide and deduced polypeptide sequences of the photosynthetic reaction center, B870 antenna and flanking polypeptides from *Rb. capsulatus. Cell* **37**: 949-957.

Chapter 6

Chlorophyll Biosynthesis II: Adventures with Native and Recombinant Enzymes

Diter von Wettstein
Departments of Crop and Soil Sciences & Genetics and Cell Biology
Washington State University, Pullman, WA 99164-6420, USA

ABSTRACT

Biochemical, genetic and molecular adventures result in never-ending, exciting discoveries in the elucidation of the chlorophyll biosynthetic pathway. This is exemplified by the discovery that Shemin's single enzyme pathway for the 5-aminolevulinate precursor of chlorophyll is only used by some photosynthetic bacteria. Cyanobacteria, algae and all plants use a multi-enzyme pathway involving a unique step of activation with a tRNA. Most fascinating is the variety of enzymes and mechanisms photosynthetic organisms employ for the same step in the pathway. This chapter highlights interesting biochemical and molecular aspects of fourteen steps in the biosynthesis of chlorophyll.

Surprises in the Biosynthesis of 5-Aminolevulinate

In chlorophyll, four pyrrole rings (designated I-IV) are ligated into a tetrapyrrole ring with a magnesium atom in the center (Fig. 1). Ring IV is esterified with phytol. Higher plants and algae use for light harvesting an additional form of chlorophyll, chlorophyll *b*, that has a formyl group instead of a methyl group in position 3. The porphyrin ring with its conjugated double bonds is assembled in the chloroplast from eight molecules of 5-aminolevulinic acid, a highly reactive non-protein amino acid (5-amino, 4-keto pentanoic acid). The bonds highlighted in red in the chlorophyll structure of Fig. 1 delineate the locations of the atoms derived from the eight 5-aminolevulinic acid carbon skeletons. An intact 5-aminolevulinate molecule can be recognized in pyrrole ring IV: The nitrogen coordinated with the magnesium atom originates from the amino group followed by four carbon atoms and the carboxyl-group, which is esterified to phytol.

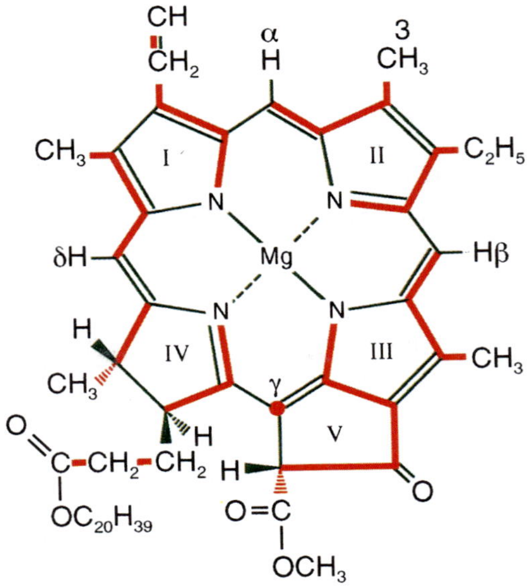

Fig. 1. The chemical structure of chlorophyll *a*. Every chlorophyll molecule is synthesized in the chloroplast from eight molecules of 5-aminolevulinic acid. The location of the atoms derived from these molecules of 5-aminolevulinic acid in the finished chlorophyll molecule are indicated by the heavy lines. Position 3, which is occupied by a methyl group in chlorophyll *a* and a formyl group in chlorophyll *b* is marked.

There are two distinct routes for the synthesis of 5-aminolevulinate, one utilizing a condensation reaction of glycine with succinyl-CoA by the enzyme 5-aminolevulinic acid synthase (cf. above) and the other a three-step pathway from glutamate that is called the C_5 pathway (Jordan, 1991). The first route is used by animals, yeast, and a number of bacteria, including the facultative phototrophic genera *Rhodobacter* and *Rhodospirillum*, the obligatory phototrophic bacteria of the *Erythrobacter* and *Methylobacterium groups*, and the non-photosynthetic bacteria such as *Agrobacterium* and *Rhizobium* (Beale, 1995). 5-aminolevulinic acid synthase is a homodimeric enzyme consisting of 41–45kDa subunits and uses pyridoxal phosphate as cofactor. A detailed review of its reaction mechanism is given by Beale (1995).

The C$_5$ pathway is characteristic of higher plants, bryophytes, cyanobacteria, and many eubacteria including the green sulfur bacteria, green nonsulfur bacteria, heliobacteria, and archaebacteria. In the phytoflagellata *Euglena gracilis*, the two pathways are used in different compartments. The C$_5$ pathway operates in chloroplasts, and is exclusively responsible for chlorophyll synthesis, while heme *a* of cytochrome c oxidase in mitochondria is formed by 5-aminolaevulinate synthase (Weinstein and Beale, 1983).

When greening plants or algae are treated with levulinate, 5-aminolevulinate accumulates because levulinate inhibits the enzyme 5-aminolevulinate dehydratase (Beale and Castelfranco, 1974a). If ^{14}C-labeled glutamate is fed together with levulinate, the radioactivity is found in the accumulated 5-aminolevulinic acid (Beale and Castelfranco, 1974b). By using glutamic acid labeled with ^{14}C in different positions as precursors, and determining the label distribution in the resulting 5-aminolaevulinic acid or the entire chlorophyll molecule, it was shown that the intact five-carbon skeleton of glutamate is incorporated into 5-aminolevulinic acid (Beale *et al.*, 1975; Meller *et al.*, 1975; Porra, 1986).

It was a great surprise when it was discovered that conversion of glutamate to 5-aminolevulinate requires glutamate to be activated at the α-carboxyl by ligation to tRNA^{Glu}. It was equally surprising that the *hemA* gene of *E. coli* (Sasarman *et al.*, 1968) encodes not 5-aminolevulinate synthase, as had been thought for 20 years, but glutamyl-tRNA^{Glu} dehydrogenase. Moreover a second *E. coli* gene, identified as giving rise upon mutation to 5-aminolevulinate auxotrophy (Powell *et al.*, 1973), turned out to be a remarkable enzyme — glutamate 1-semialdehyde aminotransferase.

As detailed in Fig. 2, glutamic acid is first activated by ligation to tRNA^{Glu} with an aminoacyl-tRNA synthetase in the presence of ATP and Mg^{++}. The activated glutamate is reduced to glutamate 1-semialdehyde in an NADPH-dependent reaction that is catalyzed by Glu-tRNA reductase. Glutamate 1-semialdehyde 2,1-aminomutase then carries out the conversion into 5-aminolevulinate (Kannangara *et al.*, 1994). All components involved in 5-aminolevulinate synthesis in greening barley are soluble, and localized in the plastid stroma (Gough and Kannangara, 1977). The components required for conversion of glutamate to 5-aminolevulinate, and subsequently to uroporphyrinogen are routinely partially purified from the stroma of greening barley plastids by Sephacryl S-300 gel filtration, and affinity chromatography, with sequentially, Cibacron Blue-Sepharose, Procion Red-agarose, and chlorophyllin- or heme-Sepharose (Wang *et al.*, 1981). This procedure separates the components into three fractions that must be combined for the conversion of glutamate to 5-aminolevulinate. Blue-Sepharose binds the ligase and the reductase, and Procion Red-agarose binds several unwanted proteins. Chlorophyllin or heme Sepharose binds the tRNA complement of the chloroplasts (Kannangara *et al.*, 1984). Glutamate-1-semialdehyde amino transferase (Kannangara and Gough, 1978) is not retained by the affinity columns and is collected in the run-off fraction. When the aminotransferase fraction is omitted from the reconstituted mixture, the product formed is glutamate-1-semialdehyde and not 5-aminolevulinate.

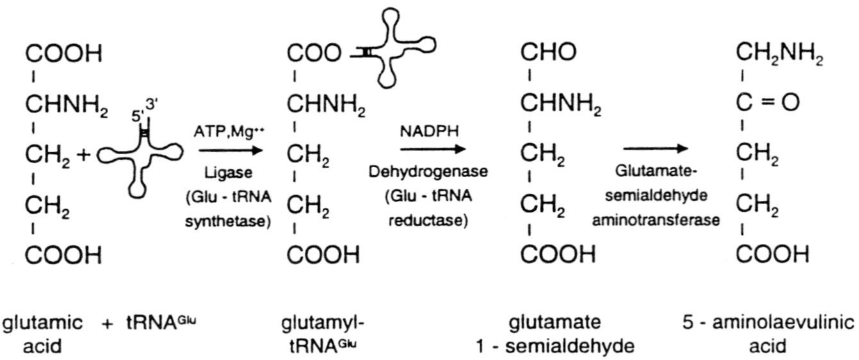

| glutamic + tRNA^{Glu} acid | glutamyl-tRNA^{Glu} | glutamate 1 - semialdehyde | 5 - aminolaevulinic acid |

Fig. 2. Biosynthesis of 5-aminolevulinate.

The Role of tRNA^{Glu} and the Glutamate-tRNA^{Glu} Ligase (Glutamyl-tRNA Synthetase)

The nucleotide sequence of the barley tRNA (Schön *et al.*, 1986) involved in 5-aminolevulinate synthesis can be fitted into the clover leaf structure. At the 3′ end, it has the CCA sequence characteristic of all tRNAs. The anticodon sequence UUC identifies it as glutamate-specific tRNA. The 3′CCA, as well as the anticodon sequence, are required for the ligation. Removal of the CCA by digestion with snake venom phosphodiesterase leads to complete loss of the tRNA′s ability to ligate glutamate. Replacing the CCA sequence with nucleotidyl transferase restores this ability (Schön *et al.*, 1986). The first position of the anticodon of barley tRNA^{Glu} is occupied by 5-methylaminomethyl-2-thiouridine. This hypermodified nucleotide can be oxidized under mild conditions with iodine and is then unable to function in the ligase reaction, but its activity can be restored by reduction with thiosulphate. tRNA^{Glu} has been found to activate glutamate in all organisms that use the C$_5$ pathway (Schneegurt and Beale, 1988), including *Chlamydomonas* (Huang *et al.*, 1984; Huang and Wang, 1986), *Chlorella* (Weinstein and Beale, 1985; Weinstein *et al.*, 1986), *Scenedesmus* (Breu and Dörnemann, 1988), *Chlorobium*, and other phototrophic bacteria (Avissar *et al.*, 1989).

In higher plants, the gene encoding the tRNA^{Glu} is encoded in chloroplast DNA, whereas the three enzymes — the aminoacyl-tRNA synthetase, the reductase, and the aminotransferase — are encoded by nuclear DNA, and are imported into the chloroplast stroma after synthesis on cytoplasmic ribosomes. This single chloroplast tRNA^{Glu} has to serve for both chlorophyll synthesis and protein synthesis on chloroplast ribosomes. The same glutamyl-tRNA synthetase charges tRNA^{Glu} for protein and 5-aminolevulinate synthesis (Bruyant and Kannangara, 1987) as is also the case in the cyanobacterium *Synechocystis* (Rieble and Beale, 1991). In the presence of ATP and glutamate, purified glutamate-tRNA synthetase, tRNA^{Glu}, and Glu-tRNA reductase can form a complex; this implies that different domains of the tRNA^{Glu} molecule recognize and bind to the two enzymes. This conclusion has been verified by studies of a *Euglena gracilis* mutant in which a cytosine of the T-loop of tRNA^{Glu} was converted to a uracil (Stange-Thomann *et al.*, 1994). The tRNA can still be charged with glutamate by the aminoacyl-tRNA synthetase, as judged by the capacity of the mutant to synthesize the large subunit of ribulose-bisphosphate carboxylase and ATP synthase on chloroplast ribosomes, but it is unable to function with the reductase in the synthesis of 5-aminolevulinate.

Although the chloroplast ligase does not discriminate between tRNA^{Glu} and tRNA^{Gln}, and it loads glutamate onto both tRNA species, the reductase recognizes glutamyl-tRNA^{Glu} exclusively. Glutamyl-tRNA^{Gln} is converted to glutaminyl-tRNA^{Gln} by a specific amidotransferase (Schön *et al.*, 1988). Chloroplast glutamyl-tRNA^{Glu} from barley, wheat, tobacco, spinach, cucumber, *Synechocystis*, and *Chlamydomonas* are equally efficient as substrates for the barley glutamyl-tRNA reductase, but *E. coli* tRNA^{Glu} is a much poorer substrate, and yeast tRNA^{Glu} is only slightly better (Willows *et al.*, 1995). A comparison of the bases conserved among tRNA^{Glu} and tRNA^{Gln} from different species and cell compartments reveals that five nucleotides are probably required for recognition by the glutamyl-tRNA synthetase. An analogous comparison of sequences that can and cannot be used by the reductase indicates that seven nucleotides are involved in recognition by the barley chloroplast glutamyl-tRNA^{Glu} reductase.

The chloroplast tRNA^{Glu} ligase comprises two identical subunits, both synthesized in the cytoplasm, each with a pre-sequence of 34 amino acids and a molecular mass of 58kDa. The subunits are then transported into the chloroplast, processed to give the mature subunits of 526 amino acids (molecular mass 54kDa), and assembled to give the functional enzyme. The barley chloroplast enzyme shows 40% amino acid sequence homology to other known glutamate-tRNA^{Glu} ligases (Andersen, 1992). Aminoacyl-tRNA synthetases ligate their amino acids to their cognate tRNAs in two steps. In the first step, the enzyme catalyses a reaction between the amino acid and Mg^{++}-ATP to give an aminoacyl-adenylate and pyrophosphate. The aminoacyl moiety is then transferred from aminoacyl-adenylate to either the 2´or the 3´OH of the tRNA, and AMP is released. The release of pyrophosphate is not observed with glutamate-tRNA^{Glu} ligase in the absence of tRNA^{Glu}, in contrast to most other aminoacyl-tRNA synthetases; the synthetase´s mechanism of catalysis is therefore likely to involve the formation of a ternary intermediate, which in turn might be associated with the glutamyl-tRNA^{Glu} reductase.

Glutamyl-tRNA^{Glu} Reductase

Two approaches have been taken to characterize the latter unique enzyme that converts glutamyl-tRNA^{Glu} to glutamate 1-semialdehyde (reviewed in Pontoppidan and Kannangara, 1994). In one approach, purification of this unstable enzyme of low abundance has been pursued, while the other centered around mutations in the *hemA* gene of the *E. coli* K12 strain, which leads to auxotrophy for 5-aminolaevulinate. In barley, unequivocal, evidence has been obtained that *hemA* encodes the glutamyl-tRNA^{Glu} reductase, and that no other protein is required for activity (Pontoppidan and Kannangara, 1994). The purified barley enzyme with a molecular mass of about 270kDa is made up of four to six identical subunits with a molecular weight of 54kDa, and has an activity for glutamatesemialdehyde synthesis of 250 pmol·min^{-1}·µg^{-1}. The determined sequence of 18 N-terminal amino acids of the mature subunits of the active enzyme is identical to the N-terminal amino acid sequence deduced from a barley cDNA clone that rescues the *E. coli hemA* mutant (Bougri and Grimm, 1996). Complementation of the *E. coli* mutant has yielded over the years, *HemA* genes from *Salmonella typhimurium*, *Bacillus subtilis*, *Synechocystis*, *Arabidopsis*, *Clostridium josui*, barley, and other species. Uncertainty reigned about the function of the *hemA* gene product for two reasons: (1) The mutant can also be rescued by the genes encoding the monomeric 5-aminolevulinate synthase from *Rhodobacter* and *Rhizobium*, the enzyme using the ubiquitous metabolites, succinyl-CoA and glycine as substrate. When the *E. coli hemA* gene was cloned and sequenced, its deduced amino acid sequence showed no homology to that of 5-aminolevulinate synthase from *Rhodobacter* and the mouse. At the same time, identification of the *E. coli hemL* gene as the structural gene for glutamate 1-semialdehyde aminotransferase, and demonstration that this enzyme as well as glutamyl-tRNA synthetase are active in the *E. coli hemA* mutant, established that *E. coli* uses the C_5 pathway. This is a good illustration that complementation of a mutant with expression of a known enzyme does not necessarily identify the enzymatic function of the cognate wild type gene. (2) While *hemA* genes from several bacteria and higher plants can complement the *E. coli hemA* mutant, and *E. coli* cells overproduce uroporphyrinogen, when they are engineered to overexpress the *hemA* gene, the purified *hemA* protein is not enzymatically active *in vitro*.

The deduced amino acid sequences of eight glutamyl-tRNA^{Glu} reductases contain several highly conserved sequence domains, including one of 27 residues that is 70% identical in all eight proteins (Vothknecht *et al.*, 1998). Analysis of the function of this domain and elucidation of the mechanism of the reduction of glutamyl tRNA^{Glu} require establishment of an expression system yielding high amounts of active enzyme. This has been achieved by expressing a barley *hemA* cDNA gene encoding the glutamyl tRNA reductase in *E. coli* joined at its amino terminal end to the glutathione *S*-transferase of *Schistosoma japonicum* (Vothknecht *et al.*, 1996). Milligram quantities of highly active enzyme could be produced. Both the fusion protein and the reductase released from it by thrombin digestion catalyzed the reduction of glutamyl tRNA^{Glu} to glutamate 1-semialdehyde with a specific activity of 120 pmol·µg^{-1}·min^{-1}. The fusion protein used tRNA^{Glu} from barley chloroplasts preferentially to *E. coli* tRNA^{Glu}, and its activity was inhibited by hemin. The released reductase migrated as a ≈60 kDa polypeptide with SDS/PAGE, while gel filtration on Superose 12 yielded an apparant molecular mass of 250kDa. The fusion protein contained heme, which could be reduced by NADPH, and oxidized by air. The mature plant enzymes contain a highly conserved extension of 31–34 amino acids at the N-terminus not present in bacterial enzymes. It was found that barley glutamyl tRNA reductases with a deletion of the 30 N-terminal amino acids have the same high specific activity as the untruncated enzymes, but are highly resistant to feed-back inhibition by heme (Vothknecht *et al.*, 1998). This peptide domain apparently interacts directly or indirectly with heme and the toxicity of the 30-amino acid peptide for *E. coli* experienced in mutant rescue and overexpression experiments can be explained by extensive heme removal from the metabolic pools that cannot be tolerated by the cell. By induced mis-sense mutations, nine amino acids have been identified in the 451-residue-long C-terminal part of the barley enzyme that, upon substitution, curtail drastically but do not eliminate entirely the catalytic activity of the enzyme. These amino acids are thus important for the catalytic reaction or tRNA binding. Availability of substantial amounts of enzyme allows one to test the hypothesis that the conversion of glutamyl-tRNA to glutamate 1-semialdehyde proceeds via azeridinone propionic acid, which is then reduced and hydrolyzed (Kannangara *et al.*, 1994).

Barley contains two *hemA* genes that have different expression patterns. One gene is not transcribed in roots, but gives a high transient transcript level during greening; a second is expressed moderately in roots and leaves (Bougri and Grimm, 1996). Light-stimulated transcription of the *Arabidopsis* genes for glutamyl-tRNA^{Glu} reductase and glutamate semialdehyde aminotransferase has been reported by Ilag *et al.* (1994).   Further experiments are necessary to determine to what extent the regulation of 5-aminolevulinate synthesis takes place, at the transcriptional and/or translational level.

The Glutamate 1-Semialdehyde Intermediate

The substrate of glutamate 1-semialdehyde aminotransferase can be chemically synthesized by three different methods, and this has facilitated studies of the enzyme. Two of the methods use N-carbobenzoxy-L-glutamate γ-benzyl ester as starting material; these two methods are quite cumbersome and give poor yields (5–10%) of (S)-glutamate 1 semialdehyde. The most convenient method starts from 4-amino-5-hexenoic acid (Gough *et al.*, 1989), also called γ-vinyl GABA, which is available as a mixture of *R* and *S* enantiomers from Merrell Dow (Cincinnati, OH USA).   An aqueous solution containing equimolar amounts of 4-amino-5-hexenoic acid and HCl is treated with ozone to give the ozonide, which is then reduced by passage through a Dowex 50X8 column in its H$^+$ form. This method gives pure (*RS*)-glutamate 1-semialdehyde in yields up to 90%.   Glutamate 1-semialdehyde hydrochloride is stable, as a solid. Our group at the Carlsberg Laboratory, and Peter Jordan and co-workers then at the University of London have analyzed the structure of glutamate 1-semialdehyde by NMR and infrared sprectroscopy (Gough *et al.*, 1989; Hoober *et al.*, 1988; Jordan, 1990). The two groups obtained identical spectra but interpreted them differently.   One interpretation is that glutamate 1-semialdehyde is a linear molecule existing as hydrate in solution. Peter Jordan´s group, however, think that the spectra represent a cyclic molecule, hydroxyaminotetrahydropyranone. On the other hand, titration of (*RS*)-glutamate 1-semialdehyde hydrochloride with NaOH indicates the presence of a free carboxyl group and a free amino group in the molecule (Kannangara *et al.*, 1994). This points to a linear structure of the molecule. Glutamate 1-semialdehyde identical to the chemically synthesized compound accumulates in greening leaves (Wang *et al.*, 1981; Kannangara and Schouboe, 1985), and in extracts of *Scenedesmus* (Breu and Dörnemann, 1988) that have been treated with 3-amino-2,3-dihydrobenzoic acid (gabaculine), a mechanism-based suicide inhibitor that blocks chlorophyll synthesis.

Glutamate 1-Semialdehyde Aminotransferase

The vitamin B_6-containing enzyme glutamate 1-semialdehyde aminotransferase transfers the amino group from the C_4 carbon of glutamate 1-semialdehyde to the C_5 carbon of 5-aminolevulinic acid with pyridoxal 5' phosphate as a cofactor and 4,5-diaminovalerate as an intermediate (Kannangara *et al.*, 1994). This enzyme is found in plants (Kannangara and Gough, 1978; Nair *et al.*, 1991; Höfgen *et al.*, 1994), algae (Wang *et al.*, 1984; Mayer *et al.*, 1987; Breu and Dörnemann, 1988; Houghton *et al.*, 1989; Jahn *et al.*, 1991), and a number of bacteria (Friedman *et al.*, 1987; Bull *et al.*, 1990; Ilag *et al.*, 1991). Cloning of the structural genes for this enzyme from barley (Grimm, 1990), *Salmonella typhimurium* (Elliot *et al.*, 1990), *Synechococcus*, *E. coli* (Grimm *et al.*, 1991; Ilag and Jahn, 1992), and *Bacillus subtilis* (Hansson *et al.*, 1991) opened the way to express this gene in *E. coli* for overproduction of the recombinant enzyme in large amounts. Co-expression of the GroEL and GroES chaperonin proteins was required for isolation of active, soluble recombinant barley glutamate 1-semialdehyde aminotransferase from *E. coli* (Berry-Lowe *et al.*, 1992), but demonstrated that the chloroplast enzyme, as to substrate specificity, kinetics, and enzyme mechanism is very similar to the recombinant enzyme of *Synechococcus* that can be expressed in *E. coli* without chaperonins, and has therefore been preferred for biochemical and molecular studies.

The enzyme has an exceptionally high affinity for *S*-glutamate 1-semialdehyde (K_m=12µM). It is irreversibly inhibited by gabaculine (3-amino-2,3-dihydrobenzoic acid), methylgabaculine, 2-hydroxy-3-amino-3,5cyclohexadiene-1-carboxylic acid, 4-amino-5-hexynoic acid (acetylenic GABA), 4-amino-5-fluoropentanoic acid, and glutamic acid 5-monohydroxamate (Gough *et al.*, 1992). These compounds inhibit chlorophyll synthesis in greening barley leaves causing accumulation of glutamate 1-semialdehyde.

In barley, in tobacco, and *Arabidopsis thaliana*, glutamate 1-semialdehyde aminotransferase is synthesized in the cytoplasm as a precursor protein of 49,540kDa, and then transferred into the chloroplast. During this import, a transit sequence of 34 amino acids is cleaved from the N-terminus. At the level of deduced amino acids, the primary structure of the plant and bacterial enzymes are remarkably similar, showing greater than 55% identity. The glutamate 1-semialdehyde aminotransferase of *Synechococcus* has 72% sequence identity with the barley enzyme. In most organisms, active glutamate 1-semialdehyde aminotransferase is a homodimer, although the isolated enzyme of *Chlamydomonas* and *Synechococcus* can function as a monomer (Jahn *et al.*, 1991; Wang *et al.*, 1984).

Glutamate 1-semialdehyde aminotransferase has a characteristic absorption spectrum that changes during enzyme catalysis depending on the binding of the pyridoxal/pyridoxamine 5′phosphate cofactor and substrate intermediates (Smith *et al.*, 1991a,b; Berry-Lowe *et al.*, 1992; Ilag and Jahn, 1992; Pugh *et al.*, 1992; Brody *et al.*, 1995). These changes have allowed significant parts of the reaction mechanism to be deduced. The recombinant native enzyme of *Synechococcus* is pale yellow with absorption maxima at 280, 338 and 418nm, and produces 5-aminolevulinate from glutamate 1-semialdehyde at the rate of 570nmol·mg^{-1}·min^{-1}. When it is exposed to 4,5-diaminovalerate, absorption increases at 338nm and decreases at 418nm (Smith *et al.*, 1991a,b). This is called the colorless enzyme. Upon treatment of the enzyme with 4,5-dioxovalerate, the enzyme turns bright yellow with the absorption decreasing at 338nm and increasing at 418nm. The yellow enzyme is sensitive to acetylenic GABA and $NaBH_4$, whereas the colorless enzyme is insensitive to these compounds. Treatment with $NaBH_4$ shifts the absorption maximum from 418nm to 338nm, thus converting the yellow enzyme into the colorless form. The yellow enzyme is converted into a pink form with a new absorption maximum at 560nm upon reaction with acetylenic GABA (Kannangara *et al.*, 1994). Proof that colorless glutamate 1-semialdehyde aminotransferase absorption at 338nm is due to pyridoxamine 5′phosphate, and the yellow absorption at 418nm to pyridoxal 5′phosphate bound to the active site Lys^{272} by a Schiff base was obtained in the following way (Brody *et al.*, 1995).

When the colorless enzyme is repeatedly diluted with Na_2HPO_4 buffer and reconcentrated, increasing amounts of pryridoxamine 5′-phosphate are released. Such a release is not observed with the yellow or pink enzyme. The absorption maximum at 338nm is therefore due to non-covalently bound pyridoxamine 5′phosphate. If the yellow enzyme is reacted with $NaBH_4$, the absorption at 418nm disappears after a few seconds and the activity of the enzyme is strongly reduced, whereas the colourless form is not affected.   Acidification made the yellow enzyme colorless, and pyridoxal 5´phosphate dissociated from the enzyme as a result of hydrolysis of the Schiff base.   Acetic acid incubation of the colorless enzyme, on the other hand, released pyridoxamine 5´phosphate, whereas the pink enzyme yielded pyridoxal 5´phosphate.   After removal of the cofactor by acidification, the apoenzyme protein had an absorption at 280nm.   The native enzyme turned out to be a mixture of protein molecules carrying either a single pyridoxal 5´phosphate or a single pyridoxamine 5´phosphate cofactor.   These results were verified by electrospray ionization mass spectrometry.   As the pyridoxamine 5´-phosphate and the Schiff base-linked pyridoxal 5´-phosphate dissociate in acetic acid, the mass spectra of the native enzyme, the yellow enzyme and the apoenzyme are identical (M_r of the protein 46,401–46,410).   Reduction of the yellow enzyme with $NaBH_4$ resulted in two mass peaks, one of M_r 46,406 corresponding to the apoenzyme, and one of M_r 46,636, i.e., a mass increase of 230 units expected for one pyridoxal 5´phosphate bound per protein molecule.   This reveals that the pyridoxal 5´phosphate is bound by an unstable Schiff base linkage in the yellow enzyme, and that $NaBH_4$ reduces the Schiff base, covalently attaching the cofactor to the protein.   Treatment of the yellow enzyme with acetylenic GABA to yield the inactive pink enzyme gave an electron spray ionization spectrum with a 46,411 apoenzyme peak, and a 123 mass units higher peak.   It was concluded that this peak originated from protein molecules each carrying a single covalently bound molecule of acetylenic GABA.

Studies with site-directed mutagenesis have indicated that Lys[272] (Lys[276] of the N-terminal extended recombinant protein) is the pyridoxal 5´-phosphate binding site (Grimm *et al.*, 1992; Ilag and Jahn, 1992).   To identify the binding site of acetylenic GABA, and the pyridoxal cofactor on the enzyme, equal amounts of the pink and $NaBH_4$-reduced yellow enzymes were digested with trypsin, the peptides separated and analyzed by on-line liquid chromatography/mass spectrometry (Brody *et al.*, 1995). The amino acid sequence of glutamate 1-semialdehyde predicts 37 tryptic peptides, 28 of which have a calculated mass above 400 and could be identified with the correct mass in both spectra.   Lys[276] is within the tryptic peptide T24.   Modifying it by covalent linkage is expected to prevent the tryptic cleavage at this position.   This will lead to a peptide combining the mass of T24, its neighbor T25, and the modifying molecule.   The predicted mass of this combined peptid modified with covalently bound pyridoxal 5´-phosphate in the yellow enzyme is 1248.4+1258.5–18.0+230.14 = 2719.0. The mass found for this peptide after reduction of the yellow enzyme with

NaBH$_4$ was 2719.4. A mass of 2616.3 was observed in the mass spectrum for the linked T24+T25 peptide modified with acetylenic GABA in the pink enzyme. Isolation of this peptide and subjecting it to electrospray ionization mass spectrometry confirmed the M$_r$ value to be 2616.6 and in close agreement with the calculated value of 2615.0. It is, therefore, evident that both pyridoxal 5´phosphate and acetylenic GABA are present covalently bound to Lys[276] after modification.

When studying the reaction mechanism, it has been observed that the *S* enantiomer of glutamate 1-semialdehyde is used preferentially over the *R* enantiomer (Smith *et al.*, 1991a, 1992), and likewise addition of (*S*)-4,5-diaminovalerate stimulates the formation of 5-aminolevulinate (Friedmann *et al.*, 1992). The enzyme also slowly converts (*R*)-glutamate 1-semialdehyde to 5-aminolevulinate in an anomalous reaction (Kannangara *et al.*, 1994). Only the pyridoxamine 5´phosphate form (colorless) of glutamate 1-semialdehyde aminotransferase is active in catalyzing the conversion of glutamate 1-semialdehyde to 5-aminolevulinate, and the reaction proceeds via the intermediate 4,5-diaminovalerate (Hoober *et al.*, 1988; Smith *et al.*, 1991a; Pugh *et al.*, 1992). The aldehyde group of glutamate 1-semialdehyde first forms a Schiff´s base with the amino group of pyridoxamine 5´-phosphate. Thereafter, a proton at the C(4´) position of pyridoxamine 5´-phosphate is transferred to the C-1 of glutamate 1-semialdehyde. 4,5-diaminovalerate and pyridoxal 5´-phosphate are thus bound to each other by a Schiff's base between the C-5 amino group of 4,5-diaminovalerate and the CHO group of pyridoxal 5´-phosphate. The ε amino group of Lys[272] attacks this Schiff's base to liberate 4,5-diaminovalerate, which immediately binds by its C-4 amino group to the CHO group of pyridoxal 5´-phosphate (a Schiff´s base). The removal of a proton from C-4 of 4,5-diaminovalerate and protonation of C(4´) of pyridoxal 5´-phosphate generates 5-aminolevulinate bound by its keto group to the amino group of pyridoxamine 5´phosphate (a Schiff´s base). Hydrolysis of this Schiff´s base liberates 5-aminolevulinate from the pyridoxamine 5´-phosphate form of the enzyme, which is thus ready for a new cycle of transamination.

In order to discover the function of the dimeric feature of the enzyme in the catalytic mechanism, Michael Hennig, Bernhard Grimm, Roberto Contestabile, Robert John, and Johan Jansonius determined the crystal structure of the dimeric recombinant glutamate 1-semialdehyde aminotransferase of *Synechococcus* (Hennig *et al.*, 1997).   Using 5 isomorphous heavy atom derivatives, the structures of the native enzyme, the form reduced with NaBH$_3$CN to attach pyridoxal 5´phosphate covalently to Lys[272] (designated Lys[273] by Hennig *et al.*), and the enzyme containing gabaculine in the substrate pocket were refined at a resolution of 2,4Å to an R-factor of 18.7%. The enzyme fold in both monomers can be divided into three domains (Fig. 3). The 70 residues of the N-terminal domain form an α-helix followed by a three-stranded antiparallel β-sheet. The cofactor binding domain comprising the fold of amino acids 70 to 326 is very similar to that of other aminotransferases comprising a central seven-stranded β-sheet with six parallel and one antiparallel strands. This characteristic structure is surrounded by several α-helices of variable length. The C-terminal domain from residue 327 to 433 is made up of a three-stranded antiparallel β-sheet covered on the outer surface with four helices, one forming the C-terminal end. This three-dimensional structure is thus similar to other vitamin B_6-containing enzymes, but major specific domains can be recognized in the glutamate 1-semialdehyde aminotransferase.

Its dimer structure is compact and the two subunits interface over a large area (Fig. 3). The active sites are located at the interface. Superposition of the polypeptide chains of subunits A and B reveal identical backbone conformations with two exceptions. The region bordered by two α-helices spanning His[153] to Thr[181] located at the interface of the two subunits and at the surface of the dimer is well defined in subunit A but disordered in subunit B, indicating a flexible region. The other significant deviation in conformation concerns a loop between residue Pro[30] to Ile[42], including Ala[33] and Val[31] close to the active site. It was shown above that the native enzyme contains a mixture of the pyridoxamine 5′phosphate and pyridoxal 5′phosphate forms of the enzyme unless the enzyme is treated to shift into one of the two forms. The crystal structure shows that each dimer contains both enzyme forms. When the location and conformation of the cofactor and Lys[272] is analyzed, the location of the phosphate group is the same in both subunits but the orientation of the pyridine rings differ by about 30°. In one subunit, the electron density between the cofactor and the lysine side chain is continuous, consistent with the protonated Schiff base between Lys[272] and pyridoxal 5′-phosphate responsible for the 418nm yellow absorption. In the other subunit, there is no continuous electron density beween the cofactor and Lys[272], demonstrating the absence of a covalent bond and indicating that the cofactor in this subunit is pyridoxamine 5′phosphate. In the crystals made with enzyme after reduction of the Schiff double bond with cyanoborohydrate, both subunits contained a covalent bond between the cofactor and the Lys[272]. However, the difference of the ordered conformation between residue 153 and 181 in the subunit originally containing pyridoxamine 5´phosphate, and the unordered conformation of this segment in the subunit containing the covalent bound cofactor remained, when both subunits displayed covalent binding of the cofactor.

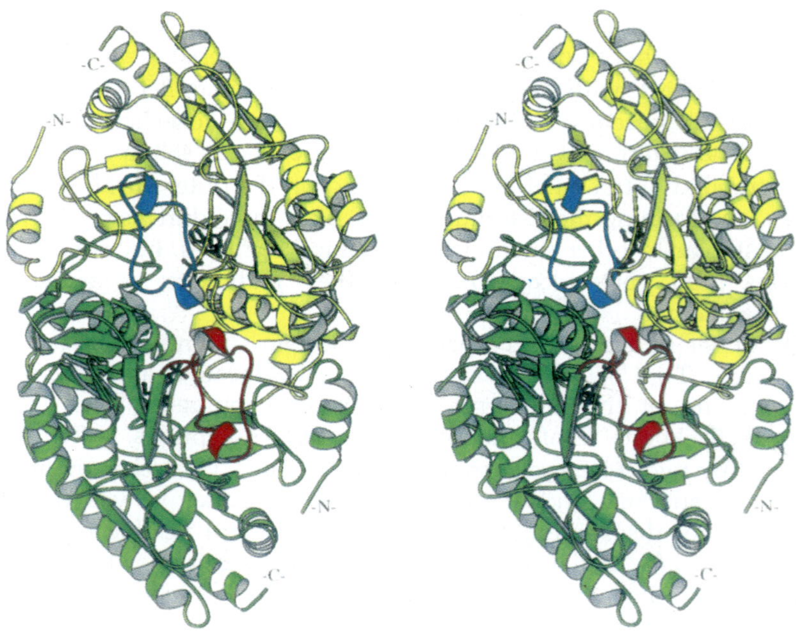

Fig. 3. Crystal structure of the dimer molecule of glutamate 1-semialdehyde aminotransferase. The cofactor in the two subunits is shown in black stick and ball. One subunit is colored yellow and the other green. The domain of residues 153–181 are shown as blue in one subunit and red in the other (C = carboxyterminus, N = aminoterminus). From Hennig *et al.*, 1997.

The substrate binding site in the three-dimensional structure was investigated by co-crystallization of the enzyme with gabaculine (3-amino-2,3-dihydrobenzoic acid), the mechanism-based suicide inhibitor that blocks chlorophyll synthesis. In the subunit containing non-covalently bound cofactor, gabaculin is bound with its carboxylate group by hydrogen bonding to the guanidinium group of Arg^{32}, an invariant residue of all glutamate 1-semialdehyde aminotransferases sequenced to date. The amino group is located close to pyridoxamine 5´phosphate and the ε-group of Lys^{272}. In the subunit containing pyridoxal 5´phosphate, gabaculin is covalently bound to the cofactor.

Several mutants of *Synechococcus* have been isolated by adapting the cyanobacterial cells to grow in increasing concentrations of gabaculine (Bull *et al.*, 1990). In one of these mutants, replacement of Met[248] with isoleucine confers gabaculine resistance to the synechococcal glutamate 1-semialdehyde aminotransferase (Grimm *et al.*, 1991). The Met[248] residue is conserved in all glutamate 1-semialdehyde aminotransferases known to date except that of *Xanthomonas campestris*, where it is replaced by valine (Murakami *et al.*, 1993). In the X-ray structure of the enzyme, the Met side-chain is extended towards the cofactor but too distant from the Lys[272] to play an important role with the true substrate, but its exchange with Ile may change the possibilities of interaction between the cofactor and gabaculine.

The binding pocket for gabaculine and, by inference, of glutamate 1-semialdehyde is funnel-shaped with a length of about 12Å, and the internal narrow (6Å) end near Arg[32], and the least buried wider (9Å) end between the side chains of Tyr[150] and Glu[406]. The pocket is lined by the side chains of the following residues: Ser[29], Val[31], Arg[32], Trp[67], Tyr[150], Ser[163], Asn[217], Met[248], Glu[406], and from the neighbouring subunit Ala[303], Gly[304], and Tyr[305]. All these residues are conserved in the 12 sequences of glutamate 1-semialdehyde aminotransferase except Met[248] as mentioned above. Glu[406] is considered as a candidate for hydrogen bonding with the aldehyde group of glutamate 1-semialdehyde and the 5-amino group of diaminovalerate. Johan Jansonius and colleagues suggest that its most important role may be its ability to exclude glutamate from the pocket by charge repulsion. The short helical domain from Ser[163] to Leu[168] blocks the entrance to the pocket in the subunit with the stable conformation of this polypeptide segment (red- and blue-marked helix in Fig. 3). The authors suggest that the subunit that displays the flexible conformation of this region may be instrumental in allowing the substrate to enter the pocket and the product to leave it. It will be of interest, if this gating mechanism operates, and ensures (i) that only one subunit is open at any one time, and (ii) that when one subunit is in the pyridoxal 5´phosphate form, the other subunit cannot open.

Three Remarkable Enzymes Generate Uroporphyrinogen III

The biosynthetic pathway of chlorophyll from 5-aminolevulinate to uroporphyrinogen III and beyond to protoporphyrin IX is, according to our present knowledge, essentially the same as that of heme in man, mouse, plants, and bacteria. Much more molecular information for plants is desirable and the present discussion endeavours to emphasize molecular results in mammals and bacteria, which might be confirmed for plants.

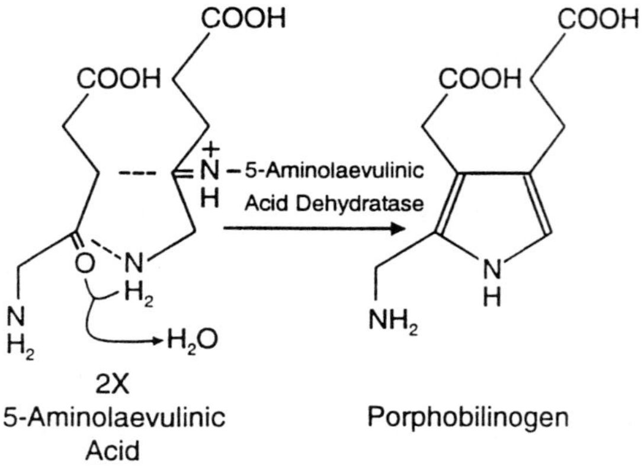

Fig. 4. Synthesis of the pyrrole ring.

The carbon-carbon and carbon-nitrogen bonds in the pyrrole ring of porphobilinogen are formed by the enzyme **5-aminolevulinic acid dehydratase** (porphobilinogen synthase), which catalyzes the assymmetric condensation of two molecules of 5-aminolevulinic acid, as shown in Fig. 4 (Jordan, 1991; Spencer and Jordan, 1994). Isotopically-labeled substrates reveal that the initially bound 5-aminolevulinic acid forms a Schiff base with its keto group and the ε-amino group of a lysine at position 247 of the enzyme (*E. coli* numbering) that, with an adjacent proline, is conserved in all 16 dehydratases whose sequence is known. These enzymes include representatives from mammals, bacteria, yeast, *Chlamydomonas* and higher plants, including spinach (Schaumburg *et al.*, 1992), pea (Boese *et al.*, 1991), tomato, and soybean. This first bound molecule gives rise to the propionic acid side chain of porphobilinogen, while the second molecule of 5-aminolevulinate (which gives rise to the acetic acid side chain) is placed in a separate substrate binding site that allows the aldol condensation between the C-3 of this second molecule and the C-4 of the first molecule to proceed. Condensation is accompanied by detachment from the enzyme's lysine nitrogen. A Schiff base between the keto group of the second 5-aminolevulinate and the amino group of the first one furnishes the pyrrole ring; the enzyme also catalyzes the tautomerization of the molecule with a stereospecific removal of a hydrogen atom at the C-2 of the pyrrole ring.

The plant enzymes and that of *Bradyrhizobium japonicum* (Chauhan and O'Brian, 1993) lack the three cysteine residues implicated in the binding of two Zn^{++} atoms that are required for activity of the mammalian and bacterial enzymes (Spencer and Jordan, 1994; Jaffe *et al.*, 1994); these have been replaced with alanine and aspartic acid residues, which signals a Mg^{++} binding site, consistent with the fact that the plant enzymes are activated by Mg^{++} but not Zn^{+-}. A second Mg^{++} binding site has been located to conserved asparagine and glutamate residues in the spinach, pea and bacterial enzymes. Interestingly, Mg^{++} can substitute for Zn^{++} in one of the two Zn^{++} binding sites of the *E. coli* enzyme (Mitchell and Jaffe, 1993; Spencer and Jordan, 1994).

By a highly innovative procedure, Ventura, Spencer, Jordan, and Timko (pers. commun. 1996) have shown that the Mg^{++} binding domain in the pea enzyme involved in catalysis can be replaced by the mammalian type Zn^{++} binding domain, and with these few amino acid changes, the pea enzyme now requires Zn^{++} for catalysis. Furthermore, this hybrid enzyme can be activated with Mg^{++}, giving a definitive proof for two Mg^{++} binding sites. The procedure used an expression system in which wild type and mutant aminolevulinic acid dehydratases complement a heme deficient *E. coli* strain.

Alternative splicing of transcripts is used in different human cells to provide, on the one hand, mRNA for the prolific production of 5-aminolevulinate dehydratase in erythrocytes, and on the other hand, a continuous house-keeping amount of the enzyme for heme production in other cells (Kaya *et al.*, 1994). It will be interesting to see, if such alternative 5-aminolevulinate dehydratase mRNA splicing is used to satisfy demands for rapid chlorophyll synthesis during greening of leaves and continuous heme synthesis in all organs.

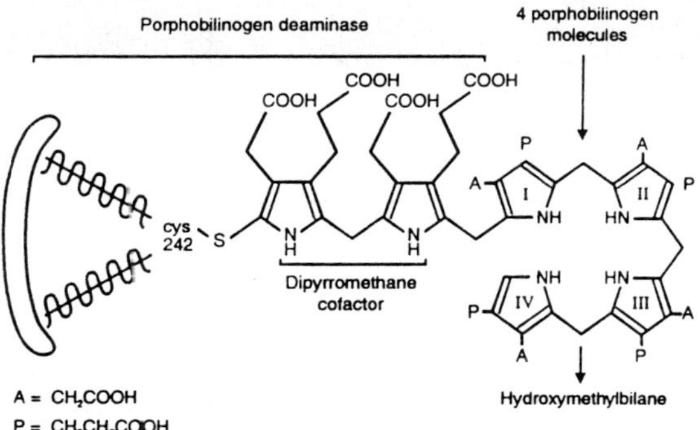

Fig. 5. Synthesis of the tetrapyrrole chain.

112   *Diter von Wettstein*

As shown in Fig. 5, the monomeric enzyme **porphobilinogen deaminase** (or 1-hydroxymethylbilane synthase), with a molecular weight of 34 to 44kDa, synthesizes its own cofactor, in the form of a dimer of porphobilinogen. It then adds four molecules of porphobilinogen by stepwise deamination to yield the tetrapyrrole 1-hydroxymethylbilane, which is also called preuroporphyrinogen (cf. Jordan, 1994). The enzyme also carries out the hydrolysis that releases the tetrapyrrole from the cofactor. Key features in the understanding of this enzyme have been obtained from the *E. coli* enzyme, whose three-dimensional crystal structure (Fig. 6) has been solved at 1.9Å resolution (Louie *et al.*, 1992).

Fig. 6. Crystal structure of *E. coli* porphobilinogen deaminase. The dipyrromethane cofactor is located in a cleft between domain 1 (blue) and domain 2 (green) and is covalently bound to Cys-242 of domain 3 (violet). From Lambert *et al.*, 1994.

Although porphobilinogen deaminases from man, mouse, rat, yeast, bacteria, and higher plants have only 15–20% overall amino acid sequence similarity, 33 individual residues involved in the catalytic and substrate binding sites are identical in all 15 enzymes analysed including those of *Euglena gracilis* (Sharif *et al.*, 1989), pea (Witty *et al.*, 1993), and *Arabidopsis*

(Jones and Jordan, 1994; EMBL X73418). The dipyrromethane cofactor is covalently bound to cysteine-242 and projects into the catalytic cleft from a loop on domain 3 (Fig. 6). The side chains of the conserved residues Arg^{132}, Arg^{155} and Lys^{83}, and the partially conserved residues Arg^{131}, Ser^{129} and Ser^{127} form salt bridges and hydrogen bonds with the acetate and propionate side chains of the cofactor. The substrate binding site is probably formed by the arginines at positions 11, 149 and 155, which may interact with the propionate and acetate side chains of the substrate. Asp^{84} is positioned such that its carboxyl side chain can interact with the amino groups of two pyrrole rings, and donate or receive protons for deamination, and C-C bond formation (Louie *et al.*, 1992). A site-directed mutant, D84E, in which this residue is changed to glutamic acid, retains less than 1% activity but can form highly stable dimer and trimer intermediate complexes (Woodcock and Jordan, 1994). Mutants D84A and D84N are inactive in preuroporphyrinogen formation but can assemble the cofactor. By contrast, mutants R131H and R132H prevent dipyrromethane assembly. This indicates that cofactor assembly and tetramerization are carried out by different reactants and provides a hint about why the cofactor remains firmly attached during repeated tetramerizations.

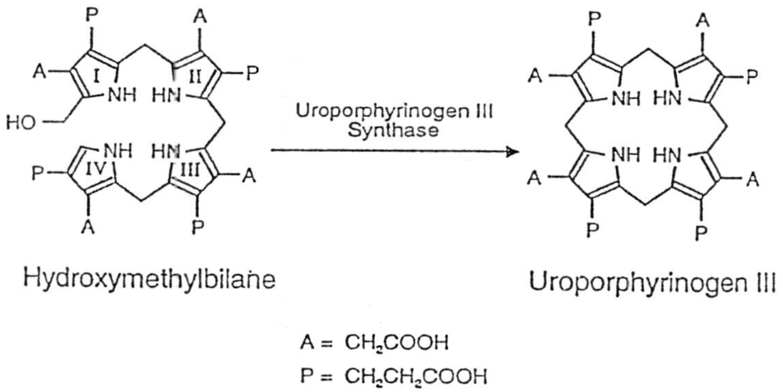

Hydroxymethylbilane

Uroporphyrinogen III Synthase

Uroporphyrinogen III

$A = CH_2COOH$

$P = CH_2CH_2COOH$

Fig. 7. Formation of the tetrapyrrole ring.

In the next step, **uroporphyrinogen (co)synthase** carries out the ring closure with isomerization of ring IV, which is thus turned around (Battersby *et al.*, 1977; Fig. 7). This can be seen in the structure of uroporphyrinogen III and the finished chlorophyll molecule by the fact that the methyl groups of ring IV and I are located in neighbouring positions within the porphyrin ring. The enzymes of man, mouse, *E. coli*, *Bacillus subtilis* (Hansson *et al.*, 1991), and *Synechococcus* 6301 (Jones *et al.*, 1994) have a molecular weight of around 30kDa and are highly diverse in primary structure. Only five residues are conserved among the nine enzymes for which the sequence is known.

Trimming of the Side Chains and Red Color Formation

Figure 8 presents the following steps of chlorophyll formation: The acetic acid side chains are shortened to methyl groups by uroporphyrinogen III decarboxylase, coproporphyrinogen III oxidase trims two of the propionic acid side chains into vinyl groups, and then the molecule is oxidized by protoporphyrinogen IX oxidase to establish the conjugated double bond system of the red-colored protoporphyrin IX (cf. Akhtar, 1994).

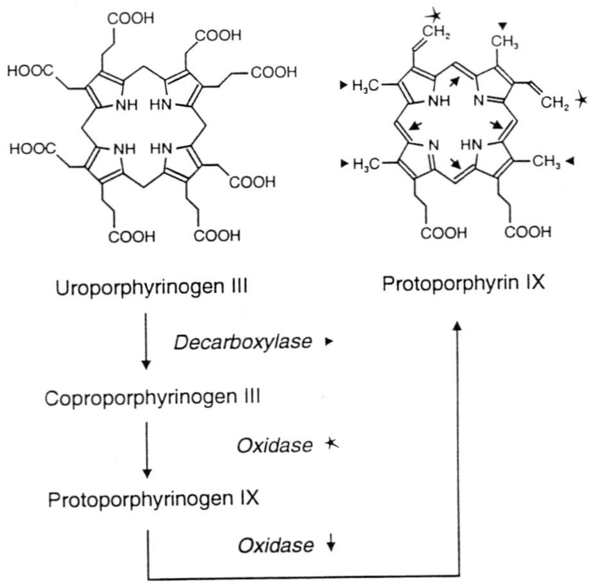

Uroporphyrinogen III

Protoporphyrin IX

Decarboxylase

Coproporphyrinogen III

Oxidase

Protoporphyrinogen IX

Oxidase

Fig. 8. Trimming of the acetate and propionate side chains and establishment of the conjugated double bond system.

Uroporphyrinogen III decarboxylases from mammalian erythrocytes and liver, yeast, *Rhodobacter sphaeroides* (Jones and Jordan, 1993), *Bacillus subtilis* (Hansson and Hederstedt, 1992), and the stroma of tobacco and barley plastids (Mock *et al.*, 1995) are monomeric proteins with a molecular weight of around 40kD. At low substrate concentrations, decarboxylation is catalyzed by the red cell and *Rhodobacter* enzymes stepwise, beginning at ring IV and proceeding clockwise through rings I and II and ending at ring III. At high substrate concentrations, decarboxylation can occur in any order. The ten primary amino acid sequences show that the eukaryotic enzymes have an overall similarity of 33% to 50%. The tobacco and barley sequences are 73% similar to the sequence of the cyanobacterial (*Synechococcus*) enzyme (Kiel *et al.*, 1992). The sequence comparisons indicate that there are three invariant domains. Their functions, and that of individual invariant amino acids, remain to be determined.

Structural genes for oxygen dependent **coproporphyrinogen III oxidases** from three higher plants, including soybean (Madsen *et al.*, 1993), barley, and tobacco (Kruse *et al.*, 1995), have been cloned. Coproporphyrinogen oxidase is synthesized as a precursor protein on cytoplasmic ribosomes and is imported into the stroma of pea chloroplasts, where it is processed into a catalytically active enzyme with a molecular weight of 39kDa. The three higher plant enzymes have 70% amino acid sequence identity, while an overall comparison from bacteria to man of the eight known primary structures shows 50% identity that is especially prominent in the C-terminal part of the oxidase. Levels of transcripts for the plant coproporphyrinogen oxidase are the same in dark-grown and greening leaves of barley; they are also constant during leaf development in tobacco but are strongly reduced in mature leaves. In the synthesis of bacteriochlorophyll by anaerobic bacteria, the oxygen requiring coproporphyrinogen III oxidase must be substituted for by an oxygen independent enzyme. A mutant in *Rhodobacter sphaeroides* that excretes coproporphyrin III has defined a gene encoding a putative anaerobic coproporphyrinogen oxidase (Coomber *et al.*, 1992) that is also present in *E. coli*, *Salmonella typhimurium*, *Pseudomonas radiora*, *Rhizobium leguminosarum*, and *Bacillus subtilis* (cf. Hansson and Hederstedt, 1994).

Protoporphyrinogen IX oxidase has been purified to homogeneity from bovine and mouse liver (cf. Akhtar, 1994) and barley etioplasts (Jacobs and Jacobs, 1987). The mammalian enzymes require molecular oxygen, use a flavin cofactor, and have a molecular mass of 65kDa, whereas the barley enzyme is 36kDa protein. Expression of the *hemY* gene of the *Bacillus subtilis hemEHY* cluster in *E. coli* results in the production of a soluble 53kDa protein that can oxidize both coproporphyrinogen III and protoporphyrinogen IX (Hansson and Hederstedt, 1994; Dailey *et al.*, 1994). Protoporphyrinogen is the last colorless intermediate in the pathway. Protoporphyrinogen IX oxidase converts the tetrapyrrole macrocycle into red protoporphyrin IX by removal of six electrons to establish the conjugated double bond system. In aerobic organisms, the electron acceptor is oxygen. The protoporphyrin IX product is the last common intermediate in the biosynthesis of heme and chlorophyll. There are two isozymes in higher plants, one located in the plastids and the other in mitochondria. Recently, cDNA clones of the genes for these two enzymes in tobacco have been obtained by complementation of an *E. coli* auxotrophic mutant in the structural gene of protoporphyrinogen oxidase (*hemG*), revealing the primary structure of the two enzymes (Lermontova *et al.*, 1997). They have 27.2 % identical amino acid residues. The gene for the chloroplast enzyme was *in vitro* transcribed and translated and the product imported into isolated chloroplasts with the aid of a transit peptide comprising 50 amino acids. The mitochondrial enzyme was also expressed and imported into mitochondria. It did not show a reduction in size after import and thus seems to lack a cleavable transit sequence. The final steps in heme synthesis thus proceed in the plastid as well as mitochondrial compartment. Recently, a chloroplast- and a mitochondrial-specific ferrochelatese were identified in *Arabidopsis* (Chow *et al.*, 1998), confirming that also the last step in heme synthesis, the insertion of iron into protoporphyrin, takes place in both organelles. Since the early steps of tetrapyrrole synthesis proceed exclusively in the chloroplasts, the protoporphyrinogen intermediate has to be safely shuttled from the chloroplast to the mitochondrion.

Heterologous expression of the genes for the protoporphyrinogen oxidases helps to verify the catalytic mechanism, which consists of three consecutive dehydrogenations and a subsequent tautomerization yielding the porphyrin through the stereospecific loss of the fourth *meso* hydrogen as a proton (Akhtar, 1994). In the three desaturations, the flavin prosthetic group acts as an acceptor of the hydride ion, whereafter the thus reduced flavin is reoxidized by O_2 for regeneration of active enzyme.

From Bright Red To Green

The chlorophyll- and bacteriochlorophyll-specific pathway from protoporphyrin IX to chlorophyll (Fig. 9) begins with the insertion of the Mg^{++} ion (Castelfranco *et al.*, 1994; Bauer *et al.*, 1993; Bollivar *et al.*, 1994b; Walker and Willows, 1997). Magnesium is inserted enzymatically into protoporphyrin IX by three separable, soluble proteins. These subunits of

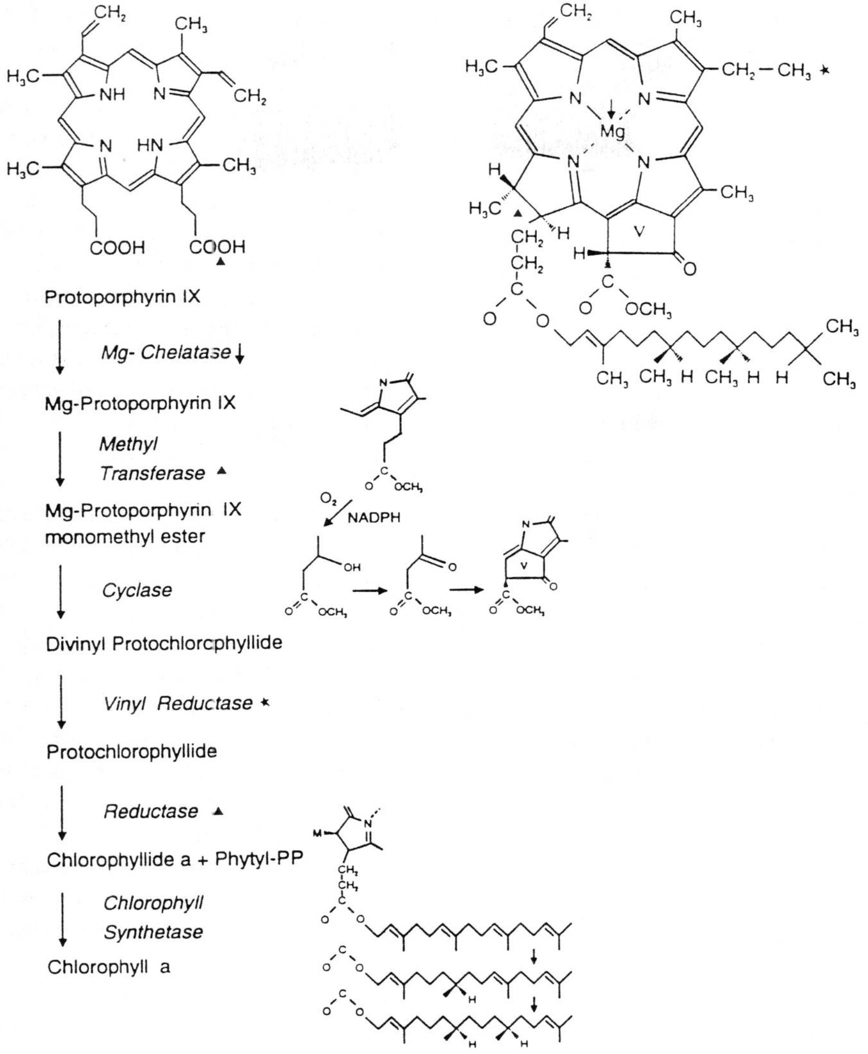

Fig. 9. Pathway from protoporphyrin IX to chlorophyll *a*.

the **magnesium chelatase** are known from *Rhodobacter* species as BchD, BchH and BchI (Gibson *et al.*, 1995; Willows *et al.*, 1996), from *Synechocystis* (ATCC 272178) as CHL D, CHL H and CHL I (Jensen *et al.*, 1996a), from barley as XAN-G, XAN-F and XAN-H (Jensen *et al.*, 1996b; Kannangara *et al.*, 1997) and from tobacco as CHL D, CHL H and CHL I (Papenbrock *et al.*, 1997). Also in pea, three separate soluble protein fractions are required for insertion of Mg^{++} into protoporphyrin (Guo *et al.*, 1998). The requirement of the products of three different genes was originally identified by mutants. In barley, mutants at the loci *Xantha-g*, *Xantha-f* and *Xantha-h* accumulate protoporphyrin IX when fed 5-aminolevulinate (von Wettstein *et al.*, 1971, 1974). Disruption of any one of the three open reading frames *bchD*, *bchI* and *bchH* of *R. capsulatus* (Bollivar *et al.*, 1994b) or *R. sphaeroides* (Coomber *et al.*, 1990; Gorchein *et al.*, 1993) by transposon insertion mutagenesis results in accumulation and secretion of protoporphyrin IX. The chelation of magnesium into protoporphyrin IX in intact cells of *R. sphaeroides* requires ATP (Gorchein, 1972, 1973), an absolute requirement also for the chelatase reaction in isolated plastids (Castelfranco *et al.*, 1979; Pardo *et al.*, 1980) and broken plastid fractions (Walker and Weinstein, 1991).

When the *bchID* gene pair and the *bchH* gene were expressed separately in *E. coli*, magnesium chelatase activity reconstituted upon combining both of the obtained soluble protein extracts from cells induced with isopropyl thio-β-D-galactoside, and by supplying ATP, Mg^{++} and protoporphyrin (Gibson *et al.*, 1995). The activity levels were high, amounting to at least $2 nmol \cdot h^{-1} \cdot mg^{-1}$ protein. The recombinant proteins BchI (40kDa) and BchH (140kDa) were purified to homogeneity while the third subunit BchD (70kDa) could not be separated entirely from the *E. coli* GroEL chaperonin (Willows *et al.*, 1996). All three purified proteins were required for Mg-chelatase activity and optimal activity was obtained with a ratio of one monomer of BchH and one dimer of BchI. BchD formed a molecular mass of about 550kDa. Expression of large amounts of the BchH in *E. coli* leads to a red color of the bacteria due to the accumulation of protoporphyrin IX. The protoporphyrin was bound specifically to BchH in an approximate molar ratio of 1:1.

The barley gene for the BchH homologue, XAN-F, contains four exons and three introns and the gene encodes a protein with an estimated molecular weight of 153kDa. The protein comprising 1380 amino acids contains a 40-amino acid-long N-terminal transit peptide for import into the chloroplast stroma. The amino acid sequence of the barley subunit displays 82% identity with that of *Antirrhinum majus* (Hudson *et al.*, 1993), and similar highly conserved values are obtained in comparison with this subunit of *Arabidopsis thaliana* (Gibson *et al.*, 1996) and *Nicotiana tabacum*

(Papenbrock *et al.*, 1997). Sequence identity to *Synechocystis* (ATCC 272178) is 66% and to *Rhodobacter* species 34%. Of the four non-leaky and one leaky *xan-f* mutants investigated, none had detectable Mg-chelatase activity in seedlings developed in the dark and illuminated for four hours, but they had wild-type amounts of Mg-protoporphyrin methyl transferase and essentially wild-type amounts of transcripts for coproporphyrinogen oxidase, and for the light-harvesting chlorophyll a/b binding protein LHCB2. Two of the *xantha-f* mutants investigated lacked in Western blots, the 150kDa protein recognized by an antibody to the BchH homologue of *Antirrhinum*, and they also lacked significant amounts of a transcript hybridizing to an appropriate probe in Northern blots. These two mutants are thus mutations in the promoter or in the intron splice sites of the gene. Two other *xan-f* mutants contained essentially wild-type levels of the transcripts and significant amounts of the 150kDa protein recognized by the antibody, and are thus to be classified as mis-sense mutations (Jensen *et al.*, 1996b).

A partial barley cDNA gene for the BchI homologue, XANH, revealed 85% identity to the *Arabidopsis* counterpart (Koncz *et al.*, 1990), and included a phosphate binding motif (GX_4GKS) common to a number of ATP binding proteins and two additional MgATPase motifs (Koonin, 1993). All four *xantha-h* mutants analyzed lacked the 42kDa protein encoded by this gene, and two of them were devoid of the cognate transcript. Due to a deletion in the *xantha-f^{38}* mutant, a shorter transcript was recognized by the cDNA probe. The analysis of the mutants in two of three genes required for Mg^{++} chelation showed that their transcriptional and translational regulation is independent of one another. The gene for the BchI subunit and its orthologues has been cloned and sequenced from fourteen different species. It shows a deduced amino acid sequence identity of 49 to 86% and the molecular weight of its product ranges from 37 to 43kDa (Walker and Willows, 1997). The higher plant subunits I are encoded in the nucleus and are synthesized with a plastid-targeting transit peptide, while the genes for this subunit in *Euglena gracilis* (Orsat *et al.*, 1992) and some other algae reside in chloroplast DNA.

The barley gene orthologous to the BchD subunit has not yet been cloned, but its functional counterpart has been identified as the XAN-G subunit by protein complementation analysis (Kannangara *et al.*, 1997). Soluble protein preparations containing more than 10 mg of protein per ml can be obtained by gentle lysis of barley plastids and *Rhodobacter* spheroplasts, and they insert Mg^{++} into deuteroporphyrin IX in the presence of ATP at a rate of 156 pmoles·20 min^{-1} and 86 pmoles·20 min^{-1}, respectively. The soluble stroma preparations of non-leaky *xantha-f, -g* and

-h mutants were inactive in the Mg-chelation reaction, but all three pairwise combination of the mutant stroma preparations reconstituted activity. Active soluble preparations from barley plastid stroma or *Rhodobacter* species can be separated into a ribosome-containing fraction and a supernatent by centrifugation at 270 000 × g. Recombinant subunit BchD from *Rhodobacter* sedimented with the ribosomes into a pellet, as did the XAN-G protein. Reconstitution of activity could be achieved by combining supernatent and pellet. Reconstitution was also possible by combining the pellet fraction of *Rhodobacter* with purified BchI and H proteins. As expected, reconstitution was not possible, e.g., by combining barley wild-type supernatant (containing subunits H and I) and mutant *xantha-g* stroma (lacking D), or wild-type pellet (containing subunit D) and *xantha-h* stroma (lacking subunit I = XAN-H). The chelation reaction is inhibited by chloroamphenicol, and it is therefore an intriguing question, if the association of the D (= XAN-G) subunit with the ribosomes is fortuitous or real, and thus involved in the regulation of chlorophyll versus protein synthesis.

Using a partial genomic clone of the *Synechocystis chlD* gene, Papenbrock *et al.* (1997) isolated an ortholog cDNA clone for this subunit of the Mg-chelatase from tobacco. The open reading frame encoded 758 amino acids with a putative plastid transit peptide of 62 residues and a molecular weight of 76.3kDa for the mature protein. Sequence identities between the *Rhodobacter* and *Synechocystis* protein were 29 and 59% respectively. The amino terminal 393 residues of the tobacco CHL D subunit (as that of the bacterial D subunits) resemble the entire CHL I/ BchI sequence (46% similarity), suggesting that the two genes have arisen by gene duplication. The unique carboxy terminal half of CHL D is linked to the amino terminal part by a domain of eight proline residues bordered by several glutamines and asparagines, and an adjacent segment of charged amino acids. These domains may be important for protein-protein interaction. The genes of the three tobacco subunits were expressed in yeast and combined extracts of the recombinant yeast strains catalyzed the chelatase reaction. Protein-protein interaction among the three subunits was studied with the two-hybrid system (Fields and Song, 1989). The *Chl D*, *Chl H*, and *Chl I* open reading frames were provided with the yeast *ADH1* (alcohol dehydrogenase) gene promoter and fused in one plasmid with the code for the DNA binding domain, and in another plasmid with that of the transcription activating domain of the GAL4 yeast transcription factor. The two plasmids were transformed together into a yeast strain carrying a *lac Z* structural gene for the β-galactosidase enzyme under the control of the *GAL 1* promoter with its target sequence for the GAL 4 transcription factor. If two subunit proteins associate, they will form a dimer and bring the DNA binding and activating domains of GAL 4 together. This assures the binding of GAL 4 to the *GAL 1* gene promoter and transcription and translation of the β-galactosidase, which acts by turning a chromogenic substrate blue. It was found that Mg-chelatase subunits D and I can form a dimer and elicit the expression of β-galactosidase enzyme.

The insertion of Mg^{++} into protoporphyrin IX is considered to proceed in two stages (Walker and Weinstein, 1994; Willows *et al.*, 1996). Subunits BchD (CHL D or XAN-G) and BchI (CHL I or XAN-H) undergo the first stage in the presence of ATP activation by complex formation. Thereafter, Mg^{++} is inserted into protoporphyrin, which involves the third subunit BchH (CHL H or XAN-F) and requires also ATP. The role of ATP in the Mg-chelatase reaction has been analyzed with purified recombinant subunits of *Rhodobacter sphaeroides* (Hansson and Kannangara, 1997). When the three subunits were analyzed individually, it was found that BchH hydrolyzed ATP (0.9 $nmole \cdot min^{-1} \cdot mg$ $protein^{-1}$) in the presence of Mg^{++}. A much higher ATPase activity (118 $nmole \cdot min^{-1} \cdot mg$ $protein^{-1}$) was obtained, when subunits D and I were combined. As mentioned above, the I subunit contains motifs characteristic for ATP binding proteins, but showed little ATPase activity by itself. Upon incubation of BchI with 4 mM ADP containing [8-^{14}C]ADP and 4mM ATP, 40% of the radioactivity appeared in ATP. Further analysis established that this transfer of phosphate from ATP to ADP by the I subunit was distinct from the ATPase activities displayed by the H subunit, and the associated I and D subunits. Binding of *o*-phenanthrolin or Co(III)-ATP-*o*-phenanthroline separately to the subunits D and H inhibited the ATPase activity of the combined subunit preparation. The lag phase observed by monitoring the formation of Mg-protoporphyrin can be eliminated by preincubation of the D and I subunits with ATP and Mg^{++}. Thus, activation of these two subunits probably involves binding of ATP. No labeling of the subunits individually or in combination occurred when they were incubated with [γ-^{32}P] ATP and Mg^{++}, excluding activation by protein phosphorylation. Sodium fluoride inhibits magnesium chelatase activity without affecting ATP hydrolysis or the ATP-to-ADP phosphate exchange reactions of the subunits. As ATP remains bound to the H subunit in the presence of NaF, this subunit is likely to contain two ATP binding sites, one sensitive and one insensitive to NaF.

In conclusion, the experiments demonstrate that ATP hydrolysis by the combined action of BchI+D is required for an activation step in the magnesium chelation, whereas ATPase activity of BchH and the phosphate exchange activity of BchI are necessary for the subsequent insertion of Mg^{++} into protoporphyrin IX. How this works is to be established by further experiments and the elucidation of the three-dimensional structure of the complexed subunits.

After insertion of the Mg^{++} ion, the **magnesium-protoporphyrin methyltransferase** esterifies the propionic side chain of ring III in preparation for the cyclization reaction that produces ring V. The S-adenosyl-L-methionine that donates the methyl group has to be imported into the chloroplast, which also contains a methyl esterase and an oxidase to degrade unused Mg-protoporphyrin IX monomethyl ester (Castelfranco *et al.*, 1994). Expression of the *bchM* gene of *Rhodobacter capsulatus* and *R. sphaeroides* in *E. coli* have identified this gene's product as the methyltransferase for Mg-protoporphyrin IX (Bollivar *et al.*, 1994a; Gibson and Hunter, 1994).

All known chlorophylls and bacteriochlorophylls possess a fifth, isocyclic ring that is formed by the cyclization of the methyl propionic acid side chain in position 6 of Mg-protoporphyrin IX monomethyl ester. The enzyme catalyzing this reaction is called **Mg-protoporphyrin monomethyl ester oxidative cyclase** or cyclase for short. Characterization of the biochemistry of the cyclase reaction was initiated by Paul Castelfranco and his co-workers (cf. Castelfranco *et al.*, 1994). First experiments with whole cucumber chloroplasts indicated that the enzyme required NADPH, in addition to Mg-protoporphyrin monomethyl ester, as substrate (Chereskin *et al.*, 1982). An *in vitro* cyclase assay was developed by hypotonic lysis of the chloroplasts at high protein concentration and centrifugation of the lysate into membrane and supernatent fractions: Both supernatant and membrane fractions were required to reconstitute cyclase activity, demonstrating that the enzyme was composed of at least two proteins (Wong and Castelfranco, 1984). This result was not surprising since the cyclization reaction involves incorporation of a keto group, a six electron oxidation, and formation of a carbon-carbon bond. In fact, it was possible to identify two stable intermediates of the cyclization reaction in the form of 13^1-hydroxy- and 13^1-oxopropionic acid methylester derivatives of Mg-protoporphyrin (Wong *et al.*, 1985; Walker *et al.*, 1988). The enzyme was diminished under anaerobic conditions, suggesting that the keto group is derived from molecular oxygen rather than water (Spiller *et al.*, 1982). Subsequently, the role of oxygen was established by incubating detached cucumber cotyledons in the presence of 5-aminolevulinate and $^{18}O_2$, and demonstrating the direct incorporation of molecular oxygen into the isocyclic ring (Walker *et al.*, 1989).

More recently, the cyclase has been assayed in *Synechocystis* and *Chlamydomonas reinhardtii* (Bollivar and Beale, 1995, 1996). In *Synechocystis*, the cyclase activity required a membrane-bound and soluble fraction, whereas in *C. reinhardtii*, attempts at fractionation were not successful. On the other hand, with cell free preparations of *C. reinhardtii*, the presence of molecular oxygen was necessary for cyclase activity. Mass spectrometry of atmospherically ^{18}O-labeled chlorophylls in *Chlorella vulgaris* (Schneegurt and Beale, 1992) and in greening maize leaves (Porra *et al.*, 1993, 1994) confirmed that, in these species, the 13^1-oxo group is derived from molecular oxygen. *Rosebacter denitrificans* (syn.*Erythrobacter* sp. Och 114), an aerobic chemotrophic bacterium, incorporated exclusively molecular oxygen into the 13^1-oxo group of its bacteriochlorophyll (Porra *et al.*, 1995, 1996). Isocyclic ring formation with molecular oxygen is thus not limited to eukaryotes.

However in *Rhodobacter sphaeroides*, where these studies were repeated with whole cells, the source of the oxygen function in the isocyclic ring was water and not molecular oxygen (Porra *et al.*, 1995). This points to a major difference in the reaction mechanism of higher plant and *Rhodobacter* cyclases. Transposon mutagenesis of the *bchE* gene of *Rhodobacter* gives rise to cells that accumulate Mg-protoporphyrin monomethyl ester, and the Bch protein with its molecular weight of 66kDa may be the multifunctional cyclase of *Rhodobacter*. We have observed, that mutations in two different genes of barley (*Xantha-l* and *Viridis-k*) lead to a block in the cyclase reaction and accumulation of Mg-protoporphyrin monomethyl ester upon forcing the pathway by supply of 5-aminolevulinate. Cloning of these two higher plant genes, and production of recombinant cyclase from these genes and the *Rhodobacter* species *bchE* genes, are required to elucidate the structure and function of the two types of cyclases.

Reduction of divinyl protochlorophyllide to monovinyl-protochlorophyllide has been inferred from product characterization (Whyte and Griffiths, 1993). As disruption of the *bchJ* gene in *Rhodobacter capsulatus* leads to accumulation of divinyl protochlorophyllide, this gene is thus a candidate for a structural gene of this enzyme (Bollivar *et al.*, 1994a).

Protochlorophyllide reduction in plants and bacteria has been studied extensively in recent years and is the topic of the next section. Briefly, the transfer of a hydrogen from NADPH to the carbon atom carrying the propionic side chain of ring IV, and a hydrogen from a protein side chain to the carbon atom carrying the methyl group by protochlorophyllide reductase, is carried out in angiosperms, as we have seen, with the aid of a photochemical reaction. In cotyledons of gymnosperms, in mosses, liver worts, algae, and photosynthetic bacteria, the reduction can be catalyzed by a very different enzyme in the dark.

The last step in the synthesis of chlorophyll *a* is catalyzed by **chlorophyll synthetase**. This enzyme esterifies the propionic acid side chain of ring IV with either phytyl-pyrophosphate, as is preferred in chloroplasts, or geranylgeranyl-pyrophosphate with subsequent reduction of the double bonds at positions 6, 10, and 14 as is preferred in etioplasts (Rüdiger *et al.*, 1980). NMR analysis of bacteriochlorophyll synthesized by *Rhodobacter sphaeroides* from 5-aminolevulinic acid labeled at the carboxyl group with ^{13}C and ^{18}O, revealed retention of both ^{18}Os in the formation of the ester bond, and thus a carboxyl-alkyl transfer mode (Akhtar, 1994). In *Rhodobacter* species, the products of two genes (*bchG* and *bchP*) are implicated in the phytolation of bacteriochlorophyll a.

After many years of discussion, there is now strong evidence that the formyl group in chlorophyll *b*, which replaces the methyl group in position 3 (cf. Fig. 1) of chlorophyll, is inserted by an oxygenase enzyme. Mass spectrometry of chlorophyll *b* synthesized in dark-grown maize leaves in the presence of either $H_2^{18}O$ or $^{18}O_2$ has documented that the oxygen in the formyl group of chlorophyll *b* is incorporated exclusively from atmospheric O_2 (Porra *et al.*, 1994), as is the case in *Chlorella vulgaris* (Schneegurt and Beale, 1992). Oxygenation of chlorophyll *a* by a single oxygenase is supported by the observation that ten independently isolated chlorophyll *b* deficient mutants in barley and fifty-four *b*-less mutants in Chlamydomonas belong to a single complementation group (Simpson *et al.*, 1985; Chunayev *et al.*, 1991).

Bacteria Invented Protochlorophyllide Reduction With and Without Light

Rhodobacter species reduce protochlorophyllide to chlorophyllide in the dark with an enzyme consisting of three polypeptides, encoded by *bchL*, *bchN*, and *bchB* (Burke *et al.*, 1993a). The elimination of the double bond in ring IV, in the course of protochlorophyllide reduction, is similar to the elimination of the double bond in ring II by chlorin reductase, which converts chlorophyllide *a* into 2-desacetyl-2 vinyl bacteriochlorophyllide *a*. Chlorin reductase comprises the products of the *bchX*, *bchY*, and *bchZ* genes. BchX and BchL share 34% amino acid sequence identity with each other, as well as 30 to 37% identity with the Fe protein (nifH) of the nitrogenase of *Acetobacter vinelandii* (Burke *et al.*, 1993b). The nucleotide binding site and the [4Fe-4S] cysteine binding ligands are especially conserved in all known BchL subunits (Koonin, 1993). In addition, *bchN* is homologous to *nifK,* the gene encoding the β-subunit of the MoFe protein of nitrogenase with its [4Fe-4S] cluster pair. The three cysteine ligands of NifK are conserved in the BchN sequences. The *bchB* product is homologous to the Vd-Fe alternative α subunit of nitrogenase (*nifD*). However, the FeMo binding ligands Cys^{275} and His^{442} are replaced by tyrosines, indicating that another cofactor reduces protochlorophyllide.

Genes encoding the three polypeptides implicated in light independent protochlorophyllide reduction have been identified in the cyanobacterium *Plectonema boryanum,* in *Synechocystis,* and in *Synechococcus* (cf.. Li *et al.,* 1993; Suzuki and Bauer, 1995). Genes encoding these polypeptides have also been identified in the chloroplast DNA of the liverwort *Marchantia polymorpha,* the chloroplast genomes of *Chlamydomonas reinhardtii* and *C. moewusii,* the red alga *Porphyra purpurea,* ferns, and gymnosperms (cf. Li *et al.,* 1993; Suzuki and Bauer, 1995), all of which are known to synthesize chlorophyll in the dark. Disruption of the *bchL* homologue of *Chlamydomonas reinhardtii* by particle gun transformation produces a "yellow in the dark" phenotype; i.e., protochlorophyllide reduction in the dark is blocked (Suzuki and Bauer, 1992). The presence of these genes has been documented in the chloroplast DNA of the ferns *Cystopteris fragilis,* and *Athyrium filix femina,* the horsetail *Equisetum arvense,* the coniferous trees *Gingko biloba* (Richard *et al.,* 1994), *Pseudotsuga menziesii, Taxus sp., Juniperus sp., Araucaria sp., Pinus contorta,* and *Picea abies.* The genes are, however, absent in the angiosperm chloroplast genome of tobacco, rice, maize, *Arabidopsis, Bougainvillea glabra,* all of which are incapable of synthesizing chlorophyll in the dark.

In contrast to lower plants and gymnosperms, angiosperms use exclusively the nuclear-encoded light and NADPH dependent protochlorophyllide oxidoreductase to synthesize chlorophyllide. The barley and *Arabidopsis* enzymes have a molecular weight of 35kDa, and are synthesized in the cytosol with transit peptides for import into etioplasts and chloroplasts. After expression of the corresponding cDNA genes in *E. coli,* the extracted enzyme reduced the substrate protochlorophyllide only in the presence of light and NADPH (Schulz *et al.,* 1989; Benli *et al.,* 1991). Monovinyl and divinyl protochlorophyllide are equally acceptable as substrates (Knaust *et al.,* 1993). Light dependent protochlorophyllide reductases from barley, oats, wheat, pea, and *Arabidopsis* are 80 to 95% identical at the amino acid sequence level (cf. Suzuki and Bauer, 1995) but bear no homology to the three peptides of the light independent protochlorophyllide reductase.

Barley contains two genes encoding light dependent protochlorophyllide oxidoreductase isoenzymes with 75% amino acid sequence identity (Holtorf *et al.*, 1995). One isoenzyme is abundantly synthesized in the dark but its synthesis declines rapidly upon illumination and greening of dark-grown seedlings. Transcription of the gene for this isoenzyme is shut off upon illumination. The gene for the other isoenzyme is constitutively transcribed in the dark and in the light, and the transcripts are translated continuously into active enzyme, which is responsible for chlorophyll synthesis during greening. Thus, it is now understandable why transcripts for the light dependent protochlorophyllide reductase as well as the protein itself were found abundant in dark-grown seedlings of bean, pea, tomato, sunflower, mustard, *Arabidopsis*, and maize, but both mRNA and enzyme levels declined when the seedlings were illuminated, and rapid chlorophyll synthesis ensued (Forreiter *et al.*, 1990; Benli *et al.*, 1991).

The isoenzyme that is abundantly synthesized in the dark and degraded in the light is designated protochlorophyllide reductase A, while the isoenzyme constitutively synthesized is named protochlorophyllide reductase B. The two enzyme precursors (44kDa) have significantly different N-terminal transit peptides of 75 and 83 amino acid residues, respectively, that provide different characteristics of import into etioplasts and chloroplasts. The method used to study the import consisted of (1) preparing *in vitro* mRNA by transcribing with T7 RNA polymerase the cDNA gene for protochlorophyllide reductase A (Schulz *et al.*, 1989), and the cDNA gene for isoenzyme B (Holtorf *et al.*, 1995), and then (2) translating the mRNA in a wheat germ system in the presence of ^{35}S methionine. The thus formed protochlorophyllide reductase A precursor actively catalyzed the formation of chlorophyllide if provided with protochlorophyllide, NADPH, and light (Reinbothe *et al.*, 1995a). For import studies of the precursor protein (Reinbothe *et al.*, 1995b), etioplasts and chloroplasts from barley, wheat, or pea were isolated in intact form by sequential continuous and step gradient density centrifugation in Percoll solutions. The isolated plastids were resuspended in import buffer and incubated for 15 minutes with the protochlorophyllide reductase precursor, ATP and other components to be studied. Thereafter, the plastids were re-isolated in Percoll gradients, treated with thermolysin for degradation of adhering protein, and analyzed for the imported protein by polyacrylamide gel electrophoresis and auto radiography. The precursor protein not sequestered by the organelles was precipitated with trichloroacetic acid and isolated from the supernatant of the reaction mixture.

Protochlorophyllide inside the etioplasts is required for the import of protochlorophyllide reductase A precursor and its processing into mature protein (Reinbothe *et al.*, 1995a). Chloroplasts do not import this enzyme precursor, unless they are supplied with protochlorophyllide, e.g., by incubation with 5-aminolevulinate, the basic precursor of protochlorophyllide. This is in contrast to other chloroplast proteins: The small subunit of ribulose bisphosphate carboxylase precursor and a dihydrofolate reductase reporter protein with an attached plastocyanin transit peptide are equally well transported into etioplasts and chloroplasts. When barley seedlings are grown under light-dark cycles, transport competence of the chloroplasts is at a minimum during the last hours of the light period and considerable at the end of the dark period when protochlorophyllide has accumulated.

If the protochlorophyllide is non-covalently bound to the reductase A precursor prior to import, ATP dependent binding to the chloroplast envelope takes place (Reinbothe *et al.*, 1995b). This binding capacity and subsequent import is abolished if the protochlorophyllide is reduced prior to adding the chlorophyllide precursor complex to the plastids. Wheat germ proplastids contain a leader peptidase that can cleave the transit peptide of the protochlorophyllide reductase *in vitro*. Experiments with an extract containing the peptidase reveal that the chlorophyllide formed and retained on the reductase A precursor causes a conformational change that not only inhibits the binding to the organelle envelope but also renders the transit peptide inaccessible to the peptidase. Binding of the precursor enzyme with its attached substrates, protochlorophyllide and NADPH to the plastid envelope, protects against trypsin proteolysis. Western blots demonstrate the presence of the 44kDa protochlorophyllide reductase A precursor on the outside of envelopes on plastids isolated from barley seedlings (Reinbothe *et al.*, 1996). The precursor can be chased into the chloroplasts, if protochlorophyllide synthesis is elicited in them. On the other hand, the precursor protein can be released from the plastid envelopes by supplying them with protochlorophyllide from the outside. Interaction with a cytosolic 70kDa heat shock protein promotes the binding of the precursor protein to the envelope. It is suggested that the precursor protein unfolds, while being translocated through the envelope, and that there are three binding sites for protochlorophyllide, one in the transit peptide and two in the mature protein part. Alternative bindings of the protochlorophyllide to the three protein domains determine which direction the unfolded protein moves across the membrane.

Light induces in developing chloroplasts a metal and ATP dependent protease activity that specifically degrades the protochlorophyll reductase A in the stroma of the plastids (Reinbothe *et al.*, 1995). An important discovery reveals that the protochlorophyllide reductase B isoenzyme precursor is imported into etioplasts and developing chloroplasts, and processed into its mature form independent of the presence of its substrate, protochlorophyllide. On the other hand, the naked reductase B precursor and the precursor B with bound chlorophyllide is as sensitive to the light-induced protease activity as are the corresponding reductase A derivatives. As long as protochlorophyllide and NADPH are bound to the reductases, they are both resistant to protease degradation. In recent experiments (Reinbothe *et al.*, 1997), the transit peptides of the precursors of the protochlorophyllide reductases A and B were exchanged with the result that the A protein is now imported into the plastids with the characteristics of the endogenous B protein and vice versa. It is thus clearly demonstrated that the amino acid sequence of the transit peptide determines if the protein is imported with attached protochlorophyllide pigment or without it. In agreement with the expectation, binding of protochlorophyllide *in vitro* can only take place to the transit peptide of the protochlorophyllide reductase A precursor. The de-etiolated mutants of *Arabidopsis thaliana* in the *DET*, *COP*, and *FUSCA* genes do not transcribe the gene for protochlorophyllide reductase A in the dark, and thus lack the safe store for protochlorophyllide during early stages of greening (Lebedev *et al.*, 1995). This makes these mutants highly sensitive to photodynamic damage. Expression of either protochlorophyllide reductase A or B in the transgenic *cop1-18* mutant with the 35S promoter of the cauliflower mosaic virus restores during seedling development in the dark, the accumulation of active protochlorophyllide reductase, and results in protection of the seedlings against photodynamic damage upon transfer to light (Sperling *et al.*, 1998). The same dramatic protection is obtained in transgenic seedlings expressing either isoenzyme A or B grown under continuous far red light which depletes active protochlorophyllide reductase in the wild type plastids and makes the seedlings highly sensitive to light (Sperling *et al.*, 1997). Obviously the two isoenzymes can substitute for each other.

In gymnosperms such as pine and spruce, the chloroplast genome encodes the light independent protochlorophyllide reductase, and the nuclear genome two light dependent protochlorophyllide reductases. The light independent enzyme is only expressed during development of the cotyledons in the dark but not during development of the primary and secondary needles, in which chlorophyll synthesis relies on the light dependent protochlorophyllide reductase. As in barley, there are two light dependent isoenzymes in each of *Pinus mugo*, *P. strobus*, and *P. taeda*; one is specifically synthesized in the dark, and the other constitutively in the dark as well as during chloroplast development in the light (Forreiter and Apel, 1993). Both isoenzymes are synthesized in the cotyledons together with the light independent enzyme.

It was thought that the light dependent protochlorophyllide oxidoreductase was an evolutionary invention of angiosperms. A seminal experiment that illustrates the power of using *Rhodobacter* mutants in the analysis of chlorophyll biosynthesis genes or enzymes from cyanobacteria and higher plants, Suzuki and Bauer (1995) demonstrated that cyanobacteria already had invented the light dependent enzyme, probably before the advent of angiosperms: A mutant of *Rhodobacter capsulatus* that is unable to reduce protochlorophyllide as a consequence of the disruption of the *bchL* gene was mated with an *E. coli* strain library containing cosmids of 19 genome equivalents of *Synechocystis* (sp. PCC6803) under the strong *Rhodobacter* PUF promoter. Some ex-conjugants selected under anaerobic light conditions permitted the synthesis of bacteriochlorophyll and photosynthesis. The *Synechocystis* gene that conferred the ability for light dependent protochlorophyllide reduction encodes a protein that is 53% identical and 73% similar to the light dependent protochlorophyllide reductase of *Arabidopsis*.

This experimental approach was further developed by Wilks and Timko (1995) into an assay for identifying substrate binding and catalytic sites in the protochlorophyllide reductase from pea. The pea enzyme was able to restore bacteriochlorophyll synthesis in the *bchN*, *bchB* or *bchL* mutants of *Rhodobacter*. By site-directed mutagenesis of a tyrosine or a lysine residue in the pea enzyme, complementation was abolished even though inactive protein was synthesized. It was suggested that in the reduction of ring IV, the proton transferred to the carbon atom carrying the methyl group is donated by the tyrosine. It has now become possible to solve the questions that James Smith had to leave suspended in 1959. The complementation and expression assays in *Rhodobacter* species will permit the identification and molecular analysis of other chlorophyll biosynthetic enzymes from higher and lower plants.

This is scientific paper 9805-18b from the College of Agriculture and Home Economics Research Center, Washington State University, Pullman, WA, USA Project 3106.

Acknowledgements: Reproduction of Figs. 1, 2, 4, 5, 7, 8 and 9 from D. von Wettstein *et al.* The Plant Cell 7, pp. 1039-1057 is by copyright permission of the American Society of Plant Physiologists. Fig. 3 is reproduced by courtesey of Professor J. N. Jansonius and Fig. 6 from Lambert *et al.*, *Structural Studies On Porphobilinogen Deaminase*. CIBA Foundation Symposium 180, pp.98, 1994 by copyright permission of John Wiley & Sons Ltd.

References

Recent reviews:

Alberti, M., Burke, D. H. and Hearst, J. H. (1995) Structure and sequence of the photosynthesis gene cluster. In *Anoxygenic Photosynthetic Bacteria*, Blankenship, R. E., Madigan, M. T. and Bauer, C. E., eds., The Hague: Kluwer, pp. 1083-1106.

Beale, S. I. (1995) Biosynthesis and structures of porphyrins and hemes. In *Anoxygenic Photosynthetic Bacteria*, Blankenship, R. E., Madigan, M. T. and Bauer, C. E., eds., The Hague: Kluwer, pp. 1083-1106.

Porra, J. R. (1997) Recent progress in porphyrin and chlorophyll biosynthesis. *Photochem. Pphotobiol.* **65:** 492-516.

Reinbothe, S. and Reinbothe, C. (1996) The regulation of enzymes involved in chlorophyll biosynthesis. *Eur. J. Biochem.* **237:** 323-343.

Rüdiger, W. (1997) Chlorophyll metabolism: from outer space down to the molecular level. *Phytochemistry* **46:** 1151-1167.

von Wettstein, D., Gough, S. and Kannangara, C. G. (1995) Chlorophyll Biosynthesis. *Plant Cell* **7:** 1039-1057.

This article:

Akhtar, M. (1994) The modification of acetate and propionate side chains during the biosynthesis of haem and chlorophylls: Mechanistic and stereochemical studies. In *The Biosynthesis of the Tetrapyrrole Pigments*, Ciba Foundation Symposium 180, Chadwick, D. J. and Ackrill, K., eds., Chichester: John Wiley & Sons, pp. 131-151.

Andersen, R. V. (1992) Characterization of a barley tRNA synthetase involved in chlorophyll biosynthesis. In *Research in Photosynthesis, Vol. 3*, Murata, N., ed., The Hague: Kluwer, pp. 27-30.

Avissar, Y. J., Ormerod, J. G. and Beale, S. I. (1989) Distribution of δ-aminolevulinic acid biosynthetic pathways among phototrophic bacterial groups. *Arch. Microbiol.* **151:** 513-519.

Battersby, A. R., McDonald, E., Williams, D. C. and Wurziger, H. K. W. (1977) Biosynthesis of the natural (type-III) porphyrins: Proof that rearrangement occurs after head to tail bilane formation. *J. Chem. Soc. Commun.* : 113-117.

Bauer, C. E., Bollivar, D. W. and Suzuki, J. Y. (1993) Genetic analyses of photopigment biosynthesis in Eubacteria: A guiding light for algae and plants. *J. Bacteriol.* **175:** 3919-3925.

Beale, S. I. (1995) Biosynthesis and structure of porphyrins and hemes. In *Anoxygenic Photosynthetic Bacteria*, Blankenship, R. E., Madigan, M. T. and Bauer, C. E., eds., The Hague: Kluwer, pp. 1083-1106.

Beale, S. I. and Castelfranco, P. A. (1974a) The biosynthesis of δ-aminolaevulinic acid in higher plants. 1. Accumulation of δ-aminolevulinic acid in greening plant tissue. *Plant Physiol.* **53:** 291-296.

Beale, S. I. and Castelfranco, P. A. (1974b) The biosynthesis of δ-aminolevulinic acid from labelled precursors in greening plant tissues. *Plant Physiol.* **53:** 297-303.

Beale, S. I., Gough, S. P. and Granick, S. (1975) Biosynthesis of δ-aminolevulinic acid from the intact carbon skeleton of glutamic acid in greening barley. *Proc. Natl. Acad. Sci. USA.* **72:** 2719-2723.

Benli, M., Schulz, R. and Apel, K. (1991) Effect of light on the NADPH-protochlorophyllide oxidoreductase of *Arabidopsis thaliana*. *Plant Mol. Biol.* **16:** 615-625.

Berry-Lowe, S. L., Grimm, B., Smith, M. A. and Kannangara C. G. (1992) Purification and characterization of glutamate 1-semialdehyde aminotransferase from barley expressed in *Escherichia coli. Plant Physiol.* **99**: 1597-1603.

Boese, Q. F., Spano, A. J., Li, J. and Timko, M. P. (1991) Aminolaevulinic acid dehydratase in pea (*Pisum sativum* L.). Identification of an unusual metal-binding domain in the plant enzyme. *J. Biol. Chem.* **266**: 17060-17066.

Bollivar, D. W. and Beale. S. I. (1995) Formation of the isocyclic ring of chlorophyll by isolated *Chlamydomonas reinhardtii* chloroplasts. *Photosynth. Res.* **43**: 113-124.

Bollivar, D. W. and Beale, S. I. (1996) The chlorophyll biosynthetic enzyme Mg protoporphyrin IX monomethyl ester oxidative cyclase. Characterization and partial purification from *Chlamydomonas reinhardtii* and *Synechocystis* sp. PCC 6803. *Plant Physiol.* **112**: 105-114.

Bollivar, D. W., Jiang, Z.-Y., Bauer, C. E. and Beale, S. I. (1994a) Heterologous expression of the *bchM* gene product from *Rhodobacter capsulatus* and demonstration that it encodes S-adenosyl-L-methionine: Mg-protoporphyrin IX methyltransferase. *J. Bacteriol.* **176**: 5290-5296.

Bollivar, D. W., Suzuki, J. Y., Beatty, J. T., Dobrowlski, J. M. and Bauer, C. E. (1994b) Directed mutational analysis of bacteriochlorophyll *a* biosynthesis in *Rhodobacter capsulatus. J. Mol. Biol.* **237**: 622-640.

Bougri, O. V. and Grimm, B. (1996) Members of a low copy gene family encoding glutamyl tRNA reductase are differentially expressed in barley. *Plant Journal* **9**: 867-878.

Breu, V. and Dörnemann, D. (1988) Formation of 5-aminolevulinate *via* glutamate-1-semialdehyde and 4,5-dioxovalerate with participation of an RNA component in *Scenedesmus obliquus* mutant C-2A´. *Biochim. Biophys. Acta* **967**: 135-140.

Brody, S., Andersen, J S., Kannangara, C. G., Meldgaard, M., Roepstorff, P. and von Wettstein, D. (1995) Characterization of the different spectral forms of glutamate 1-semialdehyde aminotransferase by mass spectrometry. *Biochemistry* **34**: 15918-15924.

Bull, A. D., Breu, V., Kannangara, C. G., Rogers, L. J. and Smith, A. J. (1990) Cyanobacterial glutamate 1-semialdehyde aminotransferase: Requirement for pyridoxamine phosphate. *Arch. Microbiol.* **154**: 56-59.

Bruyant, P. and Kannangara, C. G. (1987) Biosynthesis of δ-aminolevulinate in greening barley leaves. VIII: Purification and characterization of the glutamate-tRNA ligase. *Carlsberg Res. Commun.* **52**: 99-109.

Burke, D. H., Alberti, M. and Hearst, J. E. (1993a) *bchFNBH* bacteriochlorophyll synthesis genes of *Rhodobacter capsulatus* and identification of the third subunit of light-independent protochlorophyllide reductase in bacteria and plants. *J. Bacteriol.* **175**: 2414-2422.

Burke, D. H., Alberti, M. and Hearst, J. E. (1993b) The *Rhodobacter capsulatus* chlorin reductase - encoding locus, *bchA*, consists of three genes, *bchX, bchY* and *bchZ. J. Bacteriol.* **175**: 2407-2413.

Castelfranco, P., Walker, C. J. and Weinstein, J. (1994) Biosynthetic studies on chlorophylls: from protoporphyrin IX to protochlorophyllide. In *The Biosynthesis of the Tetrapyrrole Pigments*, Ciba Foundation Symposium 180, Chadwick, D. J. and Ackrill, K., eds., Chichester: John Wiley & Sons, pp. 194-204.

Castelfranco, P. A., Weinstein, J. D., Schwarz, S., Pardo, A. D. and Wezelman, B. E. (1979) The magnesium insertion step in chlorophyll biosynthesis. *Arch. Biochem. Biophys.* **192**: 592-598.

Chauhan, S. and O'Brian, M. R. (1993) *Bradyrhizobium japonicum* δ-aminolevulinic acid dehydratase is essential for symbiosis with soybean and contains a novel metal-binding domain. *J. Bacteriol.* **175**: 7222-7227.

Chereskin, B. M, Wong, Y. and Castelfranco, P. A. (1982) *In vitro* synthesis of the chlorophyll isocyclic ring. Transformation of magnesium-protoporphyrin IX and magnesium-protoporphyrin IX monomethyl ester into magnesium-2,4-divinyl pheoporphyrin A. *Plant Physiol.* **70**: 987-993.

Chow, K. S., Singh, D. P., Walker, A. R. and Smith, A. G. (1998) Two different genes encode ferrochelatase in Arabidopsis: mapping, expression and subcellular targeting of the precursor proteins. *Plant J.* (in press).

Chunayev, A. S., Mirnaya,O. N., Maslov, V. G. and Boschetti, A. (1991) Chlorophyll *b* and chloroxanthin-deficient mutants of *Chlamydomonas reinhardtii. Photosynthetica* 25: 291-301.

Coomber, S. A., Chaudhri, M., Connor, A., Britton, G. and Hunter, C. N. (1990) Localized transposon Tn5 mutagenesis of the photosynthetic gene cluster of *Rhodobacter sphaeroides. Mol. Microbiol.* 4: 977-989.

Coomber, S. A., Jones, R. M., Jordan, P. M. and Hunter, C. N. (1992) A putative anaerobic coproporphyrinogen III oxidase in *Rhodobacter sphaeroides* I. Molecular cloning, transposon mutagenesis and sequence analysis of the gene. *Mol. Microbiol.* 6: 3159-3169.

Dailey, T. A., Meissner, P. and Dailey, H. A. (1994) Expression of a cloned protoporphyrinogen oxidase. *J. Biol. Chem.* 269: 813-815.

Elliot, T., Avissar, Y. J., Rhie, G.- E. and Beale S. I. (1990) Cloning and sequence of the *Salmonella typhimurium hemL* gene and identification of the missing enzyme in *hemL* mutants as glutamate-1-semialdehyde aminotransferase. *J. Bacteriol.* 172: 7071-7084.

Fields, S. and Song, O. (1989) A novel genetic system to detect protein-protein interactions. *Nature* 340: 245-246.

Forreiter, C. and Apel, K. (1993) Light independent and light dependent protochlorophyllide-reducing activities and two distinct NADPH-protochlorophyllide oxidoreductase polypeptides in mountain pine (*Pinus mugo*). *Planta* 190: 536-545.

Forreiter, C., van Cleve, B., Schmidt, A. and Apel, K. (1990) Evidence for a general light-dependent negative control of NADPH-protochlorophyllide oxidoreductase in angiosperms. *Planta* 183: 126-132.

Friedmann, H. C., Thaur, R. K., Gough, S. P. and Kannangara, C. G. (1987) Δ-aminolevulinic acid formation in Methanobacterium thermoautotrophicum requires tRNA^{Glu} . *Carlsberg Res. Commun.* 52: 363-371.

Friedmann, H. C., Duban, M. E., Valasinas, A. and Frydman, B. (1992) The enantioselective participation of (S)- and (R)-diaminovaleric acids in the formation of δ-aminolevulinic acid in cyanobacteria. *Biochem. Biophys. Res. Commun.* 185: 60-68.

Gibson, L. D. C. and Hunter, C. N. (1994) The bacteriochlorophyll biosynthesis gene, *bchM*, of *Rhodobacter spheroides* encodes S-adenosyl-L-methionine: Mg-protoporphyrin IX methyltransferase. *FEBS Lett.* 352: 127-130.

Gibson, L. D. C., Willows, R. D., Kannangara, C. G., von Wettstein, D. and Hunter, C. N. (1995) Magnesium-protoporphyrin chelatase of *Rhodobacter sphaeroides*: reconstitution of activity by combining the products of the *bchH, I* and *D* genes expressed in *E. coli. Proc. Natl. Acad. Sci. USA.* 92: 1941-1944.

Gibson, L. C. D., Marrison, J. L., Leech, R. M., Jensen, P. E., Bassham, D. C., Gibson, M. and Hunter, C. N. (1996) A putative Mg chelatase subunit from *Arabidopsis thaliana* cv. C24. Sequence and transcript analysis of the gene, import of the protein into chloroplasts and *in situ* localization of the transcript and protein. *Plant Physiol.* 111: 61-71.

Gorchein, A. (1972) Magnesium protoporphyrin chelatase activity in *Rhodopseudomonas sphaeroides. Biochem. J.* 127: 97-106.

Gorchein, A. (1973) Control of magnesium protoporphyrin chelatase activity in *Rhodopseudomonas sphaeroides. Biochem. J.* 134: 833-845.

Gorchein, A., Gibson, L. C. D. and Hunter C. N. (1993) Gene expression and control of enzymes for synthesis of magnesium protoporphyrin monomethyl ester in *Rhodobacter sphaeroides. Biochem. Soc. Trans.* 21: 201S.

Gough, S. and Kannangara, C. G. (1977) Synthesis of δ-aminolevulinate by a chloroplast stroma preparation from greening barley leaves. *Carlsberg Res. Commun.* 42: 459-464.

Gough, S., Kannangara, C. G. and Bock, K. (1989) A new method for the synthesis of glutamate 1-semialdehyde. *Carlsberg Res. Commun.* **54:** 99-108.

Gough, S. P., Kannangara, C. G. and von Wettstein, D. (1992) Glutamate 1-semialdehyde aminotransferase as a target for herbicides. In *Target Assays for Modern Herbicides and Related Phytotoxic Compounds*, Böger, P. and Sandman, G., eds., Chelsea MI, Lewis Publ., pp. 21-27.

Grimm, B. (1990) Primary structure of a key enzyme in plant tetrapyrrole synthesis: Glutamate 1-semialdehyde aminotransferase. *Proc. Natl. Acad. Sci. USA.* **87:** 4169-4173.

Grimm, B., Bull, A. and Breu, V. (1991) Structural genes of glutamate 1-semialdehyde aminotransferase for porphyrin synthesis in a cyanobacterium and *Escherichia coli. Mol. Gen. Genet.* **225:** 1-10.

Grimm, B., Smith, M. A. and von Wettstein, D. (1992) The role of lys 272 in the pyridoxal 5′phosphate active site of *Synechococcus* glutamate 1-semialdehyde aminotransferase. *Eur. J. Biochem.* **206:** 579-585.

Grimm, B., Smith A. J., Kannangara, C. G. and Smith, M. (1991) Gabaculine-resistant glutamate 1-semialdehyde aminotransferase of *Synechococcus*. Deletion of a tripeptide close to the NH_2 terminus and an internal amino acid substitution. *J. Biol. Chem.* **266:** 12495-12501.

Guo, R., Luo, M. and Weinstein, J. D. (1998) Mg-chelatase from developing pea leaves: Characterization of a soluble extract from chloroplasts and resolution into three required protein fractions. *Plant. Physiol.* **116:** 605-615.

Hansson, M. and Hederstedt, L. (1992) Cloning and characterization of the *Bacillus subtilis hem EHY* gene cluster, which encodes protoheme IX biosynthetic enzymes. *J. Bacteriol.* **174:** 8081-8093.

Hansson, M. and Hederstedt, L. (1994) *Bacillus subtilis* HemY is a peripheral membrane protein essential for protoheme IX synthesis and which can oxidise coproporphyrinogen III and protoporphyrinogen IX. *J. Bacteriol.* **176:** 5962-5970.

Hansson, M. and Kannangara, C. G. (1997) ATPases and phosphate exchange activities in magnesium chelatase subunits of *Rhodobacter sphaeroides. Proc. Natl. Acad. Sci. USA.* **94:** 13351-13356.

Hansson, M., Rutberg, L., Schröder, I. and Hederstedt, L. (1991) The *Bacillus subtilis hem AXCDBL* gene cluster, which encodes enzymes of the biosynthetic pathway from glutamate to uroporphyrinogen III. *J. Bacteriol.* **173:** 2590-2599.

Hennig, M., Grimm, B., Contestabile, R., John, R. A. and Jansonius, J. N. (1997) Crystal structure of glutamate-1-semialdehyde aminomutase: An α_2-dimeric vitamin B_6-dependent enzyme with asymmetry in structure and active site reactivity. *Proc. Natl. Acad. Sci. USA.* **94:** 4866-4871.

Höfgen, R., Axelsen, K. B., Kannangara, C. G., Schüttke, J., Pohlenz, H.-D., Willmitzer, L., Grimm, B. and von Wettstein, D. (1994) A visible marker for antisense mRNA expression in plants: Inhibition of chlorophyll synthesis with a glutamate-1-semialdehyde aminotransferase antisense gene. *Proc. Natl. Acad. Sci. USA.* **91:** 1726-1730.

Holtorf, H., Reinbothe, S., Reinbothe, C., Bereza, B. and Apel, K. (1995) Two distinct NADPH-protochlorophyllide oxidoreductases are differentially expressed during the light-induced greening of dark-grown barley (Hordeum vulgare L.) seedlings. *Proc. Natl. Acad. Sci. USA.* **91:** 1726-1730.

Hoober, J. K., Kahn, A., Ash, D. E., Gough, S. P. and Kannangara, C. G. (1988) Biosynthesis of δ-aminolevulinate in greening barley leaves. IX. Structure of the substrate, mode of gabaculine inhibition, and the catalytic mechanism of glutamate 1-semialdehyde aminotransferase. *Carlsberg Res. Commun.* **53:** 11-25.

Houghton, J. D., Brown, S. B., Gough, S. P. and Kannangara, C. G. (1989) Biosynthesis of δ-aminolevulinate in *Cyanidium caldarium*: Characterization of tRNA[Glu], ligase, dehydrogenase and glutamate 1-semialdehyde aminotransferase. *Carlsberg Res. Commun.* **54**: 131-143.

Huang, D.-D. and Wang, W.-Y. (1986) Chlorophyll synthesis in *Chlamydomonas* starts with the formation of glutamyl-tRNA. *J. Biol. Chem.* **261**: 13451-13455.

Huang, D. D., Wang, W.-Y., Gough, S. P. and Kannangara, C. G. (1984) δ-aminolevulinic acid-synthesizing enzymes need an RNA moiety for activity. *Science* **225**: 1482-1484.

Hudson, A., Carpenter, R., Doyle, S. and Coen, E. S. (1993) Olive: a key gene required for chlorophyll biosynthesis in *Antirrhinum majus*. *EMBO J.* **12**: 3711-3719.

Ilag, L. L. and Jahn, D. (1992) Activity and spectroscopic properties of the *Escherichia coli* glutamate 1-semialdehyde aminotransferase and the putative active site mutant K265R. *Biochemistry* **31**: 7143-7151.

Ilag, L. L., Kumar, A. M. and Söll, D. (1994) Light regulation of chlorophyll biosynthesis at the level of 5-aminolaevulinate formation in Arabidopsis. *Plant Cell* **6**: 265-275.

Ilag, L. I., Jahn, D., Eggertson, G. and Soll, D. (1991) The *Escherichia coli hemL* gene encodes glutamate 1-semialdehyde aminotransferase. *J. Bacteriol.* **173**: 3408-3413.

Jacobs, J. M. and Jacobs, N. J. (1987) Oxidation of protoporphyrinogen to protoporphyrin, a step in chlorophyll and haem biosynthesis. *Biochem. J.* **244**: 219-224.

Jaffe, E. K., Volin, M., Myers, C. and Abrams, W. R. (1994) 5-Chloro [1,4-^{13}C] levulinic acid modification of mammalian and bacterial porphobilinogen synthase suggests an active site containing two Zn (II). *Biochemistry* **33**: 11554-11562.

Jahn, D. (1992) Complex formation between glutamyl-tRNA synthetase and glutamyl tRNA reductase during the tRNA-dependent synthesis of 5-aminolaevulinic acid in *Chlamydomonas reinhardtii*. *FEBS Lett.* **314**: 77-88.

Jahn, D., Chen, M-W. and Soll, D. (1991) Purification and functional characterization of glutamate 1-semialdehyde aminotransferase from *Chlamydomonas reinhardtii*. *J. Biol. Chem.* **266**: 161-167.

Jensen, P. E., Gibson, L. C. D., Henningsen, K. W. and Hunter, C. N. (1996a) Expression of the *chlI, chlD and chlH* genes from the cyanobacterium *Synechocystis* PCC6803 in *Escherichia coli* and demonstration that the three cognate proteins are required for magnesium-protoporphyrin chelatase activity. *J. Biol. Chem.* **271**: 16662-16667.

Jensen, P. E., Willows, R. D., Larsen Petersen, B., Vothknecht, U. C., Stummann, B. M., Kannagara, C. G., von Wettstein, D. and Henningsen, K. W. (1996b) Structural genes for Mg-chelatase subunits in barley: *Xantha-f, -g and -h*. *Molec. Gen. Genet.* **250**: 383-394.

Jones, M. C., Jenkins, J. M., Smith, A. G. and Howe, C. J. (1994) Cloning and characterization of genes for tetrapyrrole biosynthesis from the cyanobacterium *Anacystis nidulans* R2. *Plant Mol. Biol.* **24**: 435-448.

Jones, R. M. and Jordan, P. M. (1993) Purification and properties of the uroporphyrinogen decarboxylase from *Rhodobacter sphaeroides*. *Biochem. J.* **293**: 703-712.

Jones, R. M. and Jordan, P. M. (1994) Purification and properties of porphobilinogen deaminase from *Arabidopsis thaliana*. *Biochem. J.* **299**: 895-902.

Jones, M. C., Jenkins, J. M., Smith, A. G. and Howe, C. J. (1994) Cloning and characterization of genes for tetrapyrrole biosynthesis from the cyanobacterium *Anacystis nidulans* R2. *Plant Mol. Biol.* **24**: 435-448.

Jordan, P. M. (1990) Biosynthesis of 5-aminolevulinic acid and its transformation into coproporphyrinogen in animals and bacteria. In *Biosynthesis of Heme and Chlorophyll*, Dailey, H. A., ed., New York, McGraw-Hill, pp. 55-121.

Jordan, P. M. (1991) The biosynthesis of 5-aminolaevulinic acid and its transformation into uroporphyrinogen III. In *Biosynthesis of Tetrapyrroles*. New Comprehensive Biochemistry Vol. 19, Jordan, P. M., ed., Amsterdam: Elsevier, pp. 1-66.

Jordan, P. M. (1994) The biosynthesis of uroporphyrinogen III: Mechanism of action of porphobilinogen deaminase. In *The Biosynthesis of the Tetrapyrrole Pigments*, Ciba

Foundation Symposium 180, Chadwick, D. J. and Ackrill, K., eds., Chichester: John Wiley & Sons, pp. 50-64.

Kannangara, C. G. and Gough, S. P. (1978) Biosynthesis of δ-aminolevulinate in greening barley leaves: Glutamate 1-semialdehyde aminotransferase. *Carlsberg Res. Commun.* **43**: 185-194.

Kannangara, C. G. and Schouboe, A. (1985) Biosynthesis of δ-aminolevulinate in greening barley leaves VII. Glutamate 1-semialdehyde accumulation in gabaculine treated leaves. *Carlsberg Res. Commun.* **50**: 179-191.

Kannangara, C. G., Gough, S. P., Oliver, R. P. and Rasmussen, S. K. (1984) Biosynthesis of δ-aminolevulinate in greening barley leaves. VI. Activation of glutamate by ligation to RNA. *Carlsberg Res. Commun.* **49**: 417-437.

Kannangara, C. G., Vothknecht, U. C., Hansson, M. and von Wettstein, D. (1997) Magnesium chelatase: association with ribosomes and mutant complementation studies identify barley subunit Xantha-G as a functional counterpart of *Rhodobacter* subunit BchD. *Mol. Gen. Genet.* **254**: 85-92.

Kannangara, C. G., Andersen, R. V., Pontoppidan, B., Willows, R. and von Wettstein, D. (1994) Enzymic and mechanistic studies on the conversion of glutamate to 5-aminolaevulinate. In *The Biosynthesis of the Tetrapyrrole Pigments*, Ciba Foundation Symposium 180, Chadwick, D. J. and Ackrill, K., eds., Chichester: John Wiley & Sons, pp. 3-20.

Kaya, A. H., Plewinska, M., Wong, D. M., Desnick, R. J. and Wetmur, J. G. (1994) Human δ-aminolevulinate dehydratase (ALAD) gene: structure and alternative splicing of the erythroid and housekeeping mRNAs. *Genomics* **19**: 242-248.

Kiel, J. A. K.W., Ten Berge, A. M. and Venema, G. (1992) Nucleotide sequence of the Synechococcus sp. PCC7942 *hemE* gene encoding the homologue of mammalian uroporphyrinogen decarboxylase. *J. DNA Sequencing and Mapping* **2**: 415-418.

Knaust, R., Seyfried, B., Schmidt, L., Schulz, R. and Senger, H. (1993) Phototransformation of monovinyl and divinyl protochlorophyllide by NADPH: protochlorophyllide oxidoreductase of barley expressed in *Escherichia coli*. *J. Photochem. Photobiol. Ser. B.* **20**: 161-166.

Koncz, C., Mayerhofer. R., Koncz-Kalman, Z., Nawrath, C., Reiss, B., Redei, G. P. and Schell, J. (1990) Isolation of a gene encoding a novel chloroplast protein by T-DNA tagging in *Arabidopsis thaliana*. *EMBO J.* **9**: 1137-1146.

Koonin, E. V. A. (1993) A superfamily of ATPases with diverse functions containing either classical or deviant ATP-binding motifs. *J. Mol. Biol.* **229**: 1165-1174.

Kruse, E., Mock, H.-P and Grimm, B. (1995) Coproporphyrinogen III oxidase from barley and tobacco - Sequence analysis and initial expression studies. *Planta* **196**: 796-803.

Lambert, R., Brownlie, P. D., Woodcock, S. C., Louie, G. V., Cooper, J. C., Warren, M. J., Jordan, P. M., Blundell, T. L. and Wood, S. P. (1994) Structural studies on porphobilinogen deaminase. In *The Biosynthesis of the Tetrapyrrole Pigments*, Ciba Foundation Symposium 180, Chadwick, D. J. and Ackrill. K., eds., Chichester: John Wiley & Sons, pp. 97-104.

Lebedev, N., van Cleve, B., Armstrong, G. and Apel, K. (1995) Chlorophyll synthesis in a de-etiolated (det340) mutant of *Arabidopsis thaliana* without NADPH-protochlorophyllide (Pchlide) oxidoreductase (POR) A and photoactive Pchlide-F655. *Plant Cell* **7**: 2081-2090.

Lermontova, I., Kruse, E., Mock, H.-P. and Grimm, B. (1997) Cloning and characterization of a plastidal and a mitochondrial isoform of tobacco protoporphyrinogen IX oxidase. *Proc. Natl. Acad. Sci. USA.* **94**: 8895-8900.

Li, J., Goldschmidt-Clermont, M. and Timko, M. P. (1993) Chloroplast-encoded *chlB* is required for light-independent protochlorophyllide reductase activity in *Chlamydomonas reinhardtii*. *Plant Cell* **5**: 1817-1829.

Louie, G. V., Brownlie, P. D., Lambert, R., Cooper, J. B., Blundell, T. L., Wood, S. P., Warren, M. J., Woodcock, S. C. and Jordan, P. M. (1992) Structure of porphobilinogen deaminase reveals a flexible multidomain polymerase with a single catalytic site. *Nature* **359**: 33-39.

Madsen, O., Sandal, L., Sandal, N. N. and Marcker, K. A. (1993) A soybean coproporphyrinogen oxidase gene is highly expressed in root nodules. *Plant Mol. Biol.* **23**: 35-43.

Mayer, S. M., Weinstein, J. D. and Beale, S. I. (1987) Enzymatic conversion of glutamate to δ-aminolevulinate in soluble extracts of *Euglena gracilis*. *J. Biol. Chem.* **262**: 12541-12549.

Meller, E., Belkin, S. and Harel, E. (1975) The biosynthesis of δ-aminolevulinic acid in greening maize leaves. *Phytochemistry* **14**: 2399-2402.

Mitchell, L. W. and Jaffe, E. (1993) Porphobilinogen synthase from *Escherichia coli* is a Zn (II) metalloenzyme stimulated by Mg (II). *Arch. Biochem. Biophys.* **300**: 169-177.

Mock, H.-P., Trainotti, L., Kruse, E. and Grimm, B. (1995) Isolation, sequencing and expression of cDNA sequences encoding uroporphyrinogen decarboxylase from tobacco and barley. *Plant Mol. Biol.* **28**: 245-256.

Murakami, K., Korbsrisate, S., Asahara, N., Hashimoto, Y. and Murooka, Y. (1993) Cloning and characterization of the glutamate 1-semialdehyde aminomutase gene from *Xanthomonas campestris pv. phaseoli*. *Appl. Microbiol. Biotech.* **38**: 502-506.

Nair, S. P., Harwood, J. L. and John, R. A. (1991) Direct identification and quantification of the cofactor in glutamate semialdehyde aminotransferase from pea leaves. *FEBS Lett.* **283**: 4-6.

Orsat, B., Monfort, A., Chatellard, P. and Stutz, E. (1992) Mapping and sequencing of an actively transcribed *Euglena gracilis* chloroplast gene (ccsA) homologous to the *Arabidopsis thaliana* nuclear gene cs (ch42). *FEBS Lett.* **303**: 181-184.

Papenbrock, J., Gräfe, S., Kruse, E., Hänel, F. and Grimm, B. (1997) Mg-chelatase of tobacco: identification of a *ChlD* cDNA sequence encoding a third subunit, analysis of the interaction of the three subunits with the yeast two-hybrid system, and reconstitution of the enzyme activity by co-expression of recombinant CHL D, CHL H and CHL I. *Plant J.* **12**: 981-990.

Pardo, A. D., Chereskin, B. M., Castelfranco, P. A., Franceschi, V. R. and Wezelman, B. E. (1980) ATP requirement for magnesium chelatase in developing chloroplasts. *Plant Physiol.* **65**: 956-960.

Pontoppidan, B. and Kannangara, C. G. (1994) Purification and partial characterization of barley glutamyl-tRNA^{Glu} reductase, the enzyme that directs glutamate to chlorophyll biosynthesis. *Eur. J. Biochem.* **225**: 529-537.

Porra, R. J. (1986) Labelling of chlorophylls and precursors by [2-^{14}C] glycine and 2-[1-^{14}C] oxoglutarate in *Rhodopseudomonas spheroides* and *Zea mays*. *Eur. J. Biochem.* **156**: 111-121.

Porra, R. J., Schäfer, W., Cmiel, E., Kathedar, I. and Scheer, H. (1993) Derivation of the formyl group oxygen from moleular oxygen in greening leaves of a higher plant (*Zea mays*). *FEBS Lett.* **323**: 31-34.

Porra, R. J., Schäfer, W., Cmiel, E., Katheder, I. and Scheer, H. (1994) The derivation of the formyl-group oxygen of chlorophyll *b* in higher plants from molecular oxygen. Achievement of high enrichment of the 7-formyl-group oxygen from $^{18}O_2$ in greening maize leaves. *Eur. J. Biochem.* **219**: 671-679.

Porra, R. J., Schäfer, W., Katheder, I. and Scheer, H. (1995) The derivation of the 13(1)-oxo and 3-acetyl groups of bacteriochlorophyll a from water in *Rhodobacter sphaeroides* cells adapting from respiratory to photosynthetic conditions: evidence for an anaerobic pathway for the formation of isocyclic ring. E. *FEBS Lett.* **371**: 21-24.

Porra, R. J., Schäfer, W., Gad´on, N., Katheder, I., Drews, G. and Scheer, H. (1996) Origin of the two carbonyl oxygens of bacteriochlorophyll a. Demonstration of the formation of ring E in *Rhodobacter sphaeroides* and *Roseobacter denitrificans*, and a common hydratase mechanism for 3-acetyl group formation. *Eur. J. Biochem.* **239**: 85-92.

Powell, K. A., Cox, R., McConville, M. and Charles, H. P. (1973) Mutations affecting porphyrin biosynthesis in *Escherichia coli*. *Enzyme* **16**: 65-73.

Pugh, C. E., Harwood, J. L. and John, R. A. (1992) Mechanism of glutamate semialdehyde aminotransferase. Roles of diamino-intermediates and dioxo-intermediates in the synthesis of aminolevulinate. *J. Biol. Biochem.* **267**: 1584-1588.

Reinbothe, C., Apel, K. and Reinbothe, S. (1995) A light-induced protease from barley plastids degrades NADPH:protochlorophyllide oxidoreductase complexed with chlorophyllide. *Mol. Cell. Biol.* **15**: 6206-6212.

Reinbothe, C., Lebedev, N., Apel, K. and Rheinbothe, S. (1997) Regulation of chloroplast protein import through a protochlorophyllide responsive transit peptide. *Proc. Natl. Acad. Sci.USA.* **94**: 8890-8894.

Reinbothe, S., Runge, S., Reinbothe, C., van Cleve, B. and Apel, K. (1995a) Substrate-dependent transport of the NADPH:protochlorophyllide oxidoreductase into isolated plastids. *Plant Cell* **7**: 161-172.

Reinbothe, S., Reinbothe, C., Runge, S. and Apel, K. (1995b) Enzymatic product formation impairs both the chloroplast receptor-binding function as well as translocation competence of the NADPH:protochlorophyllide oxidoreductase, a nuclear-encoded plastid precursor protein. *J. Cell Biology* **129**: 299-308.

Reinbothe, S., Reinbothe, C., Neumann, D. and Apel, K. (1996) A plastid enzyme arrested in the step of precursor translocation *in vivo. Proc. Natl. Acad. Sci.USA.* **93**: 12026-12030.

Richard, M., Tremblay, C. and Bellemare, G. (1994) Chloroplastic genomes of *Gingko biloba* and *Chlamydomonas moewusii* contain a *chlB* gene encoding one subunit of light-independent protochlorophyllide reductase. *Curr. Genet.* **26**: 159-165.

Rieble, S. and Beale, S. I. (1991) Separation and partial characterization of enzymes catalyzing δ-aminolevulinic acid formation in *Synechocystis* sp PCC 6803. *Arch Biochem. Biophys.* **289**: 289-297.

Rüdiger, W., Benz, J. and Guthoff, C. (1980) Detection and partial characterization of activity of chlorophyll synthetase in etioplast *membranes. Eur. J. Biochem.* **109**: 193-200.

Sasarman, A., Surdeann, M. and Horodniceanu, T. (1968) Locus determining the synthesis of delta-aminolevulinic acid in *Escherichia coli* K-12. *J. Bacteriol.* **96**: 1882-1884.

Schaumburg, A., Schneider-Poetsch, H. A. W. and Eckerskom, C. (1992) Characterization of plastid 5-aminolevulinate dehydratase (ALAD; EC4.2.1.24) from spinach (*Spinacea oleracea* L.) by sequencing and comparison with non-plant ALAD enzymes. *Z. Naturforsch.* **47c**: 77-84.

Schneegurt, M. A. and Beale, S. I. (1988) Characterization of the RNA required for biosynthesis of δ-aminolevulinic acid from glutamate: Purification by anticodon-based affinity chromatography and determination that the UUC glutamate anticodon is a general requirement for function in ALA biosynthesis. *Plant Physiol.* **86**: 497-504.

Schneegurt, M. A. and Beale, S. I. (1992) Origin of the chlorophyll *b* formyl oxygen in *Chlorella vulgaris. Biochemistry* **31**: 11677-11683.

Schön, A., Krupp, G., Gough, S., Berry-Lowe, S., Kannangara, C. G. and Soll, D. (1986) The RNA required in the first step of chlorophyll biosynthesis is a chloroplast tRNA. *Nature* **332**: 281-284.

Schön, A., Kannangara, C. G., Gough, S. P., Soll, D. (1988) Protein biosynthesis in organelles requires misacylation of transfer RNA. *Nature* **331**: 187-190.

Schulz, R., Steinmüller, K., Klaas, M., Forreiter, C., Rasmussen, S., Hiller, C. and Apel, K. (1989) Nucleotide sequence of a cDNA coding for the NADPH-protochlorophyllide oxidoreductase (PCR) of barley (*Hordeum vulgare* L.) and its expression in *Escherichia coli. Mol. Gen. Gen.* **217**: 355-361.

Sharif, A. L., Smith, A. G. and Abell, C. (1989) Isolation and characterization of a cDNA clone for a chlorophyll synthesis enzyme from *Euglena gracilis*. The chloroplast enzyme hydroxymethylbilane synthase (porphobilinogen deaminase) is synthesized with a very long transit peptide in *Euglena. Eur. J. Biochem.* **184**: 353-359.

Simpson, D., Machold, O., Høyer-Hansen, G. and von Wettstein, D. (1985) Chlorina mutants of barley (*Hordeum vulgare* L.). *Carlsberg Res. Commun.* **50**: 223-238.

Smith, M. A., Grimm, B., Kannangara, C. G. and von Wettstein, D. (1991a) Spectral kinetics of glutamate 1-semialdehyde aminomutase of *Synechococcus*. *Proc. Natl. Acad. Sci. USA.* **88**: 9775-9779.

Smith, M. A., Kannangara, C. G., Grimm, B. and von Wettstein, D. (1991b) Characterization of glutamate 1-semialdehyde aminotransferase of *Synechococcus*. Steady State kinetics. *Eur. J. Biochem.* **202**: 749-757.

Smith, M. A., Kannangara, C. G. and Grimm, B. (1992) Glutamate 1-semialdehyde aminotransferase: anomalous enantiomeric reaction and enzyme mechanism. *Biochemistry* **31**: 11249-11254.

Spencer, P. and Jordan, P. M. (1994) 5-Aminolaevulinic acid dehydratase: characterization of the α and β metal-binding sites of the *Escherichia coli* enzyme. In *The Biosynthesis of Tetrapyrrole Pigments*, Ciba Foundation Symposium 180, Chadwick, D. J. and Ackrill, K., eds., Chichester: John Wiley & Sons, pp. 50-64.

Sperling, U., van Cleve, B., Frick, G., Apel, K. and Armstrong, G. A. (1997) Overexpression of light-dependent PORA or PORB in plants depleted of endogenous POR by far-red light enhances seedling survival in white light and protects against photooxidative damage. *Plant J.* **12**: 649-658.

Sperling, U., Franck, F., van Cleve, B., Frick, G., Apel, K. and Armstrong, G. A. (1998) Etioplast differentiation in Arabidopsis: both PORA and PORB restore the prolamellar body and photoactive protochlorophyllide-F655 to the *cop1* photomorphogenic mutant. *Plant Cell* **10**: 283-296.

Spiller, S. C., Castelfranco, A. M. and Castelfranco, P. A. (1982) Effects of iron and oxygen on chlorophyll biosynthesis. I. *In vivo* observations on iron and oxygen deficient plants. *Plant Physiol.* **69**: 107-111.

Stange-Thomann, N., Thoman, H.-U., Lloyd, A. J., Lyman, H. and Söll, D. (1994) A point mutation in *Euglena gracilis* chloroplast tRNA^{Glu} uncouples protein and chlorophyll biosynthesis. *Proc. Natl. Acad. Sci. USA.* **91**: 7947-7951.

Suzuki, J. Y. and Bauer, C. F. (1992) Light independent chlorophyll biosynthesis: Involvement of the chloroplast gene *chlL* (frxC). *Plant Cell* **4**: 929-940.

Suzuki, J. Y. and Bauer, C. E. (1995) A prokaryotic origin of light-dependent chlorophyll biosynthesis of plants. *Proc. Natl. Acad. Sci. USA.* **92**: 3749-3753.

von Wettstein, D., Henningsen, K. W., Boynton, J. E., Kannangara, C. G. and Nielsen, O. F. (1971) The genetic control of chloroplast development in barley. In *Autonomy and Biogenesis of Mitochondria and Chloroplasts*, Boardmann, N. K., Linnane, A. W. and Smillie, R. M., eds., Amsterdam: North Holland, pp. 205-223.

von Wettstein, D., Kahn, A., Nielsen, O. F. and Gough, S. P. (1974) Genetic regulation of chlorophyll synthesis analysed with mutants of barley. *Science* **184**: 800-802.

Vothknecht, U. C., Kannangara, C. G. and von Wettstein, D. (1996) Expression of catalytically active barley glutamyl tRNA^{Glu} reductase in *Escherichia coli* as a fusion protein with glutathione S-transferase. *Proc. Natl. Acad. Sci.* **93**: 9287-9291.

Vothknecht, U. C., Kannangara, C. G. and von Wettstein, D. (1998) Barley glutamyl tRNA^{Glu} reductase: Mutations affecting haem inhibition and enzyme activity. *Phytochemistry* **47**: 513-519.

Walker, C. J. and Weinstein, C. J. (1991) *In vitro* assay of the chlorophyll biosynthetic enzyme magnesium chelatase: resolution of the activity into soluble and membrane-bound fraction. *Proc. Natl. Acad. Sci. USA.* **88**: 5789-5793.

Walker, C. J. and Weinstein, C. J. (1994) The magnesium-insertion step of chlorophyll biosynthesis is a two-stage reaction. *Biochem. J.* **299**: 277-284.

Walker, C. J. and Willows, R. D. (1997) Mechanism and regulation of Mg-chelatase. *Biochem. J.* **327**: 321-333.

Walker, C. J., Mansfield, K. E., Smith, K. M. and Castelfranco, P. A. (1989) Incorporation of atmospheric oxygen into the carbonyl functionality of the protochlorophyllide isocyclic ring. *Biochem. J.* **257**: 599-602.

Walker, C. J, Mansfield, K. E., Rezzano, I. N., Hanamoto, C. M., Smith, K. M. and Castelfranco, P. A. (1988) The magnesium-protoporphyrin IX (oxidative) cyclase system. Studies on the mechanism and specificity of the reaction sequence. *Biochem. J.* **255**: 685-692.

Wang, W-Y., Gough, S. P. and Kannangara, C. G. (1981) Biosynthesis of δ-aminolevulinate in greening barley leaves. IV. Isolation of three soluble enzymes required for the conversion of glutamate to δ-aminolevulinate. *Carlsberg Res. Commun.* **46**: 243-257.

Wang, W-Y, Huang, D. D., Stachon, D., Gough, S. P. and Kannangara, C. G. (1984) Purification, characterization, and fractionation of Δ aminolevulinic acid synthesizing enzymes from light-grown *Chlamydomonas reinhardtii* cells. *Plant Physiol.* **74**: 569-575.

Weinstein, J. D. and Beale, S. I. (1983) Separate physiological roles and subcellular compartments for two tetrapyrrole biosynthetic pathways in *Euglena gracilis. J. Biol. Chem.* **258**: 6799-6807.

Weinstein, J. D. and Beale, S. I. (1985) RNA is required for enzymatic conversion of glutamate to δ-aminolevulinic acid by extracts of *Chlorella vulgaris. Arch. Biochem. Biophys.* **239**: 87-93.

Weinstein, J. D., Mayer, S. M. and Beale, S. I. (1986) Stimulation of δ-aminolevulinic acid formation in algal extracts by heterologous RNA. *Plant Physiol.* **82**: 1096-1101.

Whyte, B. J. and Griffiths, W. T. (1993) 8-vinyl reduction and chlorophyll *a* biosynthesis in higher plants. *Biochem. J.* **291**: 939-944.

Willows, R. D., Kannangara, C. G. and Pontoppidan, B. (1995) Nucleotides of tRNA[Glu] involved in recognition by barley chloroplast glutamyl-tRNA synthetase and glutamyl-tRNA reductase. *Biochem. Biophys. Acta* **1263**: 228-234.

Willows, R. D., Gibson, L. C., Kannangara, C. G., Hunter C. N. and von Wettstein, D. (1996) Three separate proteins constitute the magnesium chelatase of *Rhodobacter sphaeroides. Eur. J. Biochem.* **235:** 438-443.

Wilks, H. M. and Timko, M. P. (1995) A light-dependent complementation system for analysis of NADPH:protochlorophyllide oxidoreductase: Identification and mutagenesis of two conserved residues that are essential for enzyme activity. *Proc. Natl. Acad. USA.* **92**: 724-728.

Witty, M., Wallace-Cook, A. D. M., Albrecht, H., Spano, A. J., Michel, H., Shabanowitz, J., Hunt, D. F., Timko, M. P. and Smith, A. G. (1993) Structure and expression of chloroplast-localized porphobilinogen deaminase from pea (*Pisum sativum* L.) isolated by redundant polymerase chain reaction. *Plant Physiol.* **103**: 139-147.

Wong, Y-S. and Castelfranco, P. A. (1984) Resolution and reconstitution of Mg-Protoporphyrin IX monomethyl (oxidative) cyclase, the enzyme system responsible for the formation of the chlorophyll isocyclic ring. *Plant Physiol.* **75**: 658-661.

Wong, Y-S., Castelfranco, P. A., Goff, D. A. and Smith, K. M. (1985) Resolution and reconstitution of Mg-protoporphyrin IX monomethyl ester (oxidative) cyclase, the enzyme system responsible for the formation of the chlorophyll isocyclic ring. *Plant Physiol.* **75**: 658-661.

Woodcock, S. C. and Jordan P. M. (1994) Evidence for participation of aspartate-84 as a catalytic group at the active site of porphobilinogen deaminase obtained by site-directed mutagenesis of the *hemC* gene from *Escherichia coli. Biochemistry* **33**: 2688-2695.

Chapter 7

Discovery of the Two Parallel Pathways for Isoprenoid Biosynthesis in Plants

Hartmut K. Lichtenthaler
Botanical Institute (Plant Physiology and Biochemistry)
University of Karlsruhe, D-76128 Karlsruhe, Germany

ABSTRACT

Higher plants possess two independent pathways for isopentenyl diphosphate (IPP) and isoprenoid formation that work in different cellular compartments and deliver differential isoprenoid end products: 1) in the cytosol proceeds the classical acetate/mevalonate pathway for the biosynthesis of sterols and sesquiterpenes, and 2) in the plastids the non-mevalonate 1-deoxy-D-xylulose-5-phosphate pathway (DOXP-pathway[1]) for the biosynthesis of carotenoids, phytol, prenyl side-chains of plastoquinone-9 and phylloquinone K_1, as well as isoprene and mono- and diterpenes. The acetate/mevalonate pathway of plants had already been established in 1958. However, the DOXP-pathway was only elucidated in the past four years, although various labeling results of plastidic isoprenoids were not in agreement with the mevalonate pathway. These controversial issues and the steps that finally led to the discovery of the plastidic DOXP-pathway of IPP and isoprenoid formation are summarized in this chapter.

Introduction

Plant Prenyllipids and Terpenoids

Higher plants and algae are photosynthetically active organisms and possess various isoprenoid lipids as regular and primary products of their cell metabolism and biomembranes. These prenyllipids comprise the sterols (plus sterolglycosides and acylated sterolglycosides) bound to the

---

[1]**Abbreviations:**

DOX = 1-deoxy-D-xylulose; DOXP = 1-deoxy-D-xylulose-5-phosphate; DMAPP = dimethylallyl diphosphate; MVA = mevalonic acid; IPP = isopentenyl diphosphate.

141

cytoplasmic biomembranes (e.g., plasmalemma and tonoplast), the *plastidic* carotenoids, chlorophylls and prenylquinones (plastoquinone-9, phylloquinone K_1, α-tocoquinone and α-tocopherol) as well as the *mitochondrial* ubiquinones Q-9 and Q-10 (Goodwin, 1977; Lichtenthaler, 1977 and 1993). Sterols, carotenoids and prenyl alcohols (e.g., phytol) represent pure prenyllipids, whose carbon atoms are made up solely from isoprenic C_5-units. Mixed prenyllipids, in turn, contain an isoprenoid side-chain (prenyl chain) which is bound to a non-isoprenoid nucleus, such as a phorphyrin ring (chlorophylls *a* and *b*), a benzoquinone ring (plastoquinone-9, α-toquinone and α-tocopherol, ubiquinone-9 and -10) and a naphthoquinone ring (phylloquinone, vitamin K_1) (Goodwin, 1977; Lichtenthaler, 1977). All of the pure prenyllipids and the prenyl side-chains of mixed prenyllipids are formed by condensation of isoprenic C_5 units in a head-to-tail pattern with dimethylallyl diphosphate (DMAPP) as starting C_5 molecule to which more and more C_5 units are added from its isomer isopentenyl diphosphate (IPP).

Besides these functional prenyllipids, plants and algae can contain various other isoprenoid compounds which are regarded as secondary plant products. All of these plant isoprenoids have also been termed terpenoids whereby terpenes are counted in C_{10} carbon units. There are the hemiterpenes (C_5, e.g., isoprene), monoterpenes (C_{10}, e.g., menthol), sesquiterpenes (C_{15}, e.g., farnesol), diterpenes (C_{20}, e.g., phytol, taxol), triterpenes (C_{30}, e.g., cycloartenol, *Avena* saponins), tetraterpenes (C_{40}, e.g., β-carotene, lycopene) as well as polyterpenes (C_5H_8)n as found in rubber latex and guttapercha. The biogenetic relationship and composition of these isoprenoid/terpenoid plant products from isoprenic C_5-units derived from IPP is shown in Fig. 1.

In the past four decades, all of these plant prenyllipids and terpenoids were supposed to be synthesized via the acetate/MVA pathway of isopentenyl diphosphate (IPP) biosynthesis (Goodwin, 1965 and 1980; Fischer *et al.*, 1962; Spurgeon and Porter, 1981; Lichtenthaler *et al.*, 1982) that had been established in animals and fungi. This assumption was, however, not true since plants possess two biochemically fully different pathways for IPP formation as was shown only very recently (see the reviews Lichtenthaler *et al.*, 1997a; Lichtenthaler, 1998 and 1999). The cytosolic acetate/MVA pathway provides IPP only for sterol and sesquiterpene biosynthesis, whereas the plastids possess a separate, non-mevalonate 1-deoxy-D-xylulose-5-phosphate (DOXP) pathway of IPP formation starting from glyceraldehyde-3-phosphate and pyruvate.

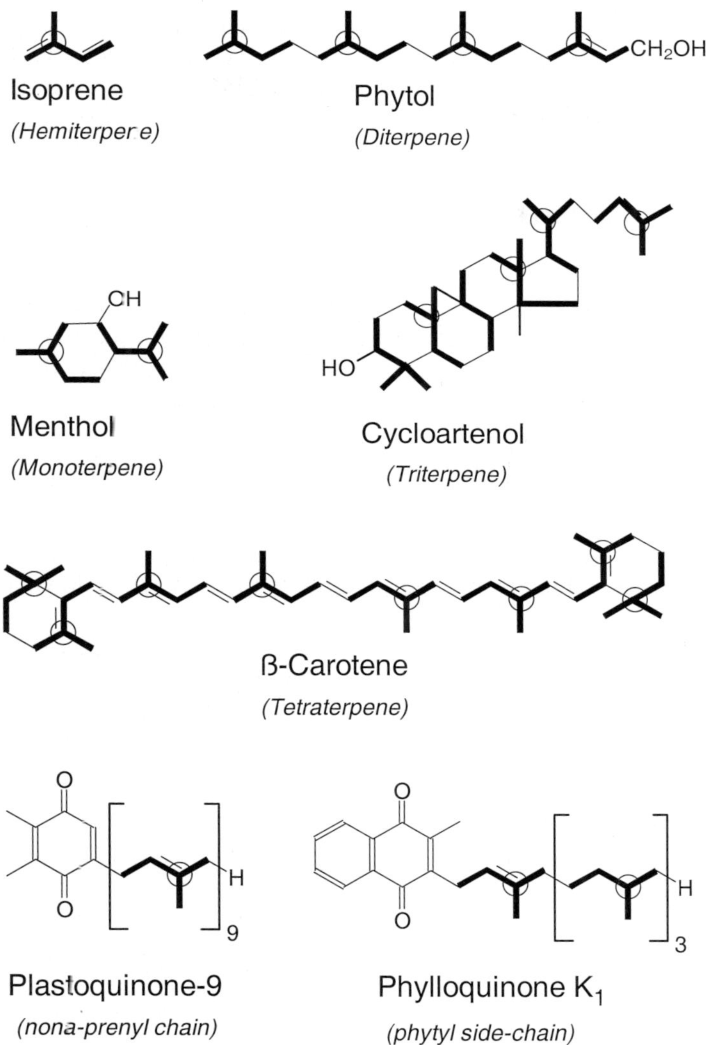

Fig. 1. Chemical structure of different isoprenoids/terpenoids of higher plants. The C_5-units derived from the biosynthetic isoprenoid precursor isopentenyldiphosphate (IPP) are shown in bold face.

Historical Aspects of IPP and Isoprenoid Biosynthesis

In the last century, O. Wallach (Nobel prize winner 1910) recognized that one or more C_5-units are found in several natural substances (Wallach, 1885 and 1887). L. Ruzicka (Nobel prize winner in 1939) and his group later dealt in great detail with the structural architecture of sterols and polyterpenes. The essential milestones in the establishment of the acetate/MVA pathway for the biosynthesis of IPP and cholesterol and also other isoprenoids, particularly in mammals and fungi, were as follows:

1) Detection of the *biogenetic isoprene rule* that the carbon skeleton is composed of C_5-units which are united by head-to-tail addition (Ruzicka *et al.*, 1922 and 1953; Ruzicka, 1932 and 1938; Eschenmoser *et al.*, 1955).
2) In 1950 K. Bloch and his group showed in biosynthetic studies using ^{14}C-acetate that cholesterol is made up from *acetate units* (Little and Bloch, 1950).
3) Then it was demonstrated that coenzyme A is required and that *acetyl-CoA* and *acetoacetyl-CoA* are the activated intermediates in cholesterol biosynthesis (Lynen *et al.*, 1951; Cornforth *et al.*, 1957).
4) In 1956 *mevalonic acid* (MVA) as essential intermediate in cholesterol biosynthesis, which is incorporated faster than acetyl-CoA, was detected (Wolf *et al.*, 1956; Tavormina *et al.*, 1956 a and b).
5) Finally, in 1958 the groups of Konrad Bloch and Feodor Lynen (Nobel prize for both scientists in 1964) demonstrated that phosphorylated compounds, such as *isopentenyl diphosphate* (IPP) and *farnesyl diphosphate*, are the intermediates in squalene and cholesterol biosynthesis (Lynen *et al.*, 1958; Chaykin *et al.*, 1958).

In these experiments, the exact position of the applied ^{13}C- or ^{14}C-label in the C-atoms of the final products squalene or cholesterol had been checked by a careful chemical degradation of these compounds. The same technique of analyzing C-atom by C-atom had also been used at that time by Melvin Calvin and his group during the establishment of the photosynthetic carbon reduction cycle after application of $^{14}CO_2$ to green algae (Bassham and Calvin, 1957; Calvin, 1960). In fact, in the 1950s and beginning 1960s the time-consuming chemical degradation of the labeled organic compounds was the usual state-of-the-art in biosynthetic studies of most laboratories in order to find out the exact position of the applied carbon label. The much faster and more elegant ^{13}C-labeling techniques combined with high resolution NMR-spectra and the sensitive GC-MS technique, which in recent years essentially helped to discover the non-mevalonate DOXP pathway of

IPP formation in plants, had yet to be developed. In addition, in the 1950s the routine methods for separation of the labeled organic compounds were column and paper chromatography. Thin layer chromatography, GC or the presently used HPLC techniques had not yet been developed.

Equipment and Plant Isoprenoid Research in 1958 and the Beginning 1960s

In 1958, I started my Ph.D. work with Professor August Seybold at the Botanical Institute of the University of Heidelberg, Germany, investigating the interrelationship between the isoprenoid chlorophylls, carotenoids and phylloquinone K_1, a naphthoquinone which, like chlorophylls, contains an isoprenoid phytyl chain. I realized that there must be a close biogenetic and biosynthetic relationship between these isoprenoid compounds, but no further research could be done at that time due to a lack of instrumentation and equipment. In these post-war years, the biochemical research possibilities in Germany and other European countries were very limited, particularly at university laboratories. The technique of applying ^{14}C-labeled precursors in the biosynthesis of bio-organic compounds including the techniques of paper chromatography and autoradiography had been started in the 1950s, not only in USA (e.g., in the investigation of the photosynthetic CO_2 reduction by Melvin Calvin and his group) but also in several European laboratories. However, it was not before the mid-1960s when this technology became available for plant physiology or at botanical institutions of universities. For this reason, I first studied the quantitative relationship between chlorophylls and phylloquinone K_1 in various green and non-green plant tissues (Lichtenthaler, 1962). Although the results of a defined quantitative correlation between both isoprenoids suggested that phylloquinone K_1 was closely associated with chlorophylls in the chloroplasts, at that time this assumption could not be verified since an ultracentrifuge for the isolation of chloroplasts was neither available in mine nor neighboring university institutes at Heidelberg.

Chlorophylls a and b and the major carotenoids of green plant tissues such as β-carotene, lutein, violaxanthin and neoxanthin were already known in the late 1950s through the work by various authors. T.W. Goodwin had performed excellent work on the identification of plant carotenoids and started to work on their biosynthesis (see Goodwin, 1980 and 1981). In 1959, Crane and his group determined ubiquinones (coenzyme Q) in plants and also detected plastoquinone-9, a benzoquinone with an isoprenoid side-chain in green plant tissue and chloroplasts (Crane, 1959; Crane *et al.*, 1960). Phylloquinone (vitamin K_1) and α-tocopherol (vitamin E) had already been established as components of green leaves. The localization of these isoprenoid quinone compounds together with chlorophylls and carotenoids in whole chloroplasts and thylakoid particles could, however, only be proved during my post-doc time (1962–1964) in Melvin Calvin's laboratory at the University of California in Berkeley (Lichtenthaler and Park, 1963; Lichtenthaler and Calvin, 1964). Due to its excellent modern equipment including ultracentrifuges and also recording spectrophotometers (e.g., Cary 14), the exact spectral identification and quantification of chlorophylls, carotenoids and the different isoprenoid quinones (including oxidized and reduced forms) in the photosynthetic membranes could be performed. In addition to this, I could demonstrate in 1963 in a biosynthetic study a fast and progressing incorporation of photosynthetically fixed $^{14}CO_2$ in *Chlorella* suspensions into all carotenoids and plastidic prenylquinones after 5 and 10 minutes of $^{14}CO_2$ exposure under the control photosynthetic standard conditions of the "Lollipop" apparatus used in M. Calvin's laboratory at those times. In fact, in those times as well as today, scientific progress depended on the availability and introduction of new instrumentation and methods.

First Studies on the Biosynthesis of Carotenoids in Fungi

After the groups of F. Lynen and K. Bloch had demonstrated the biosynthesis of cholesterol in mammals and fungi via the acetate/MVA pathway, the specific incorporation of ^{14}C-acetate at high rates into β-carotene in the fungi *Phycomyces blakesleeanus* (Lottspeich *et al.*, 1959) and *Mucor hiemalis* (Grob and Büttler, 1954) also indicated that the isoprenoid pigment ß-carotene, like cholesterol, is formed via this acetate/MVA pathway of IPP formation. This was verified by chemical degradation of the isolated carotenoid (Grob and Büttler, 1955; Lottspeich *et al.*, 1961). In contrast to higher plants, fungi do not possess chloroplasts, but some of them are able to produce carotenoids as a cytosolic activity. It has recently been re-investigated in view of the non-mevalonate DOXP pathway and has been confirmed that in such fungi both the carotenoids and sterols are, in fact, synthesized via the classical acetate/MVA pathway of IPP formation (Disch and Rohmer, 1998).

First Studies on the Biosynthesis of Carotenoids and Plastidic Isoprenoids in Plants and Algae

The first labeling studies on the biosynthesis of plant and photosynthetic isoprenoids and pigments were performed from 1958 to 1960 by the carotenoid and sterol specialist Trevor Goodwin (Fig. 2) and his group by demonstrating the incorporation of ^{14}C-label from $^{14}CO_2$, [2-^{14}C] acetate and [2-^{14}C] mevalonate in etiolated and illuminated maize seedlings (Goodwin 1958a and b) into ß-carotene of carrot root preparations (Braithwaite and Goodwin, 1960a) and into the secondary carotenoid lycopene of tomato fruit slices (Braithwaite and Goodwin, 1960b). This was complemented by the incorporation of ^{14}C-labeled acetate and MVA into chlorophylls (phytol side-chain), carotenoids and sterols of barley seedlings (Fischer *et al.*, 1962). In addition, the enzymatic synthesis of rubber from labeled mevalonate was demonstrated (Park and Bonner, 1958). Such ^{14}C-labeling studies of sterols and various kinds of plastidic isoprenoids were later extended to many other plants such as pea, lettuce, barley (Goodwin, 1965), tobacco (Threlfall and Whistance, 1971), pine (Wieckowski *et al.*, 1967) or oat (Lichtenthaler *et al.*, 1982).

Fig. 2. Prof. Trevor W. Goodwin (right), who was the first to study the biosynthesis of carotenoids and other plant isoprenoids using $^{14}CO_2$ and ^{14}C-labeled acetate and mevalonate, here shown at a plant lipid symposium held in July 1993 in Karlsruhe by Hartmut K. Lichtenthaler (left). In the middle is Paul Mazliak.

Discrepancies with the Acetate/MVA Concept of IPP Formation in Plants

These results mentioned above seemed to indicate that all isoprenoids are also synthesized in plants via the acetate/MVA pathway that had originally been established for sterols in mammals and fungi. There were, however, some discrepancies and differences in the labeling rates depending on the applied ^{14}C-substrates. Thus, plastidic isoprenoids such as β-carotene, phytol (side-chain of chlorophylls), and the side-chains of plastoquinone-9, phylloquinone K_1, or α-tocopherol were preferentially labeled from $^{14}CO_2$. In contrast, ^{14}C-acetate and ^{14}C-MVA labeled predominantly the cytosolic sterols and the mitochondrial ubiquinones. These differences were interpreted with the assumptions that i) acetate and MVA may not readily penetrate the chloroplast envelope and ii) that chloroplasts may possess an autonomous IPP biosynthesis system. Goodwin (1965) even proposed that the enzymes of IPP biosynthesis might be segregated into the cytosolic, plastidic and mitochondrial compartments of the plant cell.

Further inconsistencies with the acetate/MVA pathway came from studies using mevinolin, a fungal metabolite that is a highly specific inhibitor of HMG-CoA reductase and cholesterol biosynthesis (Bach and Lichtenthaler, 1982a). Mevinolin efficiently inhibited the biosynthesis of sterols in radish seedlings and *Sylibum* suspension cultures, whereas the accumulation of chlorophylls, carotenoids and plastidic prenylquinones was hardly or not at all affected (Bach and Lichtenthaler, 1982a and b, 1983; Döll *et al.*, 1984; Schindler *et al.*, 1985). A possible explanation for these unexpected results was the assumption that i) mevinolin might not be able to penetrate the chloroplast envelope or ii) that the target enzyme, a potential plastidic HMG-CoA reductase, could not be blocked by mevinolin. The mevinolin results clearly indicated that chloroplasts possess their own IPP biosynthesis system, but the idea that this might be possible with precursors and enzymes other than the cytosolic MVA-dependent IPP pathway was not yet acceptable by plant isoprenoid biochemists.

HMG-CoA reductase could be isolated from plants as a microsomal enzyme (Bach and Lichtenthaler, 1982a). A low activity of HMG-CoA reductase found in isolated chloroplasts (Bach *et al.*, 1976), which decreased with chloroplast purification, apparently was a contamination by microsomal HMG-CoA reductase. In fact, the existence of a plastidic HMG-CoA reductase has never been unequivocally demonstrated despite some indications in literature. Furthermore, HMG-CoA reductases are encoded in nuclear DNA by a multigene family (McGarvey and Croteau, 1965), but isoforms with a typical chloroplast leader peptide sequence have not been found (Bach, 1995). When in 1996 John Goad, a former Ph.D.

student of T. W. Goodwin, had heard from me about the novel non-mevalonate DOXP pathway of plastidic IPP formation, he informed me that he had been trying to isolate a plastidic mevalonate kinase in the 1960s for quite some time without success, and that such "negative results" were not published.

Also Gernot Schultz and co-workers had their doubts on the use of the MVA pathway of IPP formation for plastidic isoprenoids (see Heintze *et al.*, 1990 and references therein). They demonstrated in leaf segments, protoplasts and isolated chloroplasts that plastidic isoprenoid biosynthesis (β-carotene, plastoquinone-9) was not accessible to exogenously applied acetate or MVA, whereas ^{14}C-labeled CO_2, phosphoglyceric acid and pyruvate were incorporated into plastidic isoprenoids at good rates. Today we know that pyruvate and glyceraldehyde-3-phosphate are the substrates of the first enzyme of the plastidic DOXP pathway of IPP formation (DOXP synthase), which explains the labeling results of Schultz and his group. However, Schultz and co-workers concluded in 1990 that two separate metabolic pools of acetyl-CoA might exist in chloroplasts, one being filled with pyruvate via the plastidic pyruvate dehydrogenase complex and flowing into plastidic IPP biosynthesis, whereas the second being filled with acetate and used for the *de novo* fatty acid biosynthesis.

Further arguments against the existence of the acetate/MVA pathway in plastids came from H. Kleinig and co-workers (see Lütke-Brinkhaus and Kleinig, 1987 and references therein). They found that in intact or disrupted chloroplasts and chromoplasts ^{14}C-IPP was readily incorporated into β-carotene, but ^{14}C-labeled MVA, MVA-5-phosphate or MVA-5-diphosphate was not. Their results revealed that all enzymatic activities transforming MVA into IPP were not located in the plastids but in the cytoplasm only.

After the recent discovery of the non-mevalonate DOXP pathway of IPP biosynthesis for plastidic isoprenoids and prenyllipids (see below, as well as the reviews Lichtenthaler, 1998 and 1999), the controversial results mentioned above can be re-interpreted on the basis of the differential substrate requirements and enzyme equipment of the two independent IPP biosynthesis pathways of plants. The problem of the early and the later ^{14}C-labeling studies of plant and plastidic isoprenoids was that the exact position of the ^{14}C-label within the carbon skeleton of the different plant isoprenoids had either not been determined by chemical degradation or not with full certainty. For ^{14}C-labeling studies, only low amounts of the particular plant isoprenoids are required, whereas for a chemical degradation, much higher amounts are needed. Thus, final proof for the specific incorporation of the applied ^{14}C-acetate or ^{14}C-MVA into the proper position of the isoprenic C_5-units of the plastidic isoprenoids had not been made in most cases.

Discovery of the Non-mevalonate 1-deoxy-D-xylulose-5-phosphate Pathway of IPP Formation in Algae and Higher Plants

The final detection of the DOXP pathway of IPP formation in bacteria, algae and higher plants was based on the introduction of the ^{13}C-labeling technique combined with high resolution NMR-spectrometers as well as the GC-MS technique. The first impulses came from labeling studies in bacteria. Bacteria do not possess sterols, but have triterpenoid sterol surrogates — the hopanoids. In 1988, Flesch and Rohmer studied the incorporation of ^{13}C-labeled acetate into the hopanoids of the photosynthetic purple non-sulfur bacteria of *Rhodopseudomonas palustris* and *acidophila* as well as in the facultative *Methylobacterium organophilum*, and subsequently determined the position of ^{13}C-label in the hopanoid skeleton by high resolution ^{13}C-NMR spectroscopy. They found a different labeling pattern than expected according to the acetate/MVA pathway and stated that "the labeled acetate from the culture medium has not been directly incorporated into the isoprenic biosynthetic pathway." Initially this unexpected labeling pattern of hopanoids was explained by Rohmer by means of special assumptions, e.g., two different pools of acetate and/or particular metabolic activities including the glyoxylate cycle, etc. Later, Rohmer and co-workers re-investigated this phenomenon and came up with the proposal of a novel pathway for isopentenyl diphosphate (IPP) formation in eubacteria (Rohmer *et al.*, 1993).

In 1992, when I heard about this unusual isoprenoid labeling pattern in bacteria, I anticipated that this pathway might also exist in the chloroplasts of algae and higher plants, which would explain many of the controversial labeling and inhibition results regarding plastidic isoprenoids. For this reason, I invited Michel Rohmer (then Mulhouse/France) to present his data during a scientific colloquium held at our Karlsruhe University. After his lecture, we discussed in detail the possibility that chloroplasts, as endosymbionts with their bacterial genetic organization, may also possess this non-mevalonate bacterial IPP biosynthesis pathway. In our discussion, we detected various similarities between the labeling of hopanoids and plastidic isoprenoids, such as low labeling rates by ^{14}C-acetate or ^{14}C-MVA and no effects by mevinolin or other specific inhibitors of the HMG-CoA reductase on the synthesis and accumulation of these isoprenoid compounds. We immediately started a cooperation to investigate the possible existence of the non-mevalonate IPP biosynthesis pathway in algae and higher plants.

At Karlsruhe, we grew the sterile cultures of the green alga *Scenedesmus* on ^{13}C-labeled glucoses labeled at different C-atoms. Uniformly ^{13}C-labeled glucose was also applied. The ^{13}C-NMR spectra of the isolated and purified isoprenoids were determined by Rohmer and his group in Mulhouse and later in Strasbourg. In fact, these different labeling experiments clearly demonstrated that i) in *Scenedesmus* the plastidic isoprenoids β-carotene, lutein, phytol (side-chain of chlorophylls) and plastoquinone-9 (nona-prenyl side-chain) were labeled according to the alternative non-mevalonate pathway of IPP formation (Fig. 3), ii) glyceraldehyde-3-phosphate and a C_2 unit from pyruvate were the precursors of IPP and iii) the mechanism of the non-mevalonate IPP biosynthesis in *Scenedesmus* was identical to that detected in bacteria. Our first results were published in 1995 (Lichtenthaler *et al.*, 1995; Schwender *et al.*, 1995), and the full information was given in 1996 (Schwender *et al.*, 1996). The non-mevalonate labeling pattern of plastidic isoprenoids from [1-^{13}C] glucose was later confirmed for other green algae (*Chlorella*, *Chlamydomonas*), the red alga *Cyanidium*, the chrysophyte *Ochromonas* and the cyanobacterium *Synechocystis* (Lichtenthaler *et al.*, 1997b and 1998, Disch *et al.*, 1998). The only exception among algae was *Euglena*, where both, sterols and phytol, were labeled from [1-^{13}C] glucose via the classical acetate/MVA pathway of IPP formation. Thus, this new labeling approach confirmed the very early labeling studies of β-carotene in *Euglena* where the acetate/MVA pathway had also been proved by chemical degradation (Steele and Gurin, 1960).

The next step was the extension of our joint labeling studies to higher plants. Green carrot cell cultures, barley shoots and duckweed (*Lemna gibba*) plantlets were grown heterotrophically in aseptic cultures on [1-^{13}C] glucose. We found the same non-mevalonate labeling pattern of β-carotene, lutein, phytol and plastoquinone-9 (Lichtenthaler *et al.*, 1997a and b) as in green algae and several other algae groups. In contrast to plastidic isoprenoids, the cytosolic sterols were formed in higher plants via the classical acetate/MVA pathway of IPP formation (Table 1). This demonstrated that higher plants possess a compartmentation of IPP biosynthesis consisting of two independent IPP biosynthesis pathways as shown in Figs. 4 and 5. Both pathways can specifically be blocked by either the antibiotics mevinolin (Bach and Lichtenthaler, 1962a and b) or fosmidomycin (Zeidler *et al.*, 1998; Schwender *et al.*, 1999) as shown in Fig. 3. I informed Trevor Goodwin in 1996 about this DOXP pathway of IPP formation. He was particularly happy and sent his congratulations on this "major breakthrough" which explains all the former inconsistencies in plastidic isoprenoid biosynthesis.

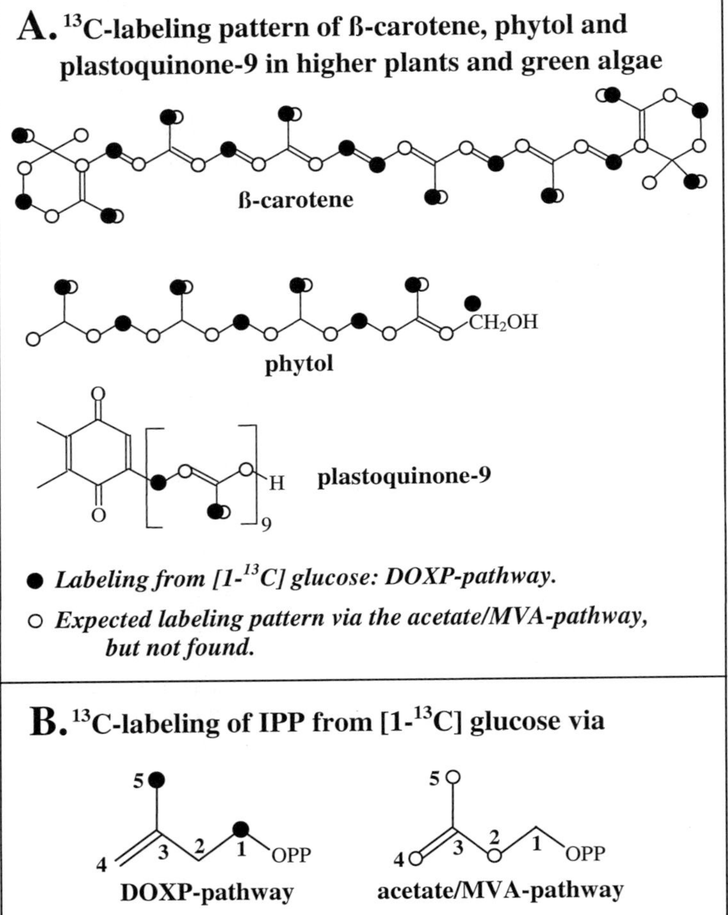

Fig. 3. **A)** ^{13}C-labeling pattern (black circles) of plastidic isoprenoids of higher plants and various algae when supplied with [1-^{13}C] glucose. **B)** Labeling pattern of C-atoms in the biosynthetic C_5-unit isopentenyl diphosphate IPP actually found (black circles) and expected according to the acetate/MVA pathway (open circles) (from Lichtenthaler *et al.*, 1997a modified).

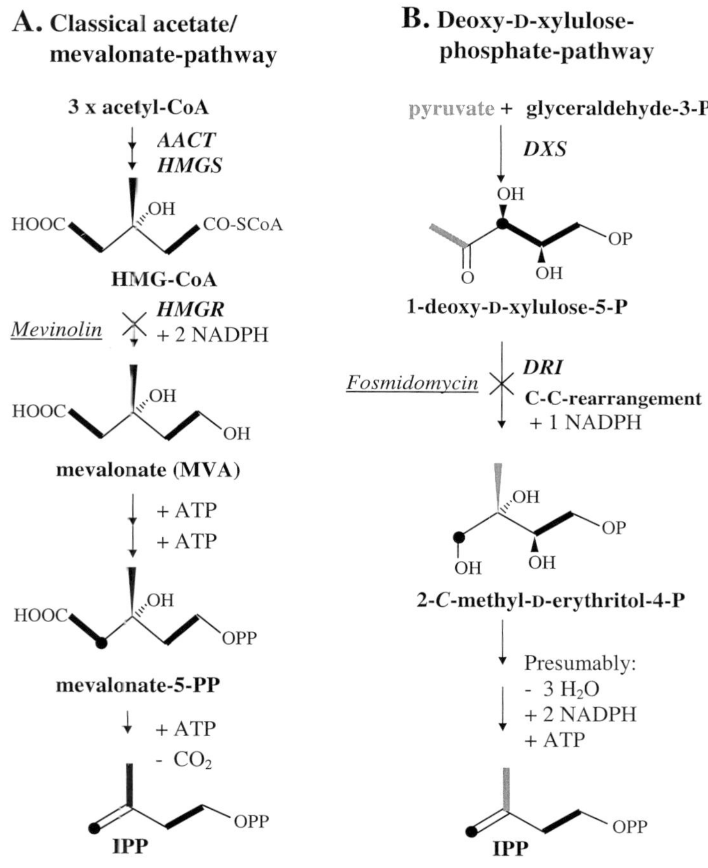

Fig. 4. Biosynthesis of isopentenyl diphosphate (IPP) from [1-^{13}C] glucose via the acetate/MVA pathway (left) and via the new non-mevalonate 1-deoxy-D-xylulose-5-phosphate (DOXP) pathway (right). The carbon-carbon fragments derived from acetate and glyceraldehyde-3-P are shown in bold face (black) and the C_2 unit from pyruvate as grey tone. The point of interaction of two antibiotics, that specifically inhibit the plants' isopentenyl diphosphate (IPP) biosynthesis, is indicated: *mevinolin* inhibits the HMG-CoA reductase (HMGR) (mevalonate pathway) and *fosmidomycin* the DOXP reductoisomerase (DRI). AACT = acetoacetyl-CoA thiolase; HMGS = HMG-CoA synthase; DXS = DOXP-synthase.

Table 1. ^{13}C-isotopic abundances (%) in the carbon atoms of cytosolic sterols and plastidic isoprenoids (β-carotene, lutein, phytol of chlorophylls, plastoquinone-9) of duckweed (*Lemna gibba*) and barley (*Hordeum vulgare*) grown on [1-^{13}C]glucose. The range of ^{13}C-abundancies of the carbon atoms corresponding to one IPP unit is given. The numbering of carbon atoms of IPP is shown in Fig. 3 (from Lichtenthaler *et al.*, 1997a).

Isoprenoid	C-1	C-2	C-3	C-4	C-5
Lemna gibba					
β-Carotene	**1.9**	1.0	0.9	1.2	**1.9**
Lutein	**2.6**	1.4	1.3	1.4	**2.8**
Phytol	**3.1**	1.4	1.5	1.6	**3.1**
Plastoquinone-9	**1.9**	1.1	1.1	1.1	**2.0**
Hordeum vulgare					
Phytol	**4.7**	1.0	1.1	1.4	**4.5**
Sitosterol	1.7	**3.8**	1.3	**3.9**	**4.9**
Stigmasterol	1.4	**3.3**	1.3	**3.2**	**3.1**

The labeling experiments of plastidic isoprenoids from [1-^{13}C] glucose indicated that 1-deoxy-D-xylulose-5-phosphate (DOXP) is the first intermediate in this non-mevalonate DOXP pathway of IPP formation in algae (Schwender *et al.*, 1996), higher plants (Lichtenthaler *et al.*, 1997a and b), and in certain eubacteria (Rohmer *et al.*, 1993). Evidence for DOXP as the first intermediate in this non-mevalonate IPP biosynthesis pathway came from the specific incorporation of deuterium (d)-labeled [1-^{2}H$_1$] deoxy-D-xylulose (d-DOX) and its methyl-glycoside (methyl-d-DOX) into the plastidic isoprenoid phytol in green algae, a red alga and higher plants as well as into isoprene of different plants (Schwender *et al.*, 1997; Zeidler *et al.*, 1997). This work was performed via high resolution NMR and GC-MS spectra in cooperation with the organic chemist Frieder W. Lichtenthaler. After this proof of DOX and DOXP as intermediates of this non-mevalonate IPP pathway, I termed this new IPP forming pathway: *1-deoxy-D-xylulose-5-phosphate pathway* (Schwender *et al.*, 1997). This name is very appropriate for this new pathway and has been used in our further publications (e.g., Lichtenthaler, 1998 and 1999; Lichtenthaler *et al.*, 1998). Additional evidence for DOXP as first intermediate came also from the independent work of other groups showing the specific incorporation of a ^{13}C-labeled DOX into β-carotene of *Catharanthus* (Arigoni *et al.*, 1997), and of a double ^{13}C-labeled DOX into the isoprenoid side-chain of ubiquinone in *E. coli* (Putra *et al.*, 1998). The incorporation of a ^{13}C-labeled DOX into a diterpenoic ginkgolide (Schwarz, 1994) and into ubiquinone of *E. coli* (Broers, 1994) had already been observed in Arigoni's laboratory in 1994, but this information became available to the scientific community only much later, as the results of these two Ph.D. theses had unfortunately never been published.

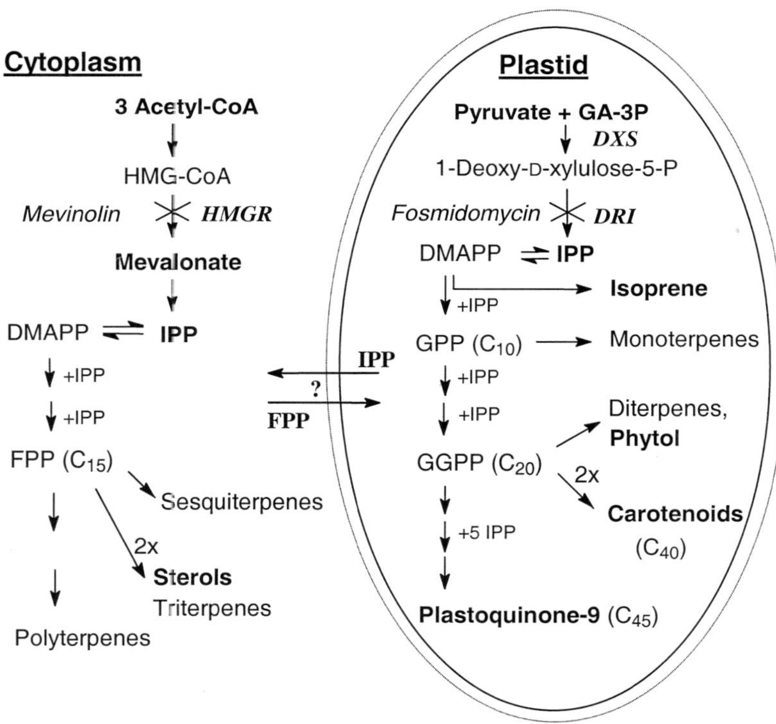

Fig. 5. Suggested compartmentation of isopentenyl diphosphate (IPP) and isoprenoid biosynthesis in higher plants between cytosol (acetate/mevalonate pathway) and plastids (DOXP pathway). The site of inhibition by *mevinolin* (cytosol) and *fosmidomycin* (plastid) is indicated. DMAPP = dimethylallyl diphosphate; GPP = geranyl diphosphate; FPP = farnesyl diphosphate; GGPP = geranylgeranyl diphosphate; HMGR = HMG-CoA reductase; DXS = DOXP synthase; DRI = DOXP reductoisomerase; GA-3P = glyceraldehyde-3-phosphate.

Concluding Remarks

The aim of this review was to present a historical view of the discovery, concepts and assumptions of the early steps of plant IPP and isoprenoid biosynthesis in the past 40 years and the many inconsistencies of plastidic IPP formation with the acetate/MVA pathway. These controversial results finally led to the detection of the non-mevalonate 1-deoxy-D-xylulose-5-phosphate (DOXP) pathway of IPP formation in photosynthetic organisms such as algae and higher plants. Two enzymes of the DOXP pathway are known today: the DOXP synthase (DXS) and the DOXP reducto-isomerase (DRI), and their genes have been or are being cloned (cf. Lichtenthaler, 1999; Schwender *et al.*, 1999). This field of plant science is developing very fast. All the details and numerous contributions on other plant terpenoids made by colleagues, who could not all be listed in this short review, are found in the reviews by Lichtenthaler *et al.*, 1997, as well as Lichtenthaler 1998 and 1999.

This discovery review of the two parallel IPP pathways of plant isoprenoid biosynthesis showed that the thoughts of plant biochemists dealing with plant isoprenoid biosynthesis were originally strongly dominated by the acetate/MVA pathway. The latter is the only IPP producing pathway in mammals and fungi, but in plants it is only one of two fully different IPP producing pathways. For many years, nobody attempted to break the unifying concept that apparently all living organisms make their IPP via the acetate/MVA pathway. This led to more and more assumptions to explain the increasing number of labeling observations that were not in agreement with the acetate/MVA pathway of IPP formation for plastidic isoprenoids. Since the plants possess the acetate/MVA pathway in the cytosol and additionally the DOXP pathway in their plastids, the subject of isoprenoid formation in plants provided greater difficulties than in animal tissue. If plants only had the DOXP and not the MVA pathway, the DOXP pathway would have been found much earlier.

The matter was further complicated by the fact that i) in *Euglena* (as an exception among algae), only the MVA pathway of IPP formation exists for the biosynthesis of sterols and plastidic isoprenoids and ii) carotenoids in fungi are, in fact, synthesized via acetate and MVA. Yet, if the exact ^{14}C-position in the carbon skeleton of plastidic isoprenoids labeled at low rates from ^{14}C-acetate or ^{14}C-MVA in plants had been determined by a correct chemical degradation (which is time-consuming work and often not a clear-cut matter), the DOXP pathway of plants could have been detected much earlier. Thus, only the ^{13}C- labeling technique in combination with ^{13}C-NMR spectra, which allows a fast and exact screening of the position of

^{13}C-label in the isoprenoid carbon skeleton, finally led to the detection of the plants' DOXP pathway of IPP formation.

All former inconsistencies with the acetate/MVA DOXP pathway in plants can now be explained by the existence of the non-mevalonate DOXP pathway in plants. Thus, the very fast labeling of plastidic isoprenoids from ^{14}CO$_2$ under photosynthetic conditions is fully explainable, as ^{14}CO$_2$ is transformed in the photosynthetic carbon cycle to glyceraldehyde-3-phosphate which is one substrate of the DOXP synthase and can readily be transformed in plastids to pyruvate as the second substrate of the DOXP synthase. A question still open is if the two IPP producing pathways of plants can cooperate as indicated in Fig. 5 by a question mark. This appears to be possible, however, only to a very low degree (cf. Lichtenthaler, 1999). The major or exclusive IPP forming pathway for plastidic isoprenoid biosynthesis is the DOXP pathway. But if some cooperation exists, the low labeling of plastidic isoprenoids from ^{14}C-acetate or ^{14}C-mevalonate mentioned above, may not only be due to an unspecific labeling but also to a partial import of some cytosolic IPP or a similar prenyl precursor into the chloroplast. This shows a further complication of the subject when two independent IPP producing pathways exist in parallel in one cell, as is the case with plants.

Acknowledgments

The work of my laboratory over the past 40 years, as included in this review, was made possible by the enthusiastic participation of my Ph.D. students and post-doctoral fellows whose names can be found in the publications cited. At this occasion I would like to thank my Ph.D. professor, the late August Seybold, Heidelberg, for introducing me to the exciting field of isoprenoid plant pigments and quinones, and the late Professor Melvin Calvin, Berkeley, for his encouraging and stimulating discussions during my post-doc time in his laboratory (1962–1964), as well as later on when he searched for isoprenoid and latex producing plants to replace gasoline by plant isoprenoids. I would also like to acknowledge the open and stimulating discussions with Professor Trevor W. Goodwin, Liverpool, on problems in carotenoid and plastidic isoprenoid biosynthesis over many years, starting in 1968 at the first International Congress on Photosynthesis in Freudenstadt, Germany. In addition, I would like to thank the organic chemists Professor Michel Rohmer and his group in Strasbourg for excellent cooperation (from 1993 to 1998) in the ^{13}C-labeling of plant isoprenoids and my elder brother, Professor Frieder Lichtenthaler, Darmstadt, and his group for synthesizing deuterium-labeled 1-deoxy-D-xylulose and measuring

NMR spectra that demonstrated its specific incorporation into the photosynthetic isoprenoids. I am thankful to Professor Paul Stumpf, Davis/California for inspiring me to write this discovery review on plant isoprenoid biosynthesis, and to Ms. Gabrielle Johnson for language assistance.

References

Arigoni, D., Sagner, S., Latzel, C., Eisenreich, W. and Bacher, A. (1997) Terpenoid biosynthesis from 1-deoxy-D-xylulose in higher plants by intra molecular skeletel rearrangement. *Proc. Natl. Acad. Sci. USA* **94**: 10600-10605.

Bach, T. J. (1995) Some aspects of isoprenoid biosynthesis in plants, a review. Lipids **30**: 191-202.

Bach, T. J., Lichtenthaler, H. K. and Retey, J. (1976) Occurrence of 3-hydroxy-3-methylglutaryl-coenzyme A (HMG-CoA) in higher plants. Abstract International Plant Lipid Symposium, p. 96. Botany II, Karlsruhe.

Bach, T. J. and Lichtenthaler, H. K. (1982a) Mevinolin, a highly specific inhibitor of microsomal 3-hydroxy-3-methyl-glutaryl-coenzyme A reductase of radish plants. *Z. Naturforsch.* **37c**: 46-50.

Bach, T. J. and Lichtenthaler, H. K. (1982b) Inhibition of mevalonate biosynthesis and plant growth by the fungal metabolite mevinolin. In *Biochemistry and Metabolism of Plant Lipids*, J. F. G. M. Wintermans and P. J. C. Kuiper, eds., pp. 515-521. Elsevier, Amsterdam.

Bach, T. J. and Lichtenthaler, H. K. (1983) Inhibition by mevinolin of plant growth, sterol formation and pigment accumulation. *Physiol. Plant.* **59**: 50-60.

Bassham, J. A. and Calvin, M. (1957) The path of carbon in photosynthesis. Prentice Hall, Englewood Cliffs, N. J.

Braithwaite, G. D. and Goodwin, T. W. (1960a) Studies on carotenogenesis. 27. Incorporation of [2-^{14}C]acetate, dl-[2-^{14}C]mevalonate and $^{14}CO_2$ into carrot-root preparations. *Biochem. J.* **76**: 194-197.

Braithwaite, G. D. and Goodwin, T. W. (1960b) Studies on carotenogenesis. 25. The incorporation of [1-^{14}C] acetate, [2-^{14}C]acetate and $^{14}CO_2$ into lycopene by tomato slices. *Biochem. J.* **76**: 1-5.

Broers, S. T. J. (1994) Ph.D. Thesis, Über die frühen Stufen der Biosynthese von Isoprenoiden in *E. coli*. Eidgenössische Technische Hochschule, Zürich.

Calvin, M. (1962) Der Weg des Kohlenstoffs in der Photosynthese. *Angew. Chemie* **74**: 165-175.

Chaykin, S., Law, J., Philipps, A. H. and Bloch, K. (1958) Phosphorylated intermediates in the synthesis of squalene. *Proceed. Natl. Acad. Sci., U.S.A.* **44**: 998-1004.

Cornforth, J. W., Gore, I. Y. and Popjak, G. (1957) Studies on the biosynthesis of cholesterol. 4. Degradation of rings C and D. *Biochem. J.* **65**: 94-109.

Crane, F. L. (1959) Internal distribution of coenzyme Q in higher plants. *Plant Physiol.* **34**: 128-131 (1959).

Crane, F. L., Ehrlich, B. and Kegel, L. P. (1960) Plastoquinone reduction in illuminated chloroplasts. *Biochem. Biophys. Res. Commun.* **3**: 37-40.

Disch, A. and Rohmer, M. (1998) On the absence of the glyceraldehyde 3-phosphate/pyruvate pathway for isoprenoid biosynthesis in fungi and yeasts. *FEMS Microbiol. Lett.* **168**: 201-208.

Disch, A., Schwender, J., Müller, C., Lichtenthaler, H. K. and Rohmer, M. (1998) Distribution of the mevalonate and glyceraldehyde phosphate/pyruvate pathways for isoprenoid biosynthesis in unicellular algae and the cyanobacterium *Synechocystis* PCC 6714. *Biochem. J.* **333**: 381-388.

Döll, M., Schindler, S., Lichtenthaler, H. K. and Bach, T. J. (1984) Differential inhibition by mevinolin of prenyllipid accumulation in cell suspension cultures of *Sylibum marianum* L. In *Structure, Function and Metabolism of Plant Lipids*. P. A. Siegenthaler and W. Eichenberger, eds., pp. 277-278. Elsevier, Dordrecht.

Eschenmoser, A., Ruzicka, L., Jeger, O. and Arigoni, D. (1955) Zur Kenntnis der Triterpene. Eine stereochemische Interpretation der biogenetischen Isoprenregel bei den Triterpenen. *Helv. Chimica Acta* **38**: 1890-1904.

Fischer, F. G., Märkl, G., Hönel, H. and Rüdiger, W. (1962) Einbau von Essigsäure und Mevalonsäure [2-14C] in Chlorophyll, Sterine und Carotinoide von Gerstenkeimlingen. *Liebigs. Ann. Chem.* **657**: 199-212.

Flesch, G. and Rohmer, M. (1988) Prokaryotic hopanoids: the biosynthesis of the bacteriopane skeleton. *Eur. J. Biochem.* **175**: 405-411.

Gershenzon, J. and Croteau, R. B. (1993) Terpenoid biosynthesis: the basic pathway and formation of monoterpenes, sesquiterpenes and diterpenes. In *Lipid Metabolism in Plants*. T. S. Moore, pp. 339-388. CRC Press, Boca Raton.

Goodwin, T. W. (1958a) Incorporation of $^{14}CO_2$, [2-^{14}C] acetate and [2-^{14}C] mevalonic acid into ß-carotene in etiolated maize seedlings. *Biochem. J.* **68**: 26-27.

Goodwin, T. W. (1958b) Studies in carotenogenesis 25. The incorporation of $^{14}CO_2$, [2-^{14}C] acetate and [2-^{14}C] mevalonic acid into ß-carotene by illuminated etiolated maize seedlings. *Biochem. J.* **70**: 612-617.

Goodwin, T. W. (1965) Regulation of terpenoid biosynthesis in higher plants. In: *Biosynthetic Pathways in Higher Plants*. J. B. Pridham and T. Swain, pp. 57-71. Academic Press, London.

Goodwin, T. W. (1977) The prenyllipids of the membranes of higher plants. In: *Lipids and Lipid Polymers in Higher Plants*. M. Tevini and H. K. Lichtenthaler, eds., pp. 29-47. Springer-Verlag, Berlin.

Goodwin, T. W. (1980) The biochemistry of carotenoids. Vol. 1, Plants. (Biosynthesis of carotenoids, p. 34-37). Chapman and Hall, London, New York.

Goodwin, T. W. (1981) Biosynthesis of plant sterols and other triterpenoids. In *Biosynthesis of Isoprenoid Compounds*, Vol. I. J. W. Porter and S. L. Spurgeon, eds., pp. 444-480. J. Wiley and Sons, New York.

Grob, E. C. and Büttler, R. (1954) Über die Biosynthese des ß-Carotins bei *Mucor hiemalis* Wehmer. *Experientia* **10**: 250-251.

Grob, E. C. and Bütler, R. (1955) Über die Biosynthese des ß-Carotins bei *Mucor hiemalis* Wehner. Die Beteiligung der Essigsäure am Aufbau der Carotinmolekel. Helv. Chim. Acta **38**: 1313-1316.

Heintze, A., Görlach, J, Leuschner, C., Hoppe, P., Hagelstein, P., Schulze-Siebert, D. and Schultz, G. (1990) Plastidic isoprenoid synthesis during chloroplast development. Change from metabolic autonomy to division-of-labor stage. *Plant Physiol.* **93**: 1121-1122.

Lichtenthaler, H. K. (1962) Vergleichende Bestimmungen der Vitamin K_1-Gehalte in Blättern. *Planta (Berl.)* **57**: 731-753.

Lichtenthaler, H. K. (1977) Regulation of prenylquinone synthesis in higher plants. In: *Lipid and Lipid Polymers in Higher plants*. M. Tevini and H. K. Lichtenthaler, eds., pp. 231-258. Springer-Verlag, Berlin.

Lichtenthaler, H. K. (1993) The plant prenyllipids including carotenoids, chlorophylls and prenylquinones. In *Lipid Metabolism in Plants*. T. S. Moore, ed., pp. 427-470. CRC Press, Boca Raton.

Lichtenthaler, H. K. (1998) The plants' 1-deoxy-D-xylulose-5-phosphate pathway for biosynthesis of isoprenoids. Fett/Lipid **100**: 128-138.

Lichtenthaler, H. K. (1999) The 1-deoxy-D-xylulose-5-phosphate pathway of isoprenoid biosynthesis in plants. *Ann. Rev. Plant Physiol. Plant Mol. Biol.* **50**: 47-65.

Lichtenthaler, H. K. and Calvin, M. (1964) Quinone and pigment composition of chloro-plasts and quantasome aggregates from *Spinacia oleracea*. *Biochim. Biophys. Acta* **79**: 30-40.

Lichtenthaler, H. K. and Calvin, M. (1964) Quinone and pigment composition of chloro-plasts and quantasome aggregates from *Spinacia oleracea. Biochim. Biophys. Acta* **79**: 30-40.

Lichtenthaler, H. K. and Park, R. B. (1963) Chemical composition of chloroplast lamellae from spinach. *Nature* **198**: 1070-1072.

Lichtenthaler, H. K., Bach, T. J. and Wellburn, A. R. (1982) Cytoplasmic and plastidic isoprenoid compounds of oat seedlings and their distinct labeling from ^{14}C-mevalonate. In *Biochemistry and Metabolism of Plant Lipids.* J. F. G. M. Wintermans and P. Kuiper, eds., pp. 489-500. Elsevier, Amsterdam.

Lichtenthaler, H. K., Schwender, J., Seemann and Rohmer, M. (1995) Carotenoid biosynthesis in green algae proceeds via a novel biosynthetic pathway. In *Photosynthesis: from Light to Biosphere.* P. Mathis, ed., pp. 115-118. Kluwer Academic Publishers, Amsterdam.

Lichtenthaler, H. K., Rohmer, M. and Schwender, J. (1997a) Two independent biochemical pathways for isopentenyl diphosphate and isoprenoid biosynthesis in higher plants. *Physiol. Plant.* **101**: 643-652.

Lichtenthaler, H. K., Schwender, J., Disch, A. and Rohmer, M. (1997b) Biosynthesis of isoprenoids in higher plant chloroplasts proceeds via a mevalonate independent pathway. *FEBS Lett.* **400**: 271-274.

Lichtenthaler, H. K., Schwender, J. and Müller, C. (1998) The 1-deoxy-D-xylulose-5-phosphate pathway for biosynthesis of carotenoids and other plastidic isoprenoids. In *Photosynthesis: Mechanisms and Effects. Proceedings of the XIth International Congress on Photosynthesis,* Vol. IV, G. Garab, ed., pp. 3215-3220. Kluwer Academic Publishers, Dordrecht.

Little, H. N. and Bloch, K. (1950) Studies on the utilization of acetic acid for the biological synthesis of cholesterol. *J. Biol. Chem.* **138**: 33-46.

Lotspeich, F. J., Krause, R. F., Lilly, V. G. and Barnett, H. L. (1959) The degradation of labeled ß-carotene. *J. Biol. Chem.* **234**: 3109-3110.

Lotspeich, F. J., Krause, R. F., Lilly, V. G. and Barnett, H. L. (1961) The degradation of labeled ß-carotene. *Fed. Proc.* **20**: 269.

Lütke-Brinkhaus, F. and Kleinig, H. (1987) Formation of isopentenyl diphosphate via mevalonate does not occur within etioplasts or etiochloroplasts of mustard (*Sinapis alba* L.) seedlings. *Planta* **171**: 401-406.

Lynen, F., Reichart, E. and Rueff, L. (1951) Zum biologischen Abbau der Essigsäure VI. "Aktivierte Essigsäure", ihre Isolierung aus Hefe und ihre chemische Natur. *Liebigs Ann. Chem.* **574**: 1-32.

Lynen, F., Eggerer, H., Henning, U. and Kessel, I. (1958) Farnesyl-pyrophosphate und 3-Methyl-Δ^3-butenyl-1-pyrophosphate, die biologischen Vorstufen des Squalens. *Angew. Chemie* **70**: 738-742.

McGarvey, D. J. and Croteau, R. (1965) Terpenoid metabolism. The Plant Cell **7**: 1015-1026.

Park, R. B. and Bonner, J. (1958) Enzymatic synthesis of rubber from mevalonic acid. *J. Biol. Chem.* **233**: 340-343.

Putra, S. R., Lois, L. M., Campos, N., Boronat, A. and Rohmer, M. (1998) Incorporation of [2,3-$^{13}C_2$]- and [2,4-$^{13}C_2$]-D-deoxyxylulose into ubiquinone of Escherichia coli via the mevalonate-independent pathway for isoprenoid biosynthesis. *Tetrahedron Lett.* **39**: 23-26.

Rohmer, M., Knani, M., Simonin, P., Sutter, B. and Sahm, H. (1993) Isoprenoid biosynthesis in bacteria: a novel pathway for early steps leading to isopentenyl diphosphate. *Biochem. J.* **295**: 517-524.

Ruzicka, L. (1932) The life and works of Otto Wallach. *J. Chem. Soc.* **1932**: 1582-1597.

Ruzicka, L. (1938) Die Architektur der Polyterpene. *Angew. Chemie* **51**: 5-11.

Ruzicka, L., Eschenmoser, A. and Heusser, H. (1953) The isoprene rule and the biogenesis of terpenic compounds. *Experientia* **9**: 357-396.

Ruzicka, L., Meyer, J. and Mingazzini, J. (1922) Über die Naphthalinkohlenwasser-stoffe Cadalin und Eudalin, zwei aromatische Grundkörper der Sesquiterpenreihe. *Helv. Chimica Acta* **5**: 345-368.

Schwarz, M. K. (1994) Terpenbiosynthese in *Ginkgo biloba*. Ph.D. Thesis, Eidgenössische Technische Hochschule, Zürich.

Schwender, J., Lichtenthaler, H. K., Seemann, M. and Rohmer, M. (1995) Biosynthesis of isoprenoid chains of chlorophylls and plastoquinone in *Scenedesmus* by a novel pathway. In *Photosynthesis. from Light to Biosphere*. P. Mathis, ed., pp. 1001-1004. Kluwer Academic Publishers, Amsterdam.

Schwender, J., Zeidler, J., Gröner, R., Müller, C. Focke, M., Braun, S., Lichtenthaler, F. W. and Lichtenthaler, H. K. (1997) Incorporation of 1-deoxy-D-xylulose into isoprene and phytol by higher plants. *FEBS Lett.* **414**: 129-134.

Schwender, J., Müller, C., Zeidler, J. and Lichtenthaler, H. K. (1999) Cloning and heterologous expression of a cDNA encoding 1-deoxy-D-xylulose-5-phosphate reductoisomerase of *Arabidopsis thaliana. FEBS Lett.* **455**: 140-144.

Spurgeon, S. L. and Porter, J. W. (1981) Introduction. In *Biosynthesis of Isoprenoid Compounds*, Vol. 1. J. W. Porter and S. L. Spurgeon, eds., pp. 1-46. J. Wiley and Sons, New York.

Steele, J. W. and Gurin, S. (1960) Biosynthesis of ß-carotene in *Euglena gracilis. J. Biol. Chem.* **235**: 2778-2785.

Tavormina, P. A and Gibbs, M. A. (1956) The metabolism of γ, γ, dihydroxy-ß-methylvaleric acid by liver homogenates. *J. Am. Chem. Soc.* **78**: 6210.

Tavormina, P. A., Gibbs, M. H. and Huff, J. W. (1956) Utilization of ß-hydroxy-ß-methyl-Δ-valerolactone in cholesterol biosynthesis. *J. Am. Chem. Soc.* **78**: 4498-4499.

Threlfall, D. R. and Whistance, G. R. (1971) Biosynthesis of isoprenoid quinones and chromanols. In *Aspects of Terpenoid Chemistry and Biochemistry*. T. W. Goodwin, ed., pp. 357-404. Academic Press, London.

Wallach, O. (1885) Zur Kenntnis der Terpene und ätherischen Öle. *Liebigs Ann. Chem.* **227**: 277-302.

Wallach, O. (1887) Zur Kenntnis der Terpene und der ätherischen Oele. *Justus Liebigs Ann. Chem.* **239**: 1-54

Wieckowski, S. and Goodwin, T. W. (1967) Incorporation o DL-[2-^{14}C] mevalonic acid lactone into ß-carotene and the phytol side-chain of chlorophyll in cotyledons of four species of pine seedlings. *Biochem. J.* **105**: 89-92.

Wolf. D. E., Hoffmann, C. H. Aldrich, P. E., Skeggs, H. R., Wright, L. D. and Folkers, K. (1956) ß-Hydroxy-ß-methyl-delta-dihydroxy-ß-methylvaleric acid (divalonic acid), a new biological factor. *J. Am. Chem. Soc.* **78**: 4499.

Zeidler, J. G., Lichtenthaler, H. K., May, H. U. and Lichtenthaler, F. W. (1997) Is isoprene emitted by plants synthesized via the novel isopentenylpyrophosphate pathway? *Z. Naturforsch.* **52c**: 15-23.

Zeidler, J. G., Schwender, J., Müller, C., Wiesner, J., Weidemeyer, C., Beck, E., Jomaa, H. and Lichtenthaler, H. K. (1998) Inhibition of the non-mevalonate 1-deoxy-D-xylulose-5-phosphate pathway of plant isoprenoid biosynthesis by fosmidomycin. *Z. Naturforsch.* **53c**: 980-986.

Chapter 8

Structure and Biosynthesis of Cellulose
Part I: Structure

Alfred D. French

Cotton Fiber Quality Research Unit, Southern Regional Research Center
1100 Robt. E. Lee Blvd., New Orleans, Louisiana 70124, USA

ABSTRACT

Studies on both the structure and mechanism of synthesis of cellulose, the world's most abundant organic compound, have been challenging and often controversial. Cellulose got its name by being the 'sugar" of plant cell walls. It was first isolated and characterized as an aggregation of glucose units by Anselme Payen 160 years ago. Knowledge of its structure has developed over time, along with many of the concepts of carbohydrate and polymer chemistry. Some major landmarks include an appreciation for the complexity of the structures of cellulose in cell walls by Balls, and that cellulosic fibers are composed of particles or micelles that we now call microfibrils. On a molecular basis, the determination by Polanyi of the unit cell for the most prevalent native form and the subsequent refinements of the unit cell and the molecular structure by Meyer and colleagues were major findings. Their reports were interspersed with the realization that the glucose monomers were rings of six atoms instead of five. That development enabled correct interpretation of the methylation data regarding the β-1,4 linkages between glucose residues. It also allowed a revision of Tollens' earlier idea that cellulose was a long chain of covalently joined glucose units rather than an aggregation held together by secondary forces. Cellulose is polymorphic. The chains apparently pack in both parallel and antiparallel arrays, and different backbone shapes are found for different derivatives and complexes. Several different forms are found for native plant systems. Elucidation of the exact details of these different forms is a current research area. To date, the best results are available for crystalline cellodextrins which have permitted full-scale, single-crystal X-ray diffraction determinations.

Cellulose structure and chemistry have long been of substantial interest because cellulose is the most prevalent biomolecule, as well as an industrially important material. Cellulose has even been tentatively identified in outer space (Hoyle and Wickramasinghe, 1977). In his review,

Preston (1939) stated that "it is becoming increasingly evident that a knowledge of the detailed structure of the cell wall will eventually throw light on many problems in botany, and will prove to be of paramount importance to the developmental anatomist, to the cytologist and to the physiologist." Nägeli (1858) expressed similar views much earlier.

Many workers and many techniques have contributed information on cellulose structure, and it is impossible to cover all of this work here. Greater detail is available in a number of other reviews, as well as in the original literature. Heuser (1936) published a series of review chapters on the history of cellulose structure in the American Dyestuff Reporter with 180 references. Preston (1939) cited seven other reviews from the period 1930-1936. Marsh and Wood (1942) wrote an entire textbook on cellulose, with a chapter on structure. A particularly comprehensive review of cellulose chemistry was published in 1943 in Ott and Spurlin's book, "Cellulose and Cellulose Derivatives" and the revised edition (Ott, Spurlin and Grafflin, 1954). A follow-on book, edited by Bikales and Segal (1971) also contains chapters with extensive historical detail that relate to developments following the publication of the Ott and Spurlin volumes. Better coverage of the work in Russia is available in a book by Nikitin (1962, 1966). Shenouda (1979) reviewed the structure of cotton cellulose. Marchessault and Sundararajan (1983) wrote an extensive review on cellulose history, as well. Another historical perspective is contained in "Polymers — The Origins and Growth of a Science", by Morawetz (1985, 1995). This work recognizes that cellulose and another natural polymer, rubber, were responsible for many of the concepts of polymer chemistry. Recently, Krassig (1993) wrote a comprehensive book on cellulose.

Anselme Payen, the Father of Cellulose Chemistry

Cellulose chemistry began with the discovery of the sugar of cell walls, or cellulose, published by the French scientist, Anselme Payen (1838) whose life spanned the years 1795–1871 (Fig. 1). His research solved a mystery regarding the variable composition of cell walls in different plants. After purifying the cell walls of various plant tissues, Payen found that the remaining material (resistant to acid, ammonia, water, alcohol and ether), had the same elemental composition as starch. In the following year, a committee of the French Academy evaluated Payen's work and the word "cellulose" was first used. In that same year, Payen isolated a crude preparation of lignin (Payen, 1839). Payen was an excellent scientist and published more than 200 papers. His first work at age 20 was on the production of synthetic borax and he went on to develop bone char for use by the sugar industry. Before the age of 31, he had studied syrup, starch and alcohol, lead for use in paint, cold drawing of iron, potatoes and economic utilization of animal carcasses. At 40, he left industrial management and took an academic position. He earned numerous honors, including the Gold Medal of the French Agriculture Society (1826), Knight of the Legion of Honor (Sweden, 1828), Member of the Agricultural Society of France and Secretary for 26 years (beginning in 1833), and Commander of the Legion of Honor (1863). He was named the "Father of Cellulose Chemistry" (Reid and Dryden, 1940). The American Chemical Society's Division of Cellulose, Paper and Textiles honors him by awarding the Anselme Payen Medal annually to the best workers in cellulose or lignin chemistry. Other biographies of Payen are available, including those by Phillips (1961) and Wise (1963).

Fig. 1. Anselme Payen (1795–1871). From the Archives of the French Academy of Science, Institute of France.

   Payen's work was the first piece of the puzzle of cellulose structure. Although he understood that cellulose could be hydrolyzed to give glucose, the nature of the aggregation of these monosaccharide units was a deep mystery and the structure of glucose itself was not at all clear. Even today, there are still unanswered questions. For example, the hydrogen bonding systems and mechanisms of swelling are not known with any certainty. When it comes to a complete understanding of the actual mechanisms of biosynthesis and hydrolysis, nothing less than a complete knowledge of the x, y and z coordinates of each atom will permit a trustworthy picture.

Other Molecules Associated With Cellulose

The matter that Payen removed from wood in the purification treatment, which varied in quantity and composition depending on the source, was mostly lignin. Of course, we know now that many other components besides lignin are found in cell walls from various plants. Among them are pectins, proteins and some waxes. Hemicelluloses are polysaccharides that are often intimately associated with cellulose in plant cell walls. These molecules are chemically similar to cellulose, but incorporate monosaccharides other than glucose, such as xylose, arabinose or galactose, in both the backbone and in side chains.

Because of these accompanying other molecules, the term cellulose does not always have the same meaning. Originally it meant the carbohydrate material remaining after a series of physical and chemical treatments were carried out on cell walls. Even now that the various chemical constituents of the cell wall are fairly well-delineated, cellulose, in some industrial usages, refers to mixtures that include hemicelluloses or other substances. Alpha-cellulose connotes a material that fails to dissolve in 17.5% sodium hydroxide at 20 °C, and other names apply to chemically diverse industrial materials that have been through specified pulping procedures. Mature cotton fibers are often informally called cellulose in the textile industry, even though they may contain as much as 10% of molecules such as pectin, waxes and proteins on a dry weight basis. For the remainder of this chapter, however, only pure, β-1,4-glucan is considered to be cellulose. The impurities in many cellulose samples hindered research, especially early on.

How Is Cellulose Held Together?

In 1859, Frémy (1859) and Payen had a heated debate, with Payen arguing that cellulose, starch and dextrin, which appear to be so different, constitute the same substance in different states of aggregation. Frémy insisted that "differences in the properties of celluloses are not due to varying states of aggregation ... but to isomeric states of these substances."

Although Payen was correct about the identity of the empirical formulae for starch, dextrins and cellulose, and Frémy was correct about them being isomers, these views were not universally adopted in their lifetimes. Bevan and Cross (1880) and Cross and Bevan (1905–1910, 1916) believed that, in the plant, the cellulose, pectin, lignin, and fatty materials merged into one another by insensible chemical gradations. They called the original complexes "ligno-cellulose", "pecto- or muco-cellulose", and "adipo- or cuto-cellulose" according to the constituents preponderating in a given plant. "Cellulose" was merely the aggregate resulting from a chemical breakdown carried to the point where the reactions generally accepted for pure cellulose were obtained. This view, indefatigably upheld and amplified, had many adherents during the period 1880 to 1920. The glucose moieties were often said to be held together by "residual forces of affinity" or "secondary valencies".

Perhaps part of the reason that little progress in structural knowledge took place in this period was that carbohydrate chemistry was just then being put on a firm foundation by Emil Fischer and others. According to Freudenberg (1933), Fischer had even pronounced that starch and cellulose were not yet ripe for study of their constitution, and it was not until after his death in 1919 that there was a renewal of effort in this direction. Rebuttal to the above pre-Payen mindset of many workers in this era eventually came from physical measurements such as optical birefringence and X-ray diffraction, as well as from chemical analyses.

Rayon, A New Form Of Cellulose

At this same time in history, the treatment of cellulose by alkali and carbon disulfide was discovered (Cross *et al.*, 1894 and 1898; Cross and Bevan, 1901). This process yields a cellulose xanthate solution that can be extruded into a coagulating bath to make regenerated, chemically pure cellulose. Extruded filaments of regenerated cellulose from the xanthate process were called artificial silk and are now called rayon, a fiber that is used for textiles and tire cord. In the late 1800s, cellulose trinitrate had been developed as a plastic, but its explosive nature made its many products, ranging from false teeth to motion picture film, somewhat hazardous. This new process of xanthation assured that cellulose would be a major industrial feedstock for many years. It also added to the questions to be answered about cellulose structure. What are the reasons for the differences between the properties of rayon fibers and those of other cellulose fibers?

The Concept of A Chain Molecule

The idea that cellulose consisted of a number of glucose residues had existed from the time of Payen, but the first recorded articulation of the concept of a linear polymer apparently came almost 60 years after Payen's discovery. Tollens (1895) proposed that cellulose is a chain of glucose molecules (Fig. 2), with a "double linking" of the carbonyl oxygen on one glucose with the hydroxyl groups on carbons 5 and 6 of the neighboring glucose unit. This linkage accommodated the prevailing view that glucose had a five-membered ring. The proposal that cellulose is a polymer was accepted by some workers early on, and the first estimate of the molecular weight of acetylated cellulose (Nastukoff, 1900) corresponded to 40 glucose residues. Progress was made in 1901 when Skraup and König (1901) isolated cellobiose octaacetate from cellulose and concluded that "cellose" is the simplest polymer of cellulose, just as maltose is the simplest polymer of starch.

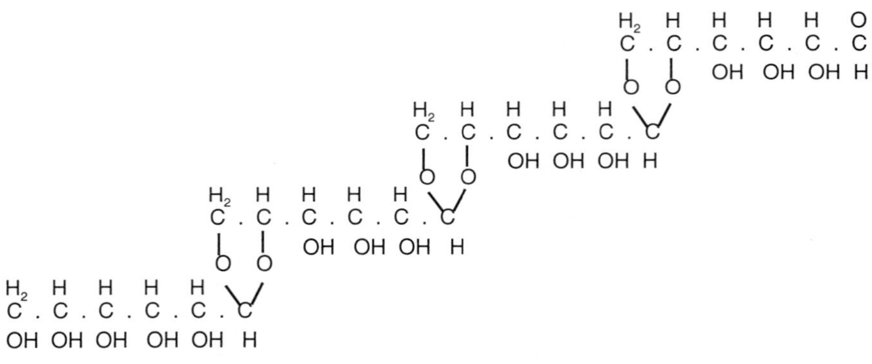

Fig. 2. Proposed structure for cellulose molecule (Tollens, 1895).

Oriented Micelles

Nishikawa and Ono (1913) produced the first known X-ray diffraction fiber diagram of cellulose (see Fig. 3 for a diffraction pattern by the author) and made a qualitative interpretation regarding the presence of oriented micelles parallel to the long axis of the fiber. Micelles are crystalline particles, typically with one long dimension compared to the other two. Ambronn

(1917) published the optical properties of stretched, denitrated cellulose nitrate film. He concluded that the birefringence could only be explained by the presence of rodlike optically anisotropic particles which tended to be oriented with their long axes parallel to the direction of stretch. Unaware of Nishikawa and Ono's work, he also proposed that it would be interesting to do X-ray diffraction studies.

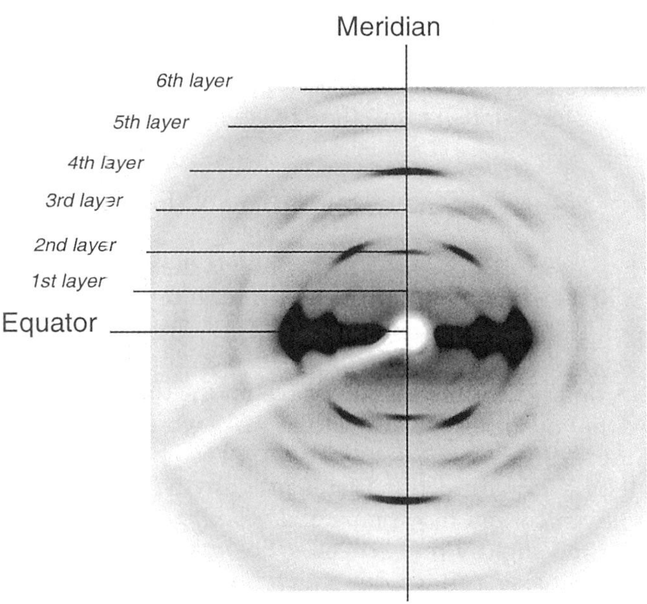

Fig. 3. X-ray diffraction pattern of ramie cellulose Iβ, showing the layer lines, the equator and the meridian. The layer lines are straight because a precession camera was used, whereas the diffraction patterns of early workers had curved layer lines because simple, flat-film cameras were used. The photograph is over-exposed. The fiber sample was parallel to the meridian. The light, diagonal marks on the lower left are from the beam stop holder.

Macrostructural Issues

One of the major problems in understanding cellulose structure and its relationships to function and properties comes from the many levels of structural organization, perhaps best exemplified by the cotton fiber. Before the concept of cellulose molecules was developed, it was realized that the fibers could be subdivided into smaller units called microfibrils. These microfibrils are in complex layers (Balls, 1919) that usually have their microfibrillar axes at an angle to the fiber axis. (In other fibers, such as flax or ramie, the microfibrillar axes are aligned with the fiber axis.)   The individual layers have different average orientations of the microfibrils, and there are reversals of the angle of orientation relative to the fiber axis at intervals along the fiber (Balls and Hancock, 1926; Anderson and Kerr, 1938). Additionally, there is a distinct primary wall (Tripp *et al.*, 1951) on the cotton fiber that has a lower cellulose content. When such a complex assembly is treated, for example, with alkali, all levels down to the individual crystalline unit cells are affected. However, the observed change in the properties may have little direct causation from the change in the crystal lattice dimensions, even though that is one of the most easily quantified changes.

Further Understanding Of The X-Ray Pattern

The unit cell is a conceptual object that contains the smallest set of structural components of a crystalline material that can be repeated to form an entire crystal by unit translations along the cell edges. In an attempt to determine the structure of a crystalline solid, the dimensions of this construct are determined from the positions of the spots on the diffraction pattern. Then, if possible, the positions of the atoms are worked out, based on the intensities of the different spots. Polanyi (1921a) determined the dimensions of a unit cell by X-ray diffraction that is roughly the same as is known today for cellulose from cotton, flax, ramie and wood: $a = 7.9$, $b = 8.45$, $c = 10.2$ Å (See Fig. 4 for the slightly revised cell). Polanyi's work was an impressive feat, for the cellulose diffraction pattern was his introduction to X-ray crystallography. He also worked out the full implications of fiber diffraction patterns, relating them to single crystal rotation diagrams (Polanyi, 1921b). A key result of this crystallography was that the size of unit cell was just big enough to contain four glucose residues.   His proposed cell was orthorhombic, with all angles between the axes of 90° while the current cell dimensions for cellulose I (see Fig. 4) correspond to a monoclinic unit cell, with an angle γ of about 96.5° between the a and b axes.   Other crystallographic conventions have been used over the years, with b being the fiber axis and the monoclinic angle being β.   Lately, the earlier convention has been preferred by those doing complete structure determination, while the convention with b as the fiber axis is often used during studies of the degree of crystallinity.

Polanyi proposed that the diffraction pattern was compatible with two alternatives: Either there were two disaccharide segments of long polysaccharide chains or there were two cyclic disaccharides. At that time, the structure commonly accepted by organic chemists for glucose had a five-membered ring, and this obscured the correct choice. The correct, six-membered ring structure would be needed for a successful interpretation. Substantial attention was given to the second alternative, and numerous speculations were published regarding cellulose structures that would fit entirely within the unit cell.

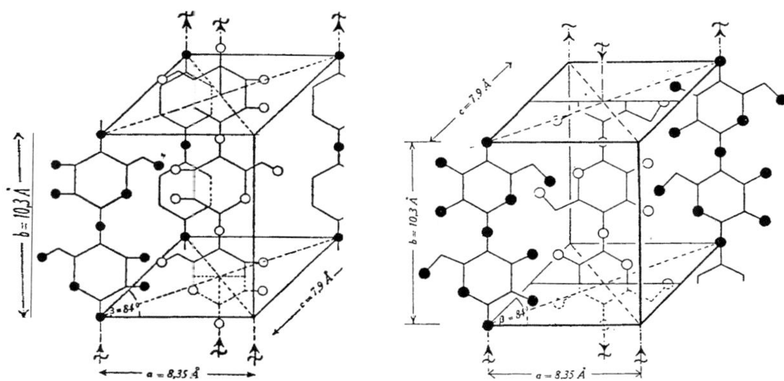

Fig. 4. The unit cell structures of cellulose I, according to Meyer and Mark (1928) (left) and Meyer and Misch (1937) (right). The dimensions of the unit cells are identical, but the chain packing in the left drawing is parallel whereas it is antiparallel in the right drawing. Some slight improvement in the stereochemistry occurred during the 1928–1937 interval. The curly line and arrow symbols represent the location of crystallographic screw axes. These 2_1 screw axes generate successive residues by the symmetry operators that state: for every point at x, y and z, there will be an identical point at $-x$, $-y$ and $z+1/2$.

Contributions of Hirst and his Colleagues, Especially Haworth

Rees and Skerret (1968) gave Sir Edmund Hirst and his colleagues much of the credit for the determination of cellulose structure. In the first of two papers cited by Rees and Skerret to support that claim, Irvine and Hirst (1923) declined to review the previous literature, because "Constitutional studies of this type are beset with unusual experimental difficulties and speculation is hazardous, but many of the investigators who have been attracted to this field have proposed structural formulae with a confidence which the experimental evidence and the present state of knowledge regarding the simple sugars do not justify. As a result, the extensive literature on the constitution of cellulose contains many conflicting statements and suggestions which make for confusion rather than enlightenment." Although this paper showed clearly that hydrolysis of trimethyl cellulose gave only the 2,3,6 methyl glucoside, as found earlier (Denham and Woodhouse, 1913), they still concluded that the cellobiose (and cellulose) linkages were between carbons 1 and 5.

By the time of the second paper cited (Haworth *et al.*, 1931) the 1,4 linkage of cellobiose was well established (Zemplén, 1926; Haworth *et al.*, 1927). Haworth (1925) realized that glucose has primarily a six-membered ring, similar to the pyran rings found in many natural products, and stated, "The elucidation of the ring structure of sugars gave a new impetus to constitutional study, and the allocation of the hexagon formula to glucose provided new interpretations of the experimental evidence bearing on the constitution of the polysaccharides."

The remaining question was whether secondary changes had occurred during acetolysis of cellobiose and some other, less recognizable disaccharide was contained in cellulose. Haworth, Hirst and Thomas's contribution in the 1931 paper was the isolation of hendecamethylcellotriose, with 11 methyl substituents, and tridecamethylcellotetraose, with 13 methyls. These oligomers give breakdown products consistent only with a 1,4 linkage. Further, the successive 1,4 linkages in these oligomers ruled out the remote possibility that 1,4 linkages alternated with some other, more easily hydrolyzed linkage.

Compatibility Of The Unit Cell With Polymeric Structures

Katz (Morawetz, 1995, p. 75-76) took on the proponents of the notion (correct in the absolute technical sense) that the molecule could not exceed the size of the crystallographic unit cell, based on his studies of rubber. In diffraction studies of long polymers, the ends are not observed for two reasons. They are a very small fraction of the total material, and the end-to-end distance is so large that it would not be observed in any conventional diffraction experiment. Instead, the regularly repeating matter within the chain gives rise to the observable diffraction pattern.

Sponsler and Dore (1926) first showed that the small unit cell of cellulose was compatible with a long chain of glucopyranosyl residues. They were able to rule out a model of cellulose based on five-membered rings and explained the unit cell dimensions based on the just-proposed, six-membered glucose ring. Strangely, their model was based on a head-to-head, tail-to-tail bonding pattern that gave alternating 1,1 and 4,4 linkages. Meyer and Mark (1928) showed that the then-new 1,4 linkage was completely consistent with the unit cell first determined by Polanyi. Freudenberg and Braun (1928), unaware of the crystallographic work, used methods of organic chemistry that same year to conclude that cellulose is not an association colloid, but instead is a long chain based on a backbone composed of cellobiose linkages.

How Are The Chains Packed In The Unit Cell?

In the same way that the idea that cellulose is a polymer required a determination of its molecular weight, knowledge that the unit cell contains two cellulose chains, each with two glucose residues, demanded a clarification of the mode of packing the chains in the unit cell. The long dimension of the unit cell, $c = \sim 10.34$ Å, is consistent with the length of a cellobiose molecule, and the c axis is parallel to the fiber axis in fibers such as flax and ramie. Also, the extremely intense diffraction spot on the equator at a spacing of about 3.9Å is compatible with the main planes of the glucose rings being spaced at half the distance between the b axes of the unit cell.

Put together, these findings gave nearly conclusive evidence that the chains were fully extended. Further, diffraction spots were observed on the meridian at the second and fourth layer lines but at best only traces of intensity are on the meridian on layer lines one and three (Fig. 3). That information is consistent with a two-fold screw axis, or at least an approximation thereto, which also leads to an extended-chain structure.

Remaining questions included the problem of what the difference is between center and corner chains that results in a two-chain unit cell. Meyer and Mark (1928) concluded that it would be difficult to conceive of a biosynthetic mechanism that would produce antiparallel chains. If antiparallel, the crystallites would have half of the reducing ends at one end of the crystal and half at the other end, so lengthening of the chain would require two different mechanisms. Therefore, they reasoned that the chains in native cellulose microcrystals would have to all be directed the same way. The difference between the chains would be a shift of the central chain along the c axis (Fig. 4). The high intensity of the fourth layer meridional reflection recommended a staggering, or shift, of one fourth of the distance along the c axis (or 2.57 Å) for the central chain relative to the chains at the four corners of the unit cell. However, Meyer (1937) and Meyer and Misch (1937) later proposed that since the cellulose chains could assemble into crystallites from solution, they must have alternating orientations in the crystalline state (Fig. 4). Even though these regenerated crystallites had the cellulose II structure (see below), the fact that they gave a diffraction pattern that is apparently identical to that of mercerized native cellulose led them to propose antiparallel structures for both polymorphs. Mercerization involves swelling, but not complete dissolution, of the fiber structure. Therefore, a link between chain orientation in cellulose I and in cellulose II seemed definite.   In the current view, their arguments for cellulose I structure are not as compelling as they seemed. For one thing, a number of natural polymers have antiparallel molecules in their native state.   For another, as we shall see below, cellulose I is currently thought to have a parallel orientation of the molecules within a given crystallite.   Finally, molecules can crystallize from solution with parallel orientations.

Cellulose Polymorphy

In the late 1920s and 1930s, the non-bonded interactions of cellulose with other molecules began to be understood. Also, various different crystalline forms (polymorphs or allomorphs) occurred after the complexing agent, such as sodium hydroxide or liquid ammonia, was removed (Fig. 5).

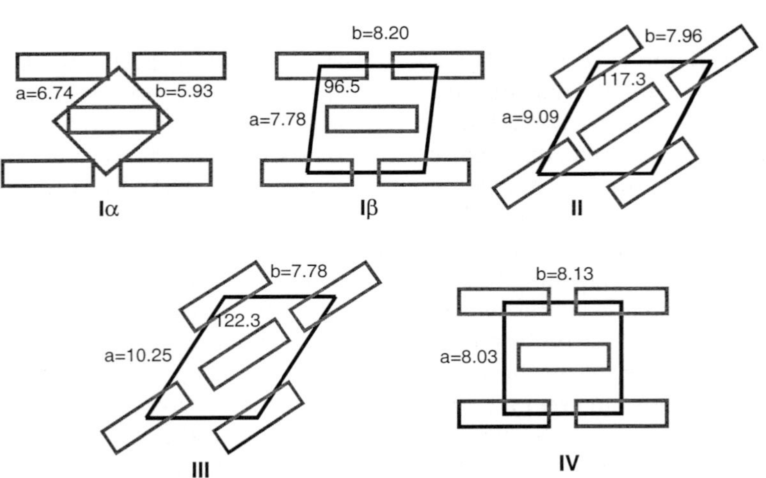

Fig. 5. The unit cells for the various crystalline forms (or polymorphs) of cellulose. The unit cell for cellulose Iα is triclinic with dimensions of a = 6.74 Å, b = 5.93 Å, c = 10.36 Å, α = 117°, β = 113° and γ = 81° (Henier, Sugiyama and Teleman, 1995). Dimensions for all of the monoclinic cells are indicated except for c, which is about 10.36 Å for all. Dimensions for cellulose Iβ were from Woodcock and Sarko (1980), for cellulose II from Stipanovic and Sarko (1976), for cellulose III from Sarko, Southwick and Hayashi (1976), and for cellulose IV from Gardiner and Sarko (1985). The rectangular boxes correspond to the end-on views of the cellulose chains (see Fig. 6). Except for the Iα structure, which has one chain per cell, all of the cells contain two chains (four glucose residues). Each of the four chains at the corners contributes, on-average, one-fourth of a cellobiose unit to the unit cell contents (the amount of the rectangle inside the cell), as well as the entire cellobiose unit positioned at the cell center.

These studies led to an understanding that the cellulose molecule could pack in different ways, and that the molecular structure, or conformation, varied to some degree in these different crystalline lattices.

In the 1840's to 1860's, Mercer treated cellulose with strong alkali and saw substantial differences in the properties of the treated cellulose, even though there was only swelling, and not dissolution, of the cellulose (see Parnell, 1886). Among the benefits of such alkali treatments, now known as "mercerization", were increased luster of the fibers and a better reception of dyes. Mercer proposed an intermediate complex with alkali, noted the shrinkage, and obtained a patent in 1850. However, the discovery of the process of mercerization has also been attributed to J. F. Persoz (Girard, 1881). Crum (1863) published micrographs, some even colored, showing the effects of sodium hydroxide solutions on cotton. The current industrial mercerization process uses lower strength sodium hydroxide solutions that have less effect on the cellulosic structure.

Mercer's work nicely set the stage for work by Andress (1929). He found distinct crystalline diffraction patterns for complexes of cellulose with alkali (see also Vieweg, 1924) in the pathway of conversion of native cellulose to the cellulose II structure. These studies with X-ray diffraction established that the cellulose underwent changes in crystal structure during the swelling with sodium hydroxide and subsequent de-swelling when rinsing with water. Perhaps the most interesting change was the formation of soda cellulose II. In this complex between cellulose and sodium hydroxide, the cellulose takes the shape of a three-fold helix instead of the normal form that more-or-less conforms to a two-fold screw-axis (Hess and Trogus, 1931a).

The Fundamental Repeating Unit

Kurt Hess and Carl Trogus were extremely prolific cellulose researchers. Among their many important contributions was the realization that the various derivatives and complexes of cellulose had crystallographic repeating distances that were multiples of 5.15Å, the factors being 2, 3, 4 and 5. This clarified that the glucose unit, rather than cellobiose, is the correct repeating unit of cellulose (Hess and Trogus, 1931b). This point has still not been universally accepted, and numerous recent papers contain a statement in their introductions that the cellobiose is the fundamental building block of the cellulose chain. This thinking comes in part from workers such as Karrer (1921) who considered anhydro-cellobiose to be repeated just twice in the cellulose molecule. Now the idea of a cellobiose repeating unit may be retained by some workers because chains in the cellulose I lattice have at least approximate two-fold symmetry, so every glucose residue is rotated roughly 180° from its predecessor. From the crystallographic point of view, however, the symmetry operator is applied to glucose itself, so glucose is the repeating unit. Other workers (Atalla, 1979) have concluded that cellobiose is the repeating unit precisely because of apparent deviations from perfect two-fold screw symmetry in the cellulose I lattice. Arguments presented by Atalla applied only to the cellulose I lattice, and it would be inappropriate to generalize to all forms of cellulose. From the biosynthetic standpoint, it is a controversial issue as to whether the chain is incremented by two residues at a time. Still, there is no evidence to suggest that alternating linkages in a free chain outside of a crystal lattice are systematically different. If hydrolysis products with even numbers of glucose residues had dominated the hydrolyzates, this would have constituted evidence for the oft-stated claim that cellobiose is the fundamental unit of the chain.

New Methods Reveal New Information

The developments of the late 1920's and early 1930's put the study of cellulose on a good foundation. Proposed improvements of structural models could be made without requiring as much speculation as had been frequent in the earlier 1920's. Still, many workers were uncomfortable regarding the concept of biopolymers. Smith (1937) wrote, "This is the so-called molecular chain of cellulose. It is the nearest approach to the usual concept of 'molecule' but cannot rightfully be so called because these chains are undoubtedly not of one definite, equal size, as are true molecules."

At the end of the 1930s, the shape of the glucose ring was verified with a single crystal study of glucosamine hydrobromide by Cox and Jeffrey (1939). This ring, like almost all glucose rings determined subsequently, has the $^4C_1$ shape, i.e., a Chair (chaise lounge) shape with carbon atom 4 high and carbon atom 1 low. The electron microscope was brought to bear on cellulose just at the start of the Second World War (Ruska and Kretschmer, 1940), and greater detail was obtained by Mühlethaler (1949). His improved detail was attributed to improved sample preparation. He was able to show the individual, very parallel fibrils in the secondary wall of ramie, and, in contrast, the loose network of the primary wall.

More On Polymorphy

When cellulose is dried after being swollen with amines or liquid ammonia, the cellulose III structure results (Bernady, 1925; Barry *et al.*, 1936). We now realize that the diffraction patterns of cellulose III differ somewhat, depending on whether the original substrate was cellulose I or II (Wellard, 1954). Therefore, the products are named III_I or III_{II}. The same applies to cellulose IV, which results from treating the starting material at high temperature (e.g., 240 °C in glycerol) (Hess, Kiessig and Gundermann, 1941). For both III_I and III_{II}, and for IV_I and IV_{II}, the unit cell dimensions and equatorial diffraction intensities are essentially the same but the non-equatorial spots have different intensities. The conversion from cellulose I to II, even in the solid state, is widely considered to be irreversible, despite reports to the contrary (Atalla and Nagel, 1974; Whitmore and Atalla, 1985). Also, cellulose II is considered to be the lowest energy form (Okamura, 1933; Rånby, 1952). Thus the cellulose I crystal structures are considered to be high-energy artifacts of the biosynthetic process. In the late 1970s, cellulose IV was found to be the native form of cellulose in the primary wall of cotton (Chanzy *et al.*, 1978) and in cultured rose cells (Chanzy *et al.*, 1979).

Around the end of the 1930s, a cellulose II pattern was observed for the native cellulose from *Halicystis* by Sisson (1938, 1941) and Sisson and Saner (1941). This observation further supported the view that the cellulose structure seen in most environments is a product of the particular developmental environment, rather than a structure that depends on inherent intramolecular characteristics.

After the Second World War, further distinctions among the various native celluloses were observed. For example, the infrared spectra for cellulose from cotton and ramie are similar to each other, but rather different from those of algal and bacterial cellulose (Marinnan and Mann, 1958). Insights into the hydrogen bonding that is widely thought to glue the solid state cellulose structures together were obtained with polarized IR spectra. This technique could discriminate among the limited number of possible hydrogen bonding arrangements, depending on which way the fiber was oriented relative to the polarized IR beam. (Liang and Marchessault, 1959a and 1959b; Marchessault and Liang, 1960). Another post-war development was the finding of the $^4C_1$ structure for glucose in solution (Reeves, 1949).

Hermans and Weidinger (1946) published a study of the hydration of cellulose. They showed that a different crystal structure (hydrate I) resulted upon the addition of about 10% water to a bone-dry, mercerized ramie sample. If cellulose crystallizes from an aqueous medium at low temperature, or if cellulose compounds containing water are decomposed, hydrate II is formed. The latter lattice is that of water cellulose (Sakurada and Okamura, 1937), which has essentially the same structure as soda cellulose IV.

Microfibril Structural Studies By Electron Microscopy

Frey-Wyssling (1955) worked out the orientational relationship of the unit cell to the edges of the microfibrils in cellulose I. The microfibrils were depicted as nearly square in cross-section, with the unit cell axes corresponding to the diagonals of the microfibrils. Frey-Wyssling suggested improvements in the positions of O6 to establish interchain hydrogen bonds parallel to the longer of the two equatorial axes, but all work was based on the "straight-chain" model of Meyer and Misch (1937) and not the more correct, stereochemically speaking, "bent-chain" model of Hermans (1949) (see Figs. 6 and 7 for modern models). As soon as the chair structure of the glucose ring had been determined, the straight-chain model could have been safely abandoned, but it persisted for some years after Hermans' book.

Mülethaler (1960) suggested that all cellulose microfibrils consist of smaller elements about 35 Å in width, corresponding to a 6 x 6 array of chains. These elementary fibrils, or protofibrils, would then aggregate to build up the microfibrils that range in size from 70 Å to 300 Å (Manley, 1971). This proposal is still controversial with regard to whether the large microfibrils of algal cellulose (with, for example, 25 x 25 chain arrays), for example, are composed of discrete smaller groupings of chains. However, a number of smaller, "sub-elementary" fibrils have been identified (Chanzy *et al.*, 1979).

Electron Diffraction

In 1958, Honjo and Watanabe (1958) applied electron diffraction to the highly crystalline microfibrils taken from the cell walls of the very large alga, *Valonia ventricosa*. They found a diffraction diagram more complex than had been previously reported, and suggested that a large unit cell was needed to explain all of the observed diffraction spots. They chose a cell that was double each of the equatorial dimensions of Meyer and Mark (1928). Its volume required eight cellobiose units to account for the observed density. In subsequent studies, this large cell was often acknowledged but it introduced unwelcome complication to structural studies, and its implications were usually ignored. It was just as well that the finding did not receive much attention, because it is now known that the extra spots are from a second crystalline phase (see below). Work by Hebert and Muller (1974) on cotton, ramie, bacterial and algal cellulose showed that the extra diffraction spots in electron diffraction patterns were visible for bacterial and algal celluloses, but not for cotton and ramie.

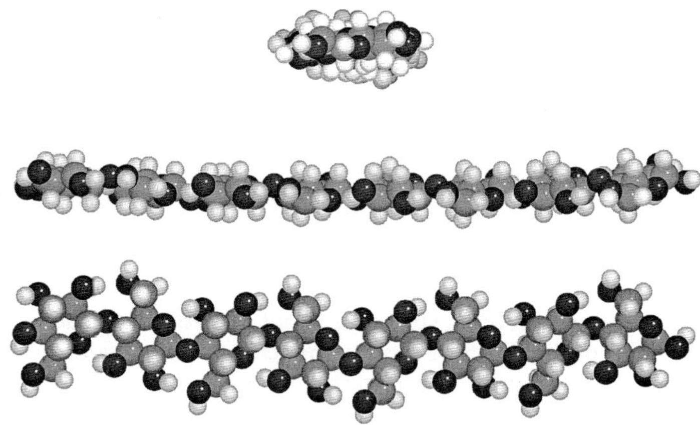

Fig. 6. Cellulose segment with eight gluocse residues, viewed (from the top down) end-on, edge-on and perpendicular to the flat, ribbon side, all in space-filling mode. This mode was energy-minimized with the molecular mechanics program MM3 and drawn with CHEM-X.

Fig. 7. Four-residue fragment of cellulose chain, with lables on atoms and dotted lines for hydrogen bonds. This basic structure is in agreement with the publications by Sarko and Muggli (1974), Gardner and Blackwell (1974), French (1978) and Woodcock and Sarko (1980), as well as the modeling studies of Aabloo *et al.* (1995), Heiner *et al.* (1995) and Kroon-Batenburg *et al.* (1996). The reducing end of the chain is to the right side. The O6 atoms are in tg positions.

Crystal Structure Studies Of Fibers And Related Small Molecules

Determination of the unit cell dimensions from the positions of the diffraction spots was an important achievement in the saga of cellulose structure. However, in most diffraction studies, that is just the first step. In principle, the quantitative positions of the atoms in a crystal can be determined from the intensities of the X-ray or electron fiber diffraction spots. Unlike the intensity data from crystals of small molecules, the evidence available from diffraction of cellulose fibers is very limited, and the problem is said to be "underdetermined". Thus, there are more variables in plausible crystal structures for cellulose than there are data to resolve them, and direct determination of its structure is not possible. Instead, a trial-and-error exploration of the possible arrangements is the most feasible approach. The different trial structures are computer models that include the x, y and z coordinates of the individual atoms. Such models are based on the idea that the glucose ring in cellulose is the same as that found in crystalline glucose and related molecules, for which very precise atomic coordinates started to become available in the late 1960's.

The First Computerized Fiber-Diffraction/Modeling Study

Mann, Roldan-Gonzalez and Wellard (1960) carefully collected X-ray intensity data from ramie cellulose. In one of the first studies to use computerized molecular modeling, Jones (1960) attempted to quantitatively reproduce those intensities by calculating them based on various trial atomic structures. However, the agreement was not good for any of the models that Jones tried, and he concluded that there were some basic problems with the proposed structure of extended cellulose chains in the literature.

Key Small Molecules

In the late 1950s and early 1960s, the advent of computers allowed small-molecule crystallographers to put away their adding machines (Jeffrey, working in England, had used business machines based on the monetary system at the time with 12 pence to the shilling and 20 shillings to the pound). Among the crystal structures determined at the dawn of this era was cellobiose (Jacobson *et al.*, 1961), establishing the O3'–O5 hydrogen bond (Fig. 7) and confirming the possibility of an approximate two-fold screw axis found by fiber diffraction for cellulose. That work was the beginning of many confirmations of the ubiquity of the $^4C_1$ shape of the glucose ring.   Improved cellobiose structures were published by Brown (1966) and by Chu and Jeffrey (1968).   Later Ham and Williams (1970) reported the crystal structure of methyl cellobioside, complexed with methanol. They inadvertently reported the coordinates for the mirror image of the molecule, causing problems for a number of workers trying to validate their modeling studies with crystallographic data.   Again, the O3'-O5 hydrogen bond was observed, despite somewhat different relative orientations of the two glucose rings.

The crystal structures of peracetylated cellobiose (Leung *et al.*, 1976) and cellotriose (Pérez and Brisse, 1977) were solved.   These molecules crystallized with orientations between the glucose rings that are in the same general range found for the cellobiose and cellobioside structures above. This was especially interesting because no conventional hydrogen bonding is possible for the acetylated structures, so it could be proposed that the geometry of the linkage between the glucose rings in cellobiose does not arise because of the formation of a hydrogen bond.   The observed space group of the trisaccharide, $P2_1$, requires a parallel packing mode for this molecule in a crystal grown from solution, refuting one of Meyer's assumptions (Meyer, 1937; Meyer and Misch, 1937).

Lots Of Fiber Diffraction Studies

The 1970s were a time of substantial growth in knowledge about cellulose structure.   The wider availability of faster computers encouraged more extensive analyses of the X-ray diffraction intensity data, based on proposed computer models of cellulose.   The X-ray intensities from *Valonia ventricosa* were collected by photodensitometry at two different laboratories and analyzed independently.   Gardner and Blackwell (1974) and Sarko and Muggli (1974) (and later, for ramie, Woodcock and Sarko, 1980) all proposed structures based on two parallel chains in the unit cell.   The reflections required only by the larger Honjo and Watanabe unit cell on the *Valonia* diagram were very weak, and it was therefore felt that they did not imply large differences from the proposed arrangement of chains in the smaller unit cell.

Gardner and Blackwell realized that there are actually two different ways that parallel chains can be packed in a monoclinic unit cell, with the reducing ends being all collected either at one end or the other.   In terms of the potential inter-chain interactions, these parallel-up and parallel-down structures are as different from each other as they are from antiparallel structures.   In the paper by Gardner and Blackwell and in the one by Woodcock and Sarko, the "most probable" structures were described as having parallel chains in the "up" direction. However, when differences in the conventions used for the unit cell axes were eliminated, it turned out that one of the two proposed structures was actually parallel-down (French and Howley, 1989). Work by French (1978) found that a reasonable accounting for the data for ramie of Mann, Roldan-Gonzalez and Wellard (1960) could have been because of the improved computer models. This solved the problem found by Jones (1960), but French found that an antiparallel model best fit the data of Mann *et al.* Thus all three of the possible packing models (up-, down- and anti-) were proposed, and it became clear that the results depended on the quality of the diffraction intensity data. Despite the variations in packing mode, all studies agreed on the basic chain conformation and hydrogen bonding system (Fig. 7).

Detailed structures were proposed for the various other polymorphs (Sarko and Muggli, 1974; Kolpak and Blackwell, 1976; Stipanovic and Sarko, 1976; Sarko, Southwick and Hayashi, 1977; Gardiner and Sarko, 1985; Kolpak *et al.*, 1978). The crystal structures of cellulose derivatives, too many to mention here, were also worked out with varying levels of detail. The underlying theme in these papers is that the chain packing in cellulose I is parallel, and the packing in III_I and IV_I is too. This common chain orientation permits the III_I and IV_I forms to be converted back to I. On the other hand, the packing in II, III_{II} and IV_{II} is antiparallel, accounting for the differences between the diffraction patterns of III_I and III_{II}, and between IV_I and IV_{II}, and for the ability of III_{II} and IV_{II} to be converted back to II. The chains are otherwise able to pack similarly, explaining why the differences in their diffraction patterns are small. Extensive studies by Pertsin *et al.* (1984, 1986) provided alternative perspectives based on the diffraction patterns of cellulose I and II.

Several structures of cellulose complexed with other molecules were worked out in this era, again based on combined model building and X-ray diffraction (Sarko *et al.*, 1987; Lee, Blackwell and Litt, 1983; Lee, Burnfield and Blackwell, 1983).

Chain Polarity And Mechanisms Of Polymorphic Conversion

The idea that the native cellulose microfibrils were composed of parallel chains became dominant, with the mercerized and regenerated (such as rayon) cellulose II structures having an antiparallel chain structure. Although fiber X-ray diffraction studies had failed to provide universally convincing evidence of a parallel structure for cellulose I, other observations with electron microscopy of different behavior of the two ends of cellulose microfibrils supported a parallel chain model. For example, Hieta, Kuga and Usuda (1984) were able to stain the microfibrils with silver, and most of the staining took place on only one end.   A similar observation of a "sharpening" of the microfibrils on only one end after action of an enzyme further supported this idea (Chanzy and Henrissat, 1985).   How then, could fully mercerized cellulose, also a cellulose II structure, be achieved from parallel cellulose I?   That was explained by noting that adjacent microfibrils, themselves composed of parallel chains, were packed antiparallel (Revol and Goring, 1983).   After the microfibrils swell, they were proposed to then merge, or "interdigitate" (Kolpak *et al.*, 1978), (Sarko, 1986).

Other proposals for the structural changes during the conversion of cellulose I to II include changes in the chain conformation.   Hayashi (Hayashi *et al.*, 1976) has championed a difference between "bent" and "bent and twisted" conformations of the backbone.   Another mechanism involves chain-folding.   Chanzy and Roche (1975) first reduced the extensive crystallinity of *Valonia* by treatment in a solution of acetic and perchloric acids and then mercerized the sample.   The resulting material had little disruption of its initial gross appearance despite developing a "shish kebab" structure with epitaxial growth of lamellar crystallites on the polymer backbone.   Another proposal was that the center chain shifts along the fiber axis during the conversion between cellulose I and II (Takai and Colvin, 1978).

A New Native Cellulose Polymorph

The new technique of Cross Polarization, Magic Angle Spinning Nuclear Magnetic Resonance (CP/MAS NMR) spectroscopy of solid state samples confirmed and expanded on some old observations regarding differences among cellulose from different sources.   Atalla and VanderHart (1984) and VanderHart and Atalla (1986) concluded from NMR data that different amounts of two forms of cellulose are present in different plants.   New names were coined to label the dominant form of cellulose in sources such as algae and bacteria (considered to be predominately Iα), and in wood, ramie, cotton, and flax (predominately Iβ).   None of the native celluloses was considered to be pure.   The various cellulose oligomers and cellulose II were also studied by CP/MAS NMR (Dudley *et al.*, 1983).

Sugiyama *et al.* (1991) studied cellulose from the alga *Microdictyon tenuius*.   They found that these microfibrils, even larger than the ones taken from the cell walls of *Valonia ventricosa*, had two different crystal structures.   This was a breakthrough.   Electron diffraction patterns from some regions of the microfibril could be indexed by a conventional, monoclinic two-chain cell similar to the Meyer and Misch (1937) cell, the Iβ form.   In other parts of the microfibril, the diffraction pattern was different, and a single-chain unit cell with triclinic symmetry sufficed. This was the Iα form. Diffraction patterns from still other parts of the same microfibril were the same as that of Honjo and Watanabe (1958) and found to represent mixtures of the Iα and Iβ fractions. Thus what was formerly thought to be a single (albeit very small) homogeneous crystal is actually heterogeneous and based on two different chain-packing arrangements. Most importantly, the single-chain unit cell proved that the chains are arranged in a parallel manner, at least in the Iα examples. Since the Iα and Iβ fractions are part of the same "crystallite", they are presumably synthesized by the same enzyme complexes and the chains in the Iβ fraction would have to be parallel, also.

Contributions from the Cellodextrins

Early on, it was observed that the cellodextrins (methyl cellotrioside and higher) gave powder diffraction patterns similar to patterns obtained from powdered cellulose II (Wellard, 1954). This similarity was the impetus for several attempts to determine the structures of crystals of these substances, starting with Poppleton and Mathieson (1968), but the small crystals and available technology were inadequate for a successful determination. Success was achieved only recently, with the reports of the crystal structures of methyl cellotrioside and cellotetraose (Raymond *et al.*, 1995a, 1995b; Gessler *et al.*, 1995). Their crystal structures clearly revealed antiparallel packing of the short chains, supporting the proposals for cellulose II from the fiber diffraction studies. However, the orientations of all of the primary alcohol groups[*] in cellotrioside and cellotetraose are "gt", different from the structures proposed earlier for cellulose II, giving a different hydrogen bonding pattern. The diffraction intensities from a previous study of cellulose II (Stipanovic and Sarko, 1976) were examined again. This time, having the primary alcohol groups of cellulose II in the same orientation as in the cellodextrins satisfied the diffraction intensities just as well as the previous, mixed "tg, gt" arrangement. Further, this proposed structure is compatible with the NMR data for cellulose II, unlike the previous result. In another report, basing a model of cellulose II on the crystal structure of cellotetraose, Raymond *et al.* (1995c) carefully determined the fiber axis spacing at 10.38 ± 0.03 Å and established the two-fold screw axis for

---

[*]   There are three staggered orientations of the primary alcohol group that could be found. The "tg" group, trans to O5 and gauche to C4 is proposed for cellulose I and shown in Figs. 6 and 7. The "gt" orientation is gauche to O5 and trans to C4. The third structure is "gg", gauche to both C4 and O5.

Fortisan cellulose II. Their R" factor for a model with all the primary alcohol groups in the gt position and the Kolpak-Blackwell (1976) data for cellulose II was similar to that of the cellulose II model by Gessler *et al.*

Leveling Off Degree Of Polymerization

Some key aspects of cellulose structure are not well worked out. For many years, workers have been aware of the "leveling off degree of polymerization" (LODP). Thus, when treated with acid, the native polymeric chains are quickly shortened to about 200 glucose residues, instead of the thousands of residues in the original condition. Regenerated cellulose II is shortened to about 55 residues (Shibazaki *et al.*, 1995). Subsequently, chain scission proceeds more slowly, and much smaller fragments result. This has been interpreted as an indication of the presence of some periodic "weak points" in the long chains. This thinking is consistent with the long-standing concept of "fringed micelles" (Nägeli, 1864). Fringed micelles are regions of well-organized three-dimensional structure that are much shorter than the molecular length. The molecules entering and leaving the micelles are the "fringes", and individual cellulose molecules visit several micelles. In between the crystalline micelles, the molecules would be amorphous and not protected by crystallinity from chemical attack. One problem with the fringed micelle concept is that the observed density of cellulose is very high (approaching 1.6 g/cm^3), and scarcely permits extensive deviation of the chain conformation from the very efficient packing found in the crystalline regions.

Another problem is that there is no direct evidence for fringed micelles in cellulose. Two methods with atomic resolution have been used to visualize the surfaces of cellulose crystallites, producing "lattice images". The first studies were carried out with electron microscopy (Kuga and Brown, 1987; Sugiyama and Okano, 1989). More recently, atomic force microscopy successfully produced images of the surfaces of cellulose crystallites (Hanley *et al.*, 1992; Kuutti *et al.*, 1995; Pesacreta *et al.*, 1997). All of these images portray native cellulose from ramie, *Valonia* and cotton as being composed of infinitely extended chains.

In particular, the cellulose microfibrils studied by lattice imaging offer no explanation for LODP. The amorphous content indicated by diffraction apparently arises from the small size of the elementary fibrils and irregular orientation of the hydroxyl groups on the surfaces. Another likely cause of the amorphous background on diffraction patterns is the distortion of microfibrils when they are in fibers. This disrupts the periodic array of atoms needed to produce sharp diffraction spots. Blackwell and Kolpak

(1975, 1976) showed that the diffraction pattern could be influenced by spacings of about 1.0 Å between the elementary fibrils that compose the microfibrils.   These spacings would correspond to occasional water molecules adhering to the surfaces of the elementary fibrils.

At one time, weak bonds in "defective" cellulose were proposed as another explanation for the LODP (Schulz, 1948).   Perhaps an ester linkage could occur instead of an ether linkage every 500 residues or so. Substantial effort has been spent trying to prove their existence, without any clear evidence in their favor (Heuser, 1952).

Folding Of Cellulose Chains

One other explanation for the leveling-off degree of polymerization is that cellulose chains might be folded, like in crystals of a number of synthetic polymers (Manley, 1964; Manley, 1971; Chang, 1974).   In the world of synthetic polymers, such as polyethylene, for example, the chains can fold into crystallites that are only about 100 Å thick, with greater length and width.   The surfaces where cellulose folds would be especially vulnerable to chemical attack.   Lamellar crystallites for cellulose triacetate II are very similar to those of the synthetics, and are almost certain to involve extensive chain folding (Rånby and Noe, 1961; Manley, 1963).   In those crystals, the distance between folds was about 180 Å.   Crystals of short chain cellulose similar to the synthetic lamellae have been observed, crystallized in the cellulose II lattice in shish-kebab fashion on microfibrils of cellulose I (Buléon *et al.*, 1977).   Similar crystals, with and without cellulose I or II skewers, were grown with a cellulose IV_{II} lattice (Buléon and Chanzy, 1980), but they probably do not involve folding because of their short lengths. When *Valonia* was mercerized (Chanzy and Roche, 1975), however, this was apparently not the case.   The folding of chains during mercerization is a means by which a parallel-to-antiparallel conversion could occur, and might account for the shrinkage first observed by Mercer.

Another example of folded chains was furnished by crystallization of cellulose from hydrazine solution (Atkins *et al.*, 1979).   These crystals were very similar to crystals grown by Buléon and Chanzy (1978), in that both had the cellulose II structure, and both had the chain axes perpendicular to the needle or ribbon axis of the crystal.   The method used in the latter case was based on dissolving cellulose triacetate with subsequent saponification and precipitation which had evolved from the work of Rånby and Noe (1961) and that of Ramesh *et al.* (1973).

When the normal environment surrounding *Acetobacter xylinium* is disrupted with the dye Tinopal, the chains form a cellulose II lattice, with electron microscopy suggesting that the chains are folded. However, under normal conditions the chains from *Acetobacter* are apparently extended, without any chain-folding (Cousins and Brown, 1995). A mutant strain of *Acetobacter xylinium* was also found to produce folded chain cellulose II (Kuga *et. al.*, 1993). Somewhat similar circumstances (with otherwise normal post-synthesis crystallization being disrupted), are thought to result in the formation of cellulose II by *Halicystis* (Sisson, 1938; Sisson and Saner, 1941).

Computer Modeling Studies

<u>Disaccharide linkage geometries</u>. Besides the experimental studies, purely computerized experiments have been employed to study cellulose structure. A favorite subject has been the prediction of the preferred conformation of the constituent disaccharide linkages. As far as we know, the first computerized study of the linkage conformation in cellobiose, along with cellulose, was carried out by Rees and Skerret (1968). In these analyses, the two glucose residues are rotated about their bonds to the glycosidic oxygen atom, and the potential energy is calculated at each increment of rotation. Initially, the calculations involved rigid individual glucose residues and glycosidic valence bond angles. Two other notable efforts of this generation were by Yathindra and Rao (1970), and Lipkind *et al.* (1985). The Russian workers used computer modeling to help interpret their NMR data.

<u>Relaxed-residue studies</u>. Currently, such calculations allow variation in all atomic coordinates so that a minimum in the energy can be reached for each increment of change (French, 1989; Dowd *et al.*, 1992; French and Dowd, 1993). A fair number of other workers have produced energy surfaces for cellobiose. Results differ somewhat, depending on the method used to calculate the energy.

One problem with the current models for cellulose I is that the NMR evidence suggests that there is some difference in the conformations of alternating linkages (Atalla, 1979). As a first approximation in the X-ray studies, the cellulose I chain has been assigned two-fold screw axis symmetry. The fiber diffraction patterns from which that conclusion was reached are not of sufficient resolution to provide a definitive determination of symmetry, and traces of the odd-order meridional diffraction spots that would contradict the two-fold screw axis are often visible. Early studies of the energy of the cellobiose linkage also suggested that the two-fold screw symmetry has an elevated (1 to 2 kcal/mol) energy, but more recent work (French and Dowd, 1993a) suggests that the two-fold screw-axis symmetry operator increases the energy by less than one kcal/mol.

<u>Ring shape calculations.</u> Another subject for calculation has been the conformation of the glucose ring. Although observed experimentally in the $^4C_1$ shape, it is useful to know how strong the energetic preferences for that shape are, so that the fraction of molecules having other ring shapes in non-crystalline environments, such as solution, can be predicted. Some of the first predictions of energies were based on simple rules (Angyal, 1963). Joshi and Rao (1979) undertook a more sophisticated analysis that also showed that the $^4C_1$ shape was preferred by a margin that would keep the contribution of other conformations to less than 0.01%.

<u>Chain-folding calculations.</u> One use for such calculations is the further examination of possible models for chain-folding. Drastic changes in the chain direction can be achieved either by large rotations about either of the bonds to the linkage oxygen atoms, or by changes from the $^4C_1$ conformation of the glucose ring into a skew form, or possibly the other chair, $^1C_4$. Some high-level quantum mechanics studies of the two chairs (Barrows *et al.*, 1995) gave a difference in their vacuum phase free energies of about 8.5 kcal/mol. This indicates that the $^1C_4$ structure would be quite improbable. A comprehensive study of all different shapes of the glucose ring indicated that the lowest-energy skew form was also much higher in energy than the always-observed $^4C_1$, based on a molecular mechanics method (Dowd *et al.*, 1994). Still, the solution properties of lightly derivatized cellulose chains are proposed to be best explained with the presence of a few non-$^4C_1$ forms (Brant and Christ, 1990).

If, because of high calculated energies, modified shapes of the glucose ring are not likely to be involved in folding, what about rotations about the linkages between the rings? If three successive linkages between $^4C_1$ rings are set into appropriate conformations, a complete reversal of direction can be accomplished. Rigid-residue calculations (Simon *et al.*, 1988) had shown an increase of 18 or 33 kcal/mol for similar shapes. Such high energies indicated that folding was unlikely. By our (French, 1989; French and Dowd, 1993a and 1993b) flexible-residue calculations, such conformations would each have an increase of about 2–3 kcal/mol over the normal, extended shape, within the range of distortions that are observed in actual crystal structures. Calculated barriers to the chain-folding conformations with flexible-residue methods are similar to those expected for polyethylene, despite earlier statements that cellulose is stiffer than synthetic polymers.

<u>Crystal lattice calculations.</u> Although the majority of the fiber diffraction studies had used crystal packing energy calculations as part of the determination, the algorithms were fairly unsophisticated compared to modern energy minimization methods. A way to check the validity of the structures from the fiber diffraction studies is to construct model miniature

crystals based on the fiber diffraction results and minimize their energy. High final energies and excessive movement during minimization would suggest that the structures proposed from fiber diffraction studies might not be entirely correct. By this method, the proposed structures for cellulose IV were found to be questionable (French *et al.*, 1993). In that paper, the model for cellulose Iβ was in error (Aabloo *et al.*, 1995).

The cellulose Iα unit cell is an ideal candidate for structural study with modeling because there is only a single chain. A fairly thorough study produced agreement between two independent modeling methods, with the chain being directed up in the cell and O6 in the "forbidden" tg position (Aabloo *et al.*, 1995). Heiner *et al.* (1995) carried out a molecular dynamics simulation of both of the cellulose I unit cells. Their "trajectory" showed a substantial amount of motion of the hydroxyl groups and primary alcohol group orientations over time. The basic structures of Iα and Iβ agreed with those of Aabloo, *et al.* Kroon-Batenburg *et al.* (1996) also studied crystal structures of cellulose polymorphs. Their results for Iα and Iβ also agreed with Aabloo *et al.*, but they proposed that mercerized cellulose might not undergo a transition from parallel to antiparallel packing. Instead the conversion from tg to gt orientations for the O6 group would result in a unit cell having dimensions similar to those of antiparallel regenerated cellulose II. A similar suggestion had been made earlier by Sakthivel *et al.* (1989).

Although numerous workers have stated that solid cellulose is mostly held together by hydrogen bonding, these and other calculations suggest that hydrogen bonding is actually a minor factor. Almost certain to exist in crystalline cellulose Iα and Iβ are intramolecular hydrogen bonds between O3' and O5. This leaves only the O6 and O2 hydroxyl groups as potential donors. With the current proposal of the O6 in the tg conformation, donating a proton to make a hydrogen bond to O2', only the O2 hydroxyl is free to donate to an inter-chain hydrogen bond. This single hydrogen bond provides only a small fraction of the lattice energy calculated for crystalline cellulose by any of several modeling systems. The rest comes from van der Waals (general dispersive) forces (French *et al.*, 1993).

Contributions From Enzyme Crystallography

One of the most important reasons for elucidating the structure of cellulose is to understand the mechanisms of enzymatic reactions. Ironically, some of the most direct evidence on cellulose chain conformation has recently been provided by crystal structures of complexes of enzymes with the cellulosic substrates. Davies *et al.* (1995) crystallized an endoglucanase from *Humicola insolens* with a cellohexaose fragment, and a structure of cellobiohydrolase I from *Trichoderma reesei* contains bound fragments of cellulose (Divne *et al.*, 1998). In this complex, the cellulose chain fragment has torsion angles at the linkages that are similar to other cellobiosyl linkages. The cellulose fragment in the cellobiohydrolase I enzyme complex has a gently curving shape.

Some Final Thoughts

The understanding of structure and biosynthesis of cellulose is unfinished. If there is any single lesson to be learnt from this historical review, it is that generalizations about cellulose structure are very hazardous beyond the (now-) simple chemical make-up of cellulose chains. Also, many assumptions have been made, and they are not always entirely correct. At times, a complete reversal of thought has been required, such as with the establishment of the pyranosic form of the glucose ring. Despite substantial interest, about half the time since cellulose was discovered passed before this critical fact was established. Cellulose in the native state has been identified to have the Iα, Iβ, II and IV$_1$ crystal structures. What was called cellulose I for many years was determined only recently to be a two-phase material, each with its own crystal lattice. While differences between the structures of cotton and algal celluloses had been indicated for many years by various experiments, the co-existence of two different lattices within the same apparently crystalline biosynthetic entity was a new development in science. We now have clear evidence that some cellulose is composed of extended chains, and equally clear evidence that some celluloses fold.

As this is written, further studies with lattice imaging are underway, using both electron microscopes or atomic force microscopy. Also, synchrotron radiation, with its extremely powerful X-ray beams, is being used to study the tiny crystals of longer cellodextrins or short cellulose molecules. More powerful NMR instruments will be applied to cellulose, and computer modeling methods are increasing in sophistication and speed at almost exponential rates. It has been 160 years since this important material was first identified. With any luck, it will be less than 160 years more before all of its major mysteries are explained.

With respect to biosynthesis, after many years of struggle, the field is now advancing at an extremely rapid rate. The identification of genes in both plants and bacteria that encode subunits of the synthase complexes has opened the way for many more sophisticated studies on how complexes are synthesized, how they assemble, and how subunits interact with each other and with other components within the cell. Our increasing ability to synthesize cellulose in isolated complexes will also undoubtedly lead to a more detailed understanding of the mechanism of polymerization and crystallization.

Studies on both structure and synthesis of cellulose have been among the most difficult and controversial areas of polysaccharide biology. Many people with strong personalities and strong opinions were involved — probably because in a field so frustrating, it took strong personalities to stick with the topic! We also re-emphasize that no single breakthrough led to the great insights we now have, and it is instructive to see how the field has progressed mostly by slow, incremental advances.

References

Aabloo, A., French, A. D., Mikelsaar, R.-H. and Pertsin, A. J. (1995) Studies of crystalline native celluloses using potential energy calculations. *Cellulose* 1: 161-168.

Ambronn, H. (1917) Ueber das Zusammenwirken von Stäbchendoppelbrechung und Eigen doppelbrechung, III. *Kolloid Z.* 20: 173-185.

Anderson, D. B. and Kerr, T. (1938) Growth and structure of cotton fiber. *Indust. and Eng. Chem.* 30: 48-54.

Andress, K. R. (1929) The X-ray diagram of mercerized cellulose. *Z. Physik. Chem.* B4: 190-206.

Angyal, S. J. (1963) The composition and conformation of sugars in solution. *Angew. Chem. Int. Ed. Eng.* 8: 157-166.

Atalla, R. H. (1979) Conformational effects in the hydrolysis of cellulose. *Advances in Chem.* 181: 55-69.

Atalla, R. H. and Nagel, S. C. (1974) Cellulose: its regeneration in the native lattice. *Science* 185: 522-523.

Atalla, R. H. and VanderHart, D. L. (1984) Native cellulose: a composite of two distinct crystalline forms. *Science* 223: 283-285.

Atkins, E. D. T., Blackwell, J. and Litt, M. H. (1979) Texture of cellulose crystallized from hydrazine. *Polymer* 20: 145-147.

Balls, W. L. (1919) The existence of daily growth rings in the cell wall of cotton hairs. *Proc. Roy. Soc.* (London) B90: 542-555.

Balls, W. L. and Hancock, H. A. (1926) Measurements of the reversing spiral in cotton hairs. *Proc. Roy. Soc.* (London) B99: 130-147.

Barrows, S. E., Dulles, F. J., Cramer, C. J., French, A. D. and Truhlar, D. G. (1995) Relative stability of alternative chair forms and hydroxymethyl conformations of β-D-glucopyranose. *Carbohydr. Res.* 276: 219-251.

Barry, A. J., Peterson, F. C. and King, A. J. (1936) X-ray studies of reactions of cellulose in non-aqueous systems. I. Interaction of cellulose and liquid ammonia. *J. Am. Chem. Soc.* 58: 333-337.

Bernardy, G. (1925) The action of ammonia on cotton cellulose. *Z. Angew. Chem.* 38: 1195-1197.

Bevan, E. J. and Cross, C. F. (1880) Chemistry of bast fibers. *J. Chem. Soc.* 38: 666-668.

Bikales, N. M. and Segal, L. (1971) *Cellulose and cellulose derivatives.* IV. Wiley Interscience, New York.

Blackwell, J. and Kolpak, F. J. (1975) The cellulose microfibril as an imperfect array of elementary fibrils. *Macromolecules* 8: 322-326.

Blackwell, J. and Kolpak, F. J. (1976) Cellulose microfibrils as disordered arrays of elementary fibrils. *J. Appl. Polymer Symp.* 28: 751-761.

Brant, D. A. and Christ, M. D. (1990) Realistic conformational modeling of carbohydrates. *ACS Symp. Ser.* **430**: 42-68.

Brown, R. M., Jr. (1996) The biosynthesis of cellulose. *J. M. S. – Pure Appl. Chem.* **A33**: 1345-1373.

Brown, R. M., Jr. and Montezinos, D. (1976) Cellulose microfibrils: visualization of biosynthetic and orienting complexes in association with the plasma membrane. *Proc. Natl. Acad. Sci. USA* **73**: 143-147.

Buléon, A. and Chanzy, H. (1978) Single crystals of cellulose II. *J. Polym. Sci.: Polym. Phys. Ed.* **16**: 833-839.

Buléon, A. and Chanzy, H. (1980) Single Crystals of cellulose IV_{II}: preparation and properties. *J. Polym. Sci.: Polym. Phys. Ed.* **18**: 1209-1217.

Buléon, A., Chanzy, H. and Roche, E. (1977) Shish kebab-like structures of cellulose. *J. Polym. Sci. Polym. Letts. Ed.* **15**: 265-270.

Chang, M. M. Y. (1974) Crystallite structure of cellulose. *J. Polym. Sci.: Polym. Chem. Ed.* **12**: 1349-1374.

Chanzy, H. and Henrissat, B. (1985) Unidirectional degradation of Valonia cellulose microcrystals subjected to celluase action. *FEBS Letters* **184**: 285-288.

Chanzy, H. D. and Roche, E. J. (1975) Fibrous Mercerization of Valonia Cellulose. *J. Polym. Sci.: Polym. Phys. Ed.* **13**: 1859-1862.

Chanzy, H., Imada, K. and Vuong, R. (1978) Electron diffraction from the primary wall of cotton fibers. *Protoplasma* **94**: 299-306.

Chanzy, H., Imada, K., Mollard, A., Vuong, R. and Barnoud, F. (1979) Crystallographic aspects of sub-elementary cellulose fibrils occurring in the wall of rose cells cultured *in vitro*. *Protoplasma* **100**: 303-316.

Chu, S. S. C. and Jeffrey, G. A. (1968) Refinement of the crystal structures of β-D-glucose and cellobiose. *Acta Crystallogr. B.* **24**: 830-838.

Cousins, S. K. and Brown, R. M., Jr. (1995) Cellulose I microfibril assembly: computational molecular mechanics energy analysis favors bonding by van der Waals forces as the initial step in crystallization. *Polymer* **36**: 3885-3888.

Cox, E. G. and Jeffrey, G. A. (1939) Crystal structure of glucosamine hydrobromide. *Nature* (London) **143**: 894-895.

Cross, C. F. and Bevan, E. J. (1901) Ueber die Cellulose-Xanthogensäure. *Ber.* **34**: 1513-1520.

Cross, C. F. and Bevan, E. J. (1905-1910) Researches on cellulose, Longman, Green & Co. London.

Cross, C. F. and Bevan, E. J. (1916) *Cellulose*, 3rd edition. Longman, Green & Co. New York.

Cross, C. F., Bevan, E. J. and Beadle, C. (1898) Thiokohlensäureester der Cellulose. *Ber.* **26**: 1090-1097.

Cross, C. F., Bevan, E. J. and Beadle, C. (1894) U. S. Pat. 520, 770.

Crum, W. (1863) On the manner in which cotton unites with colouring matter. *J. Chem. Soc.* **16**: 404-414 plus 8 plates.

Davies, G. J., Tolley, S. P., Henrissat, B., Hjort, C. and Schülein, M. (1995) Structures of oligosaccharide-bound forms of the endoglucanase V from *Humicola insolens* at 1.9Å resolution. *Biochemistry* **34**: 16210-16220.

Denham, W. S. and Woodhouse, H. (1913) The methylation of cellulose. *J. Chem. Soc.* **103**: 1735-1742.

Divne, C., Ståhlberg, J., Teeri, T. T. and Jones, T. A. (1998) High-resolution crystal structures reveal how a cellulose chain is bound in the 50 Ångstrom long tunnel of cellobiohydrolase I from *Trichoderma reesei*. *J. Mol. Biol.* **275**: 309-325.

Dowd, M. K., French, A. D. and Reilly, P. J. (1992) Conformational analysis of the anomeric forms of sophorose, laminarabiose and cellobiose using MM3. *Carbohydr. Res.* **233**: 15-34.

Dowd, M. K., French, A. D. and Reilly, P. J. (1994) Modeling of aldopyranosyl ring puckering with MM3(92). *Carbohydr. Res.* **264**: 1-19.

Dudley, R. L., Fyfe, C. A., Stephanson, P. J., Deslandes, Y., Hamer, G. K. and Marchessault, R. H. (1983) High-resolution ^{13}C CP/MAS NMR spectra of solid cellulose oligomers and the structure of cellulose II. *J. Am. Chem. Soc.* **105**: 2469-2472.

Frémy, E. (1859) Remarques à l'occasion d'une communication de M. Payen sur les tissues des vegetaux. *Compt. Rend.* **48**: 325-360.

French, A. D. (1978) The crystal structure of native ramie cellulose. *Carbohydr. Res.* **61**: 67-80.

French, A. D. (1989) Computer models of cellulose. *Cellulose and Wood - Chemistry and Technology*. Schuerch, C.,ed., John Wiley and Sons., Inc., New York, pp. 103-118.

French, A. D. and Dowd, M. K. (1993a) Conformational analysis of cellobiose with MM3, in "Biosynthesis, structure and organization", Part 1, Chap. 8, 51-56.

French, A. D. and Dowd, M. K. (1993b) Explorations of disaccharide conformations by molecular mechanics. *J. Mol. Struct. (Theochem)* **286**: 183-201.

French, A. D. and Howley, P. S. (1989) Comparisons of structures proposed for cellulose, in *Cellulose and Wood - Chemistry and Technology*. Schuerch, C., ed., John Wiley and Sons., Inc. New York, pp. 159-167.

French, A. D., Miller, D. P. and Aabloo, A. (1993) Miniature crystal models of cellulose polymorphs and other carbohydrates. *Int. J. Biol Macromol.* **15**: 30-36.

Freudenberg, K. (1933) *Tannin, Cellulose, Lignin*. Springer, Berlin, p. 91.

Freudenberg, K. and Braun, E. (1928) Nachtrag zu der Mitteilung über Methylcellulose: Zugleich 6. Mitteilung über Lignin und Cellulose. *Ann.* **461**: 130-131.

Frey-Wyssling, A. (1955) On the crystal structure of cellulose I. *Biochim. Biophys. Acta* **18**: 166-168.

Gardiner, E. S. and Sarko, A. (1985) Packing analysis of carbohydrates and polysaccharides. 16. The crystal structures of celluloses IV$_I$ and IV$_{II}$. *Can. J. Chem.* **63**: 173-180.

Gardner, K. H. and Blackwell, J. (1974) The structure of native cellulose. *Biopolymers* **13**: 1975-2001.

Gessler, K., Krauss, N., Steiner, T., Betzel, C., Sarko, A. and Saenger, W. (1995) β-D-Cellotetraose hemihydrate as a structural model for cellulose II. An X-ray diffraction study. *J. Am. Chem. Soc.* **117**: 11397-11406.

Girard, A. (1881) Mémoire sur l'hydrocellulose et ses dérivés. *Ann. Chim. Phys.* **24**: 337.

Ham, J. T. and Williams, D. G. (1970) The crystal and molecular structure of methyl β–cellobioside-methanol. *Acta Crystallogr. Sect. B.* **37**: 1373-1383.

Hanley, S. J., Giasson, J., Revol, J.-F. and Gray, D. G. (1992) Atomic force microscopy of cellulose microfibrils: comparison with transmission electron microscopy. *Polymer* **33**: 4639-4642.

Haworth, W. N. (1925) Revision of the structural formula of dextrose. *Nature* **116**: 430.

Haworth, W. N., Long, C. W. and Plant, J. H. G. (1927) Constitution of the disaccharides. Part XVI. Cellobiose. *J. Chem. Soc.* 2809-2814.

Haworth, W. N., Hirst, E. L. and Thomas, H. A. (1931) Polysaccharides. Part VII. Isolation of octamethyl cellobiose, hendecamethyl cellotriose, and a methylated cellodextrin (cellotetrose?) as crystalline products of the acetolysis of cellulose derivatives. *J. Chem. Soc.* 824-829.

Hayashi, J., Yamada, T. and Kimura, K. (1976) The change of the chain conformation of cellulose from type I to type II. *Appl. Polym. Symp.* **2**: 713-727.

Hebert, J. J. and Muller, L. L. (1974) An electron diffraction study of the crystal structure of native cellulose. *J. Appl. Polym. Sci.* **18**: 3373-3377.

Heiner, A. P., Sugiyama, J. and Teleman, O. (1995) Crystalline cellulose Iα and Iβ studied by molecular dynamics simulation. *Carbohydr. Res.* **273**: 207-223.

Hermans, P. H. (1949) Physics and chemistry of cellulose fibres, Elsevier, New York.

Hermans, P. H. and Weidinger, A. (1946) The hydrates of cellulose. *J. Coll. Sci.* **1**: 185-193.

Hess, K. and Trogus, C. (1931a) X-ray studies of cellulose derivatives. VIII. The alkali celluloses. *Z. Physik. Chem.* (Leipzig) **B11**: 381-408.

Hess, K. and Trogus, C. (1931b) The fiber periods of cellulose derivatives. X. Röntgenographic studies on cellulose derivatives. *Z. physik. Chem.* Bodenstein-festband. 385-391.

Hess, K., Kiessig, H. and Gundermann, R. (1941) *Z. Phys. Chem.* **8**: 64.

Heuser, E. (1936) The nature of cellulose - an historical review. *American Dyestuff Reporter* **25**: 55-60; 80-83; 116-120; 135-137; 315-318; 338-339.

Heuser, E. (1952) The present status of the chemistry of cellulose. *TAPPI* **35**: 481-489.

Hieta, K., Kuga S. and Usuda, M. (1984) Electron staining of reducing ends evidences a parallel-chain structure in *valonia* cellulose. *Biopolymers* **23**: 1807-1810.

Honjo, G. and Watanabe, M. (1958) Examination of cellulose fibre by the low-temperature specimen method of electron diffraction and elecctron microscopy. *Nature* **181**: 326-328.

Hoyle, F. and Wickramasinghe, N. C. (1977) Polysaccharides and infrared spectra of galactic sources. *Nature* **268**: 610-612.

Irvine, J. C. and Hirst, E. L. (1923) The constitution of polysaccharides. Part VI. The molecular structure of cotton cellulose. *J. Chem. Soc.* **123**: 518-532.

Jacobson, R. A., Wunderlich, J. A. and Lipscomb, W. N. (1961) The crystal and molecular structure of cellobiose. *Acta Crystallogr.* **14**: 598-607.

Jones, D. W. (1960) Crystalline modifications of cellulose. Part V. A crystallographic study of ordered molecular arrangements. *J. Polym. Sci.* **42**: 173-188.

Joshi, N. V. and Rao, V. S. R. (1979) Flexibility of the pyranose ring in α- and β-glucoses. *Biopolymers* **18**: 2993-3004.

Karrer, P. (1921) Alkali cellulose and the decomposition of cellulose. *Cellulosechemie* **2**: 125-128.

Kolpak, F. J. and Blackwell, J. (1976) Determination of the structure of cellulose II. *Macromolecules* **9**: 273-278.

Kolpak, F. J., Weih, M. and Blackwell, J. (1978) Mercerization of cellulose: 1. Determination of the structure of mercerized cotton. *Polymer* **19**: 123-131.

Krassig, H. A. (1993) Cellulose structure, accessibility, and reactivity. Gordon and Breach Science Publishers, USA.

Kroon-Batenburg, L. M. J., Bouma, B., and Kroon, J. (1996) Stability of cellulose structures studied by MD simulations. Could mercerized cellulose II be parallel? *Macromolecules* **29**: 5695-5699.

Kuga, S. and Brown, R. M., Jr. (1987) Lattice imaging of ramie cellulose. *Polym. Commun.*, **28**: 311-314.

Kuga, S., Takagi, S. and Brown, R. M., Jr. (1993) Native folded-chain cellulose II. *Polymer* **34**: 3293-3297.

Kuutti, L., Peltonen, J., Pere, J. and Teleman, O. (1995) Identification and surface structure of crystalline cellulose studied by atomic force microscopy. *J. Microscopy* **178**: 1-6.

Lee, D. M., Blackwell, J. and Litt, M. H. (1983) Structure of a cellulose II-Hydrazine Complex. *Biopolymers* **22**: 1383-1399.

Lee, D. M., Burnfield, K. E. and Blackwell, J. (1983) Structure of a cellulose II-ethylenediamine complex. *Biopolymers* **23**: 111-126.

Leung, F., Chanzy, H. D., Pérez, S. and Marchessault, R. H. (1976) Crystal structure of β-D-acetyl cellobiose, $C_{28}H_{38}O_{19}$. *Canad. J. Chem.* **54**: 1365-1371.

Liang, C. Y. and Marchessault, R. H. (1959a) Infrared spectra of crystalline polysaccharides. 1. Hydrogen bonds in native celluloses. *J. Polym. Sci.* **37**: 385-395.

Liang, C. Y. and Marchessault, R. H. (1959b) Infrared spectra of crystalline polysaccharides. II. Native celluloses in the region from 640 to 1700 cm^{-1}. *J. Polym. Sci.* **39**: 269-278.

Lipkind, G. M., Shashkov, A. S. and Kochetkov, N. K. (1985) Nuclear Overhauser effect and conformational states of cellobiose in aqueous solution. *Carbohydr. Res.* **141**: 191-197.

Manley, R. St. John (1963) Growth and morphology of single crystals of cellulose triacetate. *J. Polym. Sci.: Part A.* **1**: 1875-1892.

Manley, R. St. John (1964) Fine structure of native cellulose microfibrils. *Nature* **204**: 1155-1157.

Manley, R. St. John (1971) Molecular morphology of cellulose. *J. Polym. Sci. Part A-2*, **9**: 1025-1059.

Mann, J., Roldan-Gonzalez, L. and Wellard, H. J. (1960) Crystalline modifications of cellulose. Part IV. Determination of X-ray intensity data. *J. Polym. Sci.* **42**: 165-171.

Marchessault, R. H. and Liang, C. Y. (1960) Infrared spectra of crystalline polysaccharides. III. Mercerized cellulose. *J. Polym. Sci.* **43**: 71-84.

Marchessault, R. H. and Sundararajan, P. R. (1983) Cellulose. In *The polysaccharides*, Vol. 2, Aspinall, G. O., ed., Academic Press, New York, pp. 12-95.

Marrinan, H. J. and Mann, J. (1958) Crystalline modifications of cellulose. Part II. A study with plane-polarized infrared radiation. *J. Polym. Sci.* **21**: 357-370.

Marsh, J. T. and Wood, F. C. (1942) *An introduction to the chemistry of cellulose*. Chapman and Hall, London, p. 51-60 plus two plates.

Meyer, K. H. (1937) Über den Bau des krystallisierten Antiels der Cellulose, V. Mitteil. *Ber.* **70B**: 266-274.

Meyer, K. H. and Mark, H. (1928) Über den Bau des kristallisierten Antieils der Cellulose. *Ber.* **B61**: 593-614.

Meyer, K. H. and Misch, L. (1937) Positions des atomes dans le nouveau modéle spatial de la cellulose. *Helv. Chim. Acta* **20**: 232-244.

Morawetz, H. (1985) *Polymers - the origins and growth of a science*. John Wiley, New York, and Morawetz, H. (1995) Dover.

Mühlethaler, K. (1949) Electron micrographs of plant fibers. *Biochim. Biophys. Acta* **3**: 15-25.

Mühlethaler, K. (1960) Die Feinstruktur der Zellulosemikrofibrillen. *Z. Schweiz. Forstr.* **30**: 55.

Nägeli, C. von (1858) *Die Stärkekörner. Morphologische, physiologische, chemisch-physicalische und systematisch-botanische monographie*. Friedrich Schulthess, Zurich.

Nägeli, C. von (1864) Über den inneren Bau der vegetabilische Zellmembranen. *Sitzber. Kgl. Bayer. Akad. Wiss. Munchen* **1**: 282-326.

Nastukoff, A. (1900) Ueber einige Oxycellulosen und über das Molekulargewicht der Cellulose. *Ber.* **33**: 2237-2243.

Nikitin, N. I. (1962) Khimiya drevesiny i tsellyulozy. Izdatel'stvo Akademii Nauk SSSR, Moskva-Leningrad. (1966) The chemistry of cellulose and wood (translation into English). Daniel Davey and Co., New York, p. 41-61.

Nishikawa, S. and Ono, S. (1913) Transmission of X-rays through fibrous, lamellar and granular substances. *Proc. Tokyo, Math. Phys. Soc.* **7**: 131-138.

Okamura, S. (1933) Einwirkung von Lauge auf native und merzerisierte Cellulose. *Naturwiss.* **21**: 393-394.

Ott, E. and Spurlin, H. M. (1943) "Cellulose and cellulose derivatives", Interscience, New York and a revised edition (1954) Ott, E., Spurlin, H. M. and Grafflin, M. W., eds.

Parnell, E. A. (1986) "The life and labours of John-Mercer", Longmans, Green & Co., London, pp. 175-207, 214-216, 317.

Payen, A. (1838) Mémoire sur la composition du tissu propre des plantes et du ligneux. *Compt. Rend.* **7**: 1052-1056.

Payen, A. (1839) Composition de la matière ligneuse. *Compt. Rend.* **8**: 51-53.

Pérez, S. and Brisse, F. (1977) The crystal and molecular structure of a trisaccharide, β–cellotriose undecaacetate: 1,2,3,6-tetra-O-acetyl-4-O-[2,3,6-tri-O-acetyl-4-O-(2,3,4,6-tetra-O-acetyl-β-D-glucopyranosyl)-β-D-glucopyranosyl]-β-D-glucopyranose. *Acta Crystallogr. Sect. B.* **33**: 2578-2584.

Pertsin, A. J., Nugmanov, O. K., Marchenko, G. N. and Kitaigorodsky, A. I.(1984) Crystal structure of cellulose polymorphs by potential energy calculations. 1. Most probable models for mercerized cellulose. *Polymer* **25**: 107-114.

Pertsin, A. J., Nugmanov, O. K and Marchenko, G. N. (1986) Crystal structue of cellulose polymorphs by potential energy calculations: 2. Regenerated and native celluloses.

Polymer **27**: 597-601.

Pesacreta, T. C., Carlson, L. C. and Triplett, B. A. (1997) Atomic force microscopy of cotton fiber cell wall surfaces in air and water: quantitative and qualitative aspects. *Planta* **202**: 435-442.

Phillips, M. (1961) Anselme Payen. *Great Chemists*. Farber, E., ed., Interscience, New York, pp. 495-504. Taken from *J. Washington Acad. Sci.* **30**: 65-71 (1940).

Polanyi, M. (1921a) Die chemische Konstitution der Zellulose. *Naturwiss.* **9**: 288.

Polanyi, M. (1921b) The X-ray fiber diagram. *Z. physik* **7**: 149-80.

Poppleton, B. J. and Mathieson, A. McL. (1968) Crystal structure of β-D-cellotetraose and its relationship to cellulose. *Nature* **219**: 1046-1048.

Preston, R. D. (1964) Structural and mechanical aspects of plant cell walls with particular reference to synthesis and growth. In *The formation of wood in forest trees*, Zimmermann, M. H., ed., NY, Academic Press, pp. 169-201.

Preston, R. D. (1939) The molecular chain structure of cellulose and its botanical significance. *Biol. Rev. Cambridge Phil. Soc.* **14**: 281-313.

Ramesh, R., Patel, C. K. and Patel, R. D. (1973) Cellulose single crystals. *Makromolek. Chemie* **171**: 179-184.

Rånby, B. G. (1952) The mercerization of cellulose. I. A thermodynamic discussion. *Acta Chem. Scand.* **6**: 101-115.

Rånby, B. G. and Noe, R. W. (1961) Crystallization of cellulose and cellulose derivatives from dilute solution. I. Growth of single crystals. *J. Polym. Sci.* **51**: 337-347.

Raymond, S., Henrissat, B., Tran Qui, D., Kvick, A. and Chanzy, H. (1995a) The crystal structure of methyl β-cellotrioside monohydrate 0.25 ethanolate and its relationship to cellulose II. *Carbohydr. Res.* **277**: 209-229.

Raymond, S., Heyraud, A., Tran Qui, D., Kvick, A. and Chanzy, H. (1995b) Crystal and molecular structure of β-D-cellotetraose hemihydrate as a model of cellulose II. *Macromolecules* **28**: 2096-2100.

Raymond, S., Kvick, A. and Chanzy, H. (1995c) The structure of cellulose II: A revisit. *Macromolecules* **28**: 8422-8425.

Rees, D. A. and Skerrett, R. J. (1968) Conformational analysis of cellobiose, cellulose and xylan. *Carbohydr. Res.* **7**: 334-348.

Reeves, R. E. (1949) Cuprammonium-glycoside complexes. III. The conformation of the D-glycopyranoside ring in solution. *J. Am. Chem. Soc.* **71**: 215-217.

Reid, J. D. and Dryden, E. C. (1940) Anselme Payen - discoverer of cellulose. *Textile Colorist* **62**: 43-45.

Revol, F.-F. and Goring, D. A. I. (1983) Directionality of the fibre c-axis of cellulose crystallites in microfibrils of *Valonia ventricosa*. *Polymer* **24**: 1457-1550.

Ruska, H. and Kretschmer, M. (1940) Supermicroscopic investigation of the degradation of cellulose fibers. *Kolloid-Z.* **93**: 163-166.

Sakthivel, A., Turbak, A. F. and Young, R. A. (1989) New structural models for cellulose II derived from packing energy minimization. In *Cellulose and wood - chemistry and technology*. Schuerch, C., ed., John Wiley, New York, pp. 67-76.

Sakurada, I. and Okamura, S. (1937) Investigation on the molecular compounds of cellulose by determining the apparent specific volume along with the X-ray determination of the crystallographic elementary substances of swollen cellulose. *Kolloid-Z.* **81**: 199-208.

Sarko, A. (1986) Recent X-ray crystallographic studies of cellulose. In *Cellulose structure, modification and hydrolysis*, Young, R. A. and Rowell, R. M., eds., Wiley-Interscience, New York, pp. 29-49.

Sarko, A. and Muggli, R. (1974) Packing analysis of carbohydrates and polysaccharides. III. *Valonia* cellulose and cellulose II. *Macromolecules* **7**: 486-494.

Sarko, A., Southwick, J. and Hayashi, J. (1976) Packing analysis of carbohydrates and polysaccharides. 7. Crystal structure of cellulose III (subscript I), and its relationship to other cellulose polymorphs. *Macromolecules* **9**: 857-863.

Sarko, A., Southwick, J. and Hayashi, J. (1977) Packing analysis of carbohydrates and polysaccharides. 7. Crystal structure of cellulose III$_1$ and its relationship to other cellulose polymorphs. *Macromolecules* **7**: 857-863.

Sarko, A., Nishimura, H. and Okano, T. (1987) Crystalline alkali-cellulose complexes as intermediates during mercerization. *ACS Symp. Ser.* **340**: 169-177.

Schulz, J. (1948) Die Kinetik des Celluloseabbaus und die langperiodische Struktur del Cellulosemoleküls. *J. Polym. Sci.* **3**: 365-370.

Shenouda, S. G. (1979) The structure of cotton cellulose. In *Applied Fiber Science #3*, Happey, F., ed., Academic Press, pp. 275-309.

Shibazaki, H., Kuga, S., Onabe, F. and Brown, R. M., Jr. (1995) Acid hydrolysis behavior of microbial cellulose II. *Polymer* **36**: 4971-4976.

Simon, I., Scheraga, H. A. and Manley, R. St. John (1988) Structure of cellulose I. Low-energy conformations of single chains. *acromolecules* **21**: 983-990.

Sisson, W. A. (1938) The existence of mercerized cellulose and its orientation in *Halicystis* as indicated by X-ray diffraction analysis. *Science* **87**: 350.

Sisson, W. A. (1941) Some X-ray Observations Regarding the Membrane Structure of *Halicystis*. *Contribs. Boyce Thompson Inst.* **12**: 31-44.

Sisson, W. A. and Saner, W. R. (1941) Effect of the temperature and the concentration of sodium hydroxide on the X-ray diffraction behavior of raw and of degraded cotton. *J. Phys. Chem.* **45**: 717-730.

Skraup, Z. H. and König, J. (1901) Ueber Cellose, eine Biose aus Cellulose. *Ber.* **34**: 1115-1118.

Smith, H. D. (1937) Structure of cellulose. *Ind. and Eng. Chem.* **29**: 1081-1084.

Sponsler, O. L. and Dore, W. H. (1926) The structure of ramie cellulose as derived from X-ray data. *Colloid. Symp. Monograph.* **4**: 174-202.

Stipanovic, A. J. and Sarko, A. (1976) Packing analysis of carbohydrates and polysaccharides. 6. Molecular and crystal structure of regenerated cellulose II. *Macromolecules* **9**: 851-857.

Sugiyama, J. and Okano, T. (1989) Electron microscopic and X-ray diffraction study of cellulose III$_1$ and cellulose I. Cellulose and wood: chemistry and technology, Schuerch, C., ed., Wiley and Sons, New York, pp. 119-127.

Sugiyama, J., Vuong, R. and Chanzy, H. (1991) Electron diffraction study of two crystalline phases occurring in native cellulose from an algal cell wall. *Macromolecules* **24**: 4168-4175.

Takai, M. and Colvin, J. R. (1978) Mechanism of transition between cellulose I and cellulose ii during mercerization. *J. Polym. Sci.: Polym. Chem. Ed.* **15**: 1335-1342.

Tollens, B. (1895) See Marsh and Wood (1942), quoted from Handbuch der Kohlenhydrate, Leipzig, Barth, J. A. ed., 1914, p. 564.

Tripp, V. W., Moore, A. T. and Rollins, M. L. (1951) Some observations on the constitution of the primary wall of the cotton fiber. *Text. Res. J.* **21**: 886-894.

VanderHart, D. L. and Atalla, R. H. (1986) In *Cellulose: Structure, Modification and Hydrolysis*, Young, R. A. and Rowell, R. M., eds., Wiley-Interscience, New York, pp. 88-118.

Vieweg, W. (1924) Action of aqueous and dilute alcoholic sodium hydroxide on cellulose. *Ber.* **57B**: 1917-1921.

Wellard, H. J. (1954) Variation in the lattice spacing of cellulose. *J. Polym. Sci.* **13**: 471-476.

Whitmore, R. E. and Atalla, R. H. (1985) Factors influencing the regeneration of cellulose I from phosphoric acid. *Int. J. Biol. Macromol.* **7**: 182-186.

Wise, L. (1963) Anselme Payen. *The Paper Industry*, April, 38-39.

Woodcock, C. and Sarko, A. (1980) Packing analysis of carbohydrates and polysaccharides. 11. Molecular and crystal structure of native ramie cellulose. *Macromolecules* **13**: 1183-1187.

Yathindra, N. and Rao, V. S. R. (1970) Configurational statistics of polysaccharide chains. Part II. Cellulose. *Biopolymers* **9**: 783-790.

Zemplén, G. (1926) Degradation of the reducing bioses. I. Direct determination of the constitution of cellobiose. *Ber.* **59B**: 1254-1266.

Chapter 9

Structure and Biosynthesis of Cellulose
Part II: Biosynthesis

Deborah P. Delmer
Section of Plant Biology, University of California Davis
One Shields Ave, Davis, CA 95616, USA

ABSTRACT

In reviewing the history of research on cellulose biosynthesis, one must first acknowledge the pioneering work on the bacterium — *Acetobacter xylinum* — first initiated by Shlomo Hestrin in Israel. Through the years, *A. xylinum* has been developed as a model system for the first demonstration of *in vitro* synthesis of cellulose using UDP-glc as substrate, to identify a unique activator of the cellulose synthase cyclic diguanylic acid, to purify the first synthase, and to clone an operon of genes that encodes essential components of the synthase complex. Studies with algae and higher plants using freeze-fracture techniques have revealed multisubunit cellulose synthase complexes in the plasma membrane. However, at the molecular level, studies with plants have always lagged behind those of *A. xylinum*. Nevertheless, the work of many groups has gradually provided insight concerning the difficulties of *in vitro* synthesis and enzyme purification in plants as well as an understanding of motifs that are conserved in glycoyltransferases that use UDP-glc as substrate. Understanding of these conserved motifs was a key to the recent success in the first cloning of a gene that encodes the plant catalytic subunit of the synthase complex. Taken together, much clearer views of both the structure and mechanism of synthesis of this polymer are now emerging.

As for many complex questions in biology, understanding of the process of cellulose biosynthesis has been a long and often frustrating endeavor. It is a story in which there was no one single breakthrough that led to total insight, but rather it has served as one of those excellent examples of how science slowly builds upon discovery after small discovery. In addition, the story relates to that of understanding the structure of cellulose, and we shall try to tie those two parts of the history of cellulose together in this section as well.

Initial Studies On Cellulose Biosynthesis

The history of studies on cellulose biosynthesis really begins with Shlomo Hestrin's research using a gram-negative bacterium called *Acetobacter xylinum* that secretes large amounts of cellulose. Hestrin immigrated in 1930 from Canada to the pre-state of Israel (Mandatory Palestine) and was one of the leading scientists in the initial biochemistry group at The Hebrew University of Jerusalem. Moshe Benziman, one of the leaders in today's work on cellulose synthesis in this bacterium, was an early student of Hestrin's, and he recalls Hestrin as someone with an extremely forceful personality having intense dedication to his work on *A. xylinum*. The early work from this group established growth conditions for efficient cellulose production and also contributed much of what we know today about the physiology and metabolic pathways of *A. xylinum* (Hestrin, 1962). More important, Hestrin's work introduced this organism to the scientific community as an attractive model system for studies on cellulose synthesis.

While the work of Hestrin was in progress, another seminal event occurred — the discovery of the nucleotide sugar UDP-glucose by Leloir and co-workers in Argentina in connection with their studies on lactose degradation in yeast (Cardini *et al.*, 1954; Leloir, 1971). Leloir and his student Enrico Cabib quickly realized that UDP-glucose had unique properties in that the free energy of hydrolysis to UDP and glucose was sufficiently high to make it a prime candidate for a sugar donor in glycosyltransferase reactions. They first showed this by demonstrating the reaction for synthesis of trehalose phosphate using UDP-glc as sugar donor (Leloir and Cabib, 1953). This work was quickly followed by the discovery by this same group of sucrose synthase (SuSy) in extracts of wheat germ, the first plant enzyme found to use UDP-glc as a substrate (Cardini *et al.*, 1955). SuSy catalyzes the reaction: UDP-glucose + fructose ↔ sucrose + UDP. Because the glycosidic linkage in sucrose links the aldehyde residues of both glucose and fructose making it a non-reducing sugar, sucrose and other such non-reducing disaccharides have a much higher than usual free energy of hydrolysis. Therefore, the reaction of SuSy is freely reversible, although Leloir's group initially assumed it functioned in the synthesis of sucrose. With their subsequent discovery of sucrose phosphate synthase (Leloir and Cardini, 1955), this notion was questioned, and this latter enzyme is now known to be the primary route for synthesis of sucrose in plants. SuSy is much more abundant in non-photosynthetic tissues where it is believed to function primarily in the degradation of sucrose (see references in Delmer and Amor, 1995). In retrospect, it is fascinating that the discoveries of UDP-glucose, now known to be the substrate for cellulose synthesis, and SuSy, recently also implicated in the process (Amor *et al.*, 1995), played such key roles in our early understanding of glycosyltransferase reactions in general. In even more general terms, it is worth noting that the overall productivity and importance of all the pioneering work that emerged from the Leloir group in the 1950s was truly remarkable and set the foundation for all subsequent work on polysaccharide synthesis.

Within a very short period following these discoveries, UDP-glucose was found to serve as a substrate for a number of glycosyltransferase reactions. Thus, it was no surprise when Glaser (1958) also tested this substrate for cellulose synthesis using membrane preparations from *A. xylinum*. His success represents the first demonstration of *in vitro* synthesis of cellulose from any organism. However, the rates achieved were far below those observed *in vivo*, and the study of *in vitro* synthesis languished with no further progress until the early 1980s.

Meanwhile another Israeli (although relocated to the University of California at Berkeley) named Zev Hassid decided to take up the challenge of studying cellulose synthesis in plants. According to anecdotal stories from those days, David Feingold and Liz Neufeld, students in Hassid's laboratory, were asked to take up this project. Using mung bean membranes as source of enzyme and UDP-glucose as substrate, Feingold and Neufeld quickly showed massive synthesis of β-glucan — the problem was that the linkage was not 1,4 as expected for cellulose, but rather 1,3 that is characteristic of a wound polymer in plants called callose. Apparently, Hassid just couldn't believe this, and the poor students were sent back to repeat their results innumerable times before Hassid finally became convinced, and the first report of callose and not cellulose synthesis from UDP-glucose in plants was published (Feingold *et al.*, 1958). However, Hassid did not give up, and George Barber and Alan Elbein were then sent to work to see if some other nucleotide sugar might be the true substrate for cellulose synthesis. Surprisingly, they did find synthesis of β-1,4-glucan from GDP-glc using mung bean membranes, thus claiming the first *in vitro* synthesis of cellulose in a plant system (Barber *et al.*, 1964). Although this and subsequent work (Chambers and Elbein, 1970) were quite reproducible, through the years, it has emerged that the synthesis observed may really have represented a system for synthesis of glucomannan that contains β-1,4 glc linkages (Villemez and Heller, 1970 and also reviewed by Delmer, 1977 and 1983). Unfortunately, one still sees statements in textbooks that GDP-glc is the substrate for cellulose synthesis when in fact all of the evidence now points to UDP-glc as substrate. (It is noteworthy that all these young scientists who worked with Hassid have gone on to be leaders in the field of glycosyltransferase biology, although none continued for long to work on cellulose synthesis.)

We relate the story of Hassid and callose synthesis primarily because it sadly was repeated again and again in laboratories around the world in the next 20–30 years. Innumerable papers were published that claimed *in vitro* synthesis of cellulose from UDP-glucose in plants when, in fact, the product was really callose. Unfortunately, few of these attempts were performed using the rigorous structural characterization that Hassid demanded of his students. One of the primary efforts of one of our laboratories (DPD) in the early years was to re-examine many of these claims in detail (reviewed in Delmer 1983 and 1987). Throughout the years, the groups of Hans Meier in Switzerland, Gerhard Franz and Heinrich Kauss in Germany, Moshe Benziman in Israel, Gordon Machlachlan in Canada, Bruce Stone in Australia, Don Northcote in England, and Peter Ray, Alan Elbein, Bruce Wasserman, Nick Carpita, Malcolm Brown, Debby Delmer, and others in the U.S., struggled with *in vitro* synthesis of cellulose, callose, and a mixed β–1,3-1,4 glucan of plants. Until recently, the primary reward from most of these efforts was a good characterization of the callose synthase of plants that is now known to be localized in the plasma membrane, to use UDP-glc as substrate, and to be activated by a β-glucoside and micromolar levels of Ca^{2+} (reviewed by Delmer and Amor, 1995). Activation by Ca^{2+} may explain why this enzyme is activated *in vivo* by wounding, a condition that may also result in inhibition of cellulose synthesis. In fact, both Delmer (1987) and Jacob and Northcote (1985) have suggested that the callose and cellulose synthases may share common subunits, a question that remains unanswered.

What Does a Cellulose Synthase Complex Look Like?

The biochemistry of cellulose synthesis languished on the back burner throughout the 1970's and 1980's, but scientists did speculate on what a cellulose synthase complex might be like and where it would be located. Since cellulose is not found inside the plant cell, it was assumed (correctly) that the enzyme must be on the plasma membrane. Preston, with his "ordered granule hypothesis" (1964) was the first to invoke the concept that a synthase complex must be composed of many catalytic subunits in association, that function to polymerize glucan chains that would then self-associate to form the microfibril. By this logic, the size and shape of such a complex would then dictate the characteristics of the microfibril formed.

Preston's vision was first realized in 1979 with the publication by R. Malcolm Brown, Jr. and his student David Montezinos (Brown and Montezinos, 1976) of a paper showing remarkable multisubunit complexes in the plasma membranes of the cellulose-producing alga *Oocystis solitaria*. These were visualized by the then-relatively new technique of freeze-fracturing of the plasma membrane. On the E-face of the plasma membrane, they observed what are now called linear terminal complexes (TC's) at the ends of microfibril imprints. Such TC's have now been observed in many of the algae that are characterized by synthesis of highly-crystalline microfibrils that, in *Valonia* for example, can be composed of up to 1,000 glucan chains. Using Preston's logic, this means that, in *Valonia*, there must be, at a minimum, around 1,000 catalytic subunits in one complex! This is perhaps the most amazing macromolecular complex of proteins in nature. Very soon after this, Brown's laboratory, with the work of graduate student Suzette Mueller, discovered that higher plants may have nine different types of putative complexes in their plasma membrane — the so-called rosettes that fracture to the E-face and terminal globules that may fit inside the hole in the rosette in the P-face (Mueller and Brown, 1980). Proof that these smaller, solitary complexes in higher plants really were cellulose synthase complexes was a bit less obvious than for the large, linear complexes found in the algae, but, through the years, the finding of such rosettes in clustered arrays at the ends of microfibrils in the alga *Microcystis denticulata* (Giddings *et al.*, 1980), their increased density at sites of high rates of cellulose synthesis (Herth, 1985), and very recently, their possible isolation in preparations synthesizing cellulose *in vitro* in mung beans (Kudlicka and Brown, 1997) and their disappearance at high temperature in a temperature-sensitive cellulose-deficient mutant of *Arabidopsis thaliana* (Arioli *et al.*, 1998), all argue strongly that these are indeed synthase complexes.

For those wishing to see examples of these complexes, Brown (1996) provides good examples along with a very clear review of their structure and probable functional organization. Speculating on how such complexes may function provides a good opportunity to see how important the understanding of cellulose structure becomes in studying biosynthesis. As described above, we now conclude that the glucan chains within a native cellulose microfibril having the cellulose I structure are all oriented in parallel. This leads to the conclusion that all chains within a complex grow by a similar mechanism, and this issue is discussed in more detail in the following section. Since cellulose I is the thermodynamically metastable form, it must be somehow "locked in" by the process of simultaneous polymerization and crystallization. There are a few rare instances in which cellulose II structures are synthesized in nature (see Sisson, 1938; Brown, 1996). In these cases, the synthase complexes most likely still synthesize chains from the non-reducing end in a parallel array. However, if the subunits within the complex are less ordered or exist as solitary catalytic units, then chain association may be hindered. This would allow for chain folding or other more subtle changes in residue associations that would lead to the cellulose II structure. Alternatively, chain synthesis and release from the catalytic subunits might be completely uncoupled, a situation that would also favor the more stable association of chains characteristic of cellulose II.

Another issue to address is what may lead to the production of the two different allomorphs Iα and Iβ. In general, organisms with linear complexes produce a higher proportion of Iα, while those with rosettes produce microfibrils more rich in Iβ. However, as mentioned previously, none of the native celluloses are a pure form of either allomorph, and so one cannot simply invoke general complex shape as being the determinant, and much more work will be necessary to understand controlling factors. There does seem to be a generalization that complex size (or aggregation state) and shape determines the overall size and shape of the microfibrils produced, and also that those algae with large linear TCs produce the most highly-crystalline microfibrils. Also, it seems that one can generalize further to say that the more crystalline the product, the higher the degree of polymerization of the chains. However, since microfibrils are longer than the individual chains within them, this means that chain initiation and termination are occurring many times during the synthesis of one microfibril. How this is regulated is not at all understood at this time.

Are Lipid Intermediates Involved in the Synthetic Pathway?

As described above, while biochemists were struggling to make cellulose *in vitro*, cellulose chemists were still engaged in ongoing debates about the structure of cellulose. Most relevant to biosynthesis was the question of the directionality of the chains — were they parallel or anti-parallel? If anti-parallel, how could the microfibrils grow from one end? If parallel, and if they did grow from one end, which end? These questions were particularly pertinent in the decades of the 70s and 80s when biochemists were engaged in heated debate about the possible role of lipid intermediates in cellulose synthesis; this is so because synthesis via such intermediates proceeds via the reducing end, whereas direct polymerization from the nucleotide sugar proceeds by addition to the non-reducing end of the growing polymer. Ross Colvin, a very imaginative but controversial Canadian, was, along with Hestrin, one of the early workers with *A. xylinum*, and Colvin was the first to propose involvement of lipid intermediates in cellulose biosynthesis. He claimed that cellulose was synthesized in the medium surrounding the bacterium *A. xylinum* and that some lipid carrier was involved (Colvin,

1980). We still recall with amusement some of the heated discussions between Colvin, Benziman, and Brown, the major workers with *A. xylinum*, particularly at one bombastic cellulose conference at Syracuse University in 1983. In the end, at least some of Colvin's work was proven wrong — synthesis certainly occurs via complexes anchored in a row along one side of the bacterium (Brown *et al.*, 1976), and to date, there is still no strong evidence for such intermediates in this organism.

At the same time, all of us engaged in biosynthesis with plant systems were also looking for the polyprenol types of lipid intermediates that were emerging as being important in glycoprotein synthesis and in cell wall biosynthesis in bacteria. In addition to being relevant to the question of directionality of chain growth, the concept of lipid intermediates was attractive to some because of the oft-stated concept that the true repeating unit in cellulose is really cellobiose and not glucose. As stated earlier, Hess and Trogus (1931) showed early on that the repeating unit of cellulose (from a chemical standpoint) is glucose. Even so, on average the alternating residues are rotated about $180°$ with respect to each other in the native form, similar to the linkage in cellobiose. This led to speculation that one active site could not polymerize successive glucose residues in their alternating orientations. One solution would be to use a lipid intermediate in which cellobiose was donated to the growing chain. Much more recently, our growing understanding of sequences involved in substrate binding and catalysis led to the alternative suggestion that there are multiple sites for binding of UDP-glucose to the catalytic subunit, allowing addition in different orientations (Saxena *et al.*, 1995; Koyama *et al.*, 1997). However, there may really be no need to invoke such complicated mechanisms. As mentioned previously, there is a fairly large degree of freedom of rotation about the glycosidic linkage (up to about $120°$), and therefore, the sugar could be added in one orientation but then eventually relax into the native orientation after polymerization.

Although there were hints of lipid intermediates here and there, particularly the work of Pont Lezica's group on the alga Prototheca (Hopp *et al.*, 1978), none of those studies has been seriously reproduced, and we still have no firm evidence for their role in cellulose synthesis in algae or plants. But the story is not really over, as we have recently reviewed (Kawagoe and Delmer, 1997a). In 1995, Matthysse and colleagues, based upon both genetic and biochemical evidence, proposed a scheme for the bacterium *Agrobacterium tumefaciens* in which UDP-glucose serves as substrate for the synthesis of lipid-linked cellodextrins (Matthysse *et al.*, 1995a,b). The final step in cellulose synthesis was then proposed to involve the action of a cellulase acting in reverse direction to catalyze a transglycosylation of cellobiose from the lipid intermediate to the growing glucan chain. There is little to argue with the results presented, but we do note that, in the past few years, no further studies on this system have appeared. However, the idea that a cellulase might function in synthesis of cellulose keeps resurfacing. Maclachlan's group in Canada were the first to talk about this (Maclachlan, 1977) having observed that cellulase treatments stimulated glucan synthesis *in vitro* (now known most likely to have been mostly callose) and also *in vivo* (probably cellulose). In addition to Matthysse's finding of a cellulase gene in an operon related to cellulose synthesis in *A. tumefaciens*, there is also one that is upstream of the synthase operon in *A. xylinum* (Volman *et al.*, 1995). Recently, a new plant gene was discovered that encodes a cellulase with one membrane-spanning domain anchored in the plasma membrane, placing the catalytic domain in the periplasm (Brummell *et al.*, 1997). A mutation in a similar gene has now been shown to cause a severely altered phenotype reminiscent of cellulose-deficient mutants of *Arabidopsis* (Nicol *et al.*, 1998). We suggest that, rather than implicating a role for lipid intermediates, the cellulase may well perform an editing function. In order to function efficiently, all subunits must catalyze polymerization at nearly equal rates. If one catalytic subunit gets ahead or behind the others, then that chain may need to be terminated to keep the process proceeding smoothly. Thus, a cellulase that recognizes non-crystalline, disrupted regions in microfibrils and is situated in proximity to the site of synthesis might function in such an editing mode. Benziman's group has offered a similar speculation about the role of cellulase in *A. xylinum* (Volman *et al.*, 1995). However, this is just speculation, and one should still keep an open mind about the exact pathway of synthesis even today. Nevertheless, many chemists now agree that the chains of native cellulose are parallel, and the exciting new results of Sugiyama's group provide powerful evidence that growth of these parallel chains occurs from the non-reducing end, thus supporting the concept of direct polymerization from UDP-glucose (Koyama *et al.*, 1997).

Crystallization and Polymerization are Coupled Events

During these early decades, another set of studies relating the structure of cellulose to biosynthesis emerged from studies of Malcolm Brown and his graduate student Candace Haigler with additional input from Moshe Benziman who was on sabbatical in Brown's lab at the time. These workers initiated study on compounds such as the fluorescent brightener Calcofluor White which can intercalate between the chains of cellulose during synthesis and thus dramatically alter the crystallinity of the fibrillar product. These studies (Benziman *et al.*, 1980; Haigler *et al.*, 1980) and others largely summarized by Haigler (1991) provided strong evidence that polymerization and crystallization are coupled events, with crystallization being the rate-limiting process, and also offered insights into the ways chains can interact with each other.

The Role of the Cytoskeleton in Orientation of Microfibril Deposition

One of the most dramatic aspects of cellulose deposition concerns the variatious patterns in which the microfibrils can be laid down in the cell wall of algae and higher plants. Since the microfibrils are secreted to the medium by the bacteria, it is clear that, during the evolution of cellulose synthesis, a new and complex addition to the process in eucaryotes was needed to give order to the deposition within the confines of a defined cell wall matrix. In higher plants, the direction of deposition determines to a great extent the direction of expansion of the growing cell wall. As cells cease expanding and assume specialized roles, the patterns of deposition in thickening secondary walls can assume any number of arrays, the most complex perhaps being that of the spiral and convoluted arrays characteristic of cells undergoing xylem differentiation. In other algae and some plant cell types, the wall may simply thicken uniformly, but even then, the deposition occurs in layers that periodically alter orientation. Paul Green of Stanford University, one of the most original thinkers in all of modern plant biology, understood early on that control of microfibril deposition was a key event in the control of plant morphogenesis (for a review see Green and Selker, 1991).

Historically, it's worth noting that microtubules were first discovered, not in animals, but in plants by Ledbetter and Porter (1963), and it wasn't long after that when people began to see a connection between these structures and the orientation of cellulose deposition. Thus, many studies through the years have shown that in most cases, the pattern of deposition is precisely mimicked by the pattern of organization of the cortical microtubule (MT) network (reviewed by Giddings and Staehelin, 1991). From this has evolved the concept that this MT network somehow controls the path in which cellulose synthase complexes move in the fluid mosaic membrane as they spin out the fibrillar network of the wall. Studies on MTs and cellulose synthesis have also been controversial, as there are indeed some cases in which co-alignment of MTs and microfibril deposition may not occur (Emons, 1994). In addition, various models as to how the process occurs have been presented — are the synthases directly attached to the microtubules and move along them as proposed by Heath (1974), or do the microtubules merely serve as guides much like railroad tracks in which the synthase complexes wander randomly but are constrained by the distance between the tracks as favored by several other groups (see Giddings and Staehelin, 1991). Although most seem to favor the latter view, in truth, no one really knows. With the rapid advances that are emerging in studies of the plant cytoskeleton and plant synthase complexes, we predict that the relationship between the cytoskeleton and synthase complexes is going to be one of the most exciting avenues for future research in this field.

Breakthroughs in *in vitro* Synthesis; Purification of Synthases

A major breakthrough in the field of cellulose biosynthesis came with the discovery of cyclic diguanylic acid (c-di-GMP) and its role as a specific activator of the cellulose synthase in the bacterium *A. xylinum* by Moshe Benziman's group in Israel. One of us (DPD) had the good fortune to be in Benziman's laboratory on a sabbatical when this work started. We had found that addition of polyethylene glycol (PEG) stimulated glucan synthesis in plants (Delmer *et al.*, 1983), thus suggesting that it might be worth an attempt in *A. xylinum*. Building on yet another finding by Enrico Cabib's group at NIH that the β-1,3-glucan synthase of yeast was stimulated by GTP (Shematek *et al.*, 1980), Shuki Aloni, a graduate student in the group, observed remarkable stimulation of cellulose synthesis *in vitro* when both PEG and GTP were present during membrane isolation and assay. Figuring out the mode of PEG and GTP stimulation was real detective work as the answer was quite surprising (reviewed by Ross *et al.*, 1991; Volman *et al.*, 1995). PEG was precipitating an enzyme (now known to be a cyclase) that converted GTP to a specific low molecular weight compound that stimulated the enzyme (Aloni *et al.*, 1982). Peter Ross, another graduate student, with collaboration from the chemist J. Van Boom of Belgium, finally made the discovery that this activator was a very unusual compound — c-di-GMP (Ross *et al.*, 1987). To date, although c-di-GMP does seem to exist, at least in other bacteria, there is no other known function for this compound except to activate the bacterial cellulose synthases. The significance of this finding was that, with addition of c-di-GMP to the assays, cellulose could be synthesized *in vitro* at rates that approached those *in vivo*, thus paving the way for the purification of the first cellulose synthase.

In addition to the discovery of c-di-GMP, another study by Enrico Cabib's group with yeast chitin synthase was critical for paving the way for the first purification of the cellulose synthase of bacteria. This finding was that the chitin synthase of yeast could be easily purified by a technique called "product entrapment" in which the detergent-solubilized synthase was found to co-sediment with its chitin product following *in vitro* synthesis (Kang *et al.*, 1984). The groups of both Brown in Texas and Benziman in Israel successfully used this technique to purify the *A. xylinum* cellulose synthase following synthesis of cellulose *in vitro* activated by c-di-GMP (Lin and Brown, 1989). F-C Lin, a graduate student of Brown's, was also able to show that an 83 kD polypeptide present in the purified preparation could bind the substrate UDP-glc (Lin *et al.*, 1990), and sequencing of the N-terminal of this peptide led to one strategy that allowed subsequent cloning of the gene encoding the enzyme (Saxena *et al.*, 1990). Raffi Mayer, Peter Ross, and Haim Weinhouse, all in the Benziman laboratory, also showed that a 94 kD polypeptide was most likely a regulatory subunit that bound c-di-GMP (Mayer *et al.*, 1991). They have also identified the enzymes that both synthesize and degrade c-di-GMP (Volman *et al.*, 1995).

All of us who worked with plant enzymes were very hopeful that these discoveries could be translated to similar successes with the plant systems, but it proved not to be the case. To date, although there are strong hints that c-di-GMP may exist in plants (Amor *et al.*, 1991; Mayer *et al.*, 1991), no one has demonstrated activation of a cellulose synthase by this compound with plant systems. In fact, several fairly convincing proofs of *in vitro* synthesis of cellulose in plant membranes have been reported only very recently, and neither of these involved use of c-di-GMP.

One of these approaches is a series of studies in which much of the work was carried out by Likun Li and Krystyna Kudlicka from Brown's laboratory in Texas. This work has gradually evolved to the point where, using primarily mung bean membranes as a source of enzyme, one can now distinguish a cellulose synthase from a callose synthase in detergent-solubilized preparations, and although rates may not be exceedingly high, one can nevertheless clearly see production of distinct types of fibrils, one of which is characteristic of callose and the other of cellulose (see Kudlicka and Brown, 1997 and references therein).

Also recently, Meme Amor, a graduate student in one of our laboratories (DPD), made the surprising discovery that, in cotton fibers, a significant amount of the total sucrose synthase (SuSy) is associated with the plasma membrane. A model was developed in which SuSy is associated with glucan synthase complexes and efficiently channeled carbon from sucrose, via its product UDP-glc, to the glucan synthase catalytic subunit, which in turn, by polymerization of glucan, also released UDP that could be recycled and used again by SuSy (Amor *et al.*, 1995; Delmer and Amor, 1995). In the same study, immunolocalization done in collaboration with Candace Haigler's group at Texas Tech showed SuSy to be located in the fiber in helical arrays that paralleled the patterns of cellulose deposition, thus further supporting a model in which SuSy and cellulose synthase are in close association. Also supporting this was our finding that when supplied to permeabilized fibers, sucrose, as opposed to UDP-glc, served as a much more efficient substrate for *in vitro* synthesis of cellulose. Although the rates of cellulose, as opposed to callose, synthesis in this system can be extremely variable, at times the rates of cellulose synthesis from sucrose can approach those obtained *in vivo*. Although progress is clearly being made in *in vitro* synthesis for plant systems, there has not yet been sufficient purification of any complex to determine precisely the subunit composition and overall structure of plant synthase complexes using such biochemical approaches.

Identification of Genes Involved in Cellulose Synthesis

Ironically, despite the intense dedication to biochemical approaches to cellulose synthesis, for cloning of the genes involved, both for the bacteria, and more recently also for higher plants, genetic approaches provided the first breakthroughs. For *A. xylinum*, the work was driven by the desire of industry to find a way to make production of bacterial cellulose a profitable venture. Thus, Weyerhauser, the U.S. paper company, contracted the biotechnology skills of the Cetus Corporation in California who, together with collaboration from Benziman's group in Israel, set about trying to clone genes involved in cellulose synthesis in *A. xylinum*. By using wild-type DNA to complement mutants defective in cellulose synthesis, an operon of four genes was identified and sequenced and shown to be either necessary or at least important for cellulose synthesis (Wong *et al.*, 1990). Shortly thereafter, Inder Saxena, while a post-doctorate in Brown's laboratory in Texas, successfully cloned a gene that encoded the catalytic subunit using the alternate strategy of designing probes based upon the sequence of a peptide from the purified UDP-glc-binding subunit (Saxena *et al.*, 1990). This gene was the same as the first gene of the operon identified by Wong *et al.* (1990). Saxena and Brown soon after (1991) cloned the second gene homologous to the B gene of Wong *et al.*, and evidence suggests that this gene may encode the c-di-GMP-binding subunit (Mayer *et al.* 1991). More recent work from Saxena *et al.* (1994) hints that the C gene may be involved in formation of a pore in the outer membrane through which the microfibrils are secreted, and that the D gene may somehow help to control the pattern of crystallization. The Cetus/Benziman collaboration has also resulted in the cloning of other genes involved in the synthesis and degradation of c-di-GMP (Volman *et al.*, 1995).

Once again, hope was provided by these results that the corresponding genes in plants might quickly be identified using the bacterial genes as probes, but, as for the excitement generated by discovery of c-di-GMP, these hopes were not initially realized. The fact that probing cDNA libraries from plants with the full-length A genes of bacteria did not pull out homologous clones suggested that the overall homology between the plant and bacterial catalytic subunits was not going to be high, at least at the primary sequence level. But a further breakthrough came from analyses both by one of our groups (DPD) and by Brown's group in collaboration with Bernard Henrissat. These analyses involved searches for conserved motifs in the A gene of bacteria and other glycosyltransferases that bound the substrate UDP-glc or UDP-GlcNAc (Delmer and Amor, 1995; Saxena *et al.*, 1994 and 1995). Once such motifs were identified, these specific conserved sequences might indeed be expected to be found in similar plant genes, and this proved to be the case.

The breakthrough in cloning of plant genes that encode the catalytic subunit came in a collaboration between one of us (DPD) and a post-doctorate Dr. Yasushi Kawagoe at the Hebrew University and Dr. Dave Stalker and his colleagues at Calgene, Inc., a plant biotechnology company in California (Pear *et al.*, 1996). Stalker was interested in using our cDNA library prepared from the mRNA of cotton fibers harvested at the active stages of secondary wall cellulose synthesis to find promoters for genes that were highly expressed at this stage of fiber development; we were interested in finding genes involved in cellulose synthesis. Together, we reasoned that sequencing of random cDNA clones might identify clones of interest for both of these goals. After only 207 clones were sequenced, one was identified that showed extremely high expression only during active secondary wall cellulose synthesis in the fiber and showed some weak homology to the bacterial A gene. Further analysis showed that it did indeed contain all the recognized conserved motifs believed to be involved in UDP-glc binding and catalysis, and Dr. Kawagoe also showed that the recombinant protein derived from the region containing these motifs could bind UDP-glc. In retrospect, we now understand what had confounded the previous cloning attempts, as it turns out that, interspersed between these regions with the conserved motifs, are domains unique to the plant A genes. In addition, the plant genes contain a unique extended N-terminal region that contains two zinc fingers (Kawagoe and Delmer, 1997b). The functions of these regions are, at present, unknown, but may represent sites of interaction between A subunits or regions of interaction with other plant-specific proteins such as SuSy or those involved with interaction with the cytoskeleton.

The "final icing on the cake" that confirms the function of A genes in plants comes from the elegant work of Richard Williamson's laboratory and others in Australia in collaboration with Werner Herth of Germany. Williamson's group was searching for mutants impaired in cytoskeletal function, and set up a screen of mutagenized *Arabidopsis thaliana* seedlings to search for those that exhibited root-tip swelling only at high temperature (Baskin *et al.*, 1992). One such temperature-sensitive mutant, called *rsw1*, shows significantly reduced synthesis of crystalline cellulose, accumulation of a non-crystalline form of cellulose, and loss of rosette structure upon transfer to the high temperature. Recent cloning of the gene shows it to be a very close homolog to the cotton A gene, thus confirming by genetic techniques that such A genes encode the catalytic subunit and also provides strong evidence that rosettes are synthase complexes. These workers are proposing that the mutation causes a loss of interaction between the subunits within the rosette which leads to synthesis of non-crystalline cellulose (Arioli *et al.*, 1998).

Once we had identified the cotton A gene, a second gene also expressed in fibers was quickly identified (Pear *et al.*, 1996). Takahisa Hayashi's group in Kyoto has by now found several other cotton A genes (Yuri *et al.*, 1997). We had also found that, once the general A sequence was known, a vast number of A genes, as well as some more distantly related sequences could be identified amongst the expressed sequences tags (ESTs) present in databanks for *Arabidopsis* and rice (Pear *et al.*, 1996). Chris Somerville and his graduate student Sean Cutler have further analyzed many of these ESTs (Cutler and Somerville, 1997), and are now suggesting that the A gene family is a large one with possibly as many as 17 different members in *Arabidopsis*. Also exciting is the possibility that some of the other genes that have similar conserved motifs but otherwise are more distantly related (termed CSL genes for "cellulose synthase-like"), may encode for other plant glycosyltransferases (Cutler and Somerville, 1997). Thus, the fun is just beginning in which major questions can be addressed, such as why do plants have so many genes, what are the functions of the unique plant-specific insertions, and how the complexes are oriented and assembled in the membrane.

Acknowledgments

It has been impossible to document all the findings in this complex field. We offer apologies to any who feel they may have been overlooked. DPD would like to acknowledge in particular the almost continuous support she has received through the years from the U.S. Department of Energy for her work on cellulose biosynthesis. Mary An Godshall and Kanniah Rajasekaran gave editorial advice.

References

Aloni, Y., Delmer, D. P. and Benziman, M. (1982) Achievement of high rates of *in vitro* synthesis of 1,4-β-glucan: Activation by cooperative interaction of the *Acetobacter xylinum* enzyme system with GTP, polyethylene glycol, and a protein factor. *Proc. Natl. Acad. Sci. USA.* **79:** 6558-6452.

Amor, Y., Mayer, R., Benziman, M. and Delmer, D. P. (1991) Evidence for a cyclic diguanylic acid-dependent cellulose synthase in plants. *The Plant Cell* **3:** 989-995.

Amor, Y., Haigler, C. H., Johnson, S., Wainscott, M. and Delmer, D. P. (1995) A membrane-associated form of sucrose synthase and its potential role in synthesis of cellulose and callose in plants. *Proc. Natl. Acad. Sci. USA.* **92:** 9353-9357.

Arioli, T., Peng, L., Betzner, A. S., Burn, J., Wittke, W., Herth, W., Camilleri, C., Hofte, H., Plazinski, J., Birch, R., Cork, A., Glover, J., Redmond, J. and Williamson, R. E. (1998) Molecular analysis of cellulose biosynthesis in Arabidopsis. *Science* **279:** 717-720.

Barber, G. A., Elbein, A. D. and Hassid, W. Z. (1964) The synthesis of cellulose by enzyme systems from higher plants. *J. Biol. Chem.* **239:** 4056-4061.

Baskin, T. I., Betzner, A. A., Hoggart, R., Cork, A. and Williamson, R. E. (1992) Root morphology mutants in *Arabidopsis thaliana. Aust. J. Plant Physiol.* **19:** 427-437

Benziman, M., Haigler, C. H., Brown, R. M. Jr., White, A. R. and Cooper, K. M. (1980). Cellulose biogenesis: polymerization and crystallization are coupled processes in *Acetobacter xylinum. Proc. Natl. Acad. Sci. USA.* **77:** 6678-6682.

Brown, R. M., Jr. and Montezinos, D. (1976) Cellulose microfibrils: visualization of biosynthetic and orienting complexes in association with the plasma membrane. *Proc. Natl. Acad. Sci. USA.* **73:** 143-147.

Brown, R. M., Jr., Willison, J. H. M. and Richardson, C. L. (1976) Cellulose biosynthesis in *Acetobacter xylinum*: Visualization of the site of synthesis and direct measurement of the *in vivo* process. *Proc. Natl. Acad. Sci. USA.* **73:** 4565-4569.

Brown, C. J. (1966) The crystalline structure of the sugars. Part IV. A three-dimensional analysis of β-cellobiose. *J. Chem. Soc.* (A) 927-932.

Brummell, D. A., Catala, C., Lashbrook, C. C. and Bennett, A. B. (1997) A membrane-anchored E-type endo-1,4-beta-glucanase is localized on Golgi and plasma membranes of higher plants. *Proc. Natl. Acad. Sci. USA.* **94:** 4794-4799.

Cardini, C. E., Paladini, A. C., Caputto, R. and Leloir, L. F. (1954) Uridine diphosphate glucose: the coenzyme of the galactose-glucose phosphate isomerization. *Nature* **165:** 191-192.

Cardini, C. E., Leloir, L. F. and Chiriboga, J. (1955) The biosynthesis of sucrose. *J. Biol. Chem.* **214:** 149-155.

Chambers, J. and Elbein, A. D. (1970) Biosynthesis of glucans in mung bean seedlings. Formation of β-(1-4) glucans from GDP-glucose and β-(1-3) glucans from UDP-glucose. *Archives of Biochem. Biophys.* **138:** 620-631.

Colvin, J. R. (1980) The biosynthesis of cellulose. In *Plant Biochemistry*, Preiss, J., ed., NY, Academic Press, Vol. 3, pp. 543-570.

Cutler, S. and Somerville, C. (1997) Cellulose synthesis: Cloning *in silico. Current Biol.* **7:** R108-R111.

Delmer, D. P. (1977) The biosynthesis of cellulose and other plant cell wall polysaccharides. *Recent Adv. In Phytochemistry* **11:** 45-77.

Delmer, D. P. (1983) Biosynthesis of cellulose. *Adv. In Carbohydr. Chem. Biochem.* **41:** 105-153.

Delmer, D. P. (1987) Cellulose biosynthesis. *Ann. Rev. Plant Physiol.* **38:** 259-290.

Delmer, D. P. and Amor, Y. (1995) Cellulose biosynthesis. *Plant Cell* **7:** 987-1000.

Delmer, D. P., Benziman, M., Klein, A. S., Bacic, A., Mitchell, B., Weinhouse, H., Aloni, Y. and Callaghan, T. (1983) A comparison of the mechanism of cellulose biosynthesis in plants and bacteria. *J. Appl. Polym. Sci., Appl. Polym. Symp.* **37:** 1-16.

Emons, A. M. C. (1994) Winding threads around plant cells: a geometrical model for microfibril deposition. *Plant Cell Environment* **17**: 3-14.

Feingold, D. S., Neufeld, E. F. and Hassid, W. Z. (1958) Synthesis of a β-1,3 linked glucan by extracts of *Phaseolus aureus* seedlings. *J. Biol. Chem.* **233**: 783-788.

Giddings, T. H., Brower, D. L. and Staehelin, L. A. (1980) Visualization of particle complexes in the plasma membrane of *Micrasterias denticulata* associated with the formation of cellulose fibrils in primary and secondary walls. *J. Cell Biol.* **84**: 327-339.

Giddings, T. H. and Staehelin, L. A. (1991) Microtubule-mediated control of microfibril deposition: A re-examination of the hypothesis. In *The Cytoskeletal Basis for of Plant Growth and Development*, Lloyd, C. W., ed., London, Academic Press, pp. 85-99.

Glaser, L. (1958) The synthesis of cellulose in cell-free extracts of *Acetobacter xylinum. J. Biol. Chem.* **232**: 627-636.

Green, P. B. and Selker, J. M. L. (1991) Mutual alignments of cell walls, cellulose and cytoskeletons: Their role in meristems. In *The Cytoskeletal Basis for Plant Growth and Development*, Lloyd, C. W., ed., London, Academic Press, pp. 303-322.

Haigler, C. H. (1991) Relationship between polymerization and crystallization in microfibril biogenesis. In *Biosynthesis and Biodegradation of Cellulose*, Haigler, C. H. and Weimer, P. J., eds., NY, Marcel Dekker, pp. 99-124.

Haigler, C. H., Brown, R. M. Jr. and Benziman, M. (1980) Calcofluor White ST alters *in vivo* assembly of cellulose microfibrils. *Science* **210**: 903-906.

Heath, I. B. (1974) A unified hypothesis for the role of membrane bound enzyme complexes and microtubules in plant cell wall synthesis. *J. Theor. Biol.* **48**: 445-449.

Herth, W. (1985) Plasma-membrane rosettes involved in localized wall thickening during xylem vessel formation of *Lepidium sativum L. Planta* **164**: 12-21.

Hess, K. and Trogus, C. (1931) The fiber periods of cellulose derivatives. X. Röntgenographic studies on cellulose derivatives. *Z. physik. Chem.* Bodenstein-festband, 385-391.

Hestrin, S. (1962) Synthesis of polymeric homopolysaccharides. In *The Bacteria*, Gunsalus, I. C. and Stanier, R. Y., eds., NY, Academic Press, Vol. 3, pp. 373-388.

Hopp, H. E., Romero, P. A., Daleo, G. and Pont Lezica, R. (1978) Synthesis of cellulose precursors. The involvement of lipid-linked sugars. *Eur. J. Biochem.* **84**: 561-571.

Jacob, S. R. and Northcote, D. (1985) *In vitro* glucan synthesis by membranes of celery petioles: the role of the membrane in determining the linkage formed. *J. Cell. Sci. Suppl.* **2**: 1-11.

Kang, M. S., Elango, N., Mattie, E., Au-Young, J., Robbins, P. and Cabib, E. (1984) Isolation of chitin synthetase from *Saccharomyces cerevisiae*. Purification of an enzyme by entrapment in the reaction product. *J. Biol. Chem.* **259**: 14966-14972.

Kawagoe, Y. and Delmer, D. P. (1997a) Pathways and genes involved in cellulose biosynthesis. *Genetic Engineering* **19**: 63-87.

Kawagoe, Y. and Delmer, D. P. (1997b) Cotton CelA1 has a LIM-like Zn binding domain in the N-terminal cytoplasmic region. *Plant Physiol.* **114 (Suppl)**: Abstract 337, p. 85.

Koyama, M., Helbert, W., Imai, R., Sugiyama, J. and Henrissat, B. (1997) Parallel-up structure evidences the molecular directionality during biosynthesis of bacterial cellulose. *Proc. Natl. Acad. Sci. USA.* **94**: 9091-9095.

Kudlicka, K. and Brown, R. M., Jr. (1997) Cellulose and callose biosynthesis in higher plants. I. Solubilizaton and separation of (1-3)- and (1-4)-β-glucan synthase activities from mung bean. *Plant Physiol.* **115**: 643-656.

Ledbetter, M. C. and Porter, K. R. (1963) A 'microtubule' in plant cell fine structure. *J. Cell Biol.* **19**: 239-250.

Leloir, L. F. (1971) Two decades of research on the biosynthesis of saccharides. *Science* **172**: 1299-1303.

Leloir, L. F. and Cabib, I. (1953) Synthesis of trehalose. *J. Amer. Chem. Soc.* **75**: 5445-448.

Leloir, L. F. and Cardini, C. E. (1955) The biosynthesis of sucrose phosphate. *J. Biol. Chem.* **214**: 157-165.

Lin, F. C. and Brown, R. M., Jr. (1989) Purification of cellulose synthase *from Acetobacter xylinum.* In *Cellulose and Wood — Chemistry and Technology,* Scheurch, C., ed., NY, Wiley Interscience, pp. 473-492.

Lin, F. C., Brown, R. M., Jr., Drake, R. R., Jr. and Haley, B. E. (1990) Identification of the uridine-5'-diphosphoglucose (UDP-glc) binding subunit of cellulose synthase in *Acetobacter xylinum* using the photoaffinity probe 5-azido-UDP-glc. *J. Biol. Chem.* **265:** 4782-4784.

Machlachlan, G. (1977) Cellulose metabolism in growing cells. Trends in Biol. Sci. **2:** 226-228.

Matthysse, A. G., White, S. and Lightfoot, R. (1995a) Genes required for cellulose synthesis in *Agrobacterium tumefaciens. J. Bacteriol.* **177:** 1069-1075.

Matthysse, A. G., Thomas, D. O. L. and White, A. R. (1995b) Mechanism of cellulose synthesis in *Agrobacterium tumefaciens. J. Bacteriol.* **177:** 1076-1081.

Mayer, R., Ross, P., Weinhouse, H., Amikam, D., Volman, G., Ohana, P., Calhoon, R. D., Wong, H. C., Emerick, A. W. and Benziman, M. (1991) Polypeptide composition of bacterial cyclic diguanylic acid-dependent cellulose synthase and the occurrence of immunologically crossreacting proteins in higher plants. *Proc. Natl. Acad. Sci. USA.* **88:** 5472-5476.

Mueller, S. C. and Brown, R. M., Jr. (1980) Evidence for an intramembrane component associated with a cellulose microfibril synthesizing complex in higher plants. *J. Cell Biol.* **84:** 315-326.

Nicol, F., His, I., Jauneau, A., Vernhettes, S., Canut, H. and Hofte, H. (1998) A plasma membrane-bound putative endo-1,4-β-D-glucanase is required for normal wall assembly and cell elongation in Arabidopsis. *EMBO J.* **17:** 5563-5576.

Pear, J., Kawagoe, Y., Schreckengost, W., Delmer, D. P. and Stalker, D. (1996) Higher plants contain homologs of the bacterial CelA genes encoding the catalytic subunit of the cellulose synthase. *Proc. Natl. Acad. Sci. USA.* **93:** 12637-12642

Preston, R. D. (1964) Structural and mechanical aspects of plant cell walls with particular reference to synthesis and growth. In *The formation of wood in forest trees,* Zimmermann, M. H., ed., NY, Cambridge Phil. Soc. **14:** 281-313.

Ross, P., Mayer, R. and Benziman, M. (1991) Cellulose biosynthesis and function in bacteria. *Microbiol. Rev.* **55:** 35-58.

Ross, P., Weinhouse, H., Aloni, Y., Michaeli, D., Weinberger-Ohana, P., Mayer, R., Braun, S., deVroom, E., van der Marel, G. A., van Boom, J. H. and Benziman, M. (1987) Regulation of cellulose synthesis in *Acetobacter xylinum* by cyclic diguanylic acid. *Nature* **325:** 279-81.

Saxena, I. M., Lin, F. C. and Brown, R. M., Jr. (1990) Cloning and sequencing of the cellulose synthase catalytic subunit gene of *Acetobacter xylinum. Plant Mol. Biol.* **15:** 673-683.

Saxena, I. M., Lin, F. C. and Brown, R. M., Jr. (1991) Identification of a new gene in a operon for cellulose synthesis in *Acetobacter xylinum. Plant Mol. Biol.* **16:** 947-954.

Saxena, I. M., Kudlicka, K., Okuda, K. and Brown, R. M., Jr. (1994) Characterization of genes in the cellulose-synthesizing operon (acs Operon) of *Acetobacter xylinum:* Implicaton for cellulose crystallization. *J. Bacteriol.* **176:** 5735-5752.

Saxena, I. M., Brown, R. M., Jr., Fevre, M., Geremia, R. A. and Henrissat, B. (1995) Multidomain architecture of β-glycosyl transferases: Implications for mechanism of action. *J. Bacteriol.* **177:** 1419-1424.

Shematek, E. M., Braatz, J. A. and Cabib, E. (1980) Biosynthesis of the yeast cell wall. I. Preparations and properties of β-(1-3) glucan synthetase. *J. Biol. Chem.* **255:** 888-894.

Sisson, W. A. (1938) The existence of mercerized cellulose and its orientation in *Halicystis* as indicated by X-ray diffraction analysis. *Science* **87:** 350.

Villemez, C. L. and Heller, J. S. (1970) Is guanosine diphosphate-D-glucose a precursor of cellulose? *Nature* **227:** 80-81.

Volman, G., Ohana, P. and Benziman, M. (1995) Biochemistry and molecular biology of cellulose biosynthesis. *Carbohydrates in Europe* **12:** 20-27.

Wong, H. C., Fear, A. L., Calhoon, R. D., Eichinger, G. H., Mayer, R., Amikam, D., Benziman, M., Gelfand, D. H., Meade, J. H., Emerick, A. W., Bruner, R., Ben-Basat, B. A. and Tal, R. (1990) Genetic organization of the cellulose synthase operon in *Acetobacter xylinum*. *Proc. Natl. Acad. Sci. USA*. **87**: 8130-8134.

Yuri, I., Fukumi, S., and Hayashi, T. (1997) Cloning and sequencing of cotton homologs of *bcsA* encoding cellulose 4-β-glycosyltransferase. *Wood Research* **84**: 1-6.

Chapter 10

The Discovery of Starch Biosynthesis

Jack Preiss and Mirta N. Sivak
Department of Biochemistry, Michigan State University
East Lansing, MI 48824, USA

ABSTRACT

Although the basic studies in starch biosynthesis were carried out in England during the 1940s and led to the discovery of phosphorylase and Q enzyme (branching enzyme), the basis of our modern ideas originated in Argentina from the work of Luis F. Leloir and Carlos E. Cardini. During the late 1950s, they established that nucleoside diphosphate glucose was involved in the synthesis of both glycogen and starch. In this paper, we describe the events leading to these discoveries and the implications of that early research for starch biochemistry as well as for plant biochemistry in general and our understanding of human and bacterial metabolism.

1. Introduction

Starch is a mixture of two polysaccharides, amylose and amylopectin. Amylose is a relatively small (about 1,000 residues) α-1,4-glucan and is largely linear containing occasional α-1,6 branch linkages. Amylopectin is a much larger molecule with a degree of polymerization of 10^4–10^5 residues and has frequent branch points. The chemical and physical aspects of the structure of the starch granule and its components, amylose and amylopectin, have been discussed in some excellent reviews (Morrison and Karkalas, 1990; Hizukuri, 1995). A similar polysaccharide to amylopectin is bacterial or animal glycogen. Glycogen is more branched than amylopectin; the structure of animal glycogen was reviewed by Manners (1991).

Starch is present in most green plants and in practically every type of tissue; leaves, fruits, pollen grains, roots, shoots and stems. Starch is an important end product of carbon dioxide fixation during photosynthesis and in 1837 von Mohl observed them in chloroplasts under the light microscope. These grains were confirmed as starch by Boem in 1856 using the well-known iodine staining test. In 1857, Gris observed their growth during the day and disappearance at night (all the above observations are cited in a monograph by Rabinowitch, 1945). Sachs (1862, 1887), showed that in leaves, starch is formed in the day and degraded by respiration in the dark and postulated that the formation of starch is associated with photosynthesis. In a famous experiment, Sachs exposed one half of a leaf to the sun and kept the other half covered. After some time only the exposed half gave the color reaction with iodine. Pfeffer in 1873 and Godlewski in 1877 (cited in Rabinowitch, 1945) showed that starch synthesis was related to the photosynthetic process by demonstrating that starch is not formed in illuminated leaves in the absence of carbon dioxide.

When a leaf is illuminated in strong light for an hour or more, starch granules appear in the chloroplasts and this can be easily demonstrated by a number of means; electron microscopy (Badenhuizen, 1969) or iodine staining (Meyer and Gibbons, 1951). When storage tissue, seeds, tubers or fruit develop, the starch biosynthetic processes predominate. Degradation of storage tissue starch is predominant during germination or sprouting conditions. In contrast, the biosynthetic and degradative processes of leaf or photosynthetic starch is more dynamic with synthesis predominating during illumination and degradation in the dark.

The biosynthesis of α-1,4-glucans is an important process by which most living organisms accumulate energy reserves to be used when carbon skeletons and energy are not readily available from the environment. The main advantage of using polysaccharides as storage reserves is that, because of their high molecular weights and other physical properties, they have little effect on the internal osmotic pressure in the cell. The many branches of amylopectin and glycogen (α-1,4-chains attached to the main molecule by α-1,6-glucosidic linkages) tend to make the polysaccharide more soluble and is likely to facilitate the action of the different degradative enzymes, which work on the non-reducing ends. In this way, the degradative enzymes, whether amylases or phosphorylases, can act simultaneously at the many ends, speeding the conversion of the large polymers into glucose.

Despite the recognition that starch synthesis was a prominent and significant process for plants, it was not until 1960, ninety eight years after Sachs' paper (1862), that an important reaction leading towards starch synthesis was discovered (de Fekete *et al.*, 1960; Leloir *et al.*, 1961). Many studies of starch synthesis were carried out in England during the 1940s, and led to the discovery of phosphorylase and Q-enzyme (branching enzyme). However, the metabolic routes leading to glucan synthesis were elucidated after the discovery of nucleoside diphosphate sugars by Luis F.

Leloir and coworkers in the 1950s. This finding led to the conclusion that biosynthesis and degradation of glycogen and starch occur by different pathways. Because of the revolutionary nature of these discoveries, in this article we will concentrate in telling this part of the history of the discovery of starch biosynthesis.

2. Early Studies: One Enzyme for Synthesis and Degradation?

In an extensive review of all aspects of starch biochemistry written in 1958, W. J. Whelan (1958) dealt mainly with starch structure. At that time, not much was known about how starch was made in the plant and, in the section dedicated to starch metabolism, two enzymes, with their substrates and functions, are mentioned: phosphorylase and Q enzyme. The first time a polysaccharide was ever synthesized *in vitro* was in the experiments of Hanes (1940). Using a potato tuber extract, he demonstrated the formation of (1,4)-α-glucosidic linkages with glucose-1-P as a glucosyl donor [reaction (I)], the product of this reaction resembled amylose.

glucose-1-P + α-glucan primer <====> Pi + (1->4)-α-glucosyl
glucan   [I]

The enzyme catalyzing this reaction was called phosphorylase (E.C. 2.41.1).

Formation of the (1->6)-α-branch point linkages present in amylopectin and phytoglycogen is catalyzed by the branching enzyme (reaction II; E.C. 2.4.1.18; Bourne and Peat, 1945; Hobson *et al.*, 1950), also called Q enzyme.

Linear glucosyl chain of α-glucan ----> Branched chain of α-1,4 glucan
with α-1->6-linkage branch
points   [II]

At the time of publication of Whelan's review (late 1950s), the idea was that phosphorylase would catalyze both synthesis and degradation of amylose, and that the overall direction of the reaction would depend on the relative concentrations of the substrates glucose-1-P and inorganic phosphate. The presence of a glucan primer that could be elongated was essential. But, even then, there were doubts. Whelan wrote "... It may be questioned whether the *in vivo* synthesis and degradation of α-1,4 linkages are both catalyzed by the same enzyme...Could it be that phosphorylase exists in the plant solely to degrade starch by phosphorolysis and

D-enzyme to catalyze amylose synthesis?" D-enzyme (E.C. 2.4.1.25) was described elsewhere in that review as using the substrates maltotriose and higher maltodextrins to form higher maltodextrins (reaction III), its function was the disproportionation of maltodextrins.

$$\text{maltotriose} + \text{maltotriose} \longrightarrow \text{maltopentaose} + \text{glucose} \qquad [\text{III}]$$

The other big question posed by Whelan was the possible interconversions between sucrose and starch at different stages of development of the plant. Some of the questions posed by Whelan would be answered very soon.

3.   Discovery of UDP Glucose

In 1950, the first sugar nucleotide, uridine disphosphate glucose (UDPGlc) was discovered in yeast by Leloir and his colleagues (Caputto *et al.*, 1950). At the time, the main interest of the group was how disaccharides were synthesized *in vivo*. They started with the study of lactose synthesis in mammary glands but because of difficulties with the experimental system they moved into *Saccharomyces fragilis*, a yeast capable of using lactose. Eventually they looked at galactose utilization and found evidence that galactose was phosphorylated to galactose-1-phosphate (reaction IV, galactose kinase; E.C. 2.7.1.6). The study of lactose utilization led them to the discovery of glucose-1,6-bisphosphate (Caputto *et al.*, 1950) and uridine diphosphate glucose (Leloir, 1950; Leloir and Cardini, 1962a).

They prepared glucose-1-phosphate and galactose-1-phosphate by chemical synthesis (no Sigma then!) and observed that they were utilized when incubated with enzymes from galactose-adapted yeast. The disappearance of the sugar phosphate substrate was accelerated by the addition of a heated yeast extract. At first, they thought that there was only one cofactor in the heated yeast extract, but upon further investigation, they realized that there were two thermostable factors involved. One was necessary for conversion of galactose-1-phosphate into a galactose sugar nucleotide (reaction V) which was then epimerized in another reaction to UDPGlc (UDPGlc 4-epimerase; E.C. 5.1.3.2, reaction VI). The other cofactor was required for the conversion of glucose-1-phosphate into glucose-6-phosphate, that is, a change of position of the phosphate group (phosphoglucomutase, reaction VII, E.C. 2.7.5.1).

galactose + ATP ----> galactose-1-P + ADP [IV]

galactose-1-P + UDPGlc <====> UDPGal + Glc-1-P [V]

UDPGlc <====> UDPGal [VI]

Glc-1-P <====> Glc-6-P [VII]

Using the laborious procedures available at the time, i.e., separation of the barium, calcium or mercury salts, they separated the two active substances. The cofactor of the phosphoglucomutase reaction (reaction VII) was identified first, and turned out to be glucose-1,6-bisphosphate.

Studies of the other factor went more slowly. Purification by fractionation of the mercury salts and charcoal adsorption yielded concentrates that absorbed light at 260 nm. The spectrum was similar but not identical to that of adenosine. Eventually, the spectrum of uridine was published by other workers and that spectrum was identical to the one displayed by the yet to be identified cofactor. In addition to uridine the cofactor was found to contain glucose and two phosphates. Identification of glucose was carried out by paper chromatography, then a relatively new technique. The presence of uridine in a cofactor was novel because in other compounds, i.e., NAD, FAD, ATP, the nucleoside was adenosine. The occurrence of a sugar derivative combined with a nucleoside was also novel. The mechanism of action of UDPGlc became clear when Leloir and his colleagues found that the enzyme preparations active in converting galactose-1-phosphate into glucose-1-phosphate could also catalyze the opposite reaction (galactose-1-P uridylyltransferase, E.C. 2.7.7.12, reaction V).

Soon, it was found that UDPGlc could act as a glucose donor in the synthesis of glycosidic bonds in disaccharides such as trehalose and glycosides. A most important finding came with the discovery of wheat germ enzymes capable of synthesizing sucrose (reaction VIII; Cardini, Leloir and Chiriboga, 1955) and sucrose phosphate (reaction IX; Leloir and Cardini, 1953; Leloir and Cardini, 1955) using UDPGlc as glucose donor.

UDPGlc + fructose <====> sucrose + UDP [VIII]

UDPGlc + fructose-6-P -----> sucrose-6-P + UDP [IX]

Other sugar nucleotides were discovered, including UDP-N'-acetylglucosamine, UDP-N'-acetylgalactosamine and GDP mannose (later found to be involved in the synthesis of mannans), and the list has grown to include over 50 sugar nucleotides (for reviews of sugar nucleotides and their transformations, see Feingold and Avigad, 1980; Feingold, 1982). The initial discovery of sugar nucleotides and subsequently their reactions has had a powerful and positive impact in biochemistry and certainly justified the awarding of the Nobel prize to Luis Leloir in 1970.

4.   Role of Sugar Nucleotides in the Synthesis of Polysaccharides

In 1957, the group at the Campomar Foundation entered a new field of research. In those days, it was firmly believed that glycogen was formed directly from glucose-1-phosphate. But some (e.g., Earl Sutherland and Herman Niemeyer) had expressed doubts about this theory, because of its inconsistency with the effects of hormones that had been observed on glycogen metabolism. For example, epinephrine, which activates phosphorylase *in vitro*, always produces glycogenolysis *in vivo*, suggestive of a degradative (but not synthetic) role for the enzyme. In 1957, the Campomar group (Leloir and Cardini, 1957, 1962b; Leloir *et al.*, 1959) found that liver extracts could catalyze the formation of glycogen, identical to the natural product, from UDPGlc via reaction X.

$$\text{UDP-glucose} + G_n \longrightarrow \text{UDP} + G_{n+1} \qquad\qquad [X]$$

Where G_n represents a molecule of glycogen, and G_{n+1} the same molecule after the addition of a glucose residue in an α-1,4 linkage. This most important finding indicated that, at least in mammals, polysaccharide synthesis and degradation could go via two different pathways. Degradation would be through the phosphorylase reaction and synthesis of α-1,4 glucosidic bonds via glycogen synthase catalysis (E.C. 2.4.1.11).

5.   Discovery of the Starch Synthase

A logical extension of those studies was to investigate synthesis of starch in plants, for which also a dual synthetic and degradative role had been claimed for phosphorylase. UDPGlc was assayed as a glucosyl donor in starch synthesis. Using starch granules isolated from different plant materials such as potato tuber, pea seeds and maize kernels, it was shown that the starch granule as such could act as an acceptor of the radioactive glucosyl portion of the UDPGlc elongating both amylose and amylopectin in α-1,4 linkages (de Fekete *et al.*, 1960; Leloir *et al.*, 1961).   The enzyme trapped in the granule could also add the glucosyl residue from UDPGlc to exogenous maltooligosaccharides (reaction XI).   Since then, starch synthase activity has been reported to be present in many plant extracts (see reviews by Preiss and Levi, 1980, and Preiss and Sivak, 1996, 1998).

Later on, Recondo and Leloir (1961) using chemically synthesized sugar nucleotides in studies of the enzyme specificity showed that another sugar nucleotide, adenosine diphosphoglucose (ADPGlc) was a better glucosyl donor than UDPGlc.   It was later concluded that ADPGlc was likely to be the *in vivo* glucosyl donor for starch synthase, rather than UDPGlc (the glucosyl donor for glycogen synthesis in mammals and yeast) as ADPGlc was ten times more active than UDPGlc.

ADP(UDP) glucose + α-glucan ---> ADP(UDP) +

$$(1\text{->}4)\text{-}α\text{-glucosyl-glucan} \qquad [XI]$$

The above reaction is catalyzed by starch synthase (E.C. 2.4.1.21).

The sugar nucleotides UDP glucose and ADP glucose can be synthesized in plants by either a pyrophosphorylase type reaction (XIIa and XIIb, Espada, 1962) or via a reversal of the sucrose synthase reaction [reaction XIII, Cardini *et al.*, 1955, and de Fekete and Cardini, 1964]

$$α\text{-glucosyl-1-P} + ATP \mathrel{<\!\!====\!\!>} ADPGlc + PPi \qquad [XIIa]$$

$$α\text{-glucosyl-1-P} + UTP \mathrel{<\!\!====\!\!>} UDPGlc + PPi \qquad [XIIb]$$

$$Sucrose + ADP \mathrel{<\!\!====\!\!>} fructose + ADP\ glucose \qquad [XIII]$$

The ADP glucose pyrophosphorylase was first isolated from wheat flour by Espada (1962) at the Campomar Foundation, and he acknowledged the guidance of Leloir.   Recondo *et al.* (1963) later showed that ADPGlc was a natural product as it was found in corn grains.   It was also subsequently found in bacterial cells.

Using ADPGlc as glucosyl donor allowed the identification of a soluble starch synthase enzyme similar in action to that included in the starch granule.   It was however, specific for ADPGlc and inactive with UDPGlc. Much research followed, including that by Carlos Cardini, Rosalia Frydman, Clara Krisman, Juana Tandecarz, Jack Preiss and Takashi Akazawa.   Surprisingly, Akazawa (1994) reported that he was the discoverer of ADPGlc in plants and the proposer of its role in starch biosynthesis. A tribute to the generosity of the Buenos Aires group is that, upon Akazawa's request in 1962, Leloir's group sent him, a competitor, a sample of the chemically synthesized ADPGlc for him to use as standard, even when the Argentinians were actively doing research in exactly the same topic. For the record, Akazawa and his group (Murata *et al.*, 1963, 1964) reported his results in rice much later than Recondo and Leloir (1961), making it difficult to understand what his claims to precedence might be.

In mammalian cells, glycogen synthesis is also relatively well understood and it is important to note that the Leloir group was also the first to demonstrate glycogen synthesis from UDPGlc (Leloir and Cardini, 1957; Leloir *et al.*, 1959). Glycogen synthase however, is specific for UDP-glucose and regulated through hormonally induced post-translational protein modification (multi-phosphorylation of the glycogen synthase by various protein kinases). The glycogen synthase is also allosterically activated by glucose-6-P (for a review of these processes, refer to Preiss and Sivak, 1999). Bacteria and plants follow a slightly different route. Despite the difference in the final product (glycogen or starch), both bacteria and plants use the same glucose donor, ADP glucose, for the elongation of the α-1,4-glucosidic chain instead of UDPGlc. Moreover, in both plants and bacteria, as will be described later, the main regulatory step of glucan synthesis takes place at the level of ADP glucose synthesis.

6.   The Discoverers

Luis Federico Leloir (1983, Fig. 1) graduated as a medical doctor in 1932 in Buenos Aires, Argentina. Frustrated by how little he could help his patients due to the lack of effective treatments, he decided to advance medical knowledge as an investigator. Leloir chose Bernardo A. Houssay as his mentor, under whose direction he started his research on the role of the adrenal glands in carbohydrate metabolism. Later, Houssay would share the 1947 Nobel Prize in Physiology and Medicine with Carl and Gerti Cori for his work on the role of the pituitary gland in carbohydrate metabolism. Leloir was wise in his choice of a mentor; continuing his education in the best laboratories in the world, first in Cambridge at the peak of its glory, and later with the Coris. He was also lucky as he did not have to earn a living since his family owned rich, agricultural land and, not depending on a salary to support his family, could devote himself full time to research in a country that has seldom provided a reliable living wage to its scientists. As for the rest, it was not luck but plain hard work, persistence and dedication. He also had a great ability to work as part of a group.

Fig. 1. The pioneers: Luis F. Leloir (seated) and Carlos E. Cardini.

Leloir was after all human: after he and his collaborators (J. M. Muñoz, E. Braun Menendez and J. C. Fasciolo) discovered angiotensin (they named it hypertensin and published their results in 1939), he was quite upset that others in the U.S., i.e., Irwin Page and his colleagues at Eli Lilly rediscovered and re-named the substance, even though the American group published their results later. Indeed, competitors in other countries had a great advantage: the Argentinian government was usually working against its own scientists, e.g., they fired Houssay because he had signed a letter supporting democracy. This affront lead to the break up of the research group that discovered angiotensin. Deprived of a work place, Leloir left for the U.S. to work with Carl and Gerti Cori in St. Louis and later with David Green in New York. On his return to Buenos Aires, he started a research group that would eventually develop into the Campomar Foundation. After an additional round of humiliations inflicted on Houssay and the others by the government, a solution arose, thanks to the generosity of a private citizen, Jaime Campomar, who in 1946 made possible the start of a (very) modest research institute that had the major advantage of being independent of government abuse. So the experiments that culminated in the discovery of the sugar nucleotides and so many other amazing scientific feats were not performed within the well-equipped, cozy surroundings of a big laboratory but in a small, dilapidated four-room house with little equipment. The facilities were minimal, and the instrumentation almost non-existent. According to the remembrances of Paladini (1995), the first post-doctoral student of Leloir and Cardini, the "cold room" was a butcher's ice box while the dark room, needed for the examination of paper chromatograms under UV light was the only bathroom in the house. We refer the reader to the description by Paladini (1995) of the first fraction collector used in Argentina, made out of a toy Leloir played with as a child.

The research group was rather small and heterogeneous. Even before the creation of the Campomar Foundation, Leloir had attracted some collaborators, Ranwell Caputto, a biochemist with post-doctoral experience obtained in Cambridge and the microbiologist Raul Trucco. Once the Instituto de Investigaciones Bioquimicas was founded, Carlos Cardini, N. Mittelman and A. Paladini (the youngest and first post-doc) came into the group. Later, Enrico Cabib, Horacio Pontis, Eduardo Recondo and others joined the group, which had become a magnet for the brightest young Argentinians interested in scientific research.

As for Cardini, Leloir's collaborator during the crucial years of discovery, his background was mostly in organic chemistry and analytical biochemistry. After major contributions from 1938 to 1946 to the newly established University of Tucuman, he was forced to resign from his job by the Peronist government. Back in Buenos Aires, he was instrumental in the creation of the Campomar Foundation (Jaime Campomar was his brother in law) and the naming of Leloir as his director. But when Cardini joined the new institute, he only accepted a second rank position and never mentioned his role in the creation of the institute. Cardini was an exceptional scholar who loved anonymity. He was also modest, generous and charitable. Carlos Cardini's original research covered a vast field of studies: the discovery of the coenzyme of phosphoglucomutase, glucose-

1,6-bis-P, the enzymatic transformation of galactose into glucose, the isolation of uridine diphosphate glucose and uridine diphosphate N'-acetylglucosamine, glucosamine biosynthesis and the biosynthesis of sucrose, glycogen, glucosides and starch. Figure 1 is a reproduction of a photo of the two eminent scientists working together.

Leloir and Cardini learnt early in their career never to trust the government. The privately funded Fundacion Campomar later accepted government contributions and became affiliated with the University of Buenos Aires, but Leloir, Cardini and the others always retained a certain degree of independence (increased, of course, by the notoriety that only a Nobel Prize can give). Their independence allowed for the acceptance of government-blacklisted investigators (including one of us, MNS) into the institute. It also gave their Institute the continuity necessary to concentrate in creative science rather than survival. Thanks to its founders, the institute became an island of political, religious and racial tolerance uncommon in the country.

7. Starch Biosynthesis Now

Due to the initial findings of the Leloir group, the fields of both animal glycogen synthesis and plant starch synthesis became attractive as a study subject to a number of research groups. Many of the synthetic steps have been explored in great detail but much more remains to be done. Of interest, is the fact that glycogen synthesis in bacteria follows the same pathway as observed for starch synthesis with the sugar nucleotide donor being ADPGlc. It is important to note that the pioneering studies by Oliver Nelson's group in the 1966 indicated that at least in maize, the ADPGlc pathway was the dominant if not the sole pathway to starch (see review of Nelson and Pan, 1995). A null ADPGlc pyrophosphorylase mutant in *Arabidopsis thaliana* (Lin *et al.*, 1988), plus the experiments by Willmitzer's group in Germany using anti-sense mRNA towards potato ADPGlc pyrophosphorylase (Müller Röber *et al.*, 1992) also indicated that for those plants, the route towards starch synthesis was solely through ADPGlc pyrophosphorylase.

An important regulatory phenomenon for starch synthesis actually came from the studies on the regulation of bacterial glycogen synthesis, where it was found that for the enteric ADPGlc pyrophosphorylases, fructose-1,6-bis-P was a potent allosteric activator (see Preiss and Romeo, 1989, 1994). 5'-Adenylate was a potent inhibitor capable of reversing or antagonizing the Fru-bis-P activation. That these phenomena were physiologically pertinent was borne out by the isolation of mutants having

ADPGlc pyrophosphorylases altered in their allosteric properties. If the mutant enzyme had higher affinity for the activator or lower affinity for the inhibitor, the mutant accumulated glycogen at a faster rate.

Similar allosteric mutants (Ball *et al.*, 1991) have been isolated in *Chlamydomonas reinhardtii* and in maize endosperm (Giroux *et al.*, 1996). Almost all plant ADPGlc pyrophosphorylases are sensitive to 3-phosphoglycerate activation and inhibition by phosphate. 3-P-glycerate activation could be reversed by phosphate and 3-P-glycerate could reverse Pi inhibition. Thus the plant and algal ADPGlc pyrophosphorylase activity is regulated by the [3-PGA]/[Pi] ratio. The *Chlamydomonas* mutant was starch-deficient and contained an ADPGlc pyrophosphorylase defective in 3-P-glycerate activation. Conversely, the maize endosperm enzyme has normal 3PGA activation but is less sensitive to Pi inhibition. The maize endosperm mutant accumulates 15% more starch than the normal maize. Thus, the allosteric activation and inhibition effects seen *in vitro* seem to be operative also in the *in vivo* situation. The Monsanto group in St. Louis using an allosteric mutant ADPGlc pyrophosphorylase gene isolated from *Escherichia coli* have transformed some plants (potato tuber, tomato fruit, tobacco callus), and have succeeded in increasing starch content (Stark *et al.*, 1992). These results are in support of the concept that ADPGlc pyrophosphorylase is a rate-controlling enzyme and that the allosteric properties are important. Transformation of the plants with the normal *E. coli* enzyme did not increase the starch levels.

Although the initial findings of the starch biosynthetic enzymes occurred 39 years ago, there is still no precise description of how synthesis of the starch granule is initiated, how amylopectin and amylose are formed, or why starch granules from different species differ in their size, number per cell and composition.

A generally accepted structure of amylopectin is the cluster model structure as postulated by Hizukuri (1986, 1995) and is shown in Fig. 2. An important feature of the model is the α-1,6 linkage branch points present in clusters at certain sites in the amylopectin structure with occurrence of B chains of varying average sizes. They are designated as B1, about 19 glucose units long, B2, about 41, B3, about 69, and B4, about 104–115 glucose units. The number of B3 and B4 chains is very small compared to the quantity of B2 and B3 chains. The B1 chains have only one cluster area extension while the B2, B3 and B4 chains have two, three and four cluster area extensions, respectively. These areas are on the average, separated by 39 to 44 glucose units (Hizukuri, 1986).

Findings of interest are that the starch biosynthetic enzymes, starch synthase and branching enzyme exist in multiple forms and these isozymes are products of different genes (see reviews of Preiss and Sivak, 1996, 1999). What is still not known are the particular roles that the starch synthases (SS) and branching enzyme (BE) isozymes have in formation of the starch granule and amylopectin structure. The different size and number of starch granules per cell from different species may be related to the SS specificities, with respect to chain elongation and size transfer of glucose chain units by BE, and at what glucosyl residues of the B chains as well as the C chain are the α-1,6 bonds formed upon transfer.

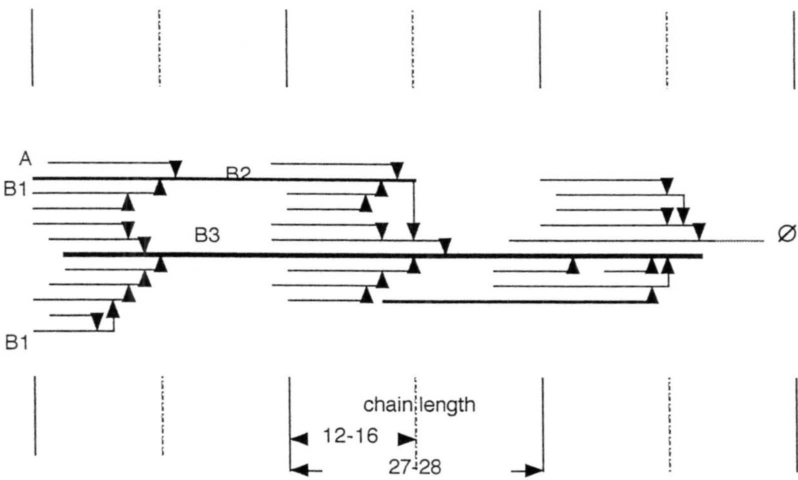

Fig. 2. The model of Hizukuri for amylopectin, showing A, B$_1$, B$_2$, B$_3$ (the very long B$_4$ chains are not illustrated) is the one more broadly accepted. "A" indicates A chains, whereas "B1", "B2" and "B3" are the B chains; the C chain has the only reducing end group, ø, in the polysaccharide. The B$_3$ chains are longer than the B$_2$ chains, which are longer than the B$_1$ chains. The B$_2$, B$_3$ and B$_4$ chains extend into 2, 3 and 4 cluster regions, respectively. The average chain lengths are 19 for B$_1$, 41 for B$_2$, 69 for B$_3$ and 104 for B$_4$ chains. The shortest chain length is for the A chains, which have no branch points.

Dr. Yasuhito Takeda, while on sabbatical leave in Jack Preiss' laboratory in 1988 showed, *in vitro,* that isozyme BEI transfers long chains (DP 40 to > 75) while isozyme BEII transfers shorter size chains (DP 6 to 14). Amylose is the preferred substrate for BEI while amylopectin is the preferred substrate for BEII. Thus, BEI may be more involved in synthesis of the interior B chains while BEII is involved in the synthesis of exterior A and B1 chains. However, it must be pointed out that most probably in the *in vivo* situation, amylose is not the physiological substrate for either branching enzyme I or II, and most probably, synthesis of the branched amylopectin occurs via continual elongation of A and B chains by the starch synthases and then branching by the BE isozymes.

The catalytic properties of the BE isozyme observed *in vitro*, however, reflect to some extent the BE *in vivo* catalysis as the maize isozymes were expressed in *E. coli*. Analysis of product made in the transgenic bacterium did indeed indicate that maize BE I transferred longer chains than BEII and BEII transferred shorter chains (Guan *et al.*,1995).

A synthetic pathway leading to amylopectin and amylose synthesis can be postulated. This pathway is based, to some extent, on the data obtained *in vitro* with the maize BE isozymes by Dr. Takeda (extended by Dr. H. Guan in Preiss' laboratory), and on data obtained by Steven Ball's group of the University of Lille, France (using *Chlamydomonas reinhardtii* mutants affected in the soluble starch synthase II and granule-bound starch synthase).

The *Chlamydomonas* mutant deficient in soluble SSII had only 20–40% of the wild-type starch content and the amylose fraction of the starch increased from 25% to 55% (Fontaine *et al.*, 1993). The structure of its amylopectin was different, with more short chains of 2 DP to 7 DP, and less intermediate size chains of 8 DP to 60 DP. These data suggest that SSII is involved in the synthesis of the intermediate size chains (mainly B chains) in amylopectin. The higher amylose contents could be caused by the inability of the mutant SSII to make extended chains for BE to work on.

Steven Ball's group also showed that the starch granule-bound starch synthase (GBSS) mutants of *C. reinhardtii* displayed the same loss of the amylose fraction as seen for the the *waxy* mutations of many plants (maize, rice and barley endosperm and sorghum and potato tuber) with little effect on the quantity of amylopectin fraction. Thus, GBSS is considered to play a major role in amylose synthesis.

The *C. reinhardtii* double mutants defective in both soluble SSII and a GBSS had a starch content of only 2%–16% of the wild-type (Maddelein *et al.*, 1994). The amount of starch present in the double mutant was dependent on the extent of the GBSS deficiency with the almost null GBSS mutant having very little starch. On this basis, it was suggested that GBSS is important not only for amylose synthesis but also for synthesis of amylopectin. It is possible that GBSS deficiency affects amylopectin synthesis when SSII activity is also low. SSII activity might substitute to some extent for GBSS activity in the synthesis of the amylopectin internal structure, but when it also becomes limiting in the GBSS-deficient background, then amylopectin synthesis becomes limiting.

Phase

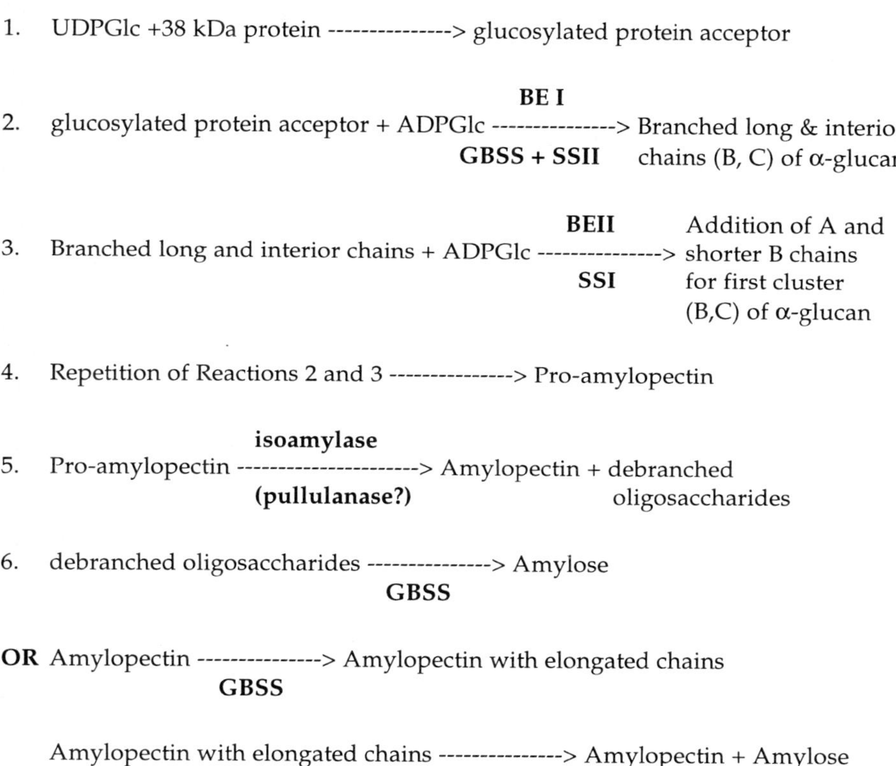

1. UDPGlc +38 kDa protein --------------> glucosylated protein acceptor

                                        BE I
2. glucosylated protein acceptor + ADPGlc --------------> Branched long & interior
                         GBSS + SSII   chains (B, C) of α-glucan

                                        BEII   Addition of A and
3. Branched long and interior chains + ADPGlc --------------> shorter B chains
                                        SSI   for first cluster
                                                     (B,C) of α-glucan

4. Repetition of Reactions 2 and 3 --------------> Pro-amylopectin

                         isoamylase
5. Pro-amylopectin ---------------------> Amylopectin + debranched
                      (pullulanase?)   oligosaccharides

6. debranched oligosaccharides ---------------> Amylose
                         GBSS

OR Amylopectin ---------------> Amylopectin with elongated chains
             GBSS

   Amylopectin with elongated chains ---------------> Amylopectin + Amylose

Fig. 3. Proposed synthesis of amylopectin and amylose. Phase 1 — Initiation of α-glucan synthesis via synthesis of a glucoprotein glucosyl acceptor. Phase 2 — Formation of the interior structure of the ultimate amylopectin product by GBSS, SSII and BEI. Phase 3 — Formation of the external part of the cluster stucture (exterior A and B chains) by SSI and BEII. Phase 4 — Continual repetition of phases 2 and 3 reactions to form the highly branched pro-amylopectin. Phase 5 — Debranching of pro-amylopectin to form amylopectin which can now crystallize and the debranched chains from amylopectin. Phase 6 — Formation of amylose by GBSS elongation or by extension of some of the long chains of amylopectin (which when cleaved by a debranching or branching enzyme become the amylose fraction).

The studies of the *Chlamydomonas* mutants provide evidence for the involvement of the GBSS not only in amylose but also in amylopectin synthesis. On the basis of the observations of the Ball group and the studies on the oligosaccharide transfer by the BEs, a possible route for amylopectin and amylose synthesis can be proposed and is shown in Fig. 3.

Tandecarz and Cardini (1978) have described a system which comprises at least two enzymatic reactions in which proplastid membranes from potato tuber, glucosylates a membrane protein at a serine or threonine residue using UDPGlc. This glucoprotein product, a 38 kDa protein, in turn, is used as an acceptor for a long chain of glucoses sequentially added in a α-1,4 bond using either ADP glucose or glucose-1-P as donors with starch synthase or phosphorylase, respectively. This 38 kDa system has been further characterized and has been purified; the potato enzyme catalyzes its own glycosylation (Ardila and Tandecarz, 1992). The reaction requires Mn^{2+} and thus the reaction is similar to the self-glycosylation carried out by glycogenin in mammalian systems. The studies on mammalian glycogenin have been carried out by Dr. Whelan's group in Miami and Dr. Peter Roach's group in Indianapolis and are reviewed by Preiss and Sivak (1999). Although the enzymatic formation of the putative glucosyl-protein has been further demonstrated in maize endosperm (Rothschild and Tandecarz, 1994), there is still not much information available on the fate of the putative glucan protein or on how ubiquitous these reactions are in plants.

After formation of the unbranched maltodextrin-protein primer of undetermined size, high rates of polysaccharide formation may occur at the surface of the developing starch granule, where GBSS, soluble SSII and branching enzyme I interact with the glucosylated protein primer to form a branched polysaccharide having both long and intermediate size chains.

The suggestion of a phase 2 in Fig. 3 is supported by the studies of the polysaccharide structures observed in the *C. reinhardtii* mutants deficient in SSII and GBSS as well as the study on the amylose extender (*ae*) mutants of rice and maize which are defective in BEII activity. BEII-deficient mutants have altered oligosaccharides with fewer branches and longer size branched chains.

In phase 3, soluble SSI and BE II are postulated to be responsible for the synthesis of the A and exterior B chains to complete formation of the first cluster region in the branched polysaccharide. Phase 4 is essentially repetition of phases 2 and 3 and the outcome is a highly branched α-glucan termed pro-amylopectin. This highly α-branched glucan is water-soluble and non-crystalline and may be similar to phytoglycogen in structure but of a higher molecular size similar to amylopectin. In phase 5, a debranching enzyme (isoamylase) debranches the pro-amylopectin to form an amylopectin fraction capable of crystallization. In phase 6, the chains, liberated by debranching action of the pro-amylopectin, may be used as primers by GBSS to form amylose. Amylose synthesis may occur only in the starch granule internal region and thus, only GBSS would be involved because it may be the only or predominant starch synthase present at the site of amylose synthesis. Branching enzyme activity is quite restricted with respect to the internal regions of the starch granule and thus, the amylose would only be slightly branched. Alternatively, the slight branching observed in the amylose fraction may have occurred prior to the debranching phase 5. Debranching of the pro-amylopectin may have liberated primer oligosaccharide already having some branch chains.

Another possibility as proposed by the group of Steven Ball on the basis of *in vitro* studies (van de Wal *et al.*, 1998) is that the synthesis of amylose occurs by extension of very long size chains, elongated by GBSS catalysis directly on the amylopectin molecule. Direct cleavage of the chains occur to give directly amylose. The nature of the cleavage whether by debranching enzyme, amylase or branching enzyme is unknown.

The postulation of a water-soluble pro-amylopectin is based on the studies of the *sugary* 1 mutation of maize endosperm which contains reduced amounts of amylopectin and starch granules. The mutant accumulates about 35% of its dry weight as phytoglycogen, a highly branched water-soluble polysaccharide (see Nelson and Pan, 1995). The *sugary* 1 mutation has been shown to be deficient in debranching enzyme activity. The evidence that the *sugary* 1 mutation affects the structural gene for a debranching enzyme is supported by the isolation of a cDNA of the *su* 1 gene. Its deduced amino acid sequence is most similar to a bacterial isoamylase (James *et al.*, 1995). The *su* 1 gene product debranching enzyme activity is an isoamylase (Rahman *et al.*, 1998). The specificity of the reaction needs to be studied with respect to the specificity of debranching activity that determines which α-1,6 linkages are cleaved, and which α-1,6 linkages remaining in amylopectin are resistant to debranching action. Crowding of the α-1,6 linkages in the cluster region possibly causes steric difficulties for the debranching of the linkages in the cluster region, but this remains to be shown.

The phases postulated in Fig. 3 do not have to occur in perfect sequence and the phases may have some overlap. For example, phases 2, 3, 4 may overlap; even 5 and 6. The present evidence such as intermediate products formed by starch mutants of *Chlamydomonas* and of higher plants also support the sequence of reactions shown in Fig. 3 for amylopectin and amylose biosynthesis.   Other experiments are certainly required to determine whether the scheme proposed in Fig. 3 is indeed correct.

8.   Conclusion

Leloir became interested in the then very young science of biochemistry because he thought it could clarify the intimate mechanisms of physiological processes. He followed the advice he would later give to young investigators: to work hard, become obsessed with problems that needed to be solved, avoid fashionable subjects, and follow the guide of the experiments rather than those of the printed word.

The important findings of Leloir, Cardini and their collaborators opened new avenues in the understanding of the molecular metabolism of carbohydrates. Their early work was continued by Cardini and his collaborators and by many researchers in other parts of the world, giving us a good understanding of how starch is synthesized in the plant and the mechanisms by which environmental factors affect starch accumulation. Here, we limit ourselves to relevant historical aspects and did not try to cover the many aspects of past and present research on starch metabolism and its interaction with sucrose, partition of photosynthates between organs, etc. We refer the readers to the recent reviews (Ball, 1995; Pan and Nelson, 1995; Stitt and Sonnewald, 1995; Martin and Smith, 1995; Smith *et al.*, 1997; Guan and Keeling, 1998; Preiss and Sivak, 1996, 1998) that cover these important aspects of plant metabolism.

References

Akazawa, T. (1994) Reminiscences, collaborations and reflections. *Photos. Res.* **39**: 93-113.

Ardila, F. J. and Tandecarz, J. S. (1992) Potato tuber UDP-glucose: protein transglucosylase catalyzes its own glucosylation. *Plant Physiol.* **99**: 1342-1347.

Badenhuizen, N. P. (1969) The biogenenesis of starch granules in higher plants. Appleton-Century-Crofts, New York.

Ball, S. G. (1995) Recent views on the biosynthesis of the plant starch granule. Trends in GlycoSci. *Glycotech.* **7**: 405-415.

Ball, S., Marianne, T., Dirick, L., Fresnoy, M., Delrue, B. and Decq, A. (1991) A *Chlamydomonas reinhardtii* low-starch mutant is defective for 3-phosphoglycerate activation and orthophosphate inhibition of ADP-glucose pyrophosphorylase. *Planta* **185**: 17-26.

Bourne, E. J. and Peat, S. (1945) The enzymic synthesis and degradation of starch. I. The synthesis of amylopectin. *J. Chem. Soc. London.* 877-882.

Caputto, R., Leloir, L. F., Cardini, C. E. and Paladini, A. C. (1950) Isolation of the coenzyme of the galactose phosphate-glucose phosphate transformation. *J. Biol. Chem.* **184**: 333-350.

Cardini, C. E., Leloir, L. F. and Chiriboga, J. (1955) The biosynthesis of sucrose. *J. Biol. Chem.* **214**: 149-153.

Cardini, C. E., Paladini, A. C., Caputto, R. and Leloir, L. F. (1950) Uridine diphosphate glucose: the coenzyme of the galactose-glucose phosphate isomerization. *Nature* **165**: 191-192.

Espada, J. (1962) Enzymic synthesis of adenosine diphosphate glucose from glucose-1-phosphate and adenosine triphosphate. *J. Biol. Chem.* **237**: 3577- 3581.

Fekete, M. A. R. and Cardini, C. E. (1964) Mechanism of glucose transfer from sucrose into the starch granule of sweet corn. *Arch. Biochem. Biophys.* **104**: 173-184.

Fekete, M. A. R., de Leloir, L. F. and Cardini, C. E. (1960) Mechanism of starch biosynthesis. *Nature* **197**: 918-920.

Feingold, D. S. (1982) Aldo (and Keto) Hexoses and Uronic acids. In *The Encyclopedia of Plant Physiology*, New Series, Vol. 13A. Plant carbohydrates I., Loewus, F. A. and Tanner, W., eds., pp. 3-76. Springer-Verlag, Berlin.

Feingold, D. S. and Avigad, G. (1980) Sugar nucleotide transformations in plants. In *The Biochemistry of Plants*, Vol. 3. Carbohydrates: structure and function, Preiss, J., ed., pp.102-170. Academic Press, New York.

Fontaine, T., D'Hulst, C., Maddelein, M.-L., Routier, F., Pepin, T. M., Decq, A., Wieruszeski, J.-M., Delrue, B., Van Den Koornhuyse, N., Bossu, J.-P., Fournet, B. and Ball, S. (1993) Toward an understanding of the biogenesis of the starch granule: evidence that *Chlamydomonas* soluble starch synthase controls the synthesis of intermediate size glucans of amylopectin. *J. Biol. Chem.* **268**: 16223-16230.

Giroux, M. J., Shaw, J., Barry, G., Cobb, B. G., Greene, T., Okita, T. and Hannah, L. C. (1996) A single gene mutation that increases maize-seed weight. *Proc. Nat. Acad. Sci. USA.* **93**: 5824-5829.

Guan, H. P. and Keeling, P. L. (1998) Starch biosynthesis: Understanding the functions and interactions of multiple isozymes of starch synthase and branching enzyme. Trends in GlycoSci. *Glycotech.* **10**: 307-319.

Guan, H. P., Kuriki, T., Sivak, M. N. and Preiss, J. (1995) Maize branching enzyme catalyzes the synthesis of glycogen-like polysaccharide in glgB deficient *Escherichia coli*. *Proc. Natl. Acad. Sci. USA.* **92**: 964-967.

Hanes, C. S. (1940) The reversible formation of starch glucose 1-phosphate catalyzed by potato phosphorylase. *Proc. Roy. Soc. London, Ser. B.* **129**: 174-208.

Hizukuri, S. (1986) Polymodal distribution of the chain lengths of amylopectins and its significance. *Carbohydr. Res.* **147**: 342-347.

Hizukuri, S. (1995) Starch: analytical aspects. In *Carbohydrates in Food*, Eliasson, A-C., ed., pp. 347-429. Marcel Dekker, Inc., New York.

Hobson, P. N., Whelan, W. J. and Peat, S. (1950) The enzymic synthesis and degradation of starch. X. The phosphorylase and Q enzyme of broad bean. The Q enzyme of wrinkled pea. *J. Chem. Soc. London.* 3566-3573.

James, M. G., Robertson, D. S. and Meyers, A. M. (1995) Characterization of the maize gene *sugary* 1, a determinant of starch composition in kernels. *Plant Cell.* **7**: 417-429.

Leloir, L. F. (1950) Uridine diphosphate glucose: the coenzyme of the galactose-glucose phosphate isomerization. *Nature* **165**: 191-192.

Leloir, L. F. (1983) Far away and long ago. *Ann. Rev. Biochem.* **32**: 1-15.

Leloir, L. F. and Cardini, C. E. (1953) The biosynthesis of sucrose. *J. Amer. Chem. Soc.* **75**: 6084-6085.

Leloir, L. F. and Cardini, C. E. (1955) The biosynthesis of sucrose phosphate. *J. Biol. Chem.* **214**: 157-162.

Leloir, L. F. and Cardini, C. E. (1957) The biosynthesis of glycogen from uridine diphosphate glucose. *J. Amer. Chem. Soc.* **79**: 6340.

Leloir, L. F. and Cardini, C. E. (1962a) Uridine Nucleotides. The Enzymes, 2nd edition, Vol. 2, pp. 32-61, Academic Press, N.Y.

Leloir, L. F. and Cardini, C. E. (1962b) UDPG-glycogen transglucosylase. In *The Enzymes*, 2nd edition, Vol. 6, pp. 317-325. Academic Press, N.Y.

Leloir, L. F., de Fekete, M. A. R. and Cardini, C. E. (1961) Starch and oligosaccharide synthesis from uridine diphosphate glucose. *J. Biol. Chem.* **236**: 636-641.

Leloir, L. F., Olavarría, J. M., Goldemberg, S. H. and Carminatti, H. (1959) Biosynthesis of glycogen from uridine diphosphate glucose. *Arch. Biochem. Biophys.* **81**: 151-158.

Lin, T. P., Caspar, T., Somerville, C. and Preiss, J. (1988) Isolation and characterization of a starchless mutant of *Arabidopsis thaliana* (L) Henyh lacking ADPglucose pyrophosphorylase activity. *Plant Physiol.* **86**: 1131-1135.

Maddelein, M.-L., Bellanger, F., Delrue, B., Libessart, N., D'Hulst, C., van den Koornhuyse, N., Fontaine, T., Wieruszeski, J.-M., Decq, A. and Ball, S. (1994) Toward an understanding of the biogenesis of the starch granule: determination of granule-bound and soluble starch synthase functions in amylopectin synthesis. *J. Biol. Chem.* **269**: 25150-25157.

Manners, D. J. (1991) Glycogen Carbohydrate Polymers. **16**: 37-82.

Martin, C. and Smith, A. M. (1995) Starch biosynthesis. *Plant Cell* **7**: 971-985.

Meyer, K. H. and Gibbons, G. C. (1951) The present status of starch chemistry. *Advan. Enzymol.* **12**: 341-377.

Morrison, W. R. and Karkalas, J. (1990) In *Methods in Plant Biochemistry*, Dey, P. M., ed., pp. 323-352. Academic Press Ltd., London.

Müller-Röber, B., Sonnewald, U. and Willmitzer, L. (1992) Inhibition of the ADP-glucose pyrophosphorylase in transgenic potatoes leads to sugar-storing tubers and influences tuber formation and expression of tuber storage protein genes. *EMBO J.* **11**: 1229-1238.

Murata, T., Minamikawa, T. and Akazawa, T. (1963) Adenosine diphosphate glucose in rice and its role in starch synthesis. *Biochem. Biophys. Res. Commun.* **13**: 439-444.

Murata, T., Minamikawa, T., Akazawa, T. and Sugiyama, T. (1964) Isolation of adenosine diphosphate glucose from ripening rice grains and its enzymic synthesis. *Arch. Biochem. Biophys.* **166**: 371-378.

Nelson, O. and Pan, D. (1995) Starch synthesis in maize endosperm. *Ann. Rev. Plant Physiol. Plant Mol. Biol.* **46**: 475-499.

Paladini, A. C. (1995) The early days of the Instituto de Investigaciones Bioquimícas, Fundacion Campomar. In *Sucrose metabolism, biochemistry, physiology and molecular biology*, Pontis, H. G., Salerno, G. L. and Echeverria, E. J., eds., pp. 1-3. Amer. Soc. Plant Physiology, Rockville, M.D.

Preiss, J. and Levi, C. (1980) Starch biosynthesis and degradation. *In The Biochemistry of Plants*, Vol. 3. Carbohydrates: structure and function, Preiss, J., ed., pp. 371-423. Academic Press, New York.

Preiss, J. and Romeo, T. (1989) Physiology, biochemistry and genetics of bacterial glycogen synthesis. In *Advances in Bacterial Physiology*, Vol. 30, 184-238.

Preiss. J. and Romeo, T. (1994) Molecular Biology and Regulation of Bacterial Glycogen biosynthesis. In *Progress in Nucleic Acids Research and Molecular Biology*, Vol. 47, Moldave, K. and Cohn, W. E., eds., pp. 299-329. Academic Press, San Diego.

Preiss, J. and Sivak, M. N. (1996) Starch synthesis in sinks and sources. In *Photoassimilate distribution in plants and crops: Source-sink relationships*, Zamski, E. and Schaffer, A. A., eds., pp. 63-96. Marcel Dekker Inc., New York.

Preiss, J. and Sivak, M. N. (1999) Biosynthesis and Regulation of Plant Starch and Bacterial and Mammalian Glycogen Synthesis. In *Comprehensive Natural Products Chemistry*, Vol. 3, Carbohydrates and their derivatives including tannins, cellulose and related lignins, Pinto, B. M., ed., pp. 441-495. Pergamon Press. Dordrecht, Netherlands.

Rabinowitch, E. I. (1945) Photosynthesis and related processes, Vol. 1., p. 42. Interscience Pub. Inc., New York.

Rahman, A., Wong, K. S., Jane, J. L., Myers, A. M. and James, M. G. (1998) Characterization of *Su* 1 isoamylase, a determinant of storage starch structure in maize. *Plant Physiol.* **117:** 425-435.

Recondo, E. F. and Leloir, L. F. (1961) Adenosine diphosphate glucose and starch synthesis. *Biochim. Biophys. Res. Commun.* **6:** 85-88.

Recondo, E., Dankert, M. and Leloir, L. F. (1963) Isolation of adenosine diphosphate D-glucose from corn grains. *Biochem. Biophys. Res. Commun.* **12:** 85-88.

Rothschild, A. and Tandecarz, J. S. (1994) UDP-glucose:protein transglucosylase in developing maize endosperm. *Plant Sci.* **97:** 141-148.

Sachs, J. (1862) Über den einflusz des Lichtes auf die Bildung des Amylums in den chlorophyllkörnern. *Botan. Z.* **20:** 365-373.

Sachs, J. (1887) In *Lectures of the Physiology of Plants* (translated by H. Ward). Oxford Univ. Press (Clarendon), London and New York, pp. 304-325.

Smith, A. M., Denyer, K. and Martin, C. (1997) The synthesis of the starch granule. *Annu. Rev. Plant Physiol. Plant Mol. Biol.* **48:** 67-87.

Stark, D. M., Timmerman, K. P., Barry, G. F., Preiss, J. and Kishore, G. M. (1992) Role of ADP glucose Pyrophosphorylase in regulating starch levels in plant tissues. *Science* **258:** 287-292.

Stitt, M. and Sonnewald, U. (1995) Regulation of metabolism in transgenic plants. *Annu. Rev. Plant Physiol. Plant Mol. Biol.* **46:** 341-368.

Tandecarz, J. S. and Cardini, C. E. (1978) A two-step enzymatic formation of a glucoprotein in potato tuber. *Biochim. Biophys. Acta* **543:** 423-429.

van de Wal, M., D'Hulst, C., Vincken, J.-P., Buléon, A., Visser, R. and Ball, S. (1998) Amylose is synthesized *in vitro* by extension of and cleavage from amylopectin. *J. Biol. Chem.* **273:** 22232-22240.

Whelan, W. J. (1958) Starch and similar polysaccharides. In *Handbuch der Pflanzenphysiologie. Encyclopedia of Plant Physiology.* Vol. VI, pp. 154-240.

Chapter 11

Seed Storage Proteins from the 1700s to the Present

Masahiro Ogawa
Department of Life Science, Yamaguchi Prefectural University
Sakurabatake, Yamaguchi 753, Japan

Thomas W. Okita
Institute of Biological Chemistry, Washington State University
Pullman, WA 99164-6340, USA

ABSTRACT

Plant seeds are a rich food source. They contain high levels of carbohydrate, lipid and protein; nutrients that are essential for human growth and development. A major class of proteins found in seeds are the storage proteins. These proteins (usually) have no enzymatic or structural function other than to simply serve as a nitrogen, sulfur and carbon reserve for the post-germination phase of growth in the young developing seedling. As a result of their nutritional value and their importance in breadmaking, the storage proteins have been studied ever since the 1700s. In this chapter, we present a historical overview on the nature of the storage proteins from different plants with special reference to their structures, evolution, and cellular processes that are responsible for their synthesis and packaging in the cell.

Early Research On the Storage Proteins

The first documented study on storage proteins was conducted by Beccari (1745) who showed that wheat flour could be divided into two fractions, a water-soluble fraction and a water-insoluble fraction. This water-insoluble fraction or wheat gluten, as we know it today, was the first example of a plant protein. Since Beccari's study, other scientists have shown that wheat gluten was composed of a complex mixture of different types of proteins. Parmentier (1773) reported that wheat gluten was largely soluble in vinegar and partially soluble in the spirits of wine which could then be precipitated by neutralization. Much later, Einhof (1805a) found that a part of wheat gluten was extractable by hot alcohol which would precipitate upon cooling. Einhof also discovered that alcohol-soluble proteins also existed in rye (Einhof, 1805b) and barley (Einhof, 1806a) but not in legumes (Einhof, 1806b). Finally, Taddei (1819) found that wheat gluten contained an alcohol-soluble fraction and an insoluble fraction; he named the former gliadin and the latter zymom which we now know today as glutenin. Since these early studies on wheat gluten, the presence of proteins in other plant seeds was recognized. Gorham (1821) described an alcohol-soluble protein in maize and designated it as zein. Braconnot (1827) also discovered the proteins from legumes and named them legumin.

The alcohol-soluble proteins from wheat, rye, barley and maize, and legumin from some leguminous seeds were the only known proteins in the 1800s. Liebig (1841) reviewed the properties and composition of these plant proteins and asserted that they were identical to proteins of animal origin that were described much earlier. Liebig (1841) suggested that these plant proteins could be classified as vegetable albumin, plant gelatin, plant casein, and plant fibrin. We now know that these proteins are grouped according to classes of albumins, prolamins, globulins and glutelins, respectively. Ritthausen (1862) continued the work undertaken by Liebig in the 1860s and prepared highly purified seed proteins from many plant species and determined their elemental composition. Ritthausen recognized that plant proteins were not as simple as animal proteins as there were species differences in the composition, properties and proportions of plant seed proteins. Although the general properties of plant seed proteins were generally accepted by most plant physiologists, Ritthausen's conclusions on the complexity of seed proteins were debated as he prepared the proteins using dilute alkaline solutions which may have possibly altered the properties of the seed protein constituents. Ritthausen's findings and conclusions were largely overlooked and disregarded by most scientists of his time.

During this same period, the localization of seed storage proteins was first studied. Hartig (1856) found that subcellular particles isolated from various oil seeds were rich in protein and he named them "aleuron" after the Greek word for flour. Therefore the word, aleurone grain, was used as the reserve site of seed storage protein until the early 1960s when Graham *et al.* (1962) discovered that wheat developing endosperm cells also contained subcellular organella called a protein body. These protein bodies are morphologically different from aleurone grains and are the reserve site of storage proteins in starchy endosperm.

T. B. Osborne — Father of Seed Storage Proteins

Thomas B. Osborne pioneered the science of seed storage protein research which is described in detail in his monograph "The Vegetable Proteins" (published in 1924). Prior to Osborne's monograph, the concept of plant proteins was not as well-established as for animal proteins. Osborne had a high opinion of Ritthausen's works, but he thought that Ritthausen's research was not extensive enough to form definitive conclusions regarding the properties of the seed proteins. Osborne decided to re-investigate the subjects dealt by Ritthausen in order to clarify the existing uncertainties about the plant seed storage proteins.

From 1891 to 1898, Osborne studied the solubility and chemical properties of seed proteins by using solvent extraction methods and accurately determining their elemental composition. Thus, he confirmed the presence of alcohol-soluble proteins in oat, rye, and barley, and introduced the concept of alcohol-soluble plant proteins which were not present in animals. Osborne designated these alcohol-soluble proteins as "prolamin" for their high content of proline and glutamin (Osborne 1908). Osborne also verified and extended the concept of storage proteins in legumes. He found that pea storage proteins consisted of three different classes of proteins based on heat stability and solubility properties. One class was soluble in 10% NaCl and not coagulated by heating, the second class was soluble in dilute NaCl solution but coagulated by heating, while the third class was soluble in water and easily coagulated by heating. The first class of proteins was found in beans and the second was abundantly found in *Vicia faba*. He designated the three classes as legumin, vicilin, and legumelin (known as albumin today), respectively. Thus the term "legumin" and "vicilin" became the representative names of storage proteins from legumes. Based on his extensive physical and chemical analysis of seed proteins, Osborne concluded that there were four different groups of proteins in plant seeds which he categorized as albumin, globulin, prolamin and glutelin, a concept that has stood the test of time.

The Intermediate Years: 1940s to 1980s

The Legume Globulins

There were few significant advances in storage protein research in the post-Osborne's years until the late 1940s when the new biochemical analytical tools of ultracentrifugation and protein electrophoresis were employed. Danielsson (1949) pioneered the fractionation of legume storage globulins by ultracentrifugation and showed that they consisted of two major components with sedimentation coefficients of 7–8S and 11–14S. These components were later found to be·vicilin and legumin, respectively (Danielsson 1956). Protein electrophoresis techniques were also being developed at the time. MaCalla (1949) demonstrated that cereal prolamins and legume globulins had distinct electrophoretic properties. However, he was unable to resolve legumin and vicilin on the basis of their electrophoretic properties. Wolf and Briggs (1958) were the first to obtain electrophoretically pure glycinin from soybean seeds by subjecting a water-extractable protein fraction to cold treatment. Using this purified protein preparation, Wolf and Briggs (1958) were able to show that glycinin had a molecular weight of 350,000 and was composed of subunits linked by intermolecular disulfide bonds. Later studies (reviewed in Casey *et al.*, 1986) were to show that the 11S storage proteins have an oligomeric structure composed of six pairs of acidic subunits with Mr 35,000 and basic subunits with Mr 20,000 linked by a single disulfide bond. The acidic and basic subunit were products of proteolysis of a larger precursor polypeptide, a post-translational processing which occurs in all 11S globulins (Tumer *et al.*, 1981; Casey *et al.*, 1986; Shotwell and Larkins, 1989).

Unlike the more uniformed properties of the 11S storage proteins, the 7S species were found to be much more heterogeneous because the subunits displayed varying degrees of susceptibility to proteolysis and glycosylation (Casey *et al.*, 1986; Shotwell and Larkins, 1989). Moreover, the quantity of the 7S storage proteins varied considerably depending on the plant species, ranging from 30% of the total proteins in soybean and up to 50% of the total protein in *Phaseolus vulgaris*. The 7S storage proteins have native molecular weights of 130,000 to 180,000. In soybean, the 7S storage protein, β-conglycinin, is a trimer composed of three subunits, α, α′, and β with molecular weights of 57,000, 57,000 and 42,000, while the 7S phaseolin from *Phaseolus aureus* have molecular weights in the range of 45,000 to 55,000 (Casey *et al.*, 1986; Shotwell and Larkins, 1989). The pea 7S fraction was found to be much more complex in structure with major subunits at 75,000, 50,000, 30,000 and 18,000 and minor subunits of 70,000, 49,000, 34,000, 25,000, 14,000, 13,000 and 12,000 (Casey *et al.*, 1986; Shotwell and Larkins, 1989). The 70–75,000 species and 50,000 species were the unprocessed convicilin and vicilin, polypeptides respectively. Pulse-chase radioactive labeling studies (Gatehouse *et al.*, 1981; Chrispeels *et al.*, 1982) showed that the lower molecular weight proteins were proteolytic products processed at one or two internal sites (Spencer *et al.*, 1983). Unlike the 11S storage protein, the 7S subunits are glycosylated, which accounts for some of the diversity in molecular weights (Ericson and Chrispeels, 1973).

Cereal Storage Proteins

The cereal storage proteins, prolamin and glutelin, are not soluble in saline solutions, and thus analytical techniques such as ultracentrifugation and chromatography were not useful for their characterization. Hence, in the 1950s–1960s, considerable effort was made to adapt various protein electrophoresis systems to characterize the protein components of wheat gluten and explain their contribution to gluten properties. Isolating an enriched gliadin fraction and resolving them by using moving-boundary electrophoresis or by urea-starch gel electrophoresis showed that the gliadin fraction contained more than six components, designated as α, β, γ and ω (reviewed in Kasarda *et al.*, 1976), a nomenclature of gliadin polypeptides that still persists today. Analysis by SDS-polyacrylamide gel electrophoresis showed that the α- and γ-gliadin components had molecular weights ranging from 30,000–50,000 with the ω-gliadins having a somewhat larger molecular weight. Bietz and Wall (1972) demonstrated by SDS-PAGE under reducing conditions that glutenin consisted of at least 15 components of both low and high molecular weight (11,600 to 133,000). The types of high molecular subunits in the glutenin fraction could be correlated with the breakmaking qualities of the flour (Payne *et al.*, 1980).

Electrophoretic studies were also carried out on the barley and rye storage proteins by Ben Miflin, Peter Shewry and their colleagues (Miflin *et al.*, 1983). These workers were the first to recognize that despite the extreme polymorphism displayed by the prolamins from wheat, barley and rye, their proteins shared considerable amino acid composition and physical properties. Hence, they classified the complex prolamins of the Triticeae into three major sub-groups: the sulfur-rich, sulfur-poor and high molecular weight prolamins (Miflin *et al.*, 1983; Kreis *et al.*, 1985; Shewry *et al.*, 1995). The sulfur-rich prolamines consist of the wheat α/β-gliadins, γ-gliadins, and low molecular weight glutenin, barley B hordeins, and rye γ-secalins, while the sulfur-poor prolamines include wheat ω-gliadins of wheat, barley C hordein and rye ω-secalins of rye. This classification was later substantiated by the elucidation of the primary structures of these proteins by recombinant DNA techniques.

The Late 1970s to the Early 90s

Just as the introduction and employment of modern biochemical techniques i.e., ultracentrifugation, column chromatography and electrophoresis led to an increase in our knowledge on the structure of the storage proteins, so too did recent methods of molecular biology and recombinant DNA technology.

A scientist who was and remains an instrumental figure in developing the seed storage protein field as it exists today is Brian Larkins. In the 1970s, Larkins and his Ph.D. mentor Eric Davies developed protocols for the isolation of intact membrane-bound and free polysomes from plants, a system notorious for the high ribonuclease activity (Larkins and Davies, 1975).   Their protocols were utilized by many laboratories to isolate polysomes, which facilitated studies on the quantitative and qualitative aspects of protein synthesis during seed development, and which eventually led to the cloning of the storage protein gene sequences. By isolating intact polysomes and assaying them for their protein synthetic capacity, the storage proteins were shown to be translated on the endoplasmic reticulum (Larkins and Dalby, 1975).   Moreover, the *in vitro*-synthesized storage protein products were larger than the native proteins (Larkins and Hurkman, 1978; Burr *et al.*, 1978; Higgins and Spencer, 1981; Croy *et al.*, 1980; Barton *et al.*, 1982), indicating that the storage proteins were synthesized with an additional "signal peptide", a structural motif essential for protein targeting.

Genes of Legume Globulins

The introduction and use of recombinant DNA technology led to the elucidation of the structure of the 7S and 11S storage proteins. Marco *et al.* (1984) first characterized a partial glycinin cDNA sequence and showed directly that the acidic and basic subunits were derived from a larger precursor. Moreover, they showed that the soybean glycinin sequence shared considerable sequence identity to the pea legumin indicating that these 11S storage proteins evolved from a common ancestral gene.

On the basis of the differences of the precursor forms, vicilin has been divided into two groups; one group is composed of precursor polypeptides with Mr of 45,000–50,000 (vicilin), subject to proteolysis, and the second group with Mr of 70,000–75,000 (convicilin), which undergo little or no post-translational cleavage. Biochemical analysis suggested that there were at least three types of vicilin subunits α, β and γ, which could be characterized by their capacity to be proteolytically processed at two major sites, the α-β site and β-γ site (Gatehouse *et al.*, 1983). Lycett *et al.* (1983) first isolated cDNA clones encoding vicilin precursors and determined the partial DNA sequences. They classified four types of vicilin genes on the basis of the processing types: type A (no cleavage at either site), type B (cleavage at the β-γ site only) , type C (cleavage at both sites) and type D (cleavage at the α-β site only).

The best characterized members of the group of 70 kDa convicilin-like polypeptides are the soybean β-congycinin and bean (*Phaseolus vulgaris*) phaseolin. The α′ and α subunits of β-conglycinin showed a higher degree of similarility in their amino acid content (Thanh and Shibasaki, 1978). They consist of multiple components in the 7S protein complex of the mature seed, suggesting that each major subunit is encoded by a small gene family. Schuler *et al.* (1982b) first reported cDNA clones encoding α and α′ subunits of β-conglycinin, and found that the α and α′ subunits are encoded by closely related multigene families differing about 7% of their nucleotide sequence.

The availability of complete cDNA sequences for the various 11S and 7S storage proteins from several different legumes enable the direct comparison of their primary sequences. Argos *et al.* (1985) compared the primary sequences based on their amino acid physical characteristics and showed that the 11S and 7S storage proteins of soybean, pea and French bean were structurally related. They proposed a model in which the primary sequences of these proteins could be divided into three structural domains. Domain I consists of the N-terminal quarter of the storage protein and shares very little conservation among the 11S and 7S storage proteins. Domain II comprises the central region of the protein and contains a mixture of helical, β-sheet and β-turn conformations and contains a hypervariable region which shows little conservation among 11S genes. Domain III consists of the carboxyl-terminal half and is highly conserved among the 11S and 7S storage proteins. Hence, molecular analysis revealed that not only the 11S storage protein genes from both dicot and monocot plants evolved from the same ancestral gene, but that genes from the 7S storage proteins shared this evolutionary origin.

Genes of Cereal Storage Proteins

Although the cereal storage proteins displayed a very diverse polymorphism differing by size and net charge, Miflin *et al.* (1983) recognized very early that they could be classified into three sub-groups: the sulfur-rich, sulfur-poor and high molecular weight (HMW) prolamins on the basis of amino acid composition and molecular weight (Miflin *et al.*, 1983). Elucidation of the primary sequences of prolamin polypeptides by the analysis of recombinant clones supported this classification. Despite the considerable variation in their sequences, all of the sulfur-rich, sulfur-poor and HMW prolamins share two common features (Shewry *et al.*, 1995). First, their structures are composed of modular domains which differ in their amino acid compositions and conformation. Second, all of these prolamins contain at least one domain composed of repeated sequences based on one or more peptide motifs. All of these peptide motifs are highly enriched in Gln and Pro residues. The sulfur-poor prolamins have the simplest structure, consisting almost entirely an eight-amino acid peptide of (Pro-Gln-Gln)-Pro-Phe-(Pro-Gln-Gln). The HMW prolamins have a similar structure as the sulfur-free prolamins in having a large (ca. 80% of the total primary sequence) central domain composed of tandem and interspersed peptide repeats of two or three motifs again rich in Gln and Pro residues. The sulfur-rich prolamines are composed of two domains, an N-terminal domain covering about one-third to one-half of the protein, composed of one or two peptide repeats arranged either tandemly or interspersed. The similarity of these repetitive peptide motifs indicates that these proteins are related and arose from a common ancestral gene.

Kreis *et al.* (1985) showed by direct comparison of the C-terminal non-repetitive domains of the sulfur-rich prolamins that they contain three conserved regions rich in cysteine residues ranging in size between 20 and 30 residues. These three conserved regions, labeled A, B and C, were themselves related to each other suggesting that they arose by a triplication from a short ancestral peptide sequence. The A, B and C regions are also evident in the HMW prolamins although the modular organization of the structural domains of the sulfur-rich and HMW prolamins differ considerably. The A and B regions comprise the N-terminal domain while region C comprises the C-terminal domain which flank the central repetitive domain.

Vestiges of the conserved A, B and C regions are also evident in the prolamins from oat, rice and maize (Shewry *et al.*, 1995). The oat avenins and β- and γ-zeins also contain repetitive peptide sequences whereas the δ-zeins and rice prolamins are devoid of repeats and contain only unique sequences. Similarly, the A, B and C conserved regions are also evident in the 2S albumins which are composed of two polypeptides, molecular weights of 9,000 and 4,000, linked by an interchain disulfide bond. The 2S albumins are synthesized and processed much like the 11S globulins. The conservation of the A, B and C regions in a variety of prolamin polypeptides from monocotyledons and 2S albumins of dicotyledons indicate that these proteins are coded by a gene superfamily of ancient origin (Shewry *et al.*, 1995).

1990 to Today

The study of the seed storage proteins continues to be an area of active research in plant biology. Such general areas of research include the genomic organization and transcriptional regulation of the storage protein genes and protein engineering efforts to increase the nutritive value of these proteins. One research area on the storage proteins which has generated much interest by plant biologists is the molecular and cellular processes that are responsible for the packaging of these proteins within the cell in discrete subcellular organelles, called protein bodies. The storage proteins are packaged into two different protein bodies which can be classified by their morphology (Lott, 1980; Pernollet, 1978). The first type consists only of proteinaceous material surrounded by a limited membrane, formed by dilation of the endoplasmic reticulum (ER). These types of protein bodies are observed in maize, sorghum and rice. The second type, present in seed tissues of all other plants including rice, originates from the vacuole and contains several globoid and crystalloid inclusion bodies, which are embedded in a proteinaceous matrix and surrounded by a membrane. This type of protein body can be seen both in cotyledon tissue of legumes and aleurone and scutellum tissues of cereals. The biochemical and cellular processes responsible for the synthesis, intracellular transport and packaging of the storage proteins continues to be an area of active research and controversy. Questions that remain to be answered are: why are some proteins stored in the ER lumen while others are transported to the vacuole? What signals, if any, are responsible for the retention of proteins within the ER lumen or for their transport to the vacuole? How are proteins transported to the vacuole? What is the nature of the storage vacuole?

Although the cellular basis on why some storage proteins are retained in the ER lumen while others are stored in the vacuole is not known, it is clear that it is not due to the solubility properties of the storage proteins. The alcohol-soluble prolamins of maize, sorghum and rice are stored as ER intracisternal inclusion granules while those from the Triticeae are stored in the vacuole. Also unclear is how ER intracisternal inclusion granules are formed. The maize, sorghum and rice prolamins lack the C-terminal peptide motif K(H)DEL or its functional equivalents, that retain proteins within the ER lumen. Some scientists, however, have suggested that these prolamins contain other types of peptide signals that can serve in this capacity. Alternatively, the capacity of proteins to form an intracisternal inclusion granule depends on the relative rates of protein synthesis versus the relative rates of protein transport to the Golgi apparatus (Okita and Rogers, 1996). Once protein synthesis rates exceed the rates of transport from the ER lumen to the Golgi apparatus, prolamins may reach a critical concentration in the ER such that protein to protein interactions are favored, thereby forming an intracisternal protein inclusion body.

The study of the rice storage proteins may help lead to answers as to why some storage proteins are stored in the ER lumen while others are transported to the vacuole. Rice contains not only prolamins but also glutelins (Tanaka *et al.*, 1980), which are highly homologous to the 11S globulins (Casey *et al.*, 1986; Shotwell *et al.*, 1988). In contrast to the prolamins, the glutelins are stored in the vacuole, and hence, the two storage proteins are synthesized on the ER and translocated to the ER lumen where they are subsequently sequestered from one another. One mechanism that may help explain how rice is able to accommodate the synthesis of both storage proteins on a common site, the ER, but subsequently transport them into separate subcellular compartments is that it segregates the storage protein mRNAs on different ER membranes. Preliminary evidence that the storage protein mRNAs were localized on different ER membranes was obtained by Yamagata *et al.* (1986) and Kim and Okita (1988) which was later proven by Li *et al.* (1993a). Li *et al.* (1993a) demonstrated that prolamin mRNAs predominated on the ER membranes that bound the prolamin protein bodies while glutelin mRNAs were enriched on cisternal ER membranes using *in situ* hybridization of thin sections of developing endosperm at the electron microscopy level. The localization of prolamin mRNAs at sites of the ER where its cognate protein will assemble and accumulate into an intracisternal protein body accomplishes two aspects. First, it reduces potential problems in protein trafficking within the ER lumen if their mRNAs are located on different domains of the ER membrane complex. Secondly, by localizing prolamin mRNAs to a specific ER subdomain, the cognate protein is concentrated in this area and therefore may facilitate protein body formation as discussed above.

How are prolamin mRNAs localized to a specific subdomain, the protein body ER? Although several different mechanisms can account for this cellular process, the currently favored hypothesis is that these mRNAs are actively transported from the nucleus after transcription to specific ER subdomains by a cytoskeleton-mediated process. Abe *et al.* (1991) has demonstrated that zein protein bodies are enmeshed in microfilaments which was confirmed microscopically by Clore *et al.* (1996). Likewise, cytoskeleton elements, actin and tubulin, are associated with the rice prolamin protein bodies. Recently, Muench *et al.* (1998) showed that prolamin polysomes are anchored to the PB surface by a second binding site, in addition to the ribosome binding site of the protein translocation complex of the ER membrane which they suggest to be the cytoskeleton. More recent immunofluorescence microscopic results indicate an interaction between the cytoskeletal elements, microtubules and microfilaments, and the prolamin protein bodies (D. Muench, personal communication). Current research is directed at understanding the role of the cytoskeleton associated with the prolamin protein bodies in possible localization of prolamin mRNAs as well as the translation of these mRNAs.

The localization of prolamin mRNAs may not be the only mechanism for concentrating prolamin within the ER. A second mechanism is the interaction of prolamins with other proteins, specifically the chaperones. The lumenal chaperones such as BiP (immunoglobulin binding protein) are essential not only to facilitate folding of translocated proteins but also aids in transport across the ER membrane (Boston *et al.*, 1991). Interestingly, Li *et al.* (1993b) and Muench *et al.* (1997) showed that BiP is found in very large amounts in prolamin protein bodies. In fact, BiP can be found associated with the prolamin nascent polypeptide chain attached to polysomes, with intact prolamin polypeptides, and with the intracisternal protein body (Li *et al.*, 1993b). These findings suggest that BiP and other lumenal chaperones interact with prolamin to form the protein body.

Protein to protein interactions may also be responsible for protein body formation in maize. In this case, different classes of zein polypeptides may interact with each other to prevent transport to the Golgi. Based on the temporal expression and location of the different classes of zein within the intracisternal protein body, Lending and Larkins (1989) proposed a model for zein protein body formation where the β- and γ-zeins are first observed as electron-dense deposits within the rough ER. Subsequently, α-zein begin to accumulate as electron-lucent globoids within β- and γ-zeins. These α-zein inclusions enlarge and eventually fuse to form a central core with the β- and γ-zeins forming a thin layer at the periphery. Specific protein interactions among the different zein classes may be responsible for retaining these proteins within the ER lumen.

Most storage proteins are transported from the ER lumen and stored into the vacuole via the Golgi apparatus. This was first suggested by Dieckert and Dieckert (1976), who obtained electron micrographs of developing oil seeds, which showed that the electron-dense material observed in the protein bodies were similar to the electron-dense material in vesicles closely associated with the Golgi apparatus. This was later confirmed by immunocytochemical results by Craig and Goodchild (1984) who showed that the storage proteins could be detected in the ER and the Golgi apparatus. Chrispeels (1985) performed pulse-chase experiments showing the passage of the storage protein through the Golgi apparatus in *Phaseolus vulgaris* cotyledons. He succeeded in demonstrating that the labelled phytohemagglutinin that was present in the Golgi cisternae stack moved to the Golgi vesicle and from there to the protein body. In this way, Dieckert's hypothesis (1976) that the Golgi apparatus mediates the transport of the storage proteins was proved immunocytochemically and biochemically. Nearly all of the evidence on the role of the Golgi in the transport of vacuolar proteins has been obtained in legume cotyledons (Chrispeels, 1985).

In contrast to the accepted role of the Golgi apparatus in protein body formation in legumes, there has been much controversy on the formation of the wheat and barley protein bodies. Earlier studies of developing wheat (Parker and Hawes, 1982) and barley (Cameron-Mills and von Wettstein, 1980) endosperm suggested the involvement of the Golgi apparatus at least during the early to mid-stages of seed development. The presence of storage protein-containing vesicles closely associated with the Golgi apparatus in developing wheat endosperm supported this view (Kim *et al.*, 1988). Other researchers obtained microscopic evidence for protein granules surrounded by or closely associated to rough ER membranes. Miflin *et al.* (1981) suggested based on biochemical data that the storage proteins assembled within the ER lumen whereupon the protein inclusion body is released into the cytoplasm. More recently, Galili and his associates have suggested that there are two routes of protein transport from the ER to the vacuole: a Golgi-dependent pathway and a Golgi-independent pathway, with the latter being the dominant transport mechanism (Galili, 1997). Electron micrographs showing the aggregation of wheat storage proteins in the ER lumen and their subsequent transportation to the vacuole by an autophagy-like process were obtained (Levanony *et al.*, 1992). BiP was also observed within the PB in the cytoplasm as well as inside the vacuoles.

The study of the storage protein vacuoles have led to a new discovery in plant cell biology, i.e., the co-existence of two types of vacuoles. It was earlier suggested based on electron microscopic observations that the storage proteins were initially deposited as peripherally located deposits within the large central vacuole. As storage proteins continued to be synthesized and deposited, the central vacuole begins to sub-divide so that at maturity the single central vacuole is displaced by thousands of small, densely filled protein storage vacuoles (Craig *et al.*, 1979). This view of storage protein body formation was widely accepted until a few years ago. Hoh *et al.* (1995) showed that the peripheral clumps of proteins were not contained in the large central vacuole, but instead, located in a closely adjacent tube-like membrane system. The tube-like membrane system could be biochemically distinguished from the central vegetative vacuole, as the former contained α-tonoplast instinct protein (α-TIP) while the latter contained γ-TIP (Okita and Rogers 1996). Hence, the storage proteins are deposited in a membrane system that is morphologically and biochemically distinct from the large vegetative type vacuole.

References

Abe, S., You, W. and Davies, E. (1991) Protein bodies in corn endosperm are enclosed by and enmeshed in F-actin. *Protoplasma* **165**: 139-149.

Agros, P., Narayana, S. V. L. and Nielsen, N. C. (1985) Structural similarity between legumin and vicilin storage proteins from legumes. *EMBO J.* **4**: 1111-1117.

Barton, K. A., Thompson, J. F. Madison, J. T. Rosental, R., Jarvis, N. P. and Beachy, R. N. (1982) The biosynthesis and processing of high molecular weight precursors of soybean glycinin subunits. *J. Biol. Chem.* **257**: 6089-6095.

Beccari, I. B. (1745) De Frumento. De bononiensi scientiarum et artium instituto atque academia commentarii II. Part I. 122.

Bietz, J. A. and Wall, J. S. (1972) Wheat gluten subunits: Molecular weights determined by sodium dodecyl sulfate-polyacrylamide gel electrophoresis. *Cereal Chem.* **49**: 416-430.

Boston, R. S., Fontes, E. B. P., Shank, B. B. and Wrobel, R. L. (1991) Increased expression of the maize immunoglobulin binding protein homolog b-70 in three zein regulatory mutants. *Plant Cell* **3**: 497-505.

Braconnot, H. (1827) Memoire sur un principe particulier aux graines de la famille des legumineuses, et analyse des pois et des haricots. *Ann. Chim. Phys.* **34**: 68-85.

Burr, B., Burr, F. A., Rubenstein, I. and Simon, M. N. (1978) Purification and translation of zein messenger RNA from maize endosperm protein bodies. *Proc. Nat. Acad. Sci. USA.* **75**: 696-700.

Cameron-Mills, V. and von Wettstein, D. (1980) Protein body formation in the developing barley endosperm. *Carlsberg Res. Commun.* **45**: 577-594.

Casey, R., Domoney, C. and Ellis, N. (1986) Legume storage proteins and their genes.*Oxford Surveys of Plant Mol. & Cell Biol.* **3**: 1-83.

Chrispeels, M. J., Higgins, T. J. V., Craig, S. and Spencer, D. (1982) Role of the endoplasmic reticulum in the synthesis of reserve proteins and the kinetics of their transport to protein bodies in developing pea cotyledons. *J. Cell Biol.* **93**: 5-14.

Chrispeels, M. J. (1985) The role of the Golgi apparatus in the transport and post-translational modification of vacuolar (protein body) proteins. *Oxford Survey of Plant Mol. Cell Biol.* **2**: 43-68.

Clore, A., Dannenhoffer, J. and Larkins, B. A. (1996) EF1α is associated with a cytoskeleton network surrounding protein bodies in maize endosperm cells. *Plant Cell* **8**: 2003-2014.

Craig, S., Goodchild, D. J. and Hardham, A. R. (1979) Structural aspects of protein accumulation in developing pea cotyledons. I. Qualitative and quantitative changes in parenchyma cell vacuoles. *Aust. J. Plant Physiol.* **6**: 81-98.

Craig, S. and Goodchild, D. J. (1984) Periodate-acid treatment of sections permits on-grid immunogold localization of pea seed vicilin in ER and Golgi. *Protoplasma* **122**: 35-54.

Croy, R. R. D., Gatehouse, J. A., Evans, I. M. and Boulter, D. (1980) Characterisation of the storage protein subunits synthesised *in vitro* by polyribosomes and RNA from developing pea (*Pisum sativum* L.) II. Vicilin. *Planta* **148**: 57-63.

Danielsson, C. E. (1949) Seed globulins of the gramineae and leguminosae. *Biochem. J.* **44**: 387-400.

Danielsson, C. E. (1956) Plant proteins. *Ann. Rev. Plant Physiol.* **7**: 215-235.

Dieckert, J. W. and Dieckert, M. C. (1976) Production of vacuolar protein deposits in developing seeds and seed protein homology. In *Genetic Improvement of Seed Proteins*, pp. 18-51. Natl. Acad. Sci., Washington, D.C.

Einhof, H. (1805a) Chemische untersuchung der kartoffeln. *Neues Allgem. J. d. Chem.* **4**: 455-508.

Einhof, H. (1805b) Chemische analyse des roggens (Secale cereale) *Neues Allgem. J. d. Chem.* **5**: 131-153.

Einhof, H. (1806a) Chemische analyse des erbsen (*Pisum sativum*) und der reifen saubohnen (*Vicia fava*). *Neues Allgem. J. d. Chem.* **6:** 115-140.

Einhof, H. (1806b) Chemische analyse des kleinen gerste (*Hordeum vulgare*). *Neues Allgem. J. d. Chem.* **6:** 62-98.

Ericson, M. C. and Chrispeels, M. J. (1973) Isolation and characterization of glucosamine-containing storage glycoproteins from the cotyledons of *Phaseolus aureus*. *Plant Physiol.* **52:** 98-104.

Galili, G. (1997) The prolamin storage proteins of wheat and its relatives. In *Cellular and Molecular Biology of Plant Seed Development*, Larkins, B. A. and Vasil, I., eds., pp. 221-256. Kluwer Academic Publisher, Dordrecht, The Netherlands.

Gatehouse, J. A., Croy, R. R. D., Morton, H., Tyler, M. and Boulter, D. (1981) Characterisation and subunit structures of the vicilin storage proteins of pea (*Pisum sativum* L.). *Eur. J. Biochem.* **118:** 627-633.

Gatehouse, J. A., Lycett, G. W., Delauney, A. J., Croy, R. R. D. and Boulter, D. (1983) Sequence specificity of the post-translational proteolytic cleavage of vicilin, a seed storage protein of pea (*Pisum sativum* L.). *Biochem. J.* **212:** 427- 432.

Gorham, J. (1821) Analysis of Indian corn. *Quarterly J. Sci., Liter. and the Arts.* **II:** 206-208.

Graham, J. S. D., Jennings, A. C., Morton, R. K., Palk, B. A. and Raison, J. K. (1962) Protein bodies and protein synthesis in developing wheat endosperm. *Nature* **196:** 967-969.

Hartig, T. (1856) Weitere mittheilungen, das klebermehl (aleuron) betreffend. *Bot. Ztg.* **14:** 257-268; 273-281; 297-305; 313-319; 329-335.

Higgins, T. J. V. and Spencer, D. (1981) Precursor forms of pea vicilin subunits. Modification by microsomal membranes during cell-free translation. *Plant Physiol.* **67:** 205-211.

Hoh, B., Hinz, G., Jeong, B-K. and Robinson, D. G. (1995) Protein storage vacuoles form *de novo* during pea cotyledon development. *J. Cell Sci.* **108:** 299-310.

Kasadra, D. D., Bernardin, J. E. and Nimmo, C. C. (1976) The wheat proteins. In *Adavances in Cereal Science*, Vol. 1, Y. Pomeranz, ed., pp. 158-236. Am. Assos. Cereal Chem, St. Paul.

Kim, W. T. and Okita, T. W. (1988) Structure, expression, and heterogeneity of the rice seed prolamines. *Plant Physiol.* **88:** 649-655.

Kim, W. T., Franceschi, V., Krishnan, H. B. and Okita, T. W. (1988) Formation of wheat protein bodies: involvement of the Golgi apparatus in gliadin transport. *Planta* **176:** 173-182.

Kreis, M., Shewry, P. R., Ford, B. G., Ford, J. and Miflin, B. J. (1985) Structure and evolution of seed storage proteins and their genes with particular reference to those of wheat, barley and rye. In *Oxford Surveys, Plant Mol. Cell Biol.*, Vol. 2, Miflin, B. J., ed., pp. 253-317.

Larkins, B. A. and Dalby, A. (1975) *In vitro* synthesis of zein-like protein by maize polyribosomes. *Biochem. Biophys. Res. Commun.* **66:** 1048-1054.

Larkins, B. A. and Davies, E. (1975) Polyribosomes from peas. V. An attempt to characterize the total free and membrane-bound polysomal population. *Plant Physiol.* **55:** 749-756.

Larkins, B. A. and Hurkman, W. J. (1978) Synthesis and deposition of zein in protein bodies of maize endosperm. *Plant Physiol.* **62:** 256-263.

Lending, C. R. and Larkins, B. A. (1989) Changes in the zein composition of protein bodies during maize endosperm development. *Plant Cell* **1:** 1011-1023.

Levanony, H., Rubin, R., Altschuler, Y. and Galili, G. (1992) Evidence for a novel route of wheat storage proteins to vacuoles. *J. Cell Biol.* **119:** 1117-1128.

Li, X., Franceschi, V. R. and Okita, T. W. (1993a) Segregation of storage protein mRNAs on the rough endoplasmic reticulum membranes of rice endosperm. *Cell* **72:** 869-879.

Li, X., Wu, Y., Zhang, Gilikin, D.-Z., Boston, R. S., Franceschi, V. R. and Okita, T. W. (1993b) Rice prolamin protein body biogenesis: a BiP-mediated process. *Science* **262:** 1054-1056.

Liebig, J. (1841) Ueber die stickstoffhaltigen nahrungsmittel des pflanzenreichs. *Annalen.* **39:** 129-160.

Lott, J. N. A. (1980) Protein bodies. In *The Biochemistry of Plants*, Vol. 1., Marcus, A., ed., pp. 589-623.

Lycett, G. W., Delauney, A. J., Gatehouse, J. A., Gilroy, J., Croy, R. R. D. and Boulter, D. (1983) The vicilin gene family of pea (*Pisum sativum* L.): a complete cDNA coding sequence for Preprovicilin. *Nucleic Acids Res.* **11**: 2367-2380.

MaCalla, A. G. (1949) Nitrogenous constituents of plants. *Ann. Rev. Biochem.* **18**: 615-638.

Marco, Y. A., Thanh, V. H., Tumer, N. E., Scallon, B. J. and Nielsen, N. C. (1984) Cloning and structural analysis of DNA encoding an A2B1a subunit of glycinin. *J. Biol. Chem.* **259**: 13436-13441.

Miflin, B. J., Burgess, S. R. and Shewry, P. R. (1981) The development of protein bodies in the storage tissues of seeds. *J. Exp. Bot.* **32**: 199-219.

Miflin, B. J., Field, J. M. and Shewry, P. R. (1983) Cereal storage proteins and their effects on technological properties. In *Seed Proteins*, Daussant, J., Mosse, J. and Vaughan, J., eds., pp. 255-319. Academic Press, London.

Muench, D. G., Wu, Y., Zhang, Z., Li, X., Boston, R. S. and Okita, T. W. (1997) Molecular cloning, expression and subcellular localization of a BiP homolog from rice endosperm. *Plant Cell Physiol.* **38**: 404-412.

Muench, D. G., Wu, Y., Coughlan, S. J. and Okita, T. W. (1998) Evidence for a cytoskeleton-associated binding site involved in prolamine mRNA localization to the protein bodies in rice endosperm tissue. *Plant Physiol.* **116**: 559-569.

Okita, T. W. and Rogers, J. C. (1996) Compartmentation of proteins in the endomembrane system of plant cells. *Ann. Rev. Plant Physiol. Plant Mol. Biol.* **47**: 327-350.

Osborne, T. B. (1908) Our present knowledge of plant proteins. *Science* **28**: 417-427.

Osborne, T. B. (1924) The vegetable proteins. "Longmans, Green and Co.: London".

Parker, M. L. and Hawes, C. R. (1982) The Golgi apparatus in developing endosperm of wheat (*Triticum aestivum* L.). *Planta* **154**: 277-283.

Parmentier, A. A. (1773) Examin chimique des pommes de terre, dans lequel on traite des parties constituantes du bled. *Didot le Jeune*, Paris, pp. 252.

Payne, P. I., Law, C. N. and Mudd, E. E. (1980) Control by homoeologous group 1 chromosomes of the high-molecular-weight subunits of glutenin, a major protein of wheat endosperm. *Theor. Appl. Genet.* **58**: 113-120.

Pernollet, J.-C. (1978) Protein bodies of seeds: ultrastructure, biochemistry, biosynthesis and degradation. *Phytochem.* **17**: 1473-1480.

Ritthausen, H. (1862) Ueber die bestandtheile des weizenklebers. *J. Pr. Chem.* **85**: 193-212.

Schuler, M. A., Ladin, B. F., Pollaco, J. C., Freyer, G. and Beachy, R. N. (1982a) Structural sequences are conserved in the genes coding for the α, α' and β-subunits of the soybean 7S seed storage protein. *Nucleic Acids Res.* **10**: 8245-8261.

Schuler, M. A., Schmitt, E. S. and Beachy, R. N. (1982b) Closely related families of genes code for the α and α' subunits of the soybean 7S storage protein complex. *Nucleic Acids Res.* **10**: 8225-8244.

Shewry, P. R., Napier, J. A. and Tatham, A. S. (1995) Seed storage proteins: structures and biosynthesis. *Plant Cell* **7**: 945-956.

Shotwell, M. A., Afonso, C., Davies, E., Chesnut, R. S. and Larkins, B. A. (1988) Molecular characterization of oat seed globulins. *Plant Physiol.* **87**: 698-704.

Shotwell, M. A. and Larkins, B. A. (1989) The biochemistry and molecular biology of seed storage proteins. In *The Biochemistry of Plants*, Vol. 15, Marcus, E., ed., pp. 297-345. Academic Press, New York.

Spencer, D., Chandler, P. M., Higgins, T. J. V., Inglis, A. S. and Rubia, M. (1983) Sequence interrelationships of the subunits of vicilin from pea seeds (*Pisum sativum*). *Plant Mol. Biol.* **2**: 259-267.

Taddei, G. (1819) Ricerche sul glutine di frumento. *Giornale di fiscia, chemica, e storia naturale, Brugnatelli*, **2**: 360-361.

Tanaka, K., Sugimoto, T., Ogawa, M. and Kasai, Z. (1980) Isolation and characterization of two types of protein bodies in the endosperm. *Agric. Biol. Chem.* **44**: 1633-1639.

Thanh, V. H. and Shibasaki, K. (1978) Major proteins of soybean seeds. Subunit structure of β-conglycinin. *J. Agr. Food Chem.* **26**: 692-695.

Tumer, N. E., Thanh, V. H. and Nielsen, N. C. (1981) Purification and characterization of mRNA from soybean seeds. Identification of glycinin and β-conglycinin precursor. *J. Biol. Chem.* **256**: 8756-8760.

Wolf, W. J. and Briggs, D. R. (1958) Studies on the cold-insoluble fraction of the water-extractable soybean proteins. II. Factors influencing conformation changes in the 11 S component. *Arch. Biochem. Biophys.* **76**: 377-393.

Yamagata, H., Tamura, K., Tanaka, K. and Kasai, Z. (1986) Cell-free synthesis of rice prolamin. *Plant Cell Physiol.* **27**: 1419-1422.

Chapter 12

The Discovery of 2S Albumins as Abundant Storage Proteins in Seeds

Anthony H. C. Huang
Department of Botany and Plant Sciences
University of California, Riverside, CA 92521, USA

Richard J. Youle
Department of Surgical Neurology, National Institute of Health
Bethesda, MD 20014, USA

Summary

Seeds contain abundant storage proteins as reserve organic nitrogen for germination and postgerminative growth. In the early 1970s, it was thought that in seeds other than the cereals, water insoluble globulins of high molecular weights were the storage proteins, whereas water soluble albumins were metabolic proteins such as enzymes. A series of reports then established that low molecular weight albumins of 2S sedimentation coefficient represent a major group of seed storage proteins, comparable in quantity and ubiquity to the globulins. The 2S albumins are localized in the protein bodies and rapidly degraded during germination and postgerminative growth. They possess high contents of nitrogen-rich amino acid residues and the sulfur-containing methionine and cysteine as reserves of organic nitrogen and sulfur, respectively. They are the seed allergens that had been studied previously. Their high content of the essential amino acids methionine and cysteine make them desirable nutrients for human and animal feed. Subsequent studies by different laboratories have defined the structure of the 2S albumins, the proteolysis of the nascent polypeptide into the final products, the targeting signals in the flexible molecule to the protein bodies, the characteristics of their genes, the inclusion of the cereal prolamins and enzyme inhibitors in the seed 2S albumin family, and the evolution of the whole gene family. Genetic engineering of the 2S albumins has included incorporation of their genes into seed crops to enhance the essential amino acid contents, manufacture of pharmaceutical polypeptides using the abundant 2S albumin as a carrier, and employment of their active seed-specific gene promoters for different purposes.

Introduction

Seed proteins traditionally have been divided into classes based on their solubility in different solvents. Seeds of most species, with the exception of the grass family, contain mainly globulins and albumins. Globulins are insoluble in water, but soluble in salt solution, whereas albumins are water soluble. Some cereals, such as rice and oat, also contain mainly globulins and albumins, whereas other cereals contain abundant prolamins, which are soluble in alcohol solutions. In the early 1970s, it was thought that the globulins, which have high molecular weights and 11–13S and 5–7S sedimentation coefficients, were the storage proteins, whereas the albumins were metabolic proteins such as enzymes. In the late 1970s, we published a series of papers documenting that abundant albumins of low molecular weights and 2S sedimentation coefficient in seeds represent a distinct class of storage proteins, comparable in quantity and ubiquity to the globulins. These proteins have been termed the seed 2S albumins. Subsequent studies by different laboratories have drastically expanded the knowledge on the seed 2S albumins in both the basic and applied arena.

Setting up a Laboratory by a New Assistant Professor

In 1973, I (described by Tony Huang in this section) attained my Ph.D. at the University of California, Santa Cruz under the supervision of Professor Harry Beevers, and then became assistant professor of the Biology Department at the University of South Carolina. Without postdoctoral training and having been in the United States for only four years, I had to quickly adjust myself professionally and culturally. At that time, the research condition in the Biology Department was far from its present sophistication. I had to teach three formal courses per year, and the courses of general biochemistry and plant physiology carried a laboratory period without teaching assistants. In my first year, I was assigned the position of department graduate advisor. Without any staff assistance, I had to file the graduate student records, handle graduate applications within my own office, and type all correspondence. I was expected to do active research but was given only a very small (about 250 square feet) and empty laboratory, and $1,900 to buy supplies. I had to make do with the simple equipment located in an adjacent teaching laboratory. I must emphasize though, that I was in no way mistreated as every faculty member had a similar workload and laboratory setting. After one year, an assistant professor, Dr. Mike Felder, inherited a relatively large, fully equipped laboratory (about 1000 square feet) and kindly offered to share it with me. Also, I had assistance from others: First, the generous Professor Beevers allowed me to take on projects I had initiated in his laboratory, and then publishing the results with myself as the sole author, and using the University of South Carolina's address as my affiliation. Second, my wife was working as a technician in another laboratory and would come to my laboratory to help out occasionally during the evenings and weekends. Still, with the heavy teaching and service loads, plus the professional and cultural adjustments I needed to make, I had to work extremely hard, waiting for luck to arrive.

Luck finally came in the second year in the form of several excellent graduate students who joined my laboratory. One of them was Rich Youle, who eventually discovered the 2S albumins. The other was Tony Cavalieri, who is currently the vice president for research at Pioneer Hybrid International, Inc.; Bob Moreau, who later joined the USDA station in Wydmoore and has quickly moved up to the highest rank; and Dai-fang Liu, who has been developing new vaccines at Lederle-Praxis Biologicals in Rochester, New York. As I was very inexperienced then, I must admit, without any attempt to be modest, that I learned more from them than they did from me. Perhaps my major contribution to them was unintentionally setting an example of hardwork to overcome adversity. These excellent graduate students, who many years later became my good friends, jokingly referred to me as their slave driver. I, however, believed that they were self-motivated; in fact, they probably used me as an excuse to avoid embarrassing their less-driven peers.

My Ph.D. thesis was on the glyoxysomes and peroxisomes in castor bean and other species. I learnt the principles and practices of organelle studies from an excellent mentor, Professor Beevers. Being a young and motivated assistant professor, I was eager to initiate work on organelles other than the peroxisomes, preferably those in castor bean, a favorite of Professor Beevers because the endosperm is large and soft and contains only one cell type. There are two dominant organelles in the ungerminated castor bean, the protein bodies and the oil bodies, neither of which had been studied extensively. I suggested the study of the protein bodies to my first graduate student, Rich Youle.

The Path of Discovery of 2S Albumins as Abundant Seed Storage Proteins

In Fall 1974, Rich Youle, who had just graduated from Albion College in Michigan, came to South Carolina. After rotating in the laboratory of another professor for one semester, he took a second rotation in Tony's laboratory, and ended up staying for his Ph.D. thesis. He worked as a teaching assistant until his third and last year when Tony finally obtained a research grant which allowed him to work as a research assistant. He had a warm-up project of characterizing the regulatory enzyme, fructose-1,6-bisphosphatase, in castor bean (Youle and Huang, 1976a). After that project, he started working on the protein bodies in castor bean.

Seed protein bodies, because of their large size and the defined shape of the internal crystalloids, had been studied by light microscopy more than a hundred years ago (Pfeffer, 1872). The protein body in castor bean has a large protein crystalloid and several small phytin globoids embedded in an amorphous protein matrix, all enclosed by a membrane. Using aqueous media to extract the protein bodies from ungerminated seeds would quickly lead to osmolysis of the organelles. We established an organic solvent system for homogenization of the castor bean and purification of the intact protein bodies by density gradient centrifugation (Youle and Huang, 1976b). We then used a procedure of sucrose density gradient centrifugation to clearly separate the crystalloids, the phytin globoids, the matrix, and the membrane into distinct fractions. This procedure has remained the only published method on the subfractionation of protein bodies from ungerminated seeds. Whereas the crystalloids are the well-known 11S storage proteins, a finding that had been reported by the Russians (Suvorov *et al.*, 1970), the matrix contain major proteins of low molecular weight and minor proteins identified as the castor bean ricin and agglutinins. There is no overlapping of the protein constituents in the matrix and the crystalloids, indicative that the matrix proteins are not those left behind after incomplete crystallization. Similar findings were obtained independently by Ray Tully, then a graduate student in Professor Beevers' laboratory (Tully and Beevers, 1976).

In the early 1970s, storage proteins in seeds other than the cereals were believed to be only water-insoluble globulins of high molecular weight (Millerd, 1975; Ashton, 1976). The concept was that the seed cells would accumulate these proteins without interfering with the osmotic potential during seed dehydration. Water-soluble proteins, the albumins, were considered to be metabolic proteins such as enzymes, and the whereabouts of these proteins inside the cell were unknown. Some cereals, such as rice and oat, also possess mainly globulins and albumins, whereas other cereals contain abundant prolamins, which are soluble in alcohol solutions. In our study, the proteins in the matrix of the castor bean protein bodies were abundant, of low molecular weights, and not very heterogeneous by SDS-PAGE (Youle and Huang, 1976b). Could the conventional concept of seed storage proteins being only water-insoluble and of high molecular weight be wrong? Our further studies indicated that these proteins have a sedimentation coefficient of about 2S and abundant nitrogen-rich amino acid residues, and disappear rapidly during germination and post-germination growth (Youle and Huang, 1978a). We concluded that the matrix 2S albumins are seed storage proteins.

Then, we noticed that decades ago, a research group at the USDA headed by Dr. Joseph Spies had reported the presence of protein allergens in castor bean and other seeds (Spies and Coulson, 1943, Spies, 1967). The group isolated the active protein ingredients on the basis of their differential solubility in various solvents and their heat stability, and tested their allergenic properties on volunteered prisoners. We were intrigued by the similarities between our castor bean 2S albumins and Spies' allergens in amino acid composition, water solubility, and low molecular weight. It was likely that our 2S albumins were the allergens. We tested the 2S albumins for anaphylaxis effect on animals, but without success. We had little knowledge on testing allergenicity on cell cultures, animals or volunteered prisoners, and there was no knowledgeable faculty on campus to consult with (the medical school was not founded until a few years later). We tried to contact Dr. Spies at the USDA in D. C., but he had long retired and left the agency. We were stuck.

Rich spent hours in the library and found the home phone number of Dr. Spies. Dr. Spies still resided in the D. C. area and had written several books on cats. On a trip to the D. C. area for a family gathering, Tony paid a visit to Dr. Spies. Dr. Spies kindly bought Tony lunch at the elite Cosmos Club. During the lunch conversation, Dr. Spies casually mentioned that when he retired, he took with him the purified castor bean and cotton allergens and stored them in the refrigerator at home. Would Tony like to have them?

Dr. Spies purified the allergens from the extracts of ungerminated seeds on the basis of their selective solubility in different solvents and was able to obtain abundant amounts of the proteins. The abundance allowed us to work with the samples with ease. We showed that our castor bean 2S albumins and Spies' allergens were highly similar in their polypeptide profiles by SDS-PAGE, amino acid compositions, and extreme heat stability (Youle and Huang, 1978b). Antibodies raised against Spies' proteins recognized our 2S albumins as revealed by a double diffusion test (Ouchterlony test) and their ability to precipitate the 2S albumins (even without the modern, then unavailable, Protein A). We concluded that the castor bean storage 2S albumins are the seed allergens. Our work on castor bean was quickly extended to cotton seed, whose protein bodies do not have the crystalloids, but nevertheless, have a similar proportion of 11S (actually 9S and 5S) globulins and 2S albumins (Youle and Huang, 1979). Incidentally, Tony still has Spies' purified proteins for whoever wants them.

By then, we realized that Dr. Spies had studied the allergens in the seeds of many species, even though he concentrated his efforts on those in castor bean and cotton seed. We also found reports that the seeds of a few species contained abundant proteins of low molecular weights, even though these proteins were looked upon as metabolic proteins or actually labeled as "nonstorage proteins." Were we looking at a new class of abundant storage proteins of ubiquitous occurrence?

We did a survey of seeds from taxonomically diverse species for abundant 2S albumins (Youle and Huang, 1981). Some of our choices of species were based on Spies' work. We did not choose cereals which were known to contain mainly a different type of storage proteins, the low molecular weight prolamins. Little did we know that the cereal prolamins would eventually be identified as members of the seed 2S albumin family (Shewry 1995). All seeds examined contain two major groups of proteins, the traditional 11S globulins, and the 2S albumins. The 2S albumins represent 20–60% of the total seed proteins — the variation being dependent on the species. They have high mole percentage of nitrogen-rich amino acids, a characteristic of storage proteins. Unlike those of the well-established globulins, the amino acid compositions of the 2S proteins are quite variable, and many of them are also rich in the essential amino acids methionine and cysteine. The Brazil nut proteins have a stunning 17% methionine and 13% cysteine. We concluded that the 2S albumins are a major group of seed storage proteins. We also made two other deductions from the findings. First, the 2S albumins store not just organic nitrogen, but also reduced sulfur for seed germination. Second, they are superior to the classical globulins in nutrition for human and animal feeds because of their higher contents of the essential amino acids, especially methionine and cysteine.

By the end of 1997, Rich graduated. He stayed in Tony's laboratory for only three years. Without any prior research experience and carrying a heavy teaching assistantship workload for the first two years, he finished a quality thesis. The thesis also had unpublished data on the characteristics and developmental changes of enzymes that hydrolyze the protein body constituents. Rich joined the NIH as a staff fellow and has been prosperous in research. In his early career at NIH, he probed the anti-cancer potential of castor bean ricin by joining the ricin toxic subunit with antibodies raised against the surface constituents of specific cancer cells (Youle *et al.*, 1981). After Rich and the other excellent graduate students left within a one-year period, it took Tony, who had been spoilt by their talents, several years to rebuild the laboratory.

The Brazil Nut Incident

After our last paper on the survey of seed storage albumins was published, Dr. Sam Sun of the ARCO Plant Cell Research Institute called. He asked if we were cloning the gene encoding the Brazil nut proteins, which have 17% methionine and 13% cysteine. He also asked for our unpublished data on the Brazil nut proteins and some of our Brazil nut. The early 1980s was the dawn of the explosion of genetic engineering in agriculture, and most academic researchers including ourselves did not know the patentability of their findings. We sent our unpublished SDS-PAGE pictures of the Brazil nut proteins and the Brazil nut to Dr. Sun. Eventually, his group cloned the genes, and the deduced amino acid sequences of the proteins confirmed our earlier findings of the methionine and cysteine contents (Altenbach *et al.*, 1987). He was kind to call and offer co-authorship, but we declined, feeling that our contributions were too minimal. He then transformed tobacco with the Brazil nut gene, and the seed of the transformed tobacco had an enhanced methionine content. Independently, another group in Brazil-Europe was working on the same project (Ampe *et al.*, 1986).

Soon, ARCO sold the Plant Cell Research Institute to Monte Dison. Within a few years, the Institute closed its doors, and the Brazil nut genes and the related findings were acquired by Pioneer Hybrid International, Inc. Pioneer researchers were successful in putting and expressing the Brazil nut gene in soybean, and the methionine content in the transformed soybean was enhanced. Soybean proteins have a very desirable content of the essential amino acids with the exception of methionine. Soybean proteins supplemented with methionine or Brazil nut extracts have a higher nutritional value in feeding animals. A success in transforming soybean with the Brazil nut gene was supposed to be a beautiful story of genetic engineering to improve crops, until the allergenic properties of the Brazil nut proteins were realized by the industrial researchers (Nordlee *et al.*, 1996). Critics were fearful that the transformed soybean would end up being used without knowledge by people who are allergic to the Brazil nut proteins. The incident made national and international news. On March 14, 1996, the Washington Post published a six-column article in Section A, describing the incident. The news article quoted an immunologist, who helped Pioneer test the allergenic properties of the transgenic soybean, of saying "There are probably 1,000 or more proteins in Brazil nuts. Pioneer was interested in one. It's pretty unlucky for them that it turned out to be the allergen." That was an uninformed statement. The last paragraph of the short Discussion of our paper (Youle and Huang, 1981), which was written in 1979 without the foresight of the capability of modern genetic engineering, stated:

"The 2S proteins with their high cysteine content should be more valuable for human nutrition than other storage proteins (globulin) in the same seeds. Selection of plant varieties for amino acid composition of high nutritional value does not usually generate new protein species but merely shifts the relative proportions of proteins already present so that the more nutritional valuable protein species constitute a high proportion of the total protein. The 2S protein should be an important consideration in the selection of plant varieties with a high content of the essential amino acids, cysteine and methionine. Since we established earlier that the castor bean and cotton 2S proteins are potent allergens, the allergenic properties of the 2S proteins deserve consideration in the nutritional evaluation of this class of seed proteins."

It was unfortunate that the industrial researchers did not pay sufficient attention to our studies. On the other hand, the fear of the allergenic properties was overstated. There are ways to overcome the allergenic problem, such as labeling of the food products, as with the peanut and other nuts that possess potent allergens. The transformed soybean could be tagged easily with a dark color seed coat by simple breeding. There are many other avenues. Nevertheless, Pioneer dropped the Brazil nut project and moved onto genes encoding other proteins that are rich in methionine or other essential amino acids. Some researchers in Europe have continued to develop seed crops transformed with the Brazil nut genes. Currently, most of the genes that encode proteins rich in essential amino acids being explored commercially belong to the 2S albumin gene family. This is not unexpected because of the flexibility of the 2S albumins in amino acid composition. Presumably, these 2S albumins are less allergenic than the Brazil nut protein.

Other Recent Studies of the Seed 2S Albumins

Studies of the seed 2S albumins in the 1980s and 1990s have been propelled by the advancing techniques of molecular biology and genetic engineering (reviewed by Shewry *et al.*, 1995). The genes encoding many 2S albumins in various seed species have been cloned. Depending upon the plant species, the 2S albumins in a seed may contain 1–2 polypeptides (e.g., of 9kD and 3kD), which are derived from a nascent polypeptide after specific proteolysis. The structures of the proteins are much more flexible than the rigid seed globulins, and the flexibility explains the variability of the amino acid composition. The targeting signals in the 2S albumins, which have variable sequences and structures, to the protein bodies have been defined. Comparing the DNA sequences of the genes and the deduced amino acid sequences has allowed the inclusion of the cereal prolamins and many seed enzyme inhibitors to the seed 2S albumin family and the charting of the evolution of all the genes.

Manipulations of the 2S albumins via genetic engineering for commercial purposes have been active. Dozens of genes encoding seed 2S albumins with high contents of specific essential amino acids from different plant species have been used to transform seed crops that are deficient in these essential amino acids (Sun *et al.*, 1996). Pharmaceutical polypeptides have been produced successfully via genetic engineering in a pilot trial by attaching the polypeptide to a highly flexible region of the 2S albumin and by subsequently releasing the polypeptide with a specific protease (Krebbers *et al.*, 1993). The promoters of the 2S genes are active and seed-specific, such as that encoding a 2S albumin (termed napin) in *Brassica napus*, and have been used to express foreign genes in the maturing seeds for different purposes (Kridl *et al.*, 1991).

Acknowledgment

We wish to express our sincerest thanks to the Biology Department at the University of South Carolina and its many staff members who had made the discovery of the 2S albumins possible. Drs. Mike Felder, Wally Dawson, and Dave Claybrook offered plenty of technical assistance as well as moral encouragement. The latter portion of our research was supported by a grant from the National Science foundation.

References

Altenbach, S. B., Pearson, K. W., Leung, F. W. and Sun, S. S. M. (1987) Cloning and sequence analysis of a cDNA encoding a Brazil nut protein exceptionally rich in methionine. *Plant Mol. Biol.* **8**: 239-250.

Ampe, C., van Damme, J., De Castro, L. A. B., Sampaio, M. J. A. M., van Montagu, M. and Vanderkerckhove, J. (1986) The amino-acid sequence of the 2S sulfur-rich proteins from seeds of Brazil nut (*Beretholletia excelsa* H.B.K.). *Euro. J. Biochem.* **159**: 597-604.

Ashton, F. M. (1976) Mobilization of storage proteins of seeds. *Annu. Rev. Plant Physiol.* **27**: 95-117.

Krebbers, E., Da Silva Conceicao, A., Denis, M., D'Hondt, K. and Vandekerckhove, J. (1993) Modification of plant seed storage protein. In *Seed Storage Compounds: Biosynthesis, Interactions and Manipulation*, Shewry, P. R. and Stobart, A. K., eds. (Oxford University Press), pp. 317-324.

Kridl, J. C., McCarter, D. W., Rose, R. E., Scherer, D. E., Knutzon, D. S., Radke, S. E., and Knauf, V. C. (1991) Isolation and characterization of an expressed napin gene from Brassica rapa. *Seed Science Research* **1**: 209-219.

Millerd, A. (1975) Biochemistry of legume seed proteins. *Annu. Rev. Plant Physiol.* **26**: 53-72.

Nordlee, J. A, Taylor, S. L., Townsend, J. A., Thomas, L. A. and Bush, R. K. (1996) Identification of a Brazil-nut allergen in transgenic soybeans. *New England J. Medicine* **334**: 688-692.

Pfeffer, W. (1872) Underschungen uber die Proteinkorner und die Bedeutung des Asparagins beim Keimen der Samen. *Jahrb. Wiss. Bot.* **8**: 429-571.

Shewry, P. R. (1995) Plant storage proteins. *Biol. Rev.* **70**: 375-426.

Shewry, P. R., Napier, J. A. and Tatham, A. S. (1995) Seed storage proteins: Structures and Biosynthesis. *Plant Cell* **7**: 945-956.

Spies, J. R. (1967) The chemistry of allergens. XIX. On the number of antigens and the homogeneity of the isolated antigens of fraction CB-1A from castor beans. *Ann. Allergens* **25**: 29-34.

Spies, J. R. and Coulson, E. J. (1943) The chemistry of allergens. VIII. Isolation and properties of an active protein-polysaccharide fraction CB-1A from castor bean. *J. Am Chem. Soc.* **65**: 1720-1725.

Sun, S. S. M., Zuo, W., Tu, H. M. and Xiong, L. (1996) Plant proteins: engineering for improved quality. *New York Acad. Sci.* **792**: 37-49.

Suvorov, V. I., Buzulukora, N. P., Sobolov, A. M. and Sveshnikova, I. N. (1970) Structure and chemical composition of globoids from aleurone grains of castor seeds. *Fiziol. Rast.* **17**: 1223-1231.

Tully, R. E. and Beevers, H. (1976) Protein bodies of castor bean endosperm. Isolation, fractionation, and the characterization of the protein components. *Plant Physiol.* **58**: 710-716.

Youle, R. J. and Huang, A. H. C. (1976a) Development and properties of fructose 1.6-bisphosphatase in the endosperm of castor bean. *Biochem. J.* **154**: 47-652.

Youle, R. J. and Huang, A. H. C. (1976b) Protein bodies from the endosperm of castor bean. Subfractionation, protein components, lectins, and changes during germination. *Plant Physiol.* **58**: 703-709.

Youle, R. J. and Huang, A. H. C. (1978a) Albumin storage proteins in the protein bodies of castor bean. *Plant Physiol.* **61**: 13-16.

Youle, R. J. and Huang, A. H. C. (1978b) Evidence that the castor bean allergens are the albumin storage proteins in the protein bodies of castor bean. *Plant Physiol.* **61**: 1040-1042.

Youle, R. J. and Huang, A. H. C. (1979) Albumin storage proteins and allergens in cottonseed. *J. Agr. Food Chem.* **27**: 500-503.

Youle, R. J. and Huang, A. H. C. (1981) Occurrence of low molecular weight and high cysteine containing albumin storage proteins in oilseeds of diverse species. *Amer. J. Bot.* **68**: 44-48.
Youle, R. J., Murray, G. J. and Neville, D. M., Jr. (1981) Studies on the galactose binding site of ricin and the hybrid toxin man6P-ricin. *Cell* **23**: 551-559.

Chapter 13

The Discovery of Maternal Inheritance of Large Subunit of Rubisco

Shain-Dow Kung
Department of Biology
The Hong Kong University of Science and Technology
Clear Water Bay, Kowloon, Hong Kong

My involvement in the study of Rubisco (ribulose-1,5-bisphosphate carboxylase-oxygenase), goes back almost thirty years ago. In 1971, I joined Dr S G Wildman's Laboratory in UCLA. Wildman's laboratory was, at that time, heavily engaged in the study of various chemical components of chloroplasts, such as chloroplast DNA (ct-DNA) and chloroplast proteins. Therefore, my PhD training on ct-DNA in Toronto qualified me to join his group. I studied the physico-chemical properties of ct-DNA in the University of Toronto and published a couple of papers in BBA (Kung and Williams, 1968 and 1969) which caught his attention. He offered me a position as a research associate, a super post-doctoral fellow, since I had already finished two years' post-doctoral training in the Biochemistry Department, The Hospital for Sick Children, Toronto. Although my PhD degree was from the department of Botany, my work was entirely in the area of biochemistry and biophysics. The training in plant physiology, biochemistry and biophysics proved to be a commendable advantage in advancing my career.

When I arrived in Wildman's laboratory in the middle of January 1971, he assigned me to work with one of his PhD students from Hong Kong, Flossie Wong. Flossie was a brilliant student who is today a world-renowned virologist in UCSD. I worked with her on the physico-chemical aspects of tobacco ct-DNA. We used the Model E Analytic Ultracentrifuge for analysis. I was the in-house expert in operating this equipment after Dr K. K. Tewari left. However, after only a very brief period of time, I was reassigned to work on a different subject with Dr P. Thornber who just joined the department a short while ago.

Thornber was a leading expert in photosystems. He was already a well established scientist before he joined UCLA. I spent most of my first year with him and learned a great deal about chlorophyll-protein complexes. I also served as a bridge between Wildman's and Thornber's laboratories on the subject they collaborated on. Thornber was a good teacher, scientist and friend. Unfortunately, he is no longer with us but we will always remember him. In about six months in 1971, I mastered and improved the isolation and purification techniques he devised for the photosystems. More importantly, by using the improved procedure, we obtained a large quantity of pure material for further analysis. The time was the summer of 1971. At the same time, Wildman had already devised a method to study the mode of inheritance of several chloroplast components. The devise was original, unique and straightforward in concept. It required the knowledge of both plant breeding and biochemical techniques in practice. I have always thought that it was one of Wildman's many great and original ideas upon which many scientific contributions were made. He deserved much more recognition from the scientific community.

Wildman's method to study the mode of inheritance of many chloroplast components including Rubisco deserves special mention. It was based on the combination of reciprocal interspecific hybridization technique and the two-dimensional chromatography or finger printing techniques (Chan and Wildman, 1972). The basis for this approach is that if genetic information for an organellar protein (chloroplast or mitochondria) is transmitted through both the maternal or paternal lines, then it is assumed that this information is located on the chromosomal DNA residing in the nucleus. Conversely, if the mode of inheritance is strictly maternal, the coding information is contained in the ct-DNA or in the mitochondrial DNA (mt-DNA). By using this approach, it was demonstrated that the genes coding for many chloroplast proteins are inherited only from the maternal line and are thus located in the chloroplast genome (Wildman *et al.*, 1972). Similarly, genetic analysis showed that the coding information for many other proteins is transmitted biparentally and is, therefore, contained in the nuclear genome (Wildman *et al.*, 1972).

Using this approach, Wildman's laboratory in 1971 demonstrated for the first time, that the nuclear DNA contained the coding information for the small subunit (SS) of Rubisco (Kawashima and Wildman, 1972) and three proteins in the 50S subunit of chloroplast ribosomes (Bourque and Wildman, 1973). Following the similar reasoning and procedures, we also, for the first for the first time, demonstrated that the light-harvesting chlorophyll-protein complex is coded by the nuclear genome (Kung *et al.*, 1972).

I consider myself very fortunate to have published a good paper on this new subject within six months (Kung and Thronber, 1971), and another one in the following months (Kung, Thronber and Wildman, 1972). This was my second post-doctoral fellowship and this was also the second time that I was able to publish a paper within the first six months. I sincerely believe that a good start assures a successful finish.

Once I completed my first year at UCLA, Wildman again reassigned me to another project, to work on Rubisco together with a post-doctoral fellow, Dr K. Sakano from Japan. My research project was changed for the second time, and the topic became the third one I was asked to work on within fifteen months.

By the middle of 1972, I switched entirely from Photosystem to Rubisco. Though I had very little knowledge of Rubisco, I was suddenly supposed to be an expert on the subject, just because I worked in the laboratory of Wildman who is widely recognized as the father of Rubisco. I studied very hard to prepare myself for this new challenge. I learnt very quickly that Rubisco is a very exciting and interesting as well as popular subject. At the end, I managed not to disappoint myself or Wildman.

Rubisco, a photosynthetic enzyme, was first discovered by Wildman and Bonner and named fraction 1 protein because of its fast mobility under centrifugal forces (Wildman and Bonner, 1947). This protein is a major leaf component comprising more than 50% of the total soluble protein in C_3 plants (Kawashima and Wildman, 1971) and has been widely recognized by all investigators in this field as the most abundant protein on earth (Kung, 1976 and Ellis, 1981). It is an important photosynthetic enzyme catalyzing the formation of phosphoglycerate (PGA) from CO_2 and ribulose-1,5-bisphosphate (RuBP). This is the carboxylation reaction in the biosphere that leads to net CO_2 conversion into carbohydrates. This enzyme also catalyzes the oxygenation reaction of the same substrate to produce one molecule each of PGA and phosphoglycolate (Lorimer, 1981). Phosphoglycolate is a substrate for photorespiration. Thus, Rubisco is central to both the net fixation of CO_2 into carbohydrates and the oxidation of carbohydrates leading to the net loss of CO_2.

Rubisco is known as one of the many important bifunctional enzymes. It is believed by many scientists that this enzyme may be the key step in regulating both photosynthetic CO_2 fixation and photorespiratory CO_2 evolution. If it is so, the relative activities of these two processes may well determine the productivity of green plants. Thus, Rubisco has become the most intensively studied plant enzyme at organismic, cellular and molecular levels. Rubisco consists of two unequally sized subunits: the small (SS) and the large subunits (LS). The LS has the molecular weight of 5.3×10^4 and the SS of 1.4×10^4 daltons (Kawashima and Wildman, 1971). A proposed model of $L^8 S^8$ for higher plant Rubisco (Baker *et al.*, 1975) stipulates that the eight LS are situated at the four corners of a two-layered structure with the eight SS distributed on the surface.

I'll devote my attention to *Nicotiana* Rubisco from this point since my involvement on this subject was exclusively with *Nicotiana* Rubisco. There are 64 species (Goodspeed, 1954) in *Nicotiana*. In a given species, the eight LS of a Rubisco molecule could be identical, whereas there could be one to four different types of SS in the same molecule. Hirai (1977) proposed that the distribution of different SS within a molecule is at random. Therefore, the variability increases dramatically with the increasing number of types of SS. By taking advantage of this structural complexity of the subunit composition, the following experiments were conducted to determine the mode of their inheritance.

In 1970, the differential synthesis of the SS and LS of Rubisco was first studied with inhibitors. This provided the first evidence that the LS was synthesized on the 70s ribosomes whereas the SS was synthesized on the 80s ribosomes (Ellis, 1981), indicating the cytoplasmic nature of the LS. A year later, Wildman's group first crystallized *Nicotiana* Rubisco (Kawashima and Wildman, 1970). Using this pure Rubisco, a series of genetic studies was followed based on the reciprocal interspecific hybridization techniques mentioned earlier. This unique approach enabled Wildman's group to identify the chloroplast and nuclear genomes as the sites containing the coding information for the LS and SS respectively (Wildman *et al.*, 1972). Once the site of the coding information for the LS was identified, the maternal inheritance of the LS was established. The following describes how the identification was made, which unveils how the maternal inheritance of the LS of Rubisco was discovered.

Wildman's group first used the trytic peptide mapping method to identify the coding sites for the LS subunits. They found that in the trytic peptide map, the LS from *Nicotiana gossei* has one more peptide than that of *Nicotiana tabacum*. This extra peptide appeared in reciprocal F1 hybrid only when *Nicotiana gossei* was the female parent. Therefore, this extra peptide, of the LS is coded for by the chloroplast genome (Chan and Wildman, 1972). This constituted the discovery of the maternal inheritance of the LS of Rubisco.

By a similar approach, the SS of Rubsico coded for by nuclear genome was also established by Wildman group (Kawashima and Wildman, 1972). In addition, many other chloroplast protein's coding sites were also identified (Wildman *et al.*, 1972). The earlier years of the 1970s in Wildman's laboratory at UCLA were truly exciting and productive. I was fortunate to be there, at the right place and time, and on the right subject with the right people.

In the years following the identification of the sites of coding information for the LS, my involvement was largely confined to its confirmation and elucidation. In 1974, we introduced the techniques known as isoelectric focusing (electro focusing) to the study of Rubisco (Sakano *et al.*, 1974). It took a long and agonizing time to modify and to master this technique applicable to Rubisco. Once we overcame a variety of technical problems encountered, we had a simple and rapid one-step procedure on hand for the study of the genetics of the *Nicotiana* Rubisco.

The principle of isoelectric focusing is straightforward. It involves the migration due to an electric force of a protein or proteins towards the pH value where it is isoelectric in a stable pH gradient. The use of this technique for the genetic analysis of Rubisco is faster and less complicated than trytic peptide mapping.

Figure 1 presents a photograph of parallel electrofocusing of Rubisco from four *Nicotiana* species, *Nicotiana gossei*, *Nicotiana excelsior*, *Nicotiana glauca*, *Nicotiana tabacum* and the reciprocal hybrid between *Nicotiana glauca* x *Nicotiana tabacum*. It illustrates that the LS of Rubisco from all four species consists of three polypeptides, whereas the SS contains one to four peptides. In the case of *Nicotiana glauca* and *Nicotiana tabacum*, one of the three LS polypeptides of *Nicotiana glauca* protein has a different isoelectric point from that of *Nicotiana tabacum*. This difference reflects the difference in amino acid composition between the LS of these two species (Kung, 1976). In the reciprocal hybrids, the isoelectric point of the three LS polypeptides correspond to that of the female parent. This genetic analysis obtained by electrofocusing is entirely consistent with the genetic analysis of the tryptic peptides of the isolated LS from different species. Both of these methods demonstrate that the genes which code for the LS are inherited only from the maternal line and therefore are located in chloroplast genome.

With the convincing evidence, the maternal inheritance of LS of Rubisco is firmly established and further elucidated. It took less than four years from the first indication (1970) to the first identification (1971), and finally to the confirmation (1974) that the LS of Rubisco was maternally inherited. This is now a well established fact. Several chloroplast genomes have been sequenced in which genes coding for the LS of Rubisco and many other genes have been identified (Sugiura, 1998).

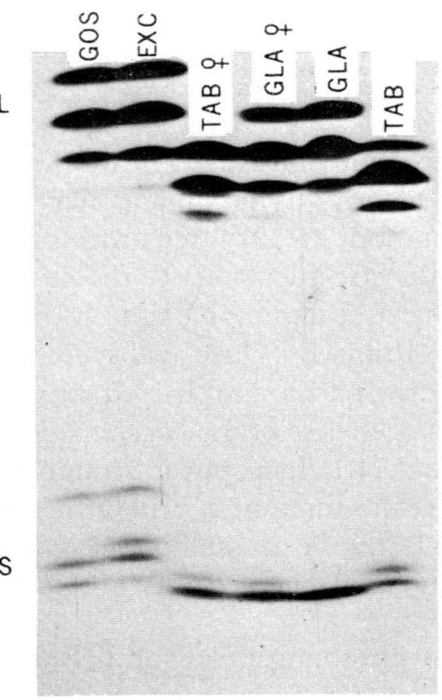

Fig. 1. Genetic analysis of *Nicotiana* fraction 1 protein by electrofocusing. The polypeptide composition of large (*L*) and small (*S*) subunits of fraction 1 protein from (left to right) *N. gossei* (*GOS*), *N. excelsior* (*EXC*), *N. tabacum* • x *N. glauca* • (*TAB* •), *N. glauca* • x *N. tabacum* • (*GLA* •), *N. glauca* (*GLA*) and *N. tabacum* (*TAB*). The large subunits consist of three polypeptides and a small subunit of one to four polypeptides.

I cannot resist mentioning that I wrote this recount of my academic career almost 25 years ago filled with human emotions. The four years that I spent in Wildman's laboratory working on Rubisco, with my esteemed colleagues: S. G. Wildman, P. Thornber, D. Bourque, J. Gray, F. Wong, K. Sakano, S. Singh, B. Hutichson, P. Chan and S. Y. Kwok, were among the most exciting of my career.

With original and innovative ideas, Wildman's group always opened new doors and pointed toward new directions for future research. It was extremely challenging working on uncertainties and converting them into certainty.

Acknowledgment

I would like to thank Winnie Cheng, Betty Law and Mecell Lee for their assistance in preparing this manuscript.

References

Baker, T. S., Eisenbery, D., Eiserling, F. A. and Weissman, L. (1975) Structure of form I crystals of D-ribulose-1,5-diphosphate carboxylase *J. Mol. Biol.* **91**: 391-399.

Bourque, D. P. and Wildman, S. G. (1973) Evidence that nuclear genes code for several chroloplast ribsomal proteins. *Biochem. Biophys. Res. Commun.* **50**: 532-37.

Chan, P. H. and Wildman, S. G. (1972) Chloroplast DNA codes for primary structure of the large subunit of Fraction 1 protein. *Biochim. Biophys. Acta.* **227**: 677-680.

Ellis, R. J. (1981) Chloroplast Proteins – Synthesis, Transport, and Assembly. *Ann. Rev. Plant Physiol.* **32**: 111-137.

Goodspeed, T. H. (1954) The Genus *Nicotiana*, Chronica botania, Waltham, Massachusetts.

Hirai, A. (1977) Random assembly of different kinds of small subunit polypeptides during formation of fraction I protein macromolecules. *Proc. Natl. Aca. Sci. USA.* **74**: 3443-3445.

Kawashima, N. and Wildman, S. G. (1970) Model of the subunit structure of fraction I protein. *Biochem. Biophys. Res. Commun.* **41**: 1463-1468.

Kawashima, N. and Wildman, S. G. (1971) *Ann. Rev. Plant Physiol.* **21**: 325-358.

Kawashima, N. and Wildman, S. G. (1972) Fraction I protein (IV) Mode of inheritance of primary structure in relation to whether chloroplast or nuclear DNA contains the code for a chloroplast protein. *Biochim. Biophys. Acta.* **262**: 42-49.

Kung S. D. and Thronber, J. P. (1971) Phtohosystem I and II chlorophyll-protein complexes of higher plant chloroplasts *Biochim, Biophys. Acta,* **253**: 285-289.

Kung, S. D. (1976) Fraction 1 protein: A Unique genetic marker. *Science* **191**: 429-434.

Kung, S. D. and Williams, J. P. (1969) Chloroplast DNA from broad bean. *Biochim. Biophys. Acta.* **195**: 434-445.

Kung, S. D., Thornber, J. P. and Wildman, S. G. (1972) Nuclear DNA codes for the photosystem II chlorophyll protein of chloroplast membrane. *FEBS Letter* **24**: 185-188.

Kung, S. D., Williams, J. P. (1968) Isolation and chloroplast free from nuclear DNA Contamination. *Biochim. Biophys. Acta.* **169**: 265-268.

Lorimer, G. H. (1981) The carboxylation and oxygenation of ribulose-1,5-biophosphate - The primary events in photosynthesis and photo-respiration. *Ann. Rev. Plant Physiol.* **32**: 349-383.

Sakano, K., Kung S. D. and Wildman, S. G. (1974) Identification of several chloroplast DNA genes which code for the large subunit of *Nicotiana* Fraction 1 protein. *Mol. Gen. Genet.* **130**: 91-97.

Sugiura, M. (1998) The discovery of the complete sequence of tobacco and rice chloroplast genomes in *Discoveries in Plant Biology*, Kung, S. D. and Yang, S. F., eds., World Scientific Publishing, Vol. II, pp. 45-60.

Wildman, S. G. and Bonner, J. (1947) Proteins of green leaves (I) isolation, enzymic properties and auxin content of spinach cytoplasmic proteins. *Arch. Biochem.* **14**: 381-413.

Wildman, S. G., Kawashima, N., Bourque, D. P., Wong, F., Singh, S., Chan, P. H., Kwok, S. Y., Sakano, K., Kung, S. D. and Thornber, J. P. (1972) Location of DNA coding for various kinds of chloroplast proteins in *Biochemistry of Gene Expression in Higher Organisms*. Pollack, J. K. and Lee, J. W., eds., Australian and New Zealand Book Co., pp. 443-456.

Chapter 14

Chloroplast Cytochromes: Discovery and Characterization

John C. Gray
Department of Plant Sciences, University of Cambridge
Downing Street, Cambridge CB2 3EA, UK

ABSTRACT

Cytochromes were first shown to be present in chloroplast membranes of higher plants in the laboratory of Robin Hill in Cambridge. Cytochromes f and b-563 were subsequently shown to be present in a cytochome bf complex, which catalyzed electron transfer between plastoquinol and plastocyanin, whereas cytochrome b-559 was located in photosystem II, where its function is still unresolved. The purification of the proteins led to the identification of the heme-binding polypeptides, and it was established that all of the chloroplast cytochromes were synthesized within the chloroplasts. The genes encoding the heme-binding polypeptides of each of the cytochromes were located in chloroplast DNA, and structural models of heme binding and transmembrane arrangement were proposed based on the protein sequences of the cytochromes. These models of the overall structures of the cytochromes will soon be tested rigorously, in the light of recent and future crystal structures of cytochrome components.

Introduction

Cytochrome was the name coined by David Keilin (1925) to describe the respiratory pigments he initially observed in insects, and subsequently in a wide variety of animal tissues, plants, yeast and bacteria. The characteristic absorption bands of cytochrome in the yellow-green part of the spectrum had previously been observed in muscle tissue by MacMunn (1886), and had been seen to change on addition of oxidizing or reducing agents, but the significance of these observations was not recognized, probably because of criticism by Hoppe-Seyler (see Keilin, 1966). Keilin's work in the Molteno Institute in Cambridge firmly established the importance of cytochrome redox reactions in respiration. However, in plants, cytochromes could be observed only in non-colored tissues, such as bulbs of shallots and garlic, and in pollen (see Keilin, 1966). Cytochromes in the green parts of plants were first observed by Yakushiji (1935), who examined leaf material after extraction of chlorophyll with acetone. However, he mistakenly attributed the absorption bands he saw to the cytochromes previously described from animals, and apparently failed to distinguish a sharp absorption band at 554–555 nm from that at 550 nm due to cytochrome *c*. This absorption band at 554–555 nm was identified as the distinctive α band of cytochrome *f* (from *frons*, Latin for leaf) by Hill and Scarisbrick (1951). Robin Hill worked in the Department of Biochemistry in Cambridge and had close links with Keilin, which arose after an exhibit of brightly colored metalloporphyrins at the opening of the new Biochemistry building in Tennis Court Road in 1924 (Bendall, 1994). The work on leaf cytochromes in Hill's laboratory was initiated in the pre-war years and provided part of Scarisbrick's doctoral thesis in 1939. Brief reports of the work were presented by Hill (1943) and Scarisbrick (1947), although full details were not published until 1951. This work clearly established that cytochrome *f* was associated with the insoluble material of chloroplasts, or thylakoid membranes as they would now be known (Hill and Scarisbrick, 1951).

Work in Robin Hill's laboratory also resulted in the first description of a plastid-located *b*-type cytochrome. It's characteristic absorption spectrum was first observed by Davenport (1952) in intact plastids isolated from etiolated barley leaves, and was photographed using a diffraction-grating camera designed by Hill (Davenport and Hill, 1952). The autoxidizable cytochrome was called cytochrome b_6, and was shown to have a redox potential of about -0.06 V (Hill, 1954). However, it was later shown to be made up of two components, subsequently called cytochrome *b*-563 and cytochrome *b*-559 (Bendall *et al.*, 1971). These components were first separated by fractionation of thylakoid membranes following digitonin treatment (Boardman and Anderson, 1967), and were shown to have different redox behaviors upon illumination of chloroplasts with red and far-red light (Cramer and Butler, 1967). Cytochrome *b*-563 was present in the digitonin-solubilized fraction of thylakoid membranes and comigrated during centrifugation in sucrose gradients with cytochrome *f* (Wessels, 1966). This comigrating material was shown to catalyze the plastoquinol-dependent reduction of plastocyanin (Wood and Bendall, 1976; Hurt and Hauska, 1981), confirming a role for the cytochromes in electron transfer between photosystems II and I. The role of the cytochrome *bf* complex in photosynthetic electron transfer and proton translocation across the thylakoid membrane is now well-established (Cramer *et al.*, 1996).

Cytochrome b-559 was not solubilized from thylakoid membranes by digitonin (Boardman and Anderson, 1967) and was later shown to be an integral component of the photosystem II reaction center core complex (Nanba and Satoh, 1987). However, there has been considerable controversy, still unresolved, over the role of cytochrome b-559 in photosynthetic electron transfer (Stewart and Brudvig, 1998), due at least in part to the existence of a variety of forms of cytochrome b-559 distinguished by their redox properties (Bendall *et al.*, 1971).

Purification of Chloroplast Cytochromes

The chloroplast cytochromes are all intrinsic membrane proteins and their extraction from the thylakoid membrane requires the use of organic solvents or detergents. Hill and Scarisbrick (1951) were able to solubilize cytochrome f by grinding leaves in ammoniacal ethanol, although the yield of cytochrome was later improved by the addition of Triton X-100 (Forti *et al.*, 1965; Bendall *et al.*, 1971). Davenport and Hill (1952) were able to purify cytochrome f after extraction from parsley leaves with ammoniacal ethanol using a specially adapted milling device for homogenizing large amounts of leaf material. The heme group was shown to be covalently attached to the polypeptide by thioether bonds that could be broken by treatment with sodium amalgam (Davenport and Hill, 1952). A variety of chromatographic or electrophoretic procedures were later employed to purify spinach cytochrome f to homogeneity (Singh and Wasserman, 1971; Nelson and Racker, 1972); this allowed the size of the cytochrome f polypeptide to be established as 31–34 kDa (Singh and Wasserman, 1971; Nelson and Racker, 1972). However, the protein behaved as an octomer in aqueous media, and aggregated further in the absence of detergents. In contrast, cytochrome f extracted from leaves of Japanese radish and spinach-mustard (*Brassica rapa* ssp. *perviridis*) behaved as a monomer (Takahashi and Asada, 1975; Matsuzaki *et al.*, 1975) and could be easily purified (Matsuzaki *et al.*, 1975). Detailed analysis of preparations of monomeric cytochrome f from cruciferous plants have provided important structural information about the protein (Gray *et al.*, 1994; Martinez *et al.*, 1994, 1996, see below).

The purification of the b-type cytochromes was not so straightforward, probably because of the ease with which heme was lost from the proteins under the harsh methods used initially for the release of the proteins from the thylakoid membrane. Sonication of ethanol-extracted thylakoid membranes in the presence of Triton X-100 and 4M urea released cytochromes b-563 and b-559 that could be further purified, although the final preparations appeared to consist of a confusing number of polypeptides (Garewal and Wasserman, 1974a, b; Stuart and Wasserman, 1973, 1975). However, the use of this extraction method on photosystem II - enriched membranes, combined with improved chromatographic separations, finally led to the isolation of cytochrome *b*-559 preparations containing two small polypeptides (Metz *et al.*, 1983). Antibodies to the purified 9 kDa polypeptide and N-terminal amino acid sequences of the 9 kDa and 4 kDa polypeptides (Widger *et al.*, 1984a, 1985) were instrumental in locating the genes in chloroplast DNA (Westhoff *et al.*, 1984). The deduced amino acid sequences of the polypeptides provided the key to understanding the structure of cytochrome *b*-559 (Herrmann *et al.*, 1983; Widger *et al.*, 1985, see below).

Cytochrome *b*-563 was also purified following sonication of ethanol-extracted thylakoid membranes in the presence of Triton X-100 and 4 M urea, and was shown to consist of a single polypeptide of about 18–20 kDa (Lach and Böger, 1977). Confirmation of the identity of the cytochrome *b*-563 polypeptide was obtained by its extraction from purified cytochrome *bf* complex with Triton X-100 and urea (Hurt and Hauska, 1982); the 23 kDa polypeptide was shown to bind two hemes (Hurt and Hauska, 1982). Antibodies to the 23 kDa polypeptide were used to locate the gene encoding cytochrome *b*-563 (Alt *et al.*, 1983; Heinemeyer *et al.*, 1984) and its co-transcription with the gene encoding subunit IV of the cytochrome complex provided an insight into the structure of the protein and its relationship to cytochrome *b* in the mitochondrial cytochrome bc_1 complex (Widger *et al.*, 1984b, see below).

Chloroplast Cytochromes in Cambridge

My interest in cytochromes started in 1974 when I was casting around for a research project for an application for a fellowship to take me back to England from Sam Wildman's laboratory at UCLA. I had spent a very productive first year in his laboratory using isoelectric focusing to study the subunits of "Rubisco" (named as a joke by David Eisenberg at Sam Wildman's retirement symposium in July 1979, see Wildman, 1998). The method for separating the subunits of rubisco by isoelectric focusing in the presence of 8 M urea had been perfected by Shain-Dow Kung just before I arrived in the laboratory, and we spent the next year applying the technique to rubisco from a variety of plants. We were able to demonstrate, on a single gel, maternal inheritance of the large subunit polypeptides and Mendelian inheritance of the small subunit polypeptides in F_1 hybrids of different *Nicotiana* species (Kung *et al.*, 1975b; Gray *et al.*, 1975). With Kevin Chen, a graduate student, we went on to examine the origins of polyploid species such as tobacco and wheat (Gray *et al.*, 1975; Chen *et al.*, 1975a), and examine the composition of rubisco in male-sterile tobacco (Chen *et al.*, 1975b) and in parasexual *Nicotiana* hybrids produced by protoplast fusion by Peter Carlson (Kung *et al.*, 1975a). This illustrated to me the value of rapid analytical methods for producing large amounts of publishable data, and more importantly, the value of *Nicotiana* hybrids for genetic analysis of chloroplast proteins.

I had decided that the basis of my research proposal was to be the genetic analysis of chloroplasts using *Nicotiana* hybrids — but which chloroplast proteins should I choose to study?   They needed to be sufficiently well characterized so that progress with the protein chemistry would be reasonably rapid, and they needed to show an interesting pattern of inheritance — ideally maternal inheritance to make the link with the newly emerging characterization of chloroplast DNA.   The chloroplast cytochromes were the obvious choice: purification methods had been described (see above), although not specifically for tobacco, and Gregory and Bradbeer (1973) had shown that the accumulation of cytochromes *f*, *b*-563 and the low potential form of *b*-559 in greening bean leaves was inhibited by chloramphenicol, suggesting that their synthesis was dependent on chloroplast protein synthesis (see below).   The idea of studying chloroplast cytochromes was reinforced by a discussion with Keith Boardman who visited Sam Wildman at exactly the right time.   The other great advantage of choosing chloroplast cytochromes was that it would lead me to Cambridge, the birthplace of cytochrome.

Derek Bendall, who obtained his doctorate with Robin Hill, was a lecturer in the Biochemistry department in Cambridge and had recently written a comprehensive review of chloroplast cytochromes for *Methods in Enzymology* (Bendall *et al.*, 1971).   He was very interested in my proposal for an SRC Research Fellowship to study the location of the genetic information for chloroplast cytochromes in his laboratory, although there was a possible problem concerning laboratory space.   There had recently been a fire in the Biochemistry department in Cambridge (due to storage of flammable solvents in an ordinary fridge) and his laboratory was being used to house those displaced from the burned-out laboratory.   To ensure that I would have a laboratory to return to if I was successful in my application for an SRC Research Fellowship, I wrote to Professor Bob Whatley at the University of Oxford to ask if he could provide laboratory space.   He was an obvious alternative choice because of his work on chloroplast development, and I was very pleased with his positive response.   However, in the end, the refurbishment at Cambridge was completed in time for me to join Derek

Bendall's laboratory in October 1975, supported by an SRC Research Fellowship.

My first task was to purify the chloroplast cytochromes, and I chose to start with cytochrome f because of the wealth of experience with this protein in Cambridge. Unfortunately, none of this experience was with tobacco, and I was forced to try a variety of methods for solubilizing the protein. Sonication of ethanol-extracted thylakoid membranes in Triton X-100 and 4M urea, based on the method devised for the extraction of b-type cytochromes (Garewal and Wasserman, 1974a), gave the best yield and allowed the subsequent purification of an octomeric form of cytochrome f (Gray, 1977). Due to the rather slow progress with the tobacco cytochrome f, I started purifying the protein from various other plants, to establish the size of the polypeptide chain. I discovered that I could extract cytochrome f in very high yield from the leaves of charlock (*Sinapis arvensis*) using butanone (methyl ethyl ketone), as described by Matsuzaki *et al.* (1975). I had a ready supply of charlock leaves from the garden of my newly-built house in Comberton. Charlock is a widespread agricultural weed with long-lived seeds which readily germinate on being brought to the surface by soil disturbance. The monomeric form of cytochrome f extracted from leaves could be easily purified to homogeneity (Gray 1977, 1978), allowing detailed physicochemical analysis, and providing material for raising antibodies (Doherty and Gray, 1979). The relative simplicity of the extraction and purification methods led me to explore cytochrome f in other cruciferous plants, including horse-radish, shepherd's purse, turnip, rape and woad (Gray, 1977, 1978). The woad was obtained from Robin Hill's garden, where he grew it as a source of indigo dye (Hill, 1975). I cycled past his house in Barton each day on my way between Comberton and the University, and occasionally stopped to talk about my experiments. When I explained how easy it was to extract cytochrome f from crucifers, he quickly suggested I try woad. The protein I purified was subsequently used for spectroscopic studies on cytochrome f in collaboration with Geoff Moore, initially at the University of Oxford and later at the University of East Anglia (Rigby *et al.*, 1988). I made the largest preparation of cytochrome f from 15.5 kg of rape leaves, yielding 133 mg of pure protein (Gray and Phillips, 1982). The leaves were collected from a farmer's field halfway between Comberton and Barton. I explained to the farmer that we would collect only one leaf from each plant so that his yield of oilseed would not be decreased too much. It took my wife, two friends who had come to visit for the weekend and myself, several hours to collect the leaves, an amount that paled into insignificance compared to the size of the field. Never again did I suggest we would collect only one leaf from a plant!

Genetics and Biosynthesis of Chloroplast Cytochromes

Although my fellowship was for a period of two years, I accepted a position as a university demonstrator in the Department of Botany at Cambridge after one year in Derek Bendall's laboratory. I continued my work with cytochrome *f*, but I now had my own laboratory and could recruit my own students and postdocs. Andrew Doherty, a calm and very talented postdoc from Belfast, was the first postdoc in my laboratory. He set out to determine the site of synthesis of cytochrome *f*. Although studies with protein synthesis inhibitors in barley (Gregory and Bradbeer, 1973) and with *Oenothera* mutants (Hallier and Heber, 1977) had indicated that cytochrome *f* was most likely synthesized in the chloroplasts, there was an element of doubt because cytochrome *f* had been detected solely by its absorbance characteristics. It was possible that one, or more, of the proteins responsible for heme addition, rather than the cytochrome *f* polypeptide, was synthesized in the chloroplast. Andrew showed conclusively that intact isolated pea chloroplasts were able to incorporate ^{35}S-methionine into the cytochrome *f* polypeptide (Doherty and Gray, 1979). We had been shown the methods for preparing intact chloroplasts capable of synthesizing proteins *in vitro* by John Ellis, at the University of Warwick, and he was very supportive of our work on chloroplast biogenesis over the years. The method worked so well that Andy Phillips, my first graduate student who started in October 1979, set out to discover if cytochrome *b*-563 or any of the other subunits of the cytochrome *bf* complex were synthesized in isolated chloroplasts. This required him to isolate the cytochrome *bf* complex from pea chloroplasts, a task that was not straightforward until we heard that Ed Hurt and Günter Hauska had been successful in solubilizing the spinach complex with a mixture of octyl glucoside and cholate (Hurt and Hauska, 1981). Andy then identified the cytochrome *b*-563 polypeptide in the pea complex by its heme-peroxidase activity, and showed that cytochrome *b*-563 and subunit IV were both synthesized in isolated chloroplasts and assembled into the cytochrome complex (Phillips and Gray, 1983, 1984a). So three of the known components of the cytochrome *bf* complex were synthesized on chloroplast ribosomes.

Following his success with the synthesis of cytochrome *f*, Andrew Doherty showed that isolated chloroplasts could also synthesize CF_o subunit III of the ATP synthase (Doherty and Gray, 1980) and he was keen to locate and characterize the genes for these components. I had shown by analyzing tryptic peptides from the purified proteins that cytochrome *f* was inherited from the maternal parent in reciprocal F_1 hybrids between *N. tabacum* and *N. glutinosa* (Gray, 1980). This provided further evidence that the gene for cytochrome *f* was likely to be located in chloroplast DNA. Fortunately, Rick Bottomley from CSIRO in Canberra visited the laboratory and told us that extracts of *Escherichia coli* were able to transcribe and translate genes in chloroplast DNA (Bottomley and Whitfeld, 1979). Andrew applied the method to cloned fragments of chloroplast DNA from wheat, obtained from Cathy Bowman and Tristan Dyer at the Plant Breeding Institute in Trumpington, just a short bicycle ride from the University, and quickly located the gene for the large subunit of rubisco (Doherty *et al.*, 1981). The method was so promising that it provided the starting point for several new graduate students. Chris Howe and Alison Huttly started to localize genes for ATP synthase in wheat and pea chloroplast DNA, respectively, and Dave Willey joined Andy Phillips in localizing genes for the cytochrome *bf* complex in pea chloroplast DNA. We were very fortunate to obtain cloned fragments of pea chloroplast DNA from Jeff Palmer (Palmer and Thompson, 1981). This was a very exciting and highly productive period in the laboratory.

However, it quickly became apparent that we were not the only laboratory interested in localizing and characterizing these genes in chloroplast DNA. Reinhold Herrmann, initially in Düsseldorf, but later in Munich, was also interested in the same genes. His approach, however, was complementary to ours. His background was in chloroplast DNA and he set up collaborations with others, such as Bill Cramer, Nathan Nelson and Günter Hauska, who could provide antibodies and expertise with chloroplast proteins. His primary approach was translation of chloroplast RNA, and of RNA hybrid-selected by cloned fragments of spinach chloroplast DNA, and identification of the products with specific antibodies (Alt *et al.*, 1983). This approach was highly successful, but the exact locations of genes could be confounded by the polycistronic nature of some transcripts, and confirmation of the gene locations was obtained by transcription and translation of cloned chloroplast DNA fragments in extracts of *E. coli* (Alt *et al.*, 1983).

The localization and characterization of the genes for subunits of the cytochrome *bf* complex developed into something of a race. We were the first to localize the gene for cytochrome *f*, in the pea chloroplast genome (Willey *et al.*, 1983), although Juliane Alt in Reinhold Herrmann's laboratory was close behind (Alt *et al.*, 1983). Andy Phillips had originally observed that cytochrome *f* was encoded within a 17.3 kb region of pea chloroplast DNA, but he left the fine dissection of the region to Dave Willey. Dave also completed the fine mapping of the gene for cytochrome *f* in wheat chloroplast DNA after Chris Howe had located the gene in a 12 kb region. He worked very hard to complete the nucleotide sequences of the pea and wheat genes for cytochrome *f* in time for the *Sixth International Congress on Photosynthesis* in Brussels in August 1983, and he was very anxious about the results on the spinach gene from Reinhold Herrmann's laboratory. However, his worries disappeared as he studied Juliane Alt's poster; part of the spinach gene sequence had been inverted, and the predicted protein sequence was incorrect and different to the sequences predicted from the pea and wheat genes (Willey and Gray, 1984). These gene sequences indicated that cytochrome *f* was synthesized as a precursor of 320 amino acid residues with an N-terminal presequence of 35 amino acid residues. The protein contained the heme-binding cysteine residues near the N-terminus and had only a single hydrophobic region near the C-terminus. Dave Willey went on to examine the orientation of the protein in pea thylakoid membranes, using proteolysis of right-side-out and inside-out vesicles to establish that the N-terminal hydrophilic heme-binding domain was located in the thylakoid lumen and a short C-terminal region of 15 amino acid residues was accessible from the stroma (Willey *et al.*, 1984). The proposed structural model for cytochrome *f* is still accepted, although a detailed structural analysis required the determination of the crystal structure of the lumen-side domain (Martinez *et al.*, 1994, 1996, see below).

The isolation of the gene for cytochrome *b*-563 and the prediction of its transmembrane structure were major successes for Reinhold Herrmann's laboratory in collaboration with Bill Cramer and Achim Trebst (Heinemeyer *et al.*, 1984; Widger *et al.*, 1984b). The gene was located in spinach chloroplast DNA using immunoprecipitation of translation products of hybrid-selected RNA, and appeared to be co-transcribed with the gene for subunit IV of the cytochrome *bf* complex (Alt *et al.*, 1983). DNA sequencing revealed that the *petB* gene encoding cytochrome *b*-563 was located about 1 kb upstream of the *petD* gene encoding subunit IV (Heinemeyer *et al.*, 1984). The really exciting feature of the sequence was the homology with mitochondrial cytochrome *b*; cytochrome *b*-563 was homologous to the N-terminal region of mitochondrial cytochrome *b*, while subunit IV was homologous to the C-terminal region (Heinemeyer *et al.*, 1984; Widger *et al.*, 1984b). Sequence comparisons of chloroplast and mitochondrial cytochromes allowed the identification of four conserved histidine residues as the ligands to the two heme groups, and led to a proposal on the transmembrane organization of the proteins (Widger *et al.*, 1984b).

Our efforts to isolate the genes for cytochrome *b*-563 and subunit IV were delayed because of an unexpected feature of the genes. Andy Phillips had located a region of the pea chloroplast genome that produced, in the *E. coli* transcription-translation system, a polypeptide of about 15 kDa immunoprecipitable by antibodies to the cytochrome *bf* complex (Phillips and Gray, 1984b). However, the polypeptide was slightly different in electrophoretic mobility to subunit IV, although DNA sequence analysis showed that the protein was homologous to the C-terminal region of mitochondrial cytochrome *b* (Phillips and Gray, 1984b). Chris Eccles, another graduate student, discovered that the N-terminal sequence of the purified cytochrome *b*-563 did not completely match the sequence predicted from the *petB* gene, although the encoded protein was homologous to the N-terminus of mitochondrial cytochrome *b*. We also found a slight discrepancy between the N-terminal sequence of the purified subunit IV and the predicted product of the *petD* open reading frame. This conundrum was later explained by the discovery of introns in both the *petB* and *petD* genes (Fukuzawa *et al.*, 1987; Tanaka *et al.*, 1987). The 5′ exons of the *petB* and *petD* genes encoded just 2 and 3 amino acid residues, respectively, and the deduced proteins synthesized from the spliced transcripts contained the N-terminal sequence we had obtained. Unfortunately, much of this work on the pea *petB* gene was never published, because Chris Eccles discovered that the prospects of becoming a Royal Air Force fighter pilot were more attractive than completing his PhD.

The combination of Reinhold Herrmann and Bill Cramer was also successful in isolating and characterizing the genes encoding cytochrome *b*-559. The synthesis of the low potential form of cytochrome *b*-559 in chloroplasts had been indicated by chloramphenicol inhibition of accumulation in greening bean leaves (Gregory and Bradbeer, 1973), and more evidence was provided by the synthesis of a 10 kDa cytochrome *b*-559 polypeptide in isolated spinach chloroplasts (Zielinski and Price, 1980). The gene for the 9 kDa polypeptide of cytochrome *b*-559 was located in spinach chloroplast DNA by immunoprecipitation of translation products of hybrid-selected RNA, using antibodies raised against the protein in Bill Cramer's laboratory (Westhoff *et al.*, 1984; Widger *et al.*, 1984a). DNA sequencing revealed a small open reading frame (*psbE*) encoding the known N-terminal amino acid sequence of the 9 kDa polypeptide of cytochrome *b*-559 (Herrmann *et al.*, 1983). However, downstream of *psbE* was another small open reading frame (*psbF*) which was suggested to encode a 4 kDa protein that would form a heterodimer with the 9 kDa polypeptide. Each polypeptide was proposed to provide a histidine residue as a ligand for the cytochrome *b*-559 heme (Herrmann *et al.*, 1983). Further N-terminal sequencing of polypeptides of the spinach cytochrome *b*-559 preparation confirmed the presence of the 4 kDa *psbF* gene product (Widger *et al.*, 1985).

Our work on cytochrome *b*-559 trailed some way behind that of Reinhold Herrmann and Bill Cramer, not least because of problems with raising antibodies to the protein. We were fortunate to obtain antibodies to barley cytochrome *b*-559 from Gunilla Hoyer-Hansen and Diter von Wettstein of the Carlsberg Laboratory in Copenhagen, and Dave Willey was able to show that a fragment of chloroplast DNA close to the cytochrome *f* gene was able to direct the synthesis of an immunoprecipitable 10 kDa protein in the *E. coli* transcription-translation system. Fine mapping and DNA sequencing of the wheat gene was carried out by Sean Hird, a graduate student, and showed the same arrangement of the *psbE* and *psbF* genes as in spinach (Hird *et al.*, 1986).

Structure of the Chloroplast Cytochromes

So, in a period of just a year or so, the chloroplast genes encoding all of the polypeptides of the chloroplast cytochromes had been isolated and sequenced, and the information was used to propose models for the structure of each of the three cytochromes (Willey *et al.*, 1984; Widger *et al.*, 1984b; Herrmann *et al.*, 1984). This was an extremely exciting and productive period of research in my laboratory and elsewhere. Subsequent experimental work has largely supported the structural features of these models, although some realignment of the transmembrane helices of cytochrome *b*-563 was necessary, resulting in an altered allocation of the histidine ligands to the two hemes. Details of the quaternary structure of cytochrome *b*-559 remain controversial, largely due to continued arguments over the stoichiometry of cytochrome *b*-559 hemes in the photosystem II reaction centre; single $\alpha\beta$ heterodimer and $\alpha_2\beta_2$ structures have been proposed (Stewart and Brudwig, 1998). Complete structural validation of the models for any of the chloroplast cytochromes has not yet been achieved. This should be achieved in the next few years as the 3D structures of the cytochrome *bf* complex and photosystem II are solved by X-ray diffraction or by electron diffraction of 2D crystals. A detailed structure of cytochrome *b*-559 should soon become available from photosystem II crystals (Rhee *et al.*, 1998), and structures of cytochrome *f* and cytochrome *b*-563 should result from structural studies on the cytochrome *bf* complex. Studies on the mitochondrial cytochrome bc_1 complex have produced detailed information on the structure and orientation of cytochrome *b* and cytochrome c_1 (Kim *et al.*, 1998; Zhang *et al.*, 1998).

A major success has been the crystal structure of the lumen-side domain of turnip cytochrome *f*, determined to a resolution of 1.96 Å, in a collaboration between the laboratories of Bill Cramer and Janet Smith at Purdue University (Martinez *et al.*, 1994, 1996). This showed the elongated nature of the molecule, explaining the somewhat anomalous behavior of the protein on gel filtration (Gray, 1978; Gray *et al.*, 1994), and resolved the question of the nature of the sixth ligand to the heme (Gray, 1992). The structure showed the unprecedented ligation of the heme iron by the α-amino group of the N-terminal tyrosine residue (Martinez *et al.*, 1994). The high-resolution structure also detected a chain of five buried water molecules within the polypeptide, leading to speculation that this might provide an exit port for protons translocated across the membrane by the cytochrome complex (Martinez *et al.*, 1996). These features, and others, are currently being explored by site-directed mutagenesis of the *petA* gene encoding cytochrome *f* in a number of laboratories. However, the structure of the lumen-side domain does not provide information on the arrangement of the transmembrane region or the stroma-side domain of cytochrome *f* and we will have to await the structure of the complete cytochrome *bf* complex to understand fully the interaction of cytochrome *f* with cytochrome *b*-563 and the other subunits of the complex.

References

Alt, J., Westhoff, P., Sears, B. B., Nelson, N., Hurt, E., Hauska, G. and Herrmann, R. G. (1983) Genes and transcripts for the polypeptides of the cytochrome b6/f complex from spinach thylakoid membranes. *EMBO J.* **2**: 979-986.

Bendall, D. S. (1994) Robin Hill. *Biog. Mem. Fell. Roy. Soc.* 143-179.

Bendall, D. S., Davenport, H. E. and Hill, R. (1971) Cytochrome components of chloroplasts of the higher plants. *Methods Enzymol.* **23**: 327-344.

Boardman, N. K. and Anderson, J. M. (1967) Fractionation of the photochemical systems of photosynthesis. II. Cytochrome and carotenoid contents of particles isolated from spinach chloroplasts. *Biochim. Biophys. Acta* **143**: 187-203.

Bottomley, W. and Whitfeld, P. R. (1979) Cell-free transcription and translation of total spinach chloroplast DNA. *Eur. J. Biochem.* **93**: 31-39.

Chen, K., Gray, J. C. and Wildman, S. G. (1975a) Fraction I protein and the origin of polyploid wheats. *Science* **190**: 1304-1306.

Chen, K., Kung, S. D., Gray, J. C. and Wildman, S. G. (1975b) Polypeptide composition of Fraction I protein from *Nicotiana glauca* and cultivars of *Nicotiana tabacum*, including a male sterile line. *Biochem. Genet.* **13**: 771-778.

Cramer, W. A. and Butler, W. L. (1967) Light-induced absorbance changes in two cytochrome *b* components in the electron-transfer system of spinach chloroplasts. *Biochim. Biophys. Acta* **143**: 332-339.

Cramer, W. A., Soriano, G. M., Ponomarev, M., Huang, D., Zhang, H., Martinez, S. E. and Smith, J. L. (1996) Some new structural aspects and old controversies concerning the cytochrome *bf* complex of oxygenic photosynthesis. *Ann. Rev. Plant Physiol. Plant Mol. Biol.* **47**: 477-508.

Davenport, H. E. (1952) Cytochrome components in chloroplasts. *Nature* **170**: 1112-1114.

Davenport, H. E. and Hill, R. (1952) The preparation and some properties of cytochrome *f*. *Proc. Roy. Soc.* B **139**: 327-345.

Doherty, A. and Gray, J. C. (1979) Synthesis of cytochrome *f* by isolated pea chloroplasts. *Eur. J. Biochem.* **98**: 87-92.

Doherty, A. and Gray, J. C. (1980) Synthesis of a dicyclohexylcarbodiimide-binding proteolipid by isolated pea chloroplasts. *Eur. J. Biochem.* **108:** 131-136.

Doherty, A., Gray, J. C., Bowman, C. M. and Dyer, T. A. (1981) Localization of genes in wheat chloroplast DNA by linked transcription-translation of cloned DNA fragments. *Biochem. Soc. Trans.* **9:** 199P.

Forti, G., Bertole, M. L. and Zanetti, G. (1965) Purification and properties of cytochrome *f* from parsley leaves. *Biochim. Biophys. Acta* **109:** 33-40.

Fukuzawa, H., Yoshida, T., Kohchi, T., Okumura, T., Sawano, Y. and Ohyama, K. (1987) Splicing of group II introns in mRNAs coding for cytochrome b_6 and subunit IV in the liverwort *Marchantia polymorpha* chloroplast genome. *FEBS Lett.* **220:** 61-66.

Garewal, H. S. and Wasserman, A. R. (1974a) Triton X-100 – 4 M urea as an extraction medium for membrane proteins. I. Purification of chloroplast cytochrome *b*–559. *Biochemistry* **13:** 4063-4071.

Garewal, H. S. and Wasserman, A. R. (1974b) Triton X-100 – 4 M urea as an extraction medium for membrane proteins. II. Molecular properties of pure cytochrome *b*–559: a lipoprotein containing small polypeptide chains and a limited lipid composition. *Biochemistry* **13:** 4072-4079.

Gray, J. C. (1977) Purification and properties of cytochrome *f* from higher plants. *Biochem. Soc. Trans.* **5:** 326-327.

Gray, J. C. (1978) Purification and properties of monomeric cytochrome *f* from charlock, *Sinapis arvensis* L. *Eur. J. Biochem.* **82:** 133-141.

Gray, J. C. (1980) Maternal inheritance of cytochrome *f* in interspecific *Nicotiana* hybrids. *Eur. J. Biochem.* **112:** 39-46.

Gray, J. C. (1992) Cytochrome *f*: structure, function and biosynthesis. *Photosynth. Res.* **34:** 359-374.

Gray, J. C. and Phillips, A. L. (1982) Isolation of cytochrome *f* and a cytochrome *b-f* complex. In *Methods in Chloroplast Molecular Biology*, M. Edelman, R. Hallick and N-H. Chua, eds, pp. 917-932. Elsevier Biomedical Press, Amsterdam.

Gray, J. C., Kung, S. D., Wildman, S. G. and Sheen, S. J. (1975) Origin of *Nicotiana tabacum* L. detected by polypeptide composition of Fraction I protein. *Nature* **252:** 226-227.

Gray, J. C., Rochford, R. J. and Packman, L. C. (1994) Proteolytic removal of the C-terminal transmembrane region of cytochrome *f* during extraction from turnip and charlock leaves generates a water-soluble monomeric form of the protein. *Eur. J. Biochem.* **223:** 481-488.

Gregory, P. and Bradbeer, J. W. (1973) Plastid development in primary leaves of *Phaseolus vulgaris*: the light-induced development of the chloroplast cytochromes. *Planta* **109:** 317-326.

Hallier, U. W. and Heber, U. W. (1977) Cytochrome *f* deficient plastome mutants of *Oenothera*. In *Photosynthetic Organelles*, special issue of *Plant Cell Physiol.*, Miyachi, S., Katoh, S., Fujita, Y. and Shibata, K., eds, pp. 257-273. Jap. Soc. Plant Physiol., Japan.

Heinemeyer, W., Alt, J. and Herrmann, R. G. (1984) Nucleotide sequence of the clustered genes for apocytochrome *b6* and subunit 4 of the cytochrome *b/f* complex in the spinach plastid chromosome. *Curr. Genet.* **8:** 543-549.

Herrmann, R. G., Alt, J., Schiller, B., Widger, W. A. and Cramer, W. A. (1983) Nucleotide sequence of the gene for apocytochrome *b*-559 on the spinach plastid chromosome: implications for the structure of the membrane protein. *FEBS Lett.* **176:** 239-244.

Hill, R. (1943) Tetrapyrrole compounds in plants. *Biochem. J.* **37:** xxiii.

Hill, R. (1954) The cytochrome *b* component of chloroplasts. *Nature* **174:** 501-503.

Hill, R. (1975) Days of visual spectroscopy. *Ann. Rev. Plant. Physiol.* **26:** 1-11.

Hill, R. and Scarisbrick, R. (1951) The haematin compounds of leaves. *New Phytol.* **50:** 98-111.

Hird, S. M., Willey, D. L., Dyer, T. A. and Gray, J. C. (1986) Location and nucleotide sequence of the gene for cytochrome *b*–559 in wheat chloroplast DNA. *Mol. Gen. Genet.* **203:** 95-100.

Hurt, E. and Hauska, G. (1981) A cytochrome f/b_6 complex of five polypeptides with plastoquinol-plastocyanin-oxidoreductase activity from spinach chloroplasts. *Eur. J. Biochem.* **117**: 591-599.

Hurt, E. and Hauska, G. (1982) Identification of the polypeptides in the cytochrome b_6/f complex of spinach chloroplasts with redox-center-carrying subunits. *J. Bioenerg. Biomemb.* **14**: 405-424.

Keilin, D. (1925) On cytochrome, a respiratory pigment, common to animals, yeast, and higher plants. *Proc. Roy. Soc. B* **98**: 312-339.

Keilin, D. (1966) *The History of Cell Respiration and Cytochrome.* Cambridge University Press, Cambridge, UK.

Kim, H., Xia, D., Yu, C-A., Xia, J-Z, Kachurin, A. M., Zhang, L., Yu, L. and Deisenhofer, J. (1998) Inhibitor binding changes domain mobility in the iron-sulfur protein of the mitochondrial $bc1$ complex from bovine heart. *Proc. Natl. Acad. Sci. USA.* **95**: 8026-8033.

Kung, S. D., Gray, J. C., Wildman, S. G. and Carlson, P. S. (1975a) Polypeptide composition of Fraction I protein from parasexual hybrid plants in the genus Nicotiana. *Science* **187**: 353-355.

Kung, S. D., Sakano, K., Gray, J. C. and Wildman, S. G. (1975b) The evolution of Fraction I protein during the origin of a new species of *Nicotiana. J. Mol. Evol.* **7**: 59-64.

Lach, H. J. and Böger, P. (1977) Some properties of plastidic cytochrome b-563. *Z. Naturforsch.* **32C**: 877-879.

MacMunn, C. A. (1886) Researches on myohaematin and the histohaematins. *Phil. Trans. Roy. Soc.* **177**: 267-298.

Martinez, S. E., Huang, D., Ponomarev, M., Cramer, W. A. and Smith, J. L. (1996) The heme redox center of chloroplast cytochrome f is linked to a buried five-water chain. *Protein Sci.* **5**: 1081-1092.

Martinez, S. E., Huang, D., Szczepaniak, A., Cramer, W. A. and Smith, J. L. (1994) Crystal structure of the chloroplast cytochrome f reveals a novel cytochrome fold and unexpected heme ligation. *Structure* **2**: 95-105.

Matsuzaki, E., Kamimura, Y., Yamasaki, T. and Yakushiji, E. (1975) Purification and properties of cytochrome f from *Brassica komatsuna* leaves. *Plant Cell Physiol.* **16**: 237-246.

Metz, J. G., Ulmer, G., Bricker, T. M. and Miles, D. (1983) Purification of cytochrome b–559 from oxygen-evolving photosystem II preparations of spinach and maize. *Biochim. Biophys. Acta* **725**: 203-209.

Nanba, O. and Satoh, K. (1987) Isolation of a photosystem II reaction center consisting of the D-1 and D-2 polypeptides and cytochrome b-559. *Proc. Natl. Acad. Sci. USA.* **84**: 109-112.

Nelson, N. and Racker, E. (1972) Partial resolution of the enzymes catalyzing photophosphorylation. X. Purification of spinach cytochrome f and its photooxidation by resolved photosystem I particles. *J. Biol. Chem.* **247**: 3848-3855.

Palmer, J. D. and Thompson, W. F. (1981) Clone banks of the mung bean, pea and spinach chloroplast genomes. *Gene* **15**: 21-26.

Phillips, A. L. and Gray, J. C. (1983) Isolation and characterization of a cytochrome b-f complex from pea chloroplasts. *Eur. J. Biochem.* **137**: 533-560.

Phillips, A. L. and Gray, J. C. (1984a) Synthesis of components of the cytochrome b-f complex by isolated pea chloroplasts. *Eur. J. Biochem.* **138**: 591-595.

Phillips, A. L. and Gray, J. C. (1984b) Location and nucleotide sequence of the gene for the 15.2 kDa polypeptide of the cytochrome b-f complex from pea chloroplasts. *Mol. Gen. Genet.* **194**: 477-484.

Rhee, K-H., Morris, E. P., Barber, J. and Kühlbrandt, W. (1998) Three-dimensional structure of the plant photosystem II reaction centre at 8Å resolution. *Nature* **396**: 283-286.

Rigby, S. E. J., Moore, G. E., Gray, J. C., Gadsby, P. M. A., George, S. J. and Thomson, A. J. (1988) N.m.r., e.p.r. and magnetic c.d. studies of cytochrome *f*. Identity of the heme axial ligands. *Biochem. J.* **256**: 571-577.

Scarisbrick, R. (1947) Haematin compounds in plants. *Ann. Rep. Chem. Soc.* **44**: 226-236.

Singh, J. and Wasserman, A. R. (1971) The use of disc gel electrophoresis with nonionic detergent in the purification of cytochrome *f* from spinach grana membranes. *J. Biol. Chem.* **246**: 3532-3541.

Stewart, D. H. and Brudvig, G. W. (1998) Cytochrome b_{559} of photosystem II. *Biochim. Biophys. Acta* **1367**: 63-87.

Stuart, A. L. and Wasserman, A. R. (1973) Purification of cytochrome b_6 a tightly bound protein in chloroplast membranes. *Biochim. Biophys. Acta* **314**: 284-297.

Stuart, A. L. and Wasserman, A. R. (1975) Chloroplast cytochrome b_6 composition as a lipoprotein. *Biochim. Biophys. Acta* **376**: 561-572.

Takahashi, M. and Asada, K. (1975) Purification of cytochrome *f* from Japanese-radish leaves. *Plant Cell Physiol.* **16**: 191-194.

Tanaka, M., Obokata, J., Chunwongse, J., Shinozaki, K. and Sugiura, M. (1987) Rapid splicing and stepwise processing of a transcript from the *psbB* operon in tobacco chloroplasts: determination of the intron sites in *petB* and *petD*. *Mol. Gen. Genet.* **209**: 427-431.

Wessels, J. S. C. (1966) Isolation of a chloroplast fragment fraction with $NADP^+$–photoreducing activity dependent on plastocyanin and independent of cytochrome *f*. *Biochim. Biophys. Acta* **126**: 581-583.

Westhoff, P., Alt, J., Widger, W. R., Cramer, W. A. and Herrmann, R. G. (1984) Localization of the gene for apocytochrome *b*–559 on the plastid chromosomes of spinach. *Plant Mol. Biol.* **4**: 103-110.

Widger, W. R., Cramer, W. A., Hermodson, M. and Herrmann, R. G. (1985) Evidence for a hetero-oligomeric structure of the chloroplast cytochrome *b*-559. *FEBS Lett.* **191**: 186-190.

Widger, W. R., Cramer, W. A., Hermodson, M., Meyer, D. and Gullifor, M. (1984a) Purification and partial sequence of the chloroplast cytochrome *b*–559. *J. Biol. Chem.* **259**: 3870-3876.

Widger, W. R., Cramer, W. A., Herrmann. R. G. and Trebst, A. (1984b) Sequence homology and structural similarity between cytochrome *b* of mitochondrial complex III and the chloroplast b_6-*f* complex: position of the cytochrome *b* hemes in the membrane. *Proc. Natl. Acad. Sci. USA.* **81**: 674-678.

Wildman, S. G. (1998) Discovery of Rubisco. In *Discoveries in Plant Biology*, Vol. 1, Kung, S. D. and Yang, S. F., eds., pp. 163-173, World Scientific, Singapore.

Willey, D. L. and Gray, J. C. (1984) Location and nucleotide sequence of the gene for cytochrome *f* in pea and wheat chloroplast DNA. In *Advances in Photosynthesis Research*, C. Sybesma, ed., Vol. IV, pp. 567-570. Martinus Nijhoff/Dr W. Junk, The Hague, Netherlands.

Willey, D. L., Auffret, A. D. and Gray, J. C. (1984) Structure and topology of cytochrome f in pea chloroplast membranes. *Cell* **36**: 555-562.

Willey, D. L., Huttly, A. K., Phillips, A. L. and Gray, J. C. (1983) Localization of the gene for cytochrome *f* in pea chloroplast DNA. *Mol. Gen. Genet.* **189**: 85-89.

Wood, P. M. and Bendall, D. S. (1976) The reduction of plastocyanin by plastoquinol-1 in the presence of chloroplasts. A dark electron transfer reaction involving components between the two photosystems. *Eur. J. Biochem.* **61**: 337-344.

Yakushiji, E. (1935) Uber das Vorkommen des Cytochroms in hoheren Pflanzen und in Algen. *Acta Phytochem., Tokyo* **8**: 325.

Zhang, Z., Huang, L., Schulmeister, V. M., Chi, Y-I., Kim, K. K., Hung, L-W., Crofts, A. R., Berry, E. A. and Kim, S-H. (1998) Electron transfer by domain movement in cytochrome bc_1. *Nature* **392**: 677-684.

Zielinski, R. E. and Price, C. A. (1980) Synthesis of thylakoid membrane proteins by chloroplasts isolated from spinach. Cytochrome b_{559} and P_{700}-chlorophyll-*a*-protein. *J. Cell Biol.* **85**: 435-445.

Chapter 15

Discovery of Plasma Membrane Proton Pumping ATPase: Our Point of View

Robert T. Leonard
Department of Plant Sciences, The University of Arizona
Tucson, AZ 8572, USA

Thomas K. Hodges
Department of Botany and Plant Pathology, Purdue University
West Lafayette, IN 47907, USA

ABSTRACT

Studies on the direct role of ATP in providing energy for the work of mineral nutrient absorption in plants led to the discovery of plasma membrane-associated adenosine triphosphatase (ATPase). The research was initiated by Tom Hodges during the late 1960s at the University of Illinois and continued during the 1970s at Purdue University. By 1972, methods for isolating plasma membrane vesicles from broken plant cells had been developed and used to characterize the associated ATPase activity. Over the next 15 years, the plant plasma membrane ATPase was shown to have many properties in common with analogous ion pumps in animal and fungal cells, including conserved DNA sequences successfully utilized for cloning plant ATPase genes. The initial hypothesis that plasma membrane ATPase directly participated in K^+ absorption by root cells was not supported by research results. Rather, this ATPase is the electrogenic proton pump that establishes and maintains a proton-free energy gradient across the plasma membrane to which energy-dependent ion and metabolite transport is coupled.

The Way Things Were!

It was 1968, for us, just thirty short years ago! The Vietnam "conflict" was raging along with frequent campus protests expressing dissent over the draft, the war, and the establishment. Robert Leonard was a graduate student at the University of Illinois taking a course on the mechanism of ion transport in plants. Tom Hodges, an assistant professor in the Department of Horticulture, was lecturing to the twenty or so students in the class. In a thorough, but clearly skeptical presentation which included original research data, Hodges conceded that some plant physiologists still believed that essential mineral nutrients were swept into the plant dissolved in the soil water that feeds the transpiration stream. That such a notion still existed was, in part, because it had not yet been possible to demonstrate that nutrient ion absorption by roots, while sensitive to inhibitors of aerobic respiration, was directly energized by cellular ATP. Indeed, it appeared that K^+ movement into root cells was passive, being largely accounted for by the electrochemical potential gradient between the inside and the outside of root cells. At this time, there was still a tendency to attribute the bulk of the electrical potential difference across the plasma membrane to a diffusion potential that required no direct expenditure of ATP. The results of Noe Higinbotham and Bud Etherton (Etherton and Higinbotham, 1960; Higinbotham *et al.*, 1970) for higher plants, and Clifford Slayman (Slayman, 1965) for *Neurospora*, were just beginning to challenge this idea. Research by Ron Poole and others showing a direct correlation between cellular ATP levels and the rate of ion uptake had not yet been completed (Poole, 1979). Unlike mammalian cells, a plasma membrane associated, ouabain-sensitive ATPase that pumped K^+ inside in exchange for Na^+ could not be convincingly demonstrated in plants. It was not even possible to obtain, from broken plant cells, a purified preparation of leaky, let alone sealed, plasma membrane vesicles. In 1968, the manner in which various substances actually traverse the outer cell membrane was a complete mystery (Pardee, 1968).

As Hodges would describe in subsequent lectures in the course, one of his graduate students, James Fisher, had just completed research with crude membrane fractions from roots identifying a Mg^{2+}-requiring, monovalent cation-stimulated, ATP-hydrolyzing activity (ATPase). The ATPase activity appeared to be present in sufficient amounts to account for energy coupling to K^+ absorption (Fisher and Hodges, 1969). Furthermore, there was a striking correlation between the amount of ion-stimulated ATPase activity in roots of various plant species and the capacity of those species to absorb monovalent cations (Fisher *et al.*, 1970). These early results also implied that plant cells used ATP as a direct source of energy for nutrient ion accumulation.

During the next two years, Leonard completed a dissertation, under the direction of John B. Hanson, on the general enhancement of ion transport that occurs following excision and washing of roots. Among the changes found to correlate to the apparent induction of ion transport activity was an increase in ion-stimulated ATPase activity (Leonard and Hanson, 1972). This led, quite logically, to postdoctoral research in the Hodges laboratory and a move with him to Purdue University.

The National Science Foundation (NSF) grant supporting the work at Purdue University was entitled "The Energetics of Ion Transport in Plants." The three years' proposal included twelve specific objectives representing, at least, 20 years of work! Objectives like determining if the ATPase formed a phosphorylated intermediate and purifying the solubilized ATPase were relatively simple compared to objective number 12 which was to determine the inheritance of ATPase(s) involved in transport. Back then, one could get such broad and all encompassing proposals funded. That's just the way things were!

How did the Search for an ATPase Begin?

As Hodges was completing his Ph.D. degree under the tutelage of Yoash Vaadia at the University of California, Davis, he developed an interest in the energetics of ion transport. A young assistant professor in the Botany Department, Joe Key, encouraged Hodges to pursue this interest by postdoctoral study with one of his mentors, John B. Hanson, at the University of Illinois. In 1962, Hodges joined the Hanson laboratory to study the energetics of Ca^{2+} transport in isolated maize mitochondria. Their initial hypothesis was that ATP formation by oxidative phosphorylation should provide the energy needed for ion accumulation into the mitochondrial matrix. Their experiments to demonstrate Ca^{2+} accumulation following ATP synthesis failed. The clue needed to solve the problem came from work on beef heart mitochondria where it was found that Ca^{2+} was accumulated if ADP was omitted. ATP synthesis in mitochondria competes with ion transport for conserved energy. Hodges went back to work, omitted the ADP from the maize mitochondrial transport assay, and the Geiger-Muller counter "glowed" announcing the successful accumulation of radioactive Ca^{2+}. There is no better feeling in science than when the predicted result is so clear that it can be seen from across the room! This was before the chemiosmotic revelations of Peter Mitchell so the result was interpreted in terms of formation in the mitochondria of a high energy phosphorylated intermediate that could be used for either ATP synthesis or for energy dependent ion accumulation. In subsequent experiments, it was shown that adding ATP to isolated mitochondria could energize Ca^{2+} accumulation, providing the first evidence in plants, albeit mitochondria, that ATP could drive ion transport across a membrane.

It was during these days (and nights) of playing with isolated mitochondria that the pioneering work of J. C. Skou on the role of ouabain sensitive, $(Na^+ + K^+)$-ATPase in coupled Na^+ and K^+ transport in animal cells was becoming well-established. In view of Skou's findings and later work by Hodges showing that ion uptake into roots was inhibited when ATP synthesis was prevented by oligomycin, it followed that the plant plasma membrane might have an ATPase that functions in energy coupling to ion transport. Hodges, now a faculty member at the University of Illinois, encouraged his student, Jim Fisher, to look for ATPase activity in root membranes and the results described above soon followed.

In 1969, Hodges returned to the Davis campus for a sabbatical leave with Ralph Stocking. The goal of the leave was to identify the plant membranes that contained ion-stimulated ATPase activity. Stocking had developed a non-aqueous method for isolating chloroplasts and his expertise in cell fractionation would be valuable for the task at hand. During the sabbatical year, Hodges tried every kind of sucrose gradient imaginable to produce an S versus RHO plot of the membrane that contained the ion-stimulated ATPase. He also developed familiarity with using marker enzymes to identify cell organelles, although, at that time, there was no known marker for either the plasma membrane or the tonoplast. He was able to show that the membrane exhibiting the cation-stimulated ATPase activity had a density greater than Golgi vesicles and less than mitochondria. A second significant result from the sabbatical leave was meeting a young Chinese graduate student by the name of Heven Sze. He convinced her to join him at Illinois to work on a Ph.D. dissertation. Later, she was among those who moved with Hodges when he joined the faculty at Purdue.

In the summer of 1971, Hodges was invited by the editors of "Methods in Enzymology" to publish a procedure for the isolation of plasma membrane vesicles from plants (Hodges and Leonard, 1974). This was a very surprising request since such a procedure had not yet been developed! Later, we learned that George Laties of UCLA had reviewed the NSF grant on this research and had made the suggestion to the editors. The NSF grant was also reviewed by Franklin M. Harold of the National Jewish Hospital and Research Center in Denver, Colorado. Harold, while not a plant biologist, was nonetheless very influential in encouraging the program director to support this work in plants even though it was clearly lagging well behind ATPase work in animals.

With the additional carrot offered by the opportunity to publish in "Methods in Enzymology," we worked long hours during the Fall of 1971 to develop a method to purify plasma membrane vesicles from oat roots.

We succeeded with just one weekend to spare! The manuscript was written over the weekend, edited, re-typed and submitted by the February 1, 1972 deadline.

Identifying Plasma Membrane Vesicles in Broken Root Cells

Progress in discovering an enzyme directly involved in transferring energy in ATP to the work of ion accumulation by roots required the ability to obtain enriched preparations of the plasma membrane. During the late 1960s and early 1970s, several laboratories were engaged in developing protocols to identify and purify plasma membrane vesicle from broken cells. There were some problems that needed to be overcome. Once cell ultrastructure is disrupted, vesicles of plasma membrane look like vesicles from other organelles (e.g., tonoplast, endoplasmic reticulum and Golgi vesicles), so a specific marker was needed to identify the plasma membrane during cell fractionation. ATPase activity, itself, was not a good choice because ATP-hydrolyzing activity was associated with several membrane systems and ways (e.g., specific inhibitors) to distinguish these activities from one another would not be developed for another ten years or so (e.g., Gallagher and Leonard, 1982).

That ATP-hydrolyzing activity was associated with a variety of cell components was inferred from studies on isolated mitochondria and chloroplasts, and from a cytochemical technique for visualizing inorganic phosphate released from various phosphate-containing substrates (Hall, 1971; Gilder and Cronshaw, 1973). In this technique, phosphate released from ATP is trapped as a lead-phosphate precipitate which is seen in electron micrographs as black deposits. Over the years, extensive efforts have been made to verify that the origin of the deposits is from a substrate-specific ATPase. It appears that the early cytochemical work probably visualized more non-specific phosphatase than substrate-specific ATPase (Katz *et al.*, 1988). However, as is sometimes the case in science, the conclusion that several organelles and membrane systems in plant cells, including the plasma membrane, contain transport ATPases was correct even though the available cytochemical technique may not have been specific enough to support it.

The method for identifying plasma membrane vesicles in cell fractions came from collaborative work between D. James Morré at Purdue University and J.-C. Roland of the University of Paris (Morré *et al.*, 1969 and Roland *et al.*, 1972). Ironically, the technique also depended on cytochemistry in combination with electron microscopy. Jim Morré began this work while on sabbatical leave in Paris in the laboratory of Professor E.

Benedetti. In the early 1960s, Benedetti and others had observed that, under the right conditions, phosphotungstic acid preferentially stained the plasma membrane of animal cells (Benedetti and Bertolini, 1963). Roland modified the technique by addition of chromic acid and used it to demonstrate the direct association of polysaccharides with the plant plasma membrane. Morré realized that this phosphotungstic acid-chromic acid staining procedure could be used to identify plasma membrane vesicles in cell fractions, and therefore, to follow them during the development of a purification procedure. While the plasma membrane staining procedure turned out to be more of an art than a science, it nonetheless proved essential for developing methods for the isolation of plasma membrane vesicles from plant tissues.

The next obvious step in the research was for Hodges to collaborate with Morré to determine if plasma membrane-rich cell fractions from roots were also enriched in ion-stimulated ATPase activity. However, for reasons that are beyond the scope of this article, this was not so obvious to the two researchers, who, at that time, were preoccupied about philosophical differences on other research matters. Fortunately, the Purdue University faculty also had Charles Bracker, an artist with the electron microscope who was brought into the collaboration.

Separating Membrane-Associated ATPases from Oat Roots

The plan was to use differential and density gradient centrifugation to separate ion-stimulated ATPase activity from broken root cells and then give the membrane fractions to Charles Bracker to determine which one contained the plasma membrane. However, recovering substrate-specific ATPase activity in plant membrane fractions was more difficult than expected. Plant cells contain soluble (e.g., acid phosphatases) and membrane-associated (e.g., nucleoside diphosphatases) enzymes that readily hydrolyze phosphate from various phosphate-containing substrates. These enzymes have a particularly ravenous appetite for ATP! In fact, once they have converted the ATP to ADP, they are very happy to convert the ADP to AMP, and so on. Since the easiest way to assay ATPase activity is to measure inorganic phosphate release from MgATP, the activities of these other enzymes provide a background of inorganic phosphate release that can mask the activity of a transport ATPase. Additionally, the vacuole contains oxidative enzymes that, when released during cell breakage, lead to the inactivation of enzymes sensitive to oxidation. Our efforts to measure ATPase activity at the end of a relatively long cell fractionation procedure were confounded by a loss of substrate-specific ATPase activity and the presence of other, more stable, but non-specific ATP hydrolyzing activities.

Success in measuring the activity of plasma membrane ATPase required the addition of a strong reductant to the medium used to homogenize plant cells and the demonstration that the ATP hydrolyzing activity being measured was substrate-specific for ATP (Leonard *et al.*, 1973). Sorting this out required more time and effort than we would now like to admit, or to discuss here. Fortunately, Hodges had a preference (probably from his days of feeding horses on the family farm) for working on oat (*Avena sativa*) rather than barley (*Hordeum vulgare*) which was more commonly used in experiments on ion transport in plants. The problem of large quantities of non-transport related ATP-hydrolyzing activities is far more severe for barley and there is good reason to believe that Hodges would have left the ion transport field sooner than he did had he selected barley seedlings as his experimental organism!

Oat roots also turned out to be a lucky choice for separating Golgi-associated nucleoside diphosphatases from plasma membrane ATPase during sucrose density gradient centrifugation. In other plant species there is more overlap between the distribution of Golgi vesicles and plasma membrane vesicles in a sucrose density gradient. This is one reason why the use of sucrose density gradients was later abandoned and replaced by an aqueous two-phase procedure that utilizes differences in membrane surface properties to produce, for a wide variety of plant tissues and species, plasma membrane preparations of greater purity (Briskin *et al.*, 1987; Larsson *et al.*, 1987).

Shades of Blue in Racks of Test Tubes

The most commonly used assays for ATPase activity measured inorganic phosphate released from ATP by various modifications of the Fiske and SubbaRow procedure (Hodges and Leonard, 1974; Briskin *et al.*, 1987). Everyone who has worked in the laboratory remembers the rhythmic sound of test tubes shaking in a warm water bath and the sight of rack upon rack of test tubes containing small volumes of different shades of blue solution. The colorimetric phosphate determination is straightforward, but because of non-enzymatic hydrolysis of inorganic phosphate from ATP, it requires the measurement of absorbance at 660 nm precisely after a timed (e.g., 30 mins) incubation with reagents. Hence, every step in the procedure from initiation of the enzymatic reaction, to quenching with acid, to addition of color-producing reagents, and even reading absorbance in a spectrophotometer was done at 20- or 30-sec intervals. The more experienced laboratory workers could cut the interval time down to just 10 seconds and could do hundreds of assays in an afternoon. All additions were made using a glass pipette with rubber bulb as automatic pipettors with plastic tips were not yet available. The stacks of test tube racks and drawers of test tubes and glass pipettes needed daily washing and acid rinsing to remove any contaminating inorganic phosphate. In order to ensure that our favorite rubber pipette bulb would be available when needed for an important experiment, we had to hide it from others in the laboratory! In the early years, absorbance was measured with an inexpensive ($300) Bausch and Lomb, Spectronic 20, spectrophotometer. Later on, we converted to more expensive ($25,000) and sophisticated dual beam spectrophotometers with automatic sampling devices. While these instruments gave rather precise absorbance figures, it was just as easy to see the peak of ATPase activity or to determine if a particular inhibitor was effective by simply looking at the shades of blue color in those racks of test tubes.

ATPase and the Plasma Membrane

After accounting for other ATP-hydrolyzing activities and loss of activity during cell fractionation, sucrose gradients revealed a peak of ion-stimulated ATPase activity at a density (1.16–1.18 g/cc) similar to that reported for the animal cell plasma membrane. Peak ATPase fractions were given to Charles Bracker to determine if they contained vesicles that stained with the phosphotungstic acid-chromic acid staining procedure. They did! In fact, for oat roots the peak ATPase fractions had "over 75% of the membranes" that stained with the plasma membrane stain. "Thus, the membrane fraction enriched in KCl-ATPase is also highly enriched in plasma membranes" (Hodges *et al.*, 1972). This provided the first evidence that plants had a plasma membrane-associated ATPase presumed to be capable of participating in absorption of inorganic ions by roots. Following these results, the emphasis of our research over the next decade was to determine if the plasma membrane ATPase directly participated in absorption of K^+.

Plasma Membrane-ATPase: Cation or Proton Pump?

Plasma membrane-ATPase requires Mg^{2+} in equal molar concentration to ATP because the substrate for the reaction is a MgATP complex. However, addition of monovalent cations ($K^+ = NH_4^+ = Rb^+ > Na^+ > Cs^+ > Li^+$) further stimulates activity by as much as 100%, depending on the plant species. In fact, the nature of the kinetic data for K^+-stimulation of ATPase activity has a remarkable resemblance to the unique, and as yet not fully understood, kinetic data for K^+ absorption into roots (Leonard and Hodges, 1973). Even though the amount of K^+-stimulation of ATPase activity was relatively small when compared to the activity measured in the presence of just MgATP, it was thought to be significant. Animal cell ion pumps such as the Ca^{2+}-ATPase of muscle and the $(Na^+ + K^+)$-ATPase of red blood cells are stimulated several fold by the ions that they transport. In the 1970s, our hypothesis was that monovalent cation stimulation of plant ATPase activity reflected a direct role for the enzyme in K^+ transport. Characterizing ion stimulation of the plasma membrane ATPase became the chore of Nelson Balke (Balke and Hodges, 1975) and Heven Sze (Sze and Hodges, 1977), the next two graduate students in the Hodges laboratory. While the relationship between the selectivity of ion stimulation of plasma membrane-ATPase and ion transport was inviting, research during the years that followed supported the now widely held view that this ATPase is an electrogenic proton pump that does not directly participate in the transport of K^+ (Fox and Guerinot, 1998).

An electrogenic pump is one that moves a net charge through a biological membrane (Poole, 1979). For a proton pump, slightly more positively charged protons are extruded from plant cells than can be compensated for by the movement of other ions (e.g., negatively charged hydroxyl ions) through membrane carriers or channels. The magnitude of the difference between cations and anions is chemically too small to measure, but it is electrically significant. The $(Na^+ + K^+)$-ATPase is electrogenic because it transports three Na^+ out for two K^+ into animal cells. The resulting electrical potential difference (cytoplasm negative relative to the outside) and chemical gradient of Na^+ (higher Na^+ outside the cell than inside) combine to form an electrochemical potential gradient, a store of energy for the work of transporting a variety of ions and metabolites. The proton pump in plants accomplishes the same thing but only pumps protons with no direct transport of a counter ion.

The initial impetus for the hypothesis that the higher plant plasma membrane ATPase is an electrogenic proton pump came from research on an analogous pump in fungal cells (Serrano, 1989; Sussman, 1994). In particular, results from studies in yeast and *Neurospora crassa* formed the background against which results in plants were compared. In fungi, the primary active transport process at the plasma membrane is the electrogenic extrusion of H^+ from cells. The fungal plasma membrane has an ATPase which is only weakly stimulated by monovalent cations, but which otherwise has biochemical characteristics similar to cation pumps of animal cells. Research characterizing the fungal ATPase greatly influenced the direction and interpretation of research on ATPases of higher plants.

Biochemical characterization of the plant ATPase during the early 1980s showed that it had structural and mechanistic features similar to animal and fungal ATPases. In particular, the enzyme is a large hydrophobic protein consisting of 100,000 Da catalytic subunits which are phosphorylated at an aspartyl residue during the course of ATP hydrolysis (Briskin and Leonard, 1982). Much of the early biochemical characterization was done by Don Briskin, then a student in Leonard's laboratory at the University of California, Riverside. For us, work on the plant ATPase came of age when Briskin spent a summer at Vanderbilt University School of Medicine in the laboratory of Robert L. Post, a pioneer in the animal ATPase field and an authority on the reaction mechanism of $(Na^+ + K^+)$-ATPase. Scientists in the medical school were fascinated by the fact that corn seedlings could be grown in a few days (even under a dark stairwell!) and can be used to isolate real biological membranes. Briskin was less than enthused by the trips to the local slaughterhouse to pick up the still warm hog kidneys for isolating $(Na^+ + K^+)$-ATPase. However, his work showed that the corn plasma membrane ATPase shared the same amino acid sequence at its active site of phosphorylation as was found for plasma membrane ATPases from a wide variety of other organisms (Walderhaug *et al.*, 1985). The plant plasma membrane ATPase had taken its place alongside real ion pumps (Briskin, 1990)!

The most compelling evidence for concluding that the plasma membrane of plant cells contains an ATP-driven proton pump came from work pioneered by Heven Sze, now at the University of Maryland, on transport in sealed membrane vesicles (Sze, 1985). Sze led the way by developing methods for isolating membrane vesicles from plant cells that were relatively sealed, and therefore, capable of retaining ATP-driven electrical and proton gradients. Her work enabled plant biologists to manipulate ATP levels in vesicles and more directly study its energetic role in ion transport. Soon after Sze's work, students (Sharman O'Neill and Alan Bennett) of Roger Spanswick at Cornell University conducted research employing phospholipid vesicles reconstituted with a solubilized ATPase preparation. They and others showed that ATP-driven H^+ transport in these reconstituted vesicles had the same characteristics as reported for plasma membrane ATPase activity (e.g., O'Neill and Spanswick, 1984). Similar results were later obtained from studies with sealed (native) plasma membrane vesicles by Briskin (now at the University of Illinois) and others (see Briskin, 1990). Hence, ATP-driven H^+ transport in plasma membrane vesicles was, in fact, catalyzed by the plasma membrane ATPase.

But, did the ATPase also directly pump K^+ in exchange for H^+, in an analogous fashion to the $(Na^+ + K^+)$-ATPase (Briskin and Hanson, 1992)? The answer is no. "The plant plasma membrane H^+-ATPase does not mediate direct K^+ transport chemically linked to ATP hydrolysis. Rather, the enzyme provides a driving force for cellular K^+ uptake by secondary mechanisms, such as K^+ channels or H^+/K^+ symporters (Briskin and Gawienowski, 1996)." All this, along with the excellent work of Serrano (Serrano, 1989), Sussman (Sussman, 1994), and Palmgren (Palmgren, 1998) on cloning plasma membrane ATPase genes and understanding the regulation of expression (Palmgren, 1998), brings the field to about where it stands today.

Today

The field has come along way since the 1960s towards characterizing the structure, mechanism, and regulation of the plasma membrane H^+-ATPase. Back then, well-respected plant biologists seriously doubted that plants needed to use ATP as the source of energy for nutrient ion absorption and the existence of a transport ATPase. Now we know that in *Arabidopsis*, for example, there are at least ten genes coding for plasma membrane ATPases alone (Sussman, 1994). The vacuole membrane has a structurally distinct H^+-ATPase (Sze *et al.*, 1992) and several membrane systems have Ca^{2+}-ATPases (Liarg *et al.*, 1997) that function to directly maintain a suitable cytosolic Ca^{2+} concentration. Plant cell membranes are full of ion-pumping ATPases!

The easy work is over! Electrogenic H^+-transport at the plasma membrane is an essential part of many physiological processes and the ATPase is a carefully regulated enzyme (Michelet and Boutry, 1995; Fox and Guerinot, 1998; Palmgren, 1998). Now, the difficult task of sorting out the details of ATPase regulation remains (e.g., Olsson *et al.*, 1998). However, only a few of the individuals mentioned in this article are still working on the plasma membrane ATPase. In fact, the field that was once so active is now rather sparsely populated. It was like a grand party initially attended and enjoyed by many, but gradually forsaken in pursuit of other interests. The party, for now at least, is over!

Hodges has actively pursued the mysteries of molecular biology and continued doing so now for over 15 years. He struggled with problems associated with genetic engineering of monocots and took a particular interest in the manipulation of transgenes through the use of site-specific recombinases such as the FLP/*FRT* system from yeast. His group developed a new method for making hybrid plants which involves the FLP-mediated excision of a male sterile gene. In addition, the FLP/*FRT* system has been shown to be valuable for production of specialty chemicals in plants. In this case, a blocking piece of DNA is used to separate a promoter from the coding region of a gene allowing, after several generations, a commercially significant increase in the quantity of seed. Then by excising the blocking DNA, such as by cross pollination with a genetic line containing the FLP gene, the previously silent gene can now be expressed, leading to accumulation of its product and easy recovery by harvesting the seeds.

As Leonard was head of an academic department, he was not able to work on ATPases for over ten years. What little time he had for research in recent years was directed towards understanding nitrate uptake and assimilation in cultured maize cells. He watched quietly as full-time research thoroughbreds feeding on molecular genetic technologies galloped past him. He thought about the enormous complexities of signal transduction pathways and families of genes regulating other genes! Thirty years ago life was so much simpler. Just find and study "the" plasma membrane ATPase!

Acknowledgments

We thank all our colleagues, those who worked in our laboratories and those who did not, for their contributions to this work. We apologize if our view of this story has failed to give you the credit that you deserve. We also thank Drs. Heven Sze and Mary Alice Webb for their comments on the manuscript. Finally, the authors acknowledge their long-standing friendship; dating from early morning carpool rides on Winter Midwest mornings, to the exchange of emails and attachments associated with writing this paper, and to those nervous moments before giving a paper at a meeting. We have always taken science seriously, but never let it seriously interfere with enjoying each other's company.

References

Balke, N. E. and Hodges, T. K. (1975) Plasma membrane adenosine triphosphatase of oat roots. Activation and inhibition by Mg^{2+} and ATP. *Plant Physiol.* **55**: 83-86.

Benedetti, E. L. and Bertolini, B. (1963) The use of the phosphotungstic acid (PTA) as a stain for the plasma membrane. *J. Royal Microscopical Society* **81**: 219-222.

Briskin, D. P. (1990) The plasma membrane H^+-ATPase of higher plant cells: biochemistry and transport function. *Biochimica et Biophysica Acta* **1019**: 95-109.

Briskin, D. P. and Gawienowski, M. C. (1996) Role of the plasma membrane H⁺-ATPase in K⁺ transport. *Plant Physiol.* **111**: 1199-1207.

Briskin, D. P. and Hanson, J. B. (1992) How does the plant plasma membrane H⁺-ATPase pump protons? *J. Exp. Bot.* **43**: 269-289.

Briskin, D. P. and Leonard, R. T. (1982) Partial characterization of a phosphorylated intermediate associated with the plasma membrane ATPase of corn roots. *Proc. Natl. Acad. Sci. USA* **79**: 6922-6926.

Briskin, D. P., Leonard, R. T. and Hodges, T. K. (1987) Isolation of the plasma membrane: Membrane markers and general principles. In *Plant Cell Membranes*, Packer, L. and Douce, R., eds., *Methods in Enzymology* **48**: 548-568.

Etherton, B. and Higinbotham, N. (1960) Transmembrane potential measurements of cells of higher plants. *Science* **131**: 409-410.

Gallagher, S. R. and Leonard, R. T. (1982) Effect of vanadate, molybdate, and azide on membrane-associated ATPase and soluble phosphatase activities of corn roots. *Plant Physiol.* **70**: 1335-1340.

Gilder, J. and Cronshaw, J. (1973) Adenosine triphosphatase in the phloem of *Curbita*. *Planta* **110**: 189-204.

Fisher, J. D. and Hodges, T. K. (1969) Monovalent ion stimulated adenosine triphosphatase from oat roots. *Plant Physiol.* **44**: 385-395.

Fisher, J. D., Hansen, D. and Hodges, T. K. (1970) Correlation between ion fluxes and ion-stimulated adenosine triphosphatase activity of plant roots. *Plant Physiol.* **46**: 812-814.

Fox, T. C. and Guerinot, M. L. (1998) Molecular biology of cation transport in plants. *Annu. Rev. Plant Physiol. Plant Mol. Biol.* **49**: 669-696.

Hall, J. L. (1971) Cytochemical localization of ATPase activity in plant root cells. *J. Microscopy* **93**: 219-225.

Higinbotham, N., Graves, J. S. and Davis, R. F. (1970) Evidence for an electrogenic ion transport pump in cells of higher plants. *J. Membrane Biol.* **3**: 210-222.

Hodges, T. K., Leonard, R. T., Bracker, C. E. and Keenan, T. W. (1972) Purification of an ion-stimulated adenosine triphosphatase from plant roots: Association with plasma membranes. *Proc. Nat. Acad. Sci. USA* **69**: 3307-3311.

Hodges, T. K. and Leonard, R. T. (1974) Purification of a plasma membrane-bound adenosine triphosphatase from plant roots. In *Biomembranes Part B*, Colowick, S. P. and Kaplan, N. O., eds., *Methods in Enzymology* **32**: 392-406.

Katz, D. B., Sussman, M. R., Mierzwa, R. J. and Evert, R. F. (1988) Cytochemical localization of ATPase activity in oat roots localizes a plasma membrane-associated soluble phosphatase, not a proton pump. *Plant Physiol.* **86**: 841-847.

Larsson, C., Widell, S. and Kjellbom, P. (1987) Preparation of high-purity plasma membranes. In *Plant Cell Membranes*, Packer, L. and Douce, R. eds., *Methods in Enzymology* **48**: 558-568.

Leonard, R. T. and Hanson, J. B. (1972) Increased membrane-bound adenosine triphosphatase activity accompanying development of enhanced solute uptake in washed corn root tissue. *Plant Physiol.* **49**: 436-440.

Leonard, R. T. and Hodges, T. K. (1973) Characterization of plasma membrane-associated adenosine triphosphatase activity of oat roots. *Plant Physiol.* **52**: 6-12.

Leonard, R. T., Hansen, D. and Hodges, T. K. (1973) Membrane-bound adenosine triphosphatase activities of cat roots. *Plant Physiol.* **51**: 749-754.

Liang, F., Cunningham, K. W., Harper, J. F. and Sze, H. (1997) ECA1 complements yeast mutants defective in Ca pumps and encodes an ER-type Ca-ATPase in *Arabidopsis thaliana*. *Proc. Natl. Acad. Sci. USA* **94**: 8579-8584.

Michelet, B. and Boutry, M. (1995) The plasma membrane H⁺-ATPase. A highly regulated enzyme with multiple physiological functions. *Plant Physiol.* **108**: 1-6.

Morré, D. J., Roland, J.-C. and Lembi, C. A. (1970) Comparisons of isolated plasma membranes from plant stems and rat liver. Proceedings Indiana Academy of Sciences for 1969. **79:** 96-106.

Olsson, A., Svennelid, F., Ek, B., Sommarin, M. and Larsson, C. (1998) A phosphothreonine residue at the C-terminal end of plasma membrane H⁺-ATPase is protected by fusicoccin-induced 14-3-3 binding. *Plant Physiol.* **118:** 551-555.

O'Neill, S. D. and Spanswick, R. M. (1984) Characterization of native and reconstituted plasma membrane H⁺-ATPase from plasma membrane of *Beta vulgaris. J. Membrane Biol.* **79:** 245-256.

Palmgren, M. G. (1998) Proton gradients and plant growth: Role of the plasma membrane H⁺-ATPase. *Adv. Bot. Res.* **28:** 1-70.

Pardee, A. B. (1968) Membrane transport proteins. *Science* **162:** 632-637.

Poole, R. J. (1979) Energy coupling for membrane transport. *Ann. Rev. Plant Physiol.* **29:** 437-460.

Roland, J.-C., Lembi, C. A. and Morré, D. J. (1972) Phosphotungstic acid-chromic acid as a selective electron-dense stain for plasma membranes of plant cells. *Stain Technology* **47:** 195-200.

Serrano, R. (1989) Structure and function of plasma membrane ATPase. *Annu. Rev. Plant Physiol. Plant Mol. Biol.* **40:** 61-94.

Slayman, C. L. (1965) Electrical properties of *Neurospora crassa.* Respiration and intracellular potential. *J. Gen. Physiol.* **49:** 93-116.

Sussman, M. R. (1994) Molecular analysis of proteins in the plant plasma membrane. *Annu. Rev. Plant Physiol. Plant Mol. Biol.* **45:** 211-234.

Sze, H. (1985) H⁺-translocating ATPases: Advances using membrane vesicles. *Annu. Rev. Plant Physiol.* **36:** 175-208.

Sze, H. and Hodges, T. K. (1977) Selectivity of alkali cation influx across the plasma membrane of oat roots. Cation specificity of the plasma membrane ATPase. *Plant Physiol.* **59:** 641-646.

Sze, H., Ward, J. M. and Lai, S. (1992) Vacuolar H⁺-ATPases from plants. *J. Bioenergetics and Biomembranes* **24:** 371-381.

Walderhaug, M. O., Post, R. L., Saccomani, G., Leonard, R. T. and Briskin, D. P. (1985) Structural relatedness of three ion-transport adenosine triphosphatases around their active sites of phosphorylation. *J. Biol. Chem.* **260:** 3852-3859.

Chapter 16

Plant Ubiquitin

Richard D. Vierstra
Department of Horticulture and
 the Cellular and Molecular Biology Program
University of Wisconsin-Madison
1575 Linden Drive, Madison, WI 53706, USA

ABSTRACT

In just the last decade, proteolysis has finally assumed its rightful place as a key regulator of cell metabolism, growth, and development. One of the central players in this process in eukaryotes is ubiquitin, a highly conserved, 76-amino-acid protein. Ubiquitin functions as a reusable signal for selective protein degradation. Via an ATP-dependent cascade of reactions, multiple ubiquitins are attached to proteins destined for breakdown; the modified proteins are then selectively degraded to amino acids by the multi-subunit 26S proteasome with release of the ubiquitin moieties in free, functional forms. Here, I provide a historical perspective of our understanding of intracellular proteolysis and ubiquitin with an emphasis on their roles in higher plants. As will be seen, determining their functional importance required the concerted effort of a number of scientists working in many model systems. Insomuch as we have only scratched the surface of this fascinating aspect of biological control, the review should be read more as a prelude than a final appraisal.

The Beginnings

"If I have seen further it is by standing on the shoulder of giants"
Sir Isaac Newton (1675)

The discovery of ubiquitin and the initial determination of its importance to plant biology was not the result of a single "Eureka!" but from a convergence of experimental results obtained by many distinguished scientists, spanning several generations. In fact, the importance of several of the seminal findings was not immediately evident, requiring some time for their significance to become clear. While I am quite sure this can be said of numerous discoveries, it is especially apt for the field of protein degradation and ubiquitin. Here, I will document the journey which led to our appreciation of proteolysis and ubiquitin as important components in cell regulation, especially as they pertain to plants. I hope to demonstrate just how convoluted science can be and provide an example of how critical it is to listen to those "giants". References that I consider key to this history are included, but the list should certainly not be viewed as comprehensive.

Living systems are in a continual state of turnover at all levels of organization from populations of organisms, to individual cells, to the molecular constituents within cells. Although it was commonly accepted in the 19th century that proteins could be degraded, especially from a zoologist's perspective of the digestive tract, it was not appreciated until the 1930s that proteins from all organisms were continually turned over. From pioneering work on the nitrogen cycle in plants (Mothes, 1933; Gregory and Sen, 1937) and subsequent studies on mammalian enzymes (Schoenheimer, 1942), the idea of a dynamic "protein cycle" emerged. This idea proposed that all proteins underwent a cycle of synthesis and degradation and that these two processes were separable and thus could be varied independently. The idea was somewhat heretical at the time because it implied superficially that cells were quite wasteful; they would devote lots of time, resources, and energy to make proteins only to degrade them later, sometimes quite quickly. Despite this conceptual breakthrough, research on protein breakdown languished for the next twenty or so years. One problem was the technical difficulties inherent in measuring turnover rates and observing how individual proteins were sequentially cleaved. Another was the discovery of DNA which shifted the focus of most biological research away from the protein cycle as a whole towards an emphasis on protein synthesis. Certainly, this shift hastened the discovery of the sophisticated mechanisms used in DNA replication, transcription, and protein translation, but it also created a prejudice that the most important regulatory processes in cells were controlled by the anabolic side of the equation.

Work on protein synthetic processes did have its benefits, mainly by helping develop tools that later would prove essential for work on proteolysis. Foremost was the identification of toxins, such as cycloheximide, chloroamphenicol and puromycin, that block protein synthesis *in vivo*. These inhibitors, when used in combination with either specific enzymatic assays or antisera, enabled researchers to determine which proteins were actively degraded. Another tool was the recently developed method of metabolic labeling, incorporating either radioactive or heavy atom-labeled (e.g., ^{2}H or ^{15}N) amino acids into proteins as a way to directly measure turnover rates (The latter exploited the same methods used to confirm semi-conservative replication of DNA). By a combination of such techniques, Robert Schimke's group was the first to document that the accumulation of an enzyme could be induced by a decreased degradation rate as well as by an increased synthetic rate (Schimke, 1964; Schimke *et al.*, 1965). Plant scientists subsequently used these labeling techniques to show that several plant enzymes (most notably nitrate reductase and phytochrome A) are rapidly degraded as well (e.g., Zielke and Filner, 1971; Quail *et al.*, 1973). Schimke also realized that shorter half-lives allowed cells to more quickly change the steady-state levels of individual proteins. And since the function/fate of any cell is ultimately determined by the abundance of specific proteins, proteolysis had to be important.

In the 1970's, the list of proteins known to be rapidly degraded grew exponentially, proving repeatedly that proteolysis is a highly selective and controlled process (see Goldberg and Dice, 1974; Goldberg and St John, 1976). Proteolysis was also shown to remove abnormal proteins and thus served an important proof-reading function for the synthetic reactions (Goldberg, 1972). But what are the proteolytic mechanisms? Despite the facts that peptide bond hydrolysis is an exergonic reaction and that most purified proteases are energy-independent, protein degradation in intact cells requires energy (Simpson, 1953). It also became clear that the turnover rates of individual proteins vary considerably from half-lives of minutes to years. How does the machinery recognize and degrade the correct proteins from the complex cellular milieu? Reasoning that structures inherent within proteins contributed to their rapid degradation, Goldberg and Dice (1974) attempted to correlate a number of general physico-chemical properties (e.g., size, charge, *in vitro* stability, and ligand binding) with half-life, but the general consensus at the time was that selectivity must be more complex. What are these degradation signals and how do the proteases recognize them? Deciphering these signals was an especially daunting task when one considers that, unlike DNA which only has four permutations (A, G, C, and T) arranged in a single three-dimensional structure (a double helix), proteins have 20 different amino acids, which can be extensively modified (e.g., phosphorylation, glycosylation, and binding of co-factors) and arranged in numerous secondary, tertiary, and quaternary structures. A likely strategy was for the cell to have a highly selective system targeting individual proteins for degradation by a relatively non-specific protease. A convergence of experimental evidence from several directions identified one important mechanism involving the protein — *ubiquitin.*

The Discovery of Ubiquitin

Ubiquitin was independently discovered by three separate groups, the first two working in areas completely unrelated to proteolysis. During a search for thymic polypeptide hormones, Goldstein *et al.* (1975) first identified ubiquitin as a small protein of 8500 molecular mass from bovine thymus. They dubbed it "ubiquitous immunopoietic polypeptide (UBIP)" or ubiquitin for short because it appeared to have hormonal activity *in vitro* (a result now discounted) and because it appeared to be ubiquitous among organisms. Using a radioimmunoassay, they detected ubiquitin in many animals, several bacteria, and even a number of plants purchased from the local grocery store (e.g., celery, eggplant, carrots, pears, and apples). Even though Goldstein *et al.* (1975) were uncertain of its biological functions, its widespread occurrence and possible immunological conservation lead them to conclude that ubiquitin must be "vital to living organisms". Subsequent more rigorous immunoassays failed to confirm the bacterial data and concomitantly questioned ubiquitin's reported presence in plants. Despite this lack of ubiquity, ubiquitin managed to keep its moniker.

Harris Busch and co-workers independently discovered ubiquitin again during a study of bovine chromosomal proteins. Upon sequencing one protein, designated A24, whose abundance differed depending on the cell type, they found that it had not one, but two N-termini (Goldknopf and Busch, 1977). One sequence corresponded to histone 2A and the other surprisingly was ubiquitin. Analysis of the complex revealed that it was formed by the covalent attachment of the C-terminal glycine of ubiquitin to a lysine residue within histone 2A (Lys^{119}) via an *isopeptide bond*. This novel linkage likely was generated by a post-translational mechanism, providing the first evidence that polypeptide tags were biologically possible.

The third group that discovered ubiquitin was a collaborative effort between the laboratories of Avram Hershko and Irwin Rose at the Technion Institute in Israel and the Fox Chase Cancer Center in Philadelphia, respectively. To determine how mammalian cells degrade intracellular proteins, several laboratories including theirs were attempting to generate *in vitro* systems that would recapitulate the properties of *in vivo* proteolysis. The best system that maintained the energy dependence and substrate specificity proved to be rabbit reticulocyte lysates (Etlinger and Goldberg, 1977; Hershko *et al.*, 1978), the same extracts commonly used for *in vitro* translation of mRNA. These lysates would degrade artificial substrates such as lysozyme in a process somehow coupled to ATP hydrolysis. Upon fractionation of the lysates, Hershko and his graduate student, Aaron Ciechanover found that the proteolytic activity could be separated by ion-exchange chromatography into two fractions, called Fraction I and II, that were active only when combined (Ciechanover *et al.*, 1978). Whereas Fraction II was a complex amalgam of proteins, Fraction I was a relatively simple mixture, mainly containing hemoglobin and another factor they called APF1 for "ATP-dependent proteolysis factor I". Preliminary characterizations revealed that APF1 was small (~9 kDa), heat stable, and insensitive to nucleases but sensitive to proteases; diagnostic of a small polypeptide. APF1 was present in most, if not all, rabbit tissues and could easily be purified using its heat tolerance as an important enrichment step (Ciechanover *et al.*, 1980). Keith Wilkinson, Michael Urban and Arthur Haas, all co-workers of Rose, noticed a marked resemblance of APF1 to ubiquitin (i.e., heat stability, size, and universal distribution). Direct comparisons of APF1 with ubiquitin subsequently confirmed that the two proteins were indeed one and the same (Wilkinson *et al.*, 1980).

Using purified preparations, Hershko, Rose, and co-workers showed that ubiquitin/APF1 was the sole factor in Fraction I necessary to restore ATP-dependent proteolysis to Fraction II (Ciechanover *et al.*, 1978, 1980). But how does ubiquitin work? Its linkage to histone 2A in the variant A24 suggested the intriguing possibility that ubiquitin became linked to targets prior to their degradation. By adding radiolabeled ubiquitin to Fraction II, Hershko *et al.* (1980) were able to generate these covalent intermediates in a test tube. In this landmark 1980 paper, they demonstrated that ubiquitin not only functioned as a proteolytic tag but that it was not destroyed in the process and thus was reusable. Furthermore, they demonstrated that both the ubiquitin conjugation and conjugate degradation steps require ATP *in vitro*, thus explaining why protein breakdown is energy-dependent *in vivo*. Hershko *et al.* (1980) also formulated the first description of the ubiquitin pathway (Fig. 1). They proposed that (1) multiple ubiquitins are bound to

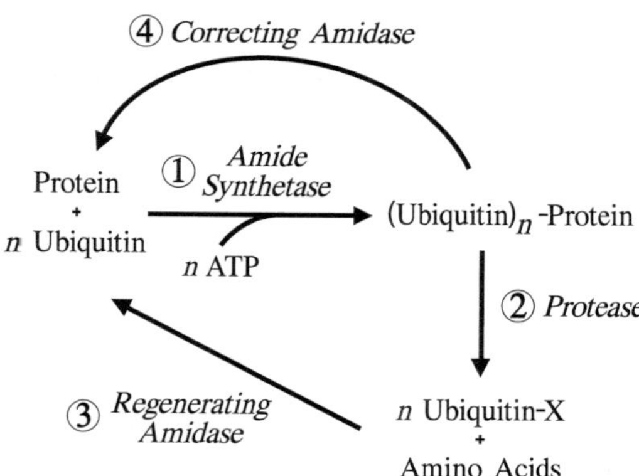

Fig. 1. Proposed sequence of events for ubiquitin-dependent proteolysis as first described by Hershko *et al.* (1980). (1) Multiple ubiquitins are first conjugated to the target protein by an ATP-dependent amide synthetase. (2) Ubiquitin-protein conjugate is degraded by a protease that breaks down the target protein and releases ubiquitin with short peptides or lysine (X) attached to the C-terminus. (3) A regenerating amidase cleaves off X to regenerate free ubiquitin. (4) A correcting amidase removes ubiquitins bound to proteins, releasing both proteins intact. [Adapted from Hershko *et al.* (1980)].

target proteins by an ATP-dependent amide synthetase reaction. (2) A protease then recognizes the ubiquitin-protein conjugate and degrades the target protein. (3) Functional ubiquitin is regenerated by an amidase that removes the lysine or short peptides that remained attached via the isopeptide bond following proteolysis of the target; and (4) a "correcting" amidase exists that can remove ubiquitin inadvertently bound to proteins. What is remarkable is that the model Hershko, Rose and co-workers developed from simple biochemical experiments is still valid today. In essence, the rest of us have spent the next 18 years attempting to fill in the details.

In the early 1980's, a flurry of papers from the Hershko/Rose collaboration began to delineate in elegant detail the enzymatic properties of the pathway, much of it described in the first comprehensive review on the topic in 1982 (Hershko and Ciechanover, 1982). Of special interest was the amide synthetase reaction since it likely represented the committed step in breakdown. They discovered that formation of the amide bond actually was orchestrated by a cascade of three reactions. The first step involved E1 or ubiquitin-activating enzyme that used ATP hydrolysis to generate an energized ubiquitin. The energized ubiquitin was passed to an E2 (or ubiquitin carrier protein/conjugating enzyme). Finally, the ubiquitin was linked to the target protein with or without the help of an E3 (or ubiquitin protein ligase) (Fig. 2). Whereas activation began with a single E1, numerous isoforms of E2 and E3 could participate in conjugation; this diversity potentially explained how the pathway could recognize the thousands of potential protein targets.

Although the *in vitro* system worked well for artificial substrates, the identities of the natural substrates were still unknown. Varshavsky and co-workers determined that some of these targets must be important cell regulators from genetic analysis of a mammalian cell line containing a temperature sensitive E1 (Ciechanover *et al.*, 1984; Finley *et al.*, 1984). At non-permissive temperatures, the mutant cells arrested in the cell cycle, activated the heat shock response, and failed to degrade almost all short-lived proteins. But what are these targets? Studies with the plant morphogenic photoreceptor *phytochrome* provided the first example.

Phytochrome is an Unstable Protein

As any respectable plant scientist knows, phytochromes (from the Greek words for "plant" and "color") are a family of photoreceptors that helps plants respond optimally to their light environment. They were identified and their physiological roles initially were resolved during three decades (1920-1950) of pioneering work by Garner and Allard and later Borthwick, Parker, and

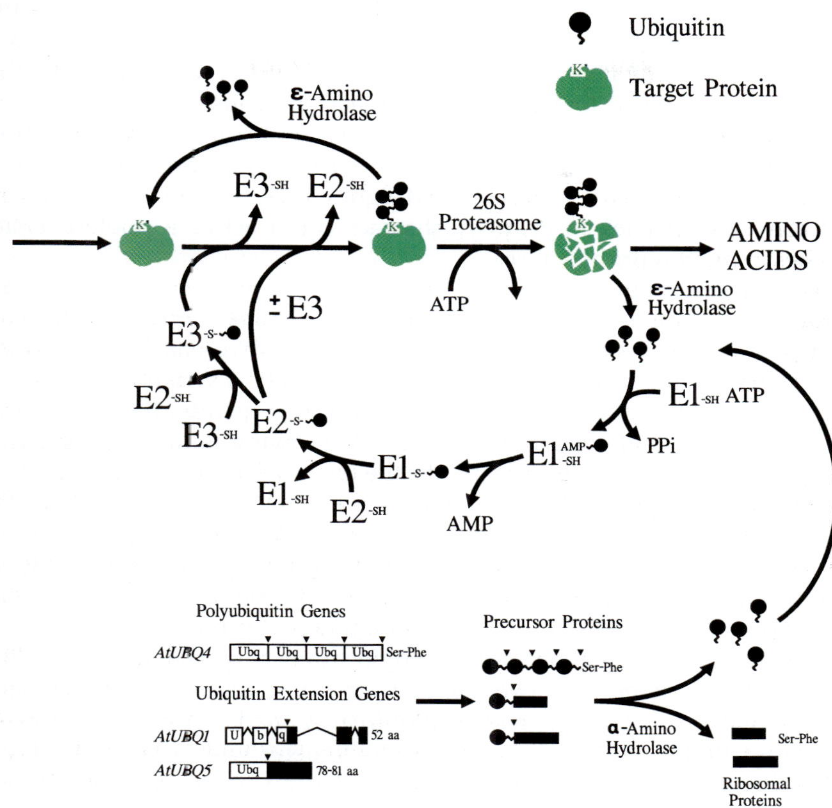

Fig. 2. The pathway for ubiquitin-dependent proteolysis as described by Vierstra (1996). The pathway begins with synthesis of ubiquitin from ubiquitin fusion genes that encode either polyubiquitin or ubiquitin extension proteins. These fusions are then processed by α-amino ubiquitin C-terminal hydrolases (or proteases) to release functional ubiquitin monomers. Several of these ubiquitin monomers are then ligated as a chain to a protein targeted for degradation using an ATP-dependent reaction sequence involving E1s, E2s, and possibly E3s. The ubiquitinated protein is either disassembled by ε-amino ubiquitin C-terminal hydrolases or degraded to amino acids and peptides by the ATP-dependent 26S proteasome with the concomitant release of free ubiquitin. The ubiquitin genes provided as examples are *AtUBQ1, 4* and *5* isolated from *Arabidopsis thaliana* [22,27]. K, lysine involved in ubiquitin attachment; Ubq, ubiquitin.

Hendricks of the USDA Research Center in Beltsville, Maryland. They were trying to determine how light triggers such important agricultural responses as seed germination, pigmentation, and flowering in plants (see Sage, 1992 for historical account). What they didn't know is that their discoveries on phytochromes would help us understand how ubiquitin works as well. Initial action spectra showed that phytochromes had an unusual property among known pigments in that they can exist in two forms, a red light absorbing form (Pr) that is biologically inactive, and a far-red light absorbing form (Pfr) that is biologically active. Because red light converts Pr to Pfr and far-red light converts Pfr back to Pr, phytochromes could act as a photoreversible switches in photomorphogenesis.

To help determine the chemical nature of this pigment, Borthwick and Hendricks recruited Karl Norris and Warren Butler, both spectroscopy experts, to design a spectrophotometer to detect subtle changes in red and far-red absorbance of phytochromes in intact, optically dense tissues. By adapting a spectrophotometer used to monitor egg quality, they designed a highly sensitive single beam instrument. After many failures using tissues from photoperiodically sensitive plants, the breakthrough came in June 1959, when instead of trying light-grown plants, the USDA team tested plants that had not yet seen light. Using etiolated turnip seedlings, Butler easily detected a pigment matching the predicted red/far-red photoreversibility for phytochromes. Subsequent analysis of crude tissue extracts revealed not only that the pigment was soluble but also that it was part protein.

The first public announcement of phytochromes' detection by the USDA group was scheduled just two months after this discovery at the Ninth International Botanical Congress in Montreal, Canada. Just for this event, Norris and Butler equipped the spectrophotometer with a large display to enable the audience to observe first hand the small photoreversible changes of phytochromes they had detected *in vivo*. With the meter and lots of etiolated maize seedlings packed in the trunk of a car, the team went off to Montreal for what they anticipated to be a big announcement. To ensure that the seedlings remained healthy during the 2-day car ride, they were kept in glass chambers and checked periodically. Unfortunately, when they finally assayed the plants during the lecture, no phytochromes were detected and the audience was left skeptical. What Butler soon realized after the fiasco was that the isoform of phytochromes abundant in etiolated seedlings (now designated as phytochrome A), is rapidly degraded as Pfr (Butler *et al.*, 1963; Butler and Lane, 1964). As it turned out, their brief illumination of the seedlings when they opened the car trunk was sufficient to convert all the Pr to Pfr during the trip; this conversion then triggered destruction of the pigment. Kinetic studies indicated that phytochrome degradation, like the breakdown of other intracellular proteins, could be blocked by respiratory inhibitors, suggesting that an energy-dependent step was involved. By density labeling, Quail *et al.* (1973) determined that the half-life of phytochrome A decreased more than 100-fold upon photoconversion to Pfr (half-life >100 hr for Pr versus 1 hr for Pfr).

A Possible Role for Ubiquitin in Phytochrome A Degradation

In 1980, I completed my Ph.D. at the Michigan State University — DOE Plant Research Laboratory and began a postdoctoral fellowship with Peter Quail at Wisconsin. My work first focused on identifying early biochemical events following Pfr formation that might give clues as to how phytochromes initiate the various photomorphogenic responses. However, I found early on that there was something wrong with the purified phytochrome that I and most other researchers were characterizing. When I rapidly extracted phytochrome A from etiolated oat seedlings by immunoprecipitation, it was approximately 6 kDa larger than that in the purified preparations (124 kDa versus 118 kDa). Further analysis revealed that this larger form was actually the intact molecule and the smaller form was generated by limited proteolysis following tissue homogenization (Vierstra and Quail, 1982). Interestingly, the proteolytic cleavage was form-specific, occurring only to Pr, the form in which phytochrome was typically purified. By identifying conditions that precluded this cleavage, I then developed a method to purify the chromoprotein intact (as Pfr with the addition of serine protease inhibitors). The remaining two years of my postdoc was spent studying the biochemical properties of the 124 kDa species, which were significantly different from the proteolytically digested form (see Vierstra and Quail, 1986).

This foray into *in vitro* proteolysis stimulated my interest in proteases and their role in protein degradation *in vivo*. As I began applying for faculty positions, I started to investigate what was known, what approaches were technically possible, and what the relevant questions were in the field. What was clear was that protein degradation in plants had not yet been studied in any detail. Although it was known that several plant proteins such as phytochrome were rapidly degraded, and that plants contained a variety of proteases, no one knew how specific proteins were broken down (Davis, 1982). A current hypothesis had the vacuole acting as the "lytic compartment", mainly because it contained most of the proteolytic activity measured *in vitro* (Matile, 1982). Akin to the animal lysosome, vacuoles were proposed to degrade most proteins regardless of the location, sometimes by engulfing entire organelles (e.g., chloroplasts), a possibility I considered unlikely. Because of this paucity of information, I saw the area ripe for study.

As my initial approach, I hoped to exploit phytochrome A to study protein turnover in plants. From the work of Butler and co-workers, I knew that the chromoprotein was stable as Pr but rapidly degraded as Pfr. As a result, the proteolytic machinery must be able to distinguish between the two forms and selectively degrade only Pfr. Moreover, the ability to manipulate the form of phytochrome *in vivo* both synchronously and non-invasively by simple light irradiations made this system ideal for kinetic studies. For the first time, one could perform accurate pulse-chase analyses *in vivo* without inhibitors or labeled amino acids, using flashes of red and far-red light to convert phytochrome A between the degradable and stable forms.

Armed with this idea, I headed off in the summer of 1983 to interview for a faculty position at Washington University in St. Louis. During the seminar explaining my future research plans, I confidently described the phytochrome system as a great new approach to study intracellular protein degradation. I outlined its many beneficial features and my idea that phytochrome became "tagged" following Pfr formation in a way that seemed to commit it for degradation. This idea was based on the work of Stone and Pratt (1979), who showed that a significant percentage of cycled phytochrome i.e., phytochrome converted from Pr to Pfr and back to Pr, was still rapidly degraded. This cycling effect implied that something happened to Pfr that could trigger degradation of the pigment even when converted back to the theoretically stable form. As I mentioned this aspect, someone from the back of the seminar room retorted, "Could it be ubiquitin?" Instantly I recognized the voice as Joe Varner's, one of the intellectual giants in plant physiology. I had never heard of ubiquitin before. So not wanting to look incompetent, I ignored the comment. Luckily for me, ubiquitin was not mentioned again for the rest of the interview.

Upon my return to Madison, I began asking others at Wisconsin what ubiquitin was. No one had heard of it, not even Peter Quail, which certainly helped my ego recover. After further research, I discovered the recently published review of Hershko and Ciechanover (1982) describing the ubiquitin proteolytic pathway. Reading it confirmed my suspicion that Varner was really onto something; ubiquitin could be the tag for phytochrome degradation! Its involvement could explain the specificity for Pfr, degradation of cycled Pr, and the energy dependence of phytochrome A turnover. Within a few months, I accepted a faculty position at Wisconsin and decided to spend some time testing this hypothesis. Luckily (and thankfully) for me, my first grant on the subject was sent to a USDA research panel headed by none other than Joe Varner. Needless to say, given his interest in the subject, the grant was funded. I was now ready to experiment!

Detection and Characterization of Plant Ubiquitin

Given that the immunoassay used by Goldstein *et al.* (1975) detected ubiquitin even when it wasn't present, the first thing I did was attempt to confirm their findings that plants actually contained this protein. Upon contacting Hershko for some rabbit ubiquitin antibodies, they directed me to Art Haas, now at the Medical College of Milwaukee. Using antibodies and the newly developed method of Western blotting, I surveyed, in the winter of 1983–1984, several plants for ubiquitin. The signals were so weak, however that I was afraid they were just a figment of my imagination. That's when Haas alerted me to a problem he had encountered with various mammalian extracts he was studying. In attempts to develop *in vitro* systems for ubiquitin conjugation, he discovered that many extracts contained one or more protease(s) that rapidly cleave ubiquitin. Although the protease(s) removed only the two C-terminal glycines, they reduced recognition by ubiquitin antibodies by almost 100-fold and completely inactivated the molecule with respect to conjugation (Haas *et al.*, 1985). As we surmised, this same proteolytic modification also was a problem in plant extracts.

After a survey of protease inhibitors that I have become familiar with during my postdoc days studying phytochrome, I came upon a cocktail that precluded this cleavage. Using these conditions, I could now easily detect ubiquitin in selected tissues from a variety of plants (Vierstra *et al.*, 1985). The cross reactivity of plant ubiquitins with human ubiquitin antibodies implied that plant proteins must be structurally related to their animal counterparts, but at the time I had no idea just how similar they were. Importantly, we also saw higher mass forms of ubiquitin in addition to the free ubiquitin monomer of 8.5 kDa after blotting crude plant extracts. These species likely represented ubiquitin-protein conjugates, providing the first indication that the ligation system was also active in plants.

Using the Western blot assay, I developed a procedure to purify ubiquitin from etiolated oats, identified as a rich source of the protein. Considering how poorly equipped my laboratory was with me being a new faculty member, I was lucky the procedure was relatively easy. In fact, the most effective purification step required only a beaker and a Bunsen burner! Simple heating of the crude extract to ~85°C for a few minutes precipitated almost all the proteins. Ubiquitin, in contrast, remained soluble and could be concentrated easily by ammonium sulfate precipitation and brought to near homogeneity by a gel filtration step (Vierstra, *et al.*, 1985). We completed the amino acid sequence of the oat protein and found it to differ from the animal counterpart by only three residues, thus making ubiquitin one of the most conserved proteins yet identified (Vierstra *et al.*, 1986). Not surprisingly, upon X-ray crystallographic analysis of oat ubiquitin in collaboration with Bill Cook, we found that the three-dimensional structures of oat and animal ubiquitins are indistinguishable (Vijay-Kumar *et al.*, 1987). Sequence analyses of the corresponding genes from a variety of species have shown since that all higher plant ubiquitins have the same peptide sequence and that this sequence differs by only one residue in algal ubiquitins (Vierstra, 1996).

To conclusively demonstrate that the ubiquitin pathway operates in plants, it remained to be proven that plants could conjugate purified ubiquitin to proteins *in vitro*. To do this, I added radiolabeled human ubiquitin (generously provided by Art Haas) to various plant extracts. Just as Hershko *et al.* (1980) had observed with reticulocyte lysates, nothing happened in the absence of ATP. But when ATP was included, the ubiquitin immediately became conjugated to other proteins, observed as a smear of radiolabeled material at the top of SDS-PAGE gels following electrophoresis of the samples (Vierstra *et al.*, 1986). Taken together, I concluded that plants contain the ubiquitin-dependent proteolytic pathway and that by all criteria investigated, it appeared to be mechanistically similar to that present in animals.

Connecting Ubiquitin to Phytochrome A Degradation

With various reagents in hand (i.e., purified ubiquitin, plant ubiquitin antibodies, the ability to detect ubiquitin conjugates by Western blotting, and an *in vitro* conjugation system), I felt ready to test the hypothesis proposed by Joe Varner that ubiquitin participated in phytochrome degradation. But was it technically feasible? Even though plants (and animals) contained an array of ubiquitin conjugates formed *in vivo*, no one had yet identified any of these natural targets, not even in animals. Maybe ubiquitination and proteolysis were so intimately connected that the ubiquitinated intermediates for any one protein never accumulated to detectable levels?

Luckily for me, two scientists had just joined my laboratory and were prepared to give it a go. One was Merten Jabben, a postdoc with extensive experience on phytochrome and the other was John Shanklin, a fiercely motivated young graduate student. We initiated a straightforward approach using highly sensitive immunoblots with phytochrome antibodies to examine what happened to phytochrome A during Pfr degradation. I had hoped to find and characterize the cleavage fragments of Pfr so that we could use this fingerprint to identify the responsible protease. After months of trial and error, and various attempts to increase the sensitivity of the blots, nothing really significant appeared. The phytochrome polypeptide just disappeared during Pfr turnover, suggesting that degradation was quite rapid and complete.

While I was ready to give up, Shanklin persevered until the spring of 1986 when he began to notice something interesting; a faint ladder of higher molecular mass forms of phytochrome A appeared when he overloaded and overdeveloped the blots (Fig. 3). The size increments of the individual rungs were consistent with the addition of one to several ubiquitins (Shanklin *et al.*, 1987). Could these be ubiquitin-phytochrome conjugates? This ladder was present at very low levels making it impossible to detect simply by blotting the crude extracts with ubiquitin antibodies. As a result, Shanklin and Jabben enriched for this ladder, using phytochrome antibodies to first immunoprecipitate the proteins from crude plant extracts. When this immunoprecipitate was subjected to Western blotting with ubiquitin antibodies the same ladder was detected, demonstrating for the first time that ubiquitinated forms of phytochrome A actually existed (Shanklin *et al.*, 1987). To our surprise, these blots detected not only one or two ubiquitins attached to phytochrome but a large array of conjugates, some potentially containing as many as 20 ubiquitins (Fig. 3). Consistent with a role in Pfr degradation, the phytochrome-ubiquitin conjugates appeared soon after Pfr formation, reached a peak when about half of the phytochrome A was degraded and then disappeared when most of the Pfr was removed.

In two subsequent papers, Shanklin and Jabben examined the kinetics of ubiquitin-phytochrome conjugate accumulation in various plant species (Jabben *et al.*, 1989a,b). In all cases, the levels of conjugates coincided with the amount of Pfr degraded. However, these conjugates were never very abundant, accumulating to only a few percent of the total protein, suggesting that the conjugates were rapidly degraded following ubiquitination. By exploiting the red/far-red light reversibility of Pfr degradation, they determined for the first time the turnover rate of a ubiquitin-protein conjugate. Whereas Pfr had a half-life of 1hr, the ubiquitin-phytochrome conjugates had half-lives of 5–10 min, exactly what one would expect for a degradation intermediate (Jabben *et al.*, 1989b). Shanklin and Jabben also tested the degradation of cycled Pr. Just as Joe Varner had predicted, this cycled Pr was ubiquitinated as well. How the ubiquitin pathway selectively recognizes Pfr is still unknown. An analysis of mutant phytochromes expressed in transgenic plants suggests that several domains are involved; one appears to serve as a recognition site for one or more ubiquitin conjugating enzymes and the other as a site that binds ubiquitins (Clough and Vierstra, 1997).

Using similar approaches, other laboratories subsequently identified a number of short-lived proteins that are also targets of the ubiquitin pathway (Rechsteiner, 1991; Hochstrasser, 1996). These include cyclins that help control the cell cycle, the αMAT2 repressor involved in switching mating type in yeast, and the p53 oncoprotein. In each case, a small percentage of the protein pool became modified with multiple ubiquitins during breakdown. Like phytochrome, many of these targets are key cellular regulators, demonstrating that the pathway really is important for developmental and metabolic control. Studies on the protease responsible for degrading

Anti-Phy

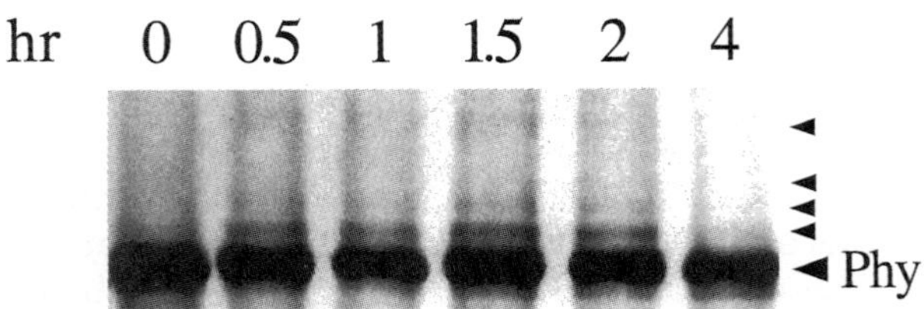

Anti-Ubq

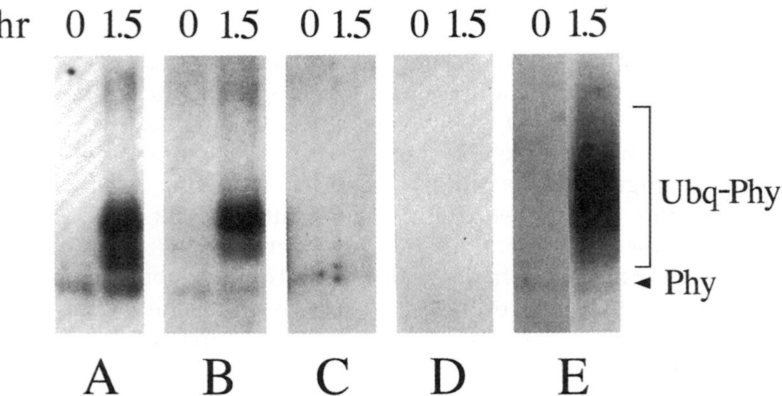

Fig. 3. Appearance of ubiquitin-phytochrome conjugates during Pfr degradation in oat seedlings. Etiolated oat seedlings were irradiated continuously with red light and at the times indicated, the tissue was homogenized and the phytochrome A was immunoprecipitated from the extract with phytochrome antibodies or non-immune antibodies. The immunoprecipitates were subjected to SDS-PAGE and Western blotting with phytochrome antibodies (**Top panel**), ubiquitin antibodies, or non-immune antibodies (**Bottom panels**). (**Top panel**) Arrowheads indicate the location of minor phytochrome polypeptides with an apparent molecular mass higher than the 124 kDa oat phytochrome A monomer. (**Bottom panels**) Panels A-E. Crude extracts were either directly used for the immunoprecipitation (E) or phytochrome was further concentrated and purified by hydroxyapatite chromatography before immunoprecipitation (A-D). Immunoprecipitation was performed with phytochrome antibodies (A, B, C, and E) or non-immune antibodies (D). Immunoprecipitates were then subjected to Western blotting with either oat ubiquitin antibodies (A, D, and E), human ubiquitin antibodies (B), or non-immune antibodies (C). The migration positions of oat phytochrome A and ubiquitin-phytochrome conjugates are indicated on the left. [Adapted from Shanklin *et al.* (1987)].

ubiquitinated proteins, the 26S proteasome, also helped explain why we as well as others failed to detect cleavage products for these targets. The protease is incredibly efficient and processively disassembles proteins into short peptides and amino acids on its own.

Wheat Germ and Arabidopsis Help Define Much of the Pathway in Plants

In the years following the detection of ubiquitinated phytochrome A, my laboratory focused on the various reactions involved in ubiquitin conjugation and conjugate degradation. Peggy Hatfield, a postdoc in the laboratory, identified wheat germ as a rich source for isolating many of the components (Hatfield and Vierstra, 1988). Using the ubiquitin-affinity method elegantly developed by Ciechanover *et al.* (1982), she, Mike Sullivan, Steve van Nocker and Pierre-Alain Girod purified several E1s, multiple E2s and enriched for an E3 activity (Sullivan *et al.*, 1990; Vierstra, 1996). The protease responsible for degrading ubiquitin conjugates (called the 26S proteasome) was detected as well as the amidases originally proposed by Hershko *et al.* (1980) to be responsible for regenerating ubiquitin and possibly correcting inadvertent conjugation [now called a ubiquitin C-terminal hydrolases or protease (Fig. 2)].

Using either antibodies against these proteins or oligonucleotide primers derived from peptide sequence, the corresponding genes to many of these factors were identified. Most of the cloning exploited *Arabidopsis thaliana* as the emerging genetic model for plants. As the cloning progressed, we began to realize that the ubiquitin proteolytic pathway was more complex than anyone first thought. For example, Thomas Burke and Judy Callis discovered 14 novelly arranged genes encoding just ubiquitin (Burke *et al.*, 1988; Callis *et al.*, 1995). More than 30 E2 genes were identified by these cloning efforts and various sequencing projects. And the 26S proteasome was found to contain more than 50 subunits that assembled into a structure almost as large as the ribosome. At the time of this writing, we knew of over 100 *Arabidopsis* genes that encode ubiquitin pathway components [or 0.5% of the total proteasome (Vierstra, 1996 and data not published)]. Clearly, the magnitude of this genetic investment alone demonstrates that the ubiquitin pathway must be critical to plant biology.

Conclusions

It has been nearly two decades since the original discovery of ubiquitin's proteolytic function in eukaryotes. During that time, a number of ubiquitin pathway targets have been identified and a plethora of factors that participate in the process have been isolated (see Hochstrasser, 1996; Vierstra,

1996). The pathway is now known to participate in almost all aspects of cell physiology and development, including the cell cycle, metabolic regulation, defense and stress responses, DNA repair, gene expression and silencing, hormone and light-regulated development in plants and the immune response in mammals, to name a few.

Despite this accumulated knowledge, many questions still remain to be answered, especially with respect to plants. At the time of this writing, phytochrome was still the only known target of the ubiquitin pathway and we still do not know how it is recognized by the conjugation machinery. Dissecting the functions of ubiquitin in plants has been hindered by the absence of relevant mutants. However, this obstacle has been solved recently with the development of facile reverse genetic strategies to identify mutations in previously known genes. Recent studies indicate that polypeptides related to ubiquitin also exist. Like ubiquitin they can be conjugated to other proteins by similar enzymatic cascades (E1s, E2s, and E3s); however, this tagging may have different functional consequences such as altering the location and/or activity of the modified protein. It is possible that our understanding of ubiquitin is only the tip of the iceberg, and that eighteen years from now we will be aware of a whole constellation of covalent polypeptide modifiers. Given the amazing array of insights that have emerged during my fourteen years working on proteolysis and plant ubiquitin, I would not rule out any possibility. I certainly concur with Martin Rechsteiner who stated that "(he) believed that ubiquitin is a lucky molecule. Almost everyone who has studied it has made fascinating observations" (Rechsteiner, 1988). I am eternally in debt to Joe Varner for making me aware of it.

Acknowledgements

This review is dedicated to the memory of **Dr. Joseph Varner** (1921–1995), Professor of Biology at Washington University, St Louis, whose vision and enthusiasm inspired a great number of plant scientists and influenced numerous aspects of plant biology, including my own. I also thank various members of my laboratory, past and present, for their dedication and hard work that made all of our results possible.

Dr. Joseph Varner (1921-1995)

References

Burke, T. J., Callis, J. A. and Vierstra, R. D. (1988) Characterization of a polyubiquitin gene in *Arabidopsis thaliana. Molec. Gen. Genetics.* **213**: 435-443.

Butler, W. L. and Lane, H. C. (1964) Dark transformations of phytochrome *in vivo. Plant Physiol.* **40**: 13-17.

Butler, W. L., Lane, H. C. and Siegleman, H. W. (1963) Nonphotochemical transformations of phytochrome *in vivo. Plant Physiol.* **38**: 514-519.

Callis, J. A., Carpenter, T. B., Sun, C.-W. and Vierstra, R. D. (1995) Structure and evolution of genes encoding polyubiquitin and ubiquitin-like proteins in *Arabidopsis thaliana* ecotype, Columbia. *Genetics* **139**: 921-939.

Ciechanover, A., Elias, S., Heller, H. and Hershko, A. (1982) Covalent affinity purification of ubiquitin activating enzyme. *J. Biol. Chem.* **257**: 2537-2542.

Ciechanover, A., Elias, S., Heller, H., Ferber, S. and Hershko, A. (1980) Characterization of the heat stable polypeptide of the ATP-dependent proteolytic system from reticulocytes. *J. Biol. Chem.* **255**: 7525-7528.

Ciechanover, A., Finley, D. and Varshavsky, A. (1984) Ubiquitin dependence of selective protein degradation demonstrated in the mammalian cell cycle mutant ts85. *Cell* **37**: 57-66.

Ciechanover, A., Hod, Y. and Hershko, A. (1978) A heat-stable polypeptide component of an ATP dependent proteolytic system from reticulocytes. *Biochem. Biophys. Res. Commun.* **81**: 1100-1105.

Clough, R. C. and Vierstra R. D. (1997) Phytochrome degradation. *Plant Cell Environ.* **20**: 713-721.

Davis, D. D. (1982) Physiological aspects of protein turnover. In *Encyclopedia of Plant Physiology, New Series,* Vol. 14A. Coulter, D. and Partier, B., eds., Springer Verlag, Berlin, pp. 189-228.

Etlinger, J. D. and Goldberg, A. L. (1977) A soluble ATP-dependent proteolytic system responsible for the degradation of abnormal proteins in reticulocytes. *Proc. Natl. Acad. Sci. USA.* **74**: 54-58.

Finley, D., Ciechanover, A. and Varshavsky, A. (1984) Thermolability of ubiquitin-activating enzyme from the mammalian cell cycle mutant ts85. *Cell* **37**: 43-55.

Goldberg, A. L. (1972) Degradation of abnormal proteins in *Escherichia coli. Proc. Natl. Acad. Sci. USA.* **69**: 422-426.

Goldberg, A. L. and Dice, J. F. (1974) Intracellular protein degradation in mammalian and bacterial cells. *Ann. Rev. Biochem.* **43**: 835-869.

Goldberg, A. L. and St. John, A. C. (1976) Intracellular protein degradation in mammalian and bacterial cells: part 2. *Ann. Rev. Biochem.* **45**: 747-803.

Goldknopf, I. L. and Busch, H. (1977) Isopeptide linkage between nonhistone and histone 2A polypeptides of chromosomal conjugate-protein A24. *Proc. Natl. Acad. Sci. USA.* **74**: 864-868.

Goldstein, G., Scheid, M., Hammerling, U., Boyse, E. A., Schlesinger, D. H. and Niall, H. D. (1975) Isolation of a polypeptide that has lymphocyte-differentiating properties and is probably represented universally in living cells. *Proc. Natl. Acad. Sci. USA.* **72**: 11-15.

Gregory, F. G. and Sen, G. K. (1937) Physiological studies in plant nutrition. VI The regulation of respiration rate to the carbohydrate and nitrogen metabolism of the barley leaf as determined by nitrogen and potassium deficiency. *Ann. Bot. (London)* **1**: 521-561.

Haas, A. L., Murphy, K. E. and Bright, P. M. (1985) The inactivation of ubiquitin accounts for the inability to demonstate ATP, dependent proteolysis in liver extracts. *J. Biol. Chem.* **260**: 4694-4703.

Hatfield, P. M. and Vierstra, R. D. (1989) Analysis of the ubiquitin-dependent proteolytic pathway in wheat germ: isolation and characterization of the "ubiquitin activating enzyme" (E1). *Biochemistry* **28**: 735-742.

Hershko, A. and Ciechanover, A. (1982) Mechanisms of intracellular protein breakdown. *Ann. Rev. Biochem.* **51**: 335-364.

Hershko, A., Ciechanover, A., Heller, H., Haas, A. L. and Rose, I. A. (1980) Proposed role of ATP in protein breakdown: conjugation of proteins with multiple chains of the polypeptide of ATP-dependent proteolysis. *Proc. Natl. Acad. Sci. USA.* **77**: 1783-1786.

Hershko, A., Heller, H., Ganoth, D. and Ciechanover, A. (1978) Mode of degradation of abnormal globin chains in rabbit reticulocytes. In *Protein Turnover and Lysosome, Function.* Academic Press, N. Y., pp. 149-169.

Hochstrasser, M. (1996) Ubiquitin-dependent protein degradation. *Annu. Rev. Genetics* **30**: 405-439.

Jabben, M., Shanklin, J. and Vierstra, R. D. (1989a) Ubiquitin-phytochrome conjugates: pool dynamics during *in vivo* phytochrome degradation. *J. Biol. Chem.* **264**: 4998-5005.

Jabben, M., Shanklin, J. and Vierstra, R. D. (1989b) Red-light accumulation of ubiquitin-phytochrome conjugates in both monocots and dicots. *Plant Physiol.* **90**: 380-385.

Matile, P.H. (1982) Protein Degradation. In *Encyclopedia of Plant Physiology, New Series,* Vol. 14A, Coulter, D. and Partier, B., eds., Springer Verlag, Berlin, pp. 169-188.

Mothes, K. (1933) Investigations on the assimilation of ammonia. *Planta* **19**: 117-138.

Quail, P. H., Schäfer, E. and Marmé, D. (1973) Turnover of phytochrome in pumpkin cotyledons. *Plant Physiol.* **52**: 128-131.

Rechsteiner, M. (1988) *Ubiquitin.* Plenum Press, N.Y., p. 346.

Rechsteiner, M. (1991) Natural substrates of the ubiquitin proteolytic pathway. *Cell* **66**: 615-618.

Sage, L. C. (1992) Pigment of the imagination: a history of phytochrome research. Academic Press, N.Y., p. 562.

Schimke, R. (1964) The importance of both synthesis and degradation in the control of arginase levels in rat liver. *J. Biol. Chem.* **239**: 3808-3817.

Schimke, R., Sweeney, E. W. and Berlin, C. M. (1965) The role of synthesis and degradation in the control of rat liver tryptophan pyrrolase. *J. Biol. Chem.* **240**: 322-331.

Schoenheimer, R. (1942) *Dynamic State of Body Constituents.* Harvard University Press, Cambridge, MA.

Shanklin, J., Jabben, M. and Vierstra, R. D. (1987) Red light-induced formation of ubiquitin-phytochrome conjugates: identification of possible intermediates of phytochrome degradation. *Proc. Natl. Acad. Sci. USA.* **84**: 359-363.

Simpson, M. V. (1953) The release of labeled amino acids from the proteins of rat liver slices. *J. Biol. Chem.* **201**: 143-154.

Stone, H. J. and Pratt, L. H. (1979) Characterization of the destruction of phytochrome in the red-absorbing form. *Plant Physiol.* **63**: 680-682.

Sullivan, M. L., Callis, J. A. and Vierstra, R. D. (1990) HPLC resolution of ubiquitin pathway enzymes from wheat germ. *Plant Physiol.* **94**: 710-716.

Vierstra, R. D. (1996) Proteolysis in plants: mechanisms and functions. *Plant Molec. Biol.* **32**: 275-302.

Vierstra, R. D. and Quail, P. H. (1982) Native phytochrome: inhibition of proteolysis yields a homogeneous monomer of 124 kilodaltons from *Avena. Proc. Natl. Acad. Sci. USA.* **79**: 5272-5276.

Vierstra, R. D. and Quail, P. H. (1986) Phytochrome: the protein. In *Photomorphogenesis in Plants,* Kendrick, R. E. and Kronenberg, G. H. M., eds., W. Junk Publishers, The Netherlands, pp. 35-60.

Vierstra, R. D., Langan, S. M. and Haas, A. L. (1985) Purification and initial characterization of ubiquitin from the higher plant, *Avena sativa. J. Biol. Chem.* **260**: 12015-12021.

Vierstra, R. D., Langan, S. M. and Schaller, G. E. (1986) The complete amino acid sequence of ubiquitin from the higher plant, *Avena sativa. Biochemistry* **25**: 3105-3108.

Vijay-Kumar, S., Bugg, C., Wilkinson, K. D., Vierstra, R. D., Hatfield, P. M. and Cook, W. J. (1987) Comparison of the three-dimensional structures of yeast and oat ubiquitin with human ubiquitin. *J. Biol. Chem.* **262**: 6396-6399.
Wilkinson, K. D., Urban, M. K. and Haas, A. L. (1980) Ubiquitin is the ATP-dependent proteolysis factor I of rabbit reticulocytes. *J. Biol. Chem.* **255**: 7529-7532.
Zielke, R. H. and Filner, P. (1971) Synthesis and turnover of nitrate reductase induced by nitrate in cultured tobacco cells. *J. Biol. Chem.* **246**: 1772-1779.

Chapter 17

Thirty Years of Fun with Antenna Pigment-Proteins and Photochemical Reaction Centers: A Tribute to the People Who Have Influenced My Career

---

J Philip Thornber
Department of Biology, University of California, Los Angeles

In Memory of Dr J Philip Thornber

We tried to invite prominent scientists who made important discoveries in plant biology to contribute to this series. It has been very unfortunate that a couple of them are no longer with us and could not share their wisdom with us.

Dr J P Thornber is one of the invitees who agreed to take up the authorship but in our tears passed away. With the kind and generous permission of Mrs Thornber (Dr Elaine Tobin) and Kluwer Academic Publisher, we are going to reprint on pages 327-346 the "Thirty years of fun with antenna pigment-proteins and photochemical reaction centers: A tribute to the people who have influenced my career". In this paper, Dr Thornber described the landmark discoveries made by himself and other photosynthesis workers. He also shared with readers his enjoyment and excitement in his research career.

Shain-Dow Kung and Shang-Fa Yang

Photosynthesis Research **44**: 3–22, 1995.
© 1995 *Kluwer Academic Publishers. Printed in the Netherlands.*

Personal perspective

Thirty years of fun with antenna pigment-proteins and photochemical reaction centers: A tribute to the people who have influenced my career

J. Philip Thornber
Department of Biology, University of California, Los Angeles, CA 90095-1606, USA

Received 13 December 1994; accepted in revised form 7 February 1995

Key words: antenna pigment-proteins, photochemical reaction centers, PAGE systems

Abstract

The author summarizes the research contributions to photosynthesis made by him, his graduate and postdoctoral students, visiting scientists and by his collaboration with other photosynthesis workers during 1964–1994. The development of isolation procedures and biochemical/biophysical characterization of antenna pigment-proteins and photochemical reaction centers are described together with the author's education and experiences as a scientific researcher. Some anecdotes hopefully add insight into what it was like to be in this area of science during the period.

Scientific research, like life, has its ups and downs. Nevertheless, I hope that the enjoyment and excitement I have had during my career comes across to the reader in my personal historical perspective of one area of photosynthesis research. I also hope that I relay the spirit of camaraderie that I have found in my research area, because it was openness and friendship with my colleagues that mainly guided my approach to scientific research. I would like students to look at my career as an indication that from simple beginnings much is possible, and that all the hard work and frustrations of their training can indeed be worthwhile. Over the past 30 years I have made many friends in photosynthesis research who had an impact on my education as a research scientist, and it is to them that I dedicate this article.

1964–1967 – Twyford Laboratories Ltd., London, England

I was 26 when I joined the staff of Twyford Laboratories in Fall 1961, immediately after finishing my PhD research in the Department of Biochemistry at Cambridge University, England. Twyford Laboratories was a basic research subsidiary of the Arthur Guinness Brewery, in London, and was patterned after the famous Carlsberg Laboratories in Copenhagen, Denmark. However, contrary to rumors about the Danish laboratories, beer was not on tap at each lab bench at Twyford. Nevertheless, we had all the benefits of food, drink and sporting facilities associated with working at a big brewery.

The Plant Biochemistry section at Twyford was headed by a protein chemist, John Leggett Bailey, a very spruce English gentleman who claimed it was possible while wearing evening dress, to do chemistry at the lab bench without ruining one's clothing; I never saw him demonstrate this. Under his direction I learned protein biochemistry techniques to add to those of cellular biochemistry I had been taught during my PhD studies on plant cell walls with Donald H. Northcote. In London, we examined the Fraction I (now known as RubisCo) and Fraction II proteins, so named by Wildman and Bonner (1947). Our studies were greatly aided by the advent of polyacrylamide gel (cyanogum) electrophoresis, first brought to our attention in an Eastman-Kodak pamphlet written by Ornstein and Davis in 1962 (see Davis 1964). But by 1965, I was to become very eager to understand how pigments were organized in photosynthetic membranes.

At that time little was known about such organization. I and others have reviewed in detail what was known (e.g., Thornber (1975) for plants; and Thornber et al. (1978b) for bacteria). Essentially, the view was that solubilization of photosynthetic membranes yielded fractions with varying chlorophyll (Chl) *a/b* ratios (for plants) and different spectral forms of bacteriochlorophyll (BChl) for purple bacteria. That different proteins might be responsible for these fractions was barely considered. Rather, it was felt, the fractions represented different large arrays of pigments associated with membrane lipids, and that excitation energy migrated through them by a random walk to a specific point where energy conversion occurred (see for example, Rabinowitch and Govindjee 1965). Proteins, while known to be present in the membrane, were thought to be more likely involved in maintaining the structure of the membrane rather than in organizing pigments therein (e.g., Criddle and Park 1964). And yet as long ago as 1938 (Smith 1938; see also Govindjee 1988), it had been shown or postulated that proteins were associated with photosynthetic pigment by Emil L. Smith and E.G. Pickels for higher plants, and E.C. Wassink and E. Katz for purple bacteria, respectively. Only in one instance had it been unambiguously demonstrated that a protein was involved in their organization: A small fraction of the total pigment of the green photosynthetic bacterium, *Chloropseudomonas ethylicum*, had been purified (and crystallized) as a water-soluble BChl *a*-protein (Olson and Romano 1962; see also Olson 1994). Whether this was an anomaly had to await future research.

An observation, made late one Friday afternoon in 1965, was to set me on a three-decade study of Chl and carotenoid organization in the many diverse photosynthetic organisms. While cleaning that week's laboratory glassware with Stuart Ridley, I noticed that the detergent we used to remove the thylakoid residue from the bottom of some centrifuge tubes produced a transparent green solution. We had previously paid little attention to this residue because we considered it a by-product of our Fraction I and II preparations. But on this day there remained some unused polyacrylamide tube gels. On a whim, I layered the green solution, after adding some sucrose to it, onto the top of the gels, switched on the power, and, to our surprise, we observed the resolution of several green bands (discs in those days) within a few minutes. I did not doubt, based on the rate at which our colorless proteins migrated under the same conditions, that we were looking at complexes of Chl with protein. A phone call to the manufacturers of the glass-washing detergent on the following Monday yielded the names of the active agents – sodium dodecyl benzene sulphonate (SDBS) and a Tween (non-ionic) detergent. Each was used to solubilize thylakoids, and it was SDBS rather than the non-ionic detergent, that yielded a soluble extract from which multiple colored bands were produced upon PAGE.

Further work on the SDBS-PAGE system by myself and Carol Smith yielded three green and one yellow band; Stuart Ridley continued the work on Fraction I protein (Ridley et al. 1967). When the polyacrylamide gels were stained with Amido Black, we found protein stains exactly coincident with the two upper green bands in the tube. These we labeled Complex I and Complex II – a terminology that inadvertently was to reflect something of their function which was determined later. We thought that the third band (Complex III) was a mixture of non-protein associated Chls and carotenoids, some of which we knew had been displaced from pigment-protein complexes by the detergent. We called this band – free pigment.

Studies of the Chl and carotenoid content in the bands dissected from the tubes of polyacrylamide, of their sedimentation coefficients, and of their amino acid compositions, left John Bailey and myself in no doubt that there were two photosynthetic pigment-protein complexes in the photosynthetic apparatus. Or, more importantly, that proteins were used to organize at least half the pigments in the photosynthetic apparatus. Many photosynthesis researchers were not so convinced, as I was to learn later. They argued that because the photosynthetic pigments were located in the lipid bilayer, detergent solubilization of the membrane resulted in the released pigments being adsorbed by the membrane structural proteins, and the bands we were observing on the gels were these artefactual complexes.

At this stage, however, we were only too eager to let the world know of our exciting discovery. I presented a 15-minute talk at a London meeting of the Biochemical Society. It was a cold Saturday morning and the room was packed with listeners for the first talks. But as the morning wore on, more and more people left the room, but I was sure it was only for coffee. Surely they would soon return to pack the room to hear about our earth-shaking discovery. When I talked towards the end of the morning, I looked up from my notes to see just five people in the large room – the chair of the session, the other two authors of the Abstract, the previous speaker and the following speaker. The chair's call for 'any questions or comments' brought nothing but silence. Much chagrined we returned to further studies.

Twyford Laboratory's Plant Biochemistry section in 1965–66 included, among others, Richard P.F. Gregory, John C. Stewart and Mark W.C. Hatton. Richard and I took advantage of Boardman and Anderson's (1964) seminal work on separation of the two photosystems; the notion of two rather than one photosystem in oxygenic organisms was only a few years old. Our group obtained evidence that Complex I, a 9S pigment-protein was derived from PS I, and Complex II, a 3S component, from PS II. Complex I had a high Chl *a/b* ratio (12) and beta-carotene represented almost its entire carotenoid content, whereas Complex II had a lower Chl *a/b* ratio (1.5–2) and xanthophylls (particularly lutein) were its major carotenoids. Amino acid analysis revealed that Complex I was unusually rich in histidine and Complex II in proline. Both contained a preponderance of non-polar amino acids, and at that period they represented the most hydrophobic analyses of a protein.

During 1964–66 the parent company (A. Guinness and Sons, Ltd.) had problems justifying the cost of Twyford Labs. to the almost total absence of any product. 'Golden handshakes' were offered to some, while others saw the writing on the wall – that the end of fundamental research at Twyford was near, under what had been idyllic support and surroundings. It was time to write up in full our research on the Chl-proteins (Thornber et al. 1967a. b), and move on. Furthermore, Ogawa et al.'s 1966 paper describing somewhat similar results had appeared. A letter to John Bailey from Werner Hirs, chairman of the Biology Department at Brookhaven National Laboratories on Long Island, New York, inquiring whether anyone in our group was interested in working at Brookhaven with John Olson on the water-soluble BChl *a*-protein of *Chloropseudomonas ethylicum* was not only timely but also fortuitous for establishing the rest of my scientific career, not that I was to know so at the time. Working in the US seemed to entail only a small diversion from having a career in the UK because I planned to be in the US for only one to two years, and would return to take up a new position at the University of Sussex, England. Little did I know of what events were to occur over the next three years that resulted in my family never again being UK residents.

1967–1970 – Brookhaven National Laboratories

The family arrived at Brookhaven in 1967 in time to experience a March blizzard; however, because we lived on site with many other visiting plant scientists (among whom were George and Joan Akoyonoglou, and Peter Schopfer), the storm added only to the general excitement of the new surroundings. Furthermore, our first child, Karen, was seven weeks old when we arrived and she kept us occupied. In the lab, John Olson taught me much about bacterial photosynthesis and about the use of biophysical approaches, of which I was totally ignorant, while I performed an in-depth biochemical analysis of the BChl *a*-protein provided to me in crystal form. Only now, on rereading Thornber and Olson (1968), as a result of being posed some questions on it by Robert E. Blankenship at a symposium honoring my retirement, am I reminded of the excellence of the analytical facilities of the Biology Department. My analysis of what is now known as the FMO protein showed it had a 'normal' amino acid composition, contained BChl *a* and an unknown, possibly an unusual carbohydrate entity; however, unlike the plant complexes it had no carotenoids and no lipids.

My long friendship with Robert Heath, also a postdoc in the Biology department as well as a native Californian and an Anglophile, started during this period. Bob was another biophysicist to whom I am grateful not only for his gentle explanation of the physics behind photosynthesis but also for the conversion of a 'Brit' to the American way of life. I think he, more than anyone else, made me realize that I could be perfectly happy having a career in the US. His insistence that there is no place on earth as great as California was particularly persuasive in my spending the major part of my career in that state.

John Olson, assisted by Betty Shaw and Frieda Engelberger, grew several purple bacteria as well as the green bacterium. Thus, after my analysis of the BChl *a*-protein, I was introduced to *Chromatium*, Strain D (later *Chr. vinosum*) and *Rhodopseudomonas* NHTC 133, the latter was to make its name later as *Rps. viridis*. Geoffrey Hind, who had been in the same Part II Biochemistry course in Cambridge as I, and Bill Siegelman were two other members of the Brookhaven department who contributed much in preparing me for a career in photosynthesis. Bill taught me how to make buckets full of hydroxyapatite that had a good flow rate for column chromatography, and provided me with ice-cream cartons packed with harvested frozen cyanobacteria. Building upon my experiences at Twyford labs I wanted to see if the techniques used by Bill to purify large quantities of biliproteins and cytochromes (Celite-reverse ammonium sulfate gradient followed by hydroxyapatite chromatography) would yield cyanobacterial Chl-proteins equivalent to those we had seen in higher plants in 1965. I devised a procedure which was so unusual it was never adopted by anyone else. It made sense to me at the time but, in retrospect, it is amazing that it worked. All the color in an SDS extract of a cyanobacterium, *Phormidium luridum*, was precipitated with ammonium sulfate and the whole mixed with celite (a filtration aid) and poured into a large column. Elution was with ammonium sulfate containing as much methanol as would just not precipitate the salt. I reasoned that the organic solvent would elute a lot of carotenoids and a proportion of the Chl *a* that in higher plants we had observed as free pigment. I believed this was a necessary step for obtaining pure products. Subsequent elution with buffer removed the salt and solvent but not the bulk of the bright green material in the column. Its elution required SDS to be added to the buffer. Since I started with masses of cells, I ended up with masses of what later was shown to be the heart of PS I, the P700-Chl *a*-protein. Its equivalence to Complex I from higher plants was revealed by its pigment content, amino acid composition, its migration on SDS-PAGE and sedimentation coefficient. John and Bill generously encouraged me to publish this paper by myself (Thornber 1969) even though they had had much impact on the work. I cannot remember whether or not I expected a Chl-containing equivalent of Complex II. There, of course was not one.

Everything now started to come together for me, particularly after attending an NSF-ASPP one-week course on 'The physical concepts and methodologies in the study of photosynthesis' held in March 1968 at Cornell University. There I met for the first time many of the leading people in the field in the US: Roderick Clayton, Bessel Kok, Kenneth Sauer, Wilse Robinson, André Jagendorf, Warren Butler, William Parson, Dan Reed, and Tom Ebrey. The series of lectures and laboratories were eye-opening for me who was a novice about the physical side of photosynthesis, and timely to say the least. Rod Clayton and Dan Reed, who had just published the first isolation of a photochemical reaction center (RC) using Triton X-100 treatment of *Rps. sphaeroides*, R-26, a carotenoidless mutant, demonstrated how to observe the photooxidation of P870 in the RC using the IR2 mode of the Cary 14R

spectrophotometer. I returned to Brookhaven full of ideas.

The next few months were among my most productive in research. It was then perfectly feasible to change the organism one studied at least once during the day. The field of pigment organization was wide open, and not particularly competitive at least for oxygenic organisms; furthermore, I was being pressed to substantiate that my studies on higher plant pigment-proteins were meaningful. I was now using the more readily available SDS, rather than SDBS, to solubilize photosynthetic membranes. In many ways it was a fortunate, but not ideal, choice, as I was to learn 20 years later. SDS added to *Rps. viridis* membranes converted the 1015 nm-antenna BChl *b* into a chlorin derivative (690 nm) but surprisingly left the RC intact, and I now knew how to look for photochemical activity therein. I passed the extract through hydroxyapatite; the chlorin was not retained but a brown band was and could be removed by 0.2 M sodium phosphate. The eluate contained mainly the RC in a photoactive form, but unlike the two RC's that had by then been purified from two carotenoid-less mutants of *Rps. sphaeroides* and *Rsp. rubrum* (Gingras and Jolchine 1969), the *Rps. viridis* component seemingly contained two cytochromes, C553 and C558, which are now known to be a single four-heme cytochrome. There are few better people with whom to study bacterial cytochromes than John Olson, and with the help of two Brookhaven summer students, a partial characterization of the first RC obtained from a wild type (i.e. carotenoid containing) organism was reported (Thornber et al. 1969).

If SDS-solubilization and hydroxyapatite chromatography yielded the *Rps. viridis* RC, would it also work for *Chromatium*, strain D? Although the answer was 'no', the fractionated material obtained proved interesting, more with respect to the antenna than the RC. Three fractions termed A, B and C were obtained (Thornber 1970). Fraction A had the longest wavelength antenna BChls (B890) bound to the reaction center (P883), which was also associated with two cytochromes (C553 and C556) (presumably also a single four-heme cytochrome). Isolation of the *Chromatium* RC was not easy (see below). Fraction B contained the B800–B850 antenna BChls. Both A and B had low-quantum-yield reversibly photo-oxidizable Bchls (Fraction A had the true functional P883 of the RC as well). The function of the photooxidizable pigments, P'890 in Fraction A and P840 in Fraction B, has never been elucidated. They remain interesting phenomena but are certainly not connected with primary reactions. Fraction C was a B800–B820 antenna component.

Researchers at the Johnson Foundation at the University of Pennsylvania (P. Leslie Dutton, Britton Chance and Don DeVault) were eager to compare their studies of the primary photochemical events in *Chromatium* chromatophores with what occurred in Fraction A when there was no membrane structure. Les Dutton was a recent arrivee also from the North of England, and this common bond was at least partly responsible for this visit. I duly arrived carrying a sample of Fraction A that was whisked away from me while I gave a lunch-time seminar. Before I left that evening, Britt asked me to glance over two abstracts he had written, with my name on them, one on proton binding to Fraction A and the other on the effect of high pressure on electron transfer in the Fraction A, for the forthcoming Biophysical Society Meeting in Los Angeles. The work had been done between the morning and evening of that day. However, in spite of this initial speed, it took nine years before the work was fully written up for a book on electron tunneling reactions (Chance et al. 1979). Les Dutton and colleagues determined that the kinetics of oxidation of cytochrome C553 by photooxidized P883 at 77 K in Fraction A were identical to those in chromatophores (Dutton et al. 1971), thereby substantiating that detergent fractionation had not altered any critical molecular arrangements on the oxidizing side of the photosystem, and that the membrane was not obligatory for the primary event to occur. To this day, Les has the flask of Fraction A used for the studies on his office windowsill. He believes that it is still 'alive'.

At least once every month of 1969 I drove to J.F. Kennedy airport and flew to some part of the US. Most flights were connected with job interviews, others with seminar or meeting presentations. While it is not so unusual for a scientist to fly that frequently now, I found it mind-boggling then because I had never traveled much. I also realized that it would be possible for me to have a career in the US in a research area that I enjoyed, and that contained many interesting and helpful scientists. Furthermore, the state of the British economy had resulted in the disappearance of the promised position at Sussex University, and three American universities were now interested in my joining their faculties: however, a barrier existed – I had a visa that required me to leave the US for two years before I could obtain resident alien status. Only Paul Boyer at UCLA, one of the three interested universities, knew how to overcome this hurdle. I happily joined the UCLA faculty in July 1970. Fifteen months later I was issued a 'green card' by the immigration authorities.

Two visits in early Fall 1969, one to William Parson at The University of Washington, Seattle, and the other a month-long visit to George Feher at University of California, San Diego, added much to my experiences during that year. Bill Parson and his then student George Case used their laser/Xenon lamp equipment to look at *Chromatium* Fraction A and the isolated *Rps. viridis* RC. They showed, while I watched, that in both preparations the two cytochromes bound to the fraction were oxidized by one and the same reaction center (P^+883 in *Chromatium* and P^+960 in *Rps viridis*). Which cytochrome was oxidized depended on the redox potential of the sample (Case et al. 1970). The results obtained in Philadelphia and Seattle, plus studies by John Olson and Olga Owens reported at the annual Biophysical Society meeting in 1970, ended a notion then prevalent that there was more than one photosystem in purple bacteria, and further strengthened the notion that photooxidation of P883 or P960 and their subsequent oxidation of the bound cytochromes is unaltered in membrane-free preparations. SDS did, however, destroy the secondary electron transfer between X (later shown to be bacteriopheophytin (BPh)) and Y (the primary quinone), which in retrospect, was likely due to SDS extracting quinones from the preparation.

With George Feher and Jim McElroy at UCSD, I attempted to improve the purification of the *Rps. sphaeroides*, R26, reaction center, the preparation of which had by then evolved so that Triton X-100 was replaced by lauryl dimethylamine oxide (LDAO). I was eager to try SDS as yet another solubilizing agent as well as the chromatographic methods I was then using at Brookhaven. The changes were of little advantage. We removed some extraneous colorless proteins but the difference was not worth altering the existing isolation procedure for the RC. Non-denaturing PAGE indicated the RC had a size of 50 ± 3 kD. George Feher has given me more credit over the years than I think I deserve for getting his group more involved in RC research. I learned much more from his group than they did from me. George would probe me until he was satisfied I had revealed everything I knew about protein purification procedures and gel electrophoresis, and only then would he evaluate the relative merits of the different approaches for purification. I was not to know at the time that we would later share not only common scientific interests but also a similar health problem, and for his advice on the latter I am also most grateful.

I was placed on the Biology Department staff for my last year (1969–70) at Brookhaven, and John Olson generously gave me William E. Dietrich Jr. as a research associate. Bill worked on the cyanobacterial Chl *a*-protein that I had isolated in bulk in 1968. He showed that it contained a functional P700 in a ratio of one per 100 Chl *a* antenna molecules. Later, a more exact extinction coefficient for P700 determined by Hiyama and Ke (1972) altered the ratio to about one per 50 Chls. Bill Dietrich also showed that in addition to Chls, carotenoid and proteins, the complex contained a quinone that was not plastoquinone but more likely a tocophoryl or polar naphtho-quinone (see below) which we thought was involved in secondary electron transfer reactions in the complex (Dietrich and Thornber 1971). Our P700 preparation was not as enriched in P700 as that (HP700) of Leo P. Vernon's group (e.g. Yamamoto and Vernon 1969) who had used Triton X-100 to solubilize their thylakoids. This difference in detergents used later influenced my approach to obtaining the P700-Chl *a*-protein from higher plant thylakoids.

My contact with students and teaching had been minimal during the 1960's, so that my participation in the Summer 1970 Marine Botany course at Woods Hole, Massachusetts, was a big change. The familiar pattern of having a seven-week-old girl when moving to the next position was maintained – Emma was born seven weeks before we went to Woods Hole. The course was headed by Bill Siegelman, and other instructors were John Olson, Frank Loewus, Francis Haxo, Bob Wilse and Bob Guillard. I think I was officially a T.A.; however, I lectured and worked with the students in the laboratories just as the instructors did. All the laboratory equipment we needed had to be brought with us. The very first evening the course prepared a batch of hydroxyapatite. It was soon discovered that we had no pipettes. This was not a problem for Bill Siegelman who went to a local store and returned with plastic straws as a substitute. Improvisation, as practiced by Bill, has been important in many aspects of my life. Three students on this course ultimately entered the field of photosynthesis – Katherine Steinbeck, Barbara Prézelin and Randall S. Alberte. Eighteen months later Randy Alberte joined me as a postdoc at UCLA.

1970–1976 – University of California, Los Angeles

Establishing my own laboratory at UCLA in Fall 1970 was not as simple as I had hoped. Paul Boyer planned that a plant biochemistry/photobiology group would occupy the seventh floor of a soon-to-be-built Molecular Biology Institute of which I was a member. I was temporarily housed in the basement of one of the four oldest buildings on campus where Jacob Biale, George Laties, David Appleman and Park Nobel, fellow members of the Botanical Sciences department, were also located. My two rooms had one sink and an assortment of tables and cupboards but little else. I began accumulating chemicals and equipment with some money provided by the department and concentrated initially on my teaching duties. In Spring 1971, I received my first extramural grant and my first graduate student (Daryl Benson), and slowly research restarted. In Fall 1971 I jumped at an offer to move into the Botany Building in which the majority of the other departmental members were located, because by then the planned Molecular Biology Institute no longer had a seventh floor; plant-oriented scientists were not then a strong selling point for fund-raising.

Samuel G. Wildman (UCLA) greatly influenced my thinking throughout this period and even after his retirement in 1978. We spent hours debating notions about chloroplast structure, plant genetics and photosynthesis. One of his postdocs, Shain-Dow Kung, and I devised a purification procedure to get higher yields of plant Complexes I and II (Kung and Thornber 1971). We renamed them the Photosystem I and II Chl-proteins; these seemed more descriptive of their function. Of course they neglected the fact that they also contained carotenoids, as Jan Anderson pointed out at a Gordon Conference that year. Hardly surprisingly in light of Sam's interest in chloroplast DNA, the three of us set about determining the site of coding of the major thylakoid protein, the PS II Chl-protein, by studying its inheritance in interspecific reciprocal hybrids of *Nicotiana tabacum* and *Nicotiana glauca*. The protein had a Mendelian mode of inheritance (Kung et al. 1972) which provided the first conclusive proof that a major thylakoid protein was coded in the nucleus. We did not know at the time whether its mRNA was translated on chloroplast or cytoplasmic ribosomes.

Randall S. Alberte's association with my research program between 1971–76 was responsible for my becoming interested in the biogenesis of the photosynthetic components of thylakoids – an interest that remains to this day. Randy was completing his Ph.D at Duke University under Aubrey Naylor's supervision in 1970–72, and spending brief periods in my laboratory before staying more permanently in Fall, 1972. His work at Duke had shown that etiolated Jack bean leaves greened under high humidity had, unlike under all other greening conditions, no lag phase in Chl accumulation, and at UCLA that SDS-PAGE showed the PS II Chl-protein appeared in the leaf before that of PS I (Alberte et al. 1972). As anticipated their appearances coincided with the onset of oxygen evolution and P700 photobleaching, respectively. Later, we observed that the Chl/P700 ratio of the P700-Chl a-protein increased from 35 to 75 during greening (Alberte et al. 1973). In retrospect this probably reflected an increase in the amount of LHC I associated with the core-complex I (CC I) (see Dreyfuss and Thornber, 1994b). There was, and still is, one aggravating aspect of greening studies that neither we nor anyone else, to the best of my knowledge, have been able to solve, and that is to get good PAGE separations of *intact* pigment-proteins of material isolated during the first hour of greening. Even after two hours of greening such separations are not routinely obtained. A solution to this problem would greatly help delineate the earliest events in assembly of pigments with their apoproteins.

Harry R. Highkin (California State University, Northridge) was a frequent visitor to the Laties/Wildman/Thornber laboratories in the 1970's. He had studied various aspects of a Chl b-less barley mutant ever since his PhD work with A.W. Frenkel (University of Minnesota) (Highkin 1950). He either attended my 'tenure seminar' in 1972 or one of the seminar courses that Sam, George Laties and I taught, and became interested in what effect the absence of Chl b might have on the PS II Chl-protein. We anticipated that the complex would have only Chl a and carotenoids. After all, the mutant was photosynthetically viable and therefore had to have a functional PS II pigment-protein, or so we thought. Of course, we observed it lacked the PS II Chl-protein (Thornber and Highkin 1974). This led to several proposals: one, we gave the wild-type complex a new name, the light-harvesting Chl a/b-protein (which I will abbreviate to LHCP for much of the rest of this article); two, we proposed its function was analogous to that of the algal biliproteins; three, we pointed out that the complex must exert a strong influence on maintaining lamellae in contact with each other; four, there must be some other component (not stable in SDS) in which the PS II RC resided. Not everything we proposed was correct;

e.g., we favored the nuclear gene coding for the LHCP apoprotein as the site of mutation and not some protein in the Chl b biosynthetic pathway. At this stage of the field's development we were unaware that there was more than one apoprotein that was associated with Chl b.

This work done in early 1973 convinced most of those who doubted the reality of photosynthetic pigment-protein complexes in higher plants that their view was incorrect. I cannot explain why it did so, but it and examination of other mutants lacking one of the pigment-proteins by different groups, plus a study of the CD spectra of thylakoids and the isolated LHCP (Gregory et al. 1972), and an Annual Review of Plant Physiology article I wrote in 1975 were primarily responsible for doing so. For a few years after its publication, researchers quoted Thornber (1975) to indicate that Chl occurred mainly in association with protein (see Thornber 1985).

Harry Highkin had to travel only a few miles to visit the UCLA labs, but Jeanette S. Brown (Department of Plant Biology, Carnegie Institution of Washington at Stanford University) made the 300 plus-mile round trip journey to UCLA weekly from February to August 1974. The availability of methods to fractionate thylakoid pigment-proteins was the 'carrot'. Jan fractionated the complexes of green and yellow-green algae and *Euglena* in collaboration with Randy Alberte. Absorbance spectra of the complexes were taken back to Stanford where C. Stacy French and Jan Brown used computer-assisted curve-fitting to obtain the red wavelength maximum of each Chl a and b spectral form in each complex. Some major influences on our thinking resulted from this collaboration: The P700-containing complex was shown to be the same in grana and stroma regions of the thylakoid. This complex and the LHCP contained the same four major Chl a spectral form of 662, 670, 677 and 684 nm in widely different kinds of plants, and indicated to us that all oxygenic organisms had essentially identical PS I complexes. However, our notion that all of the Chl b in green plants was contained in the LHCP was incorrectly reinforced. We also found that the thylakoid's content of LHCP was influenced by environmental conditions (Brown et al. 1974). Harvard Lyman (State University of New York at Stony Brook), whom I had known while at Brookhaven, spent his sabbatical leave in my laboratory in 1975–76. He worked hard but unsuccessfully to resolve the transiently stable SDS-treated pigment-proteins of *Euglena gracilis* , and thus was unable to unravel the biogenesis and turnover of the pigment-proteins when *Euglena* changed between phototrophic and heterotrophic growth.

Five graduate students were in my group in the first half of the 1970's. Daryl Benson tried to remove all of the chlorin contamination from the *Rps. viridis* RC preparation. Before he left graduate school he had had some success if sodium dithionite was included in the lamellae solubilization mixture. Terry Trosper, a post-doc, completed this study, and her studies of the NIR spectral forms of the improved RC preparation led to a conclusion that BPh *b* was involved in the primary photochemical event (Trosper et al. 1977). Lily Lin, a graduate student, obtained an RC preparation from *Chr. vinosum* by using LDAO and not SDS (Lin and Thornber 1975). Kuey-Suey Kan purified the light-harvesting Chl *a/b*-protein from SDS extracts of *Chlamydomonas reinhardtii*. The complex had an essentially identical biochemical characteristics as those of the LHCP of higher plants. She determined, using amino acid analysis, that the complex contained three Chl *a*, three Chl *b* and one xanthophyll molecule per 29 kD polypeptide (Kan and Thornber 1976). Today the pigment/protein ratios of this complex purified from non-ionic detergent extracts are larger (12–14/1), but in 1975 SDS was the solubilizing agent which extracted considerable amounts of pigment from the thylakoid complexes to yield most of the free pigment observed on PAGE. I now think that this difference in ratio is explained by one half of the LHCP's structure being more accessible to SDS than the other (see Kühlbrandt et al.'s 1994 structure) and that removal of the pigment molecules in this half resulted in the lower pigment/protein ratio in the SDS product while the change in Chl a/b ratio was small but significant (1.3–1.0) (i.e., each half of the pigment assembly is composed of both Chls *a* and *b*).

Two other graduate students, Judith A. Shiozawa and Fiona A. Hunter, studied P700-containing fractions of higher plant and cyanobacteria, respectively. Judy confirmed that SDS treatment of higher plant thylakoids destroyed P700 activity whereas the same treatment of cyanobacterial membranes was known not to do so. However, as Bill Dietrich had noted earlier, the quantum requirement for P700 photooxidation in the cyanobacterial P700-Chl *a*-protein was considerably greater than the value obtained for P700 in situ. Although the PS I Chl-protein of eukaryotes, but not prokaryotes, is spectrally altered by SDS treatment, we could conclude from Judy's subsequent studies and those of Jan Brown and Randy Alberte (above) that the heart of PS I was organized in the same manner in

all Chl *a*-containing organisms which included gymnosperms (Alberte et al. 1976). When Judy changed the solubilizing agent from SDS to Triton X-100 (see above what prompted this choice), she obtained a fraction from a higher plant plastids that contained P700 (Chl/P700 = 40–45/1) and it had the same spectral characteristics and pigment content as the SDS-prepared cyanobacterial P700-Chl *a*-protein (Shiozawa et al. 1974). In some instances (Alberte and Thornber 1978) we could lower the Chl/P700 ratio in isolated material to 20/1.

Fiona Hunter examined the subunit composition of the cyanobacterial P700-Chl *a*-protein, and observed two polypeptides of 48 and 46 kD. In view of much subsequent work on the apoproteins of this complex, it was reasonable that two similarly-sized subunits were seen, but they were at least 20 kD smaller than was subsequently determined. Perhaps proteolysis occurred in our samples. A very useful collaboration with Richard Malkin and Alan Bearden (University of California, Berkeley) developed in 1975 which clarified the differences between SDS and Triton X-100 treatments of P700 in eukaryotes and prokaryotes (Malkin et al. 1976). This investigation occurred at a time when the identity of primary acceptors in plant and bacterial photosynthesis was receiving much attention. We were able to add fuel to the proposal that an iron-sulfur center (g values 2.05, 1.94 and 1.86) was the likely primary acceptor in the PS I photoact. This was deduced because the SDS-prepared P700-Chl *a*-protein of cyanobacteria lacked the Fe-S center EPR signals whereas the equivalent Triton-prepared material contained them. Furthermore, it explained why the SDS-prepared fraction had a greater quantum requirement for P700 photooxidation.

In spring 1976, John Olson and Geoffrey Hind organized what was then an annual event – a Brookhaven Symposium in Biology. On this occasion it covered Chl-proteins, reaction centers and photosynthetic membranes. It was the first symposium in which Chl-proteins, by then accepted as authentic entities by most photosynthetic researchers, were a major topic. Somehow I managed to get most of the people in my lab to Long Island for the meeting. I believe it is very important that students starting out in an area meet the established researchers who would otherwise be just names to them. In my talk, I summarized my lab's studies on Chl-containing pigment-proteins that we had obtained since I arrived at UCLA (Thornber et al. 1977). There was much discussion about the identity of the primary electron acceptor of PS I. We had some evidence (see above) that an iron-sulfur center might be it, and I also added that Judy Shiozawa had found one molecule of vitamin K1 per P700 in her PS I preparations. We felt that it and its equivalent in cyanobacteria (tocopheryl quinone – see above) might play similar roles and that might be in the primary event of PS I, but we had no direct evidence. Later, others showed that Vitamin K1 was an early but not the immediate electron acceptor. R. Clinton Fuller had for some years been a champion of the idea that pteridines had many of the appropriate properties to fill that role. I remember his quizzing me at the end of my talk whether pheophytins, which were being touted as the putative primary acceptor in bacterial photosynthesis, could be eliminated as the PS I acceptor. I was able to state unambiguously that our active P700 fractions contained neither pheophytins nor pteridines.

It was at this Brookhaven lecture that I first proposed a model of the arrangement of pigment-proteins in the photosynthetic unit. The model was based on the 'funnel' concept of Chl spectral forms of Govindjee et al. (1967) which had recently been given more details by Gilbert Seely (Seely 1973), and on Warren Butler's fluorescence studies (e.g. Butler and Kitajima 1975). We envisaged the unit to contain four pigment-proteins. In addition to the P700-Chl *a*-protein and the light-harvesting Chl *a/b*-protein, which we estimated represented 65% of the Chl, we surmised that the rest was accounted for by, as yet unidentified, a reaction center of PS II and a PS I Chl *a*-protein. Further, we postulated that all except the Chl *a/b*-protein were common to all Chl *a*-containing organisms, and that during evolution only the nature of the major antenna component (the LHCP in green plants) had changed from biliproteins via Chl *a/c*-proteins to the Chl *a/b*-protein. For the next 18 years, we and others (e.g. Jan Anderson, David Simpson, Roberto Bassi) have been updating our versions of such models, correcting them in light of new data and adding more details. Interestingly, it was at this Brookhaven meeting that William Hopkins and Dan Hayden told me of a third higher plant pigment-protein, which they had named Complex IV and which later became known as the CP47/CP43 component (see Hayden and Hopkins 1977).

My laboratory at UCLA had expanded greatly in personnel and equipment by 1976. In fact, so much was going on in the lab that I could not follow every project in the level of detail I wanted to and also fulfill other duties expected of me. When I accepted the faculty position, I was unaware also of all the talents needed. Brookhaven Labs had pampered me, and my

British education had not adequately prepared me to be a teacher, administrator, editor, amateur psychiatrist, travel agent, repairman, accountant, fund-raiser, and movie consultant (after all, this was Los Angeles). Thus it was with considerable relief from such pressures that I took sabbatical leave in 1976–77, thanks in part to support by a Guggenheim Memorial Fellowship. I could also breathe a sigh of relief that I had completed my movie consultant duties. One of the very modern Biology textbooks had a script for a movie on Photosynthesis aimed at an undergraduate audience. I had had considerable trepidation about the script since my first reading of it. It began: 'The sun, long worshipped by ancient man as God ...' Frequent script conferences resulted in few of my suggested changes being adopted, and the project was mercifully dropped.

1976–1977 – sabbatical leave at Brookhaven and in England

The first six months of sabbatical leave were spent in John Olson's laboratory at Brookhaven. The environment provided the relaxation I needed. The deep-freeze still contained batches of organisms I had accumulated in the late 1960's, and the cold-room still had some of my solutions and hydroxyapatite I had made. This probably explains why I never insist that graduating PhD's in my laboratory throw away their refrigerator-stored materials when they leave. John was as always an inspiring host. Familiar surroundings were conducive to a fast start, I started by modifying a recently published preparative procedure for the *Rps. viridis* RC (Pucheu et al. 1976). This was in response to Jack Fajer and Dan Brune (Division of Chemical Sciences at Brookhaven) and Arthur Forman (Medical Research Center) who wanted material to study the ESR characteristics of the primary donor and acceptor. The only trouble was that they said they needed a few ml of RC's that were 1mM in P960. They said such a concentration would give excellent ESR signals. My notebook indicates that the closest I came was 3 ml of 175 μM P960 (even at that concentration the solution was the color of Guinness stout). The preparation (LDAO solubilization coupled with DEAE-cellulose chromatography and ammonium sulfate fractionation) gave an essentially chlorin-free preparation, which I measured contained per P960, four BChl *b* and two BPh molecules, two cytochromes C553 and the two cytochromes C558. It turned out that the ESR signals were so strong that the power had to be considerably reduced to get them on the screen. I shall summarize the outcome of this work in a moment.

For a short time I thought the new isolation procedure might be universally applicable to purple bacteria, and so I tried *Rsp. rubrum, Rsp. photometricum* and a second BChl *b*-containing organism, *Thiocapsa pfennigii.* RC's were isolated but in poor yield from the BChl *a*-containing organisms, and while the yield of *T. pfennigii* RC's was good, their stability was poor. This problem was solved later by Richard Seftor (see below). Nevertheless, collaborations with Fajer/Brune/Forman, and with Bill Parson/Dewey Holten/Maurice Windsor in Seattle and with Les Dutton/Roger Prince/David Tiede in Philadelphia produced exciting data on the participants in the primary photoact in BChl *b*-containing organisms. These data were published in several papers and were summarized in my symposium talk at the Fourth International Congress on Photosynthesis in Reading, England (Thornber et al. 1978a; see also Prince et al. 1977; Davis et al. 1979). Optical, ESR and pulsed laser spectroscopy of BChl *b*-containing RC's from both species showed that the charge separation process was very similar to that in BChl *a*-containing organisms. The major difference was that P^+960 was not a symmetrical dimer of BChl *b* molecules. The 'primary' electron acceptor (X) was shown to be a quinone-iron complex, and an intermediate carrier (I) between P960 and Q.Fe was a pheophytin molecule. We demonstrated spectrophotometrically that BPh was *not* generated during the isolation procedure and that illumination of the RC's at low redox potentials at room temperature resulted in the reduction of both BPh molecules in the RC (see later).

At some time during my two Brookhaven periods, I had a contretemps with *Rps. viridis.* The cells were grown on a yeast medium in 25-l bottles on large stirrers inside illuminated cabinets that were not much larger than the bottles. When the cells were harvested, their putrid smell could be detected several rooms distant. Obviously, no one liked to do the harvesting, but one day I foolishly assisted. As I removed one bottle, I succeeded in hitting and cracking its bottom against the stirrer, and 25 l of cells slowly emptied over me, soaking my clothes and shoes. Did I smell! To make it bearable to have me around, Betty and Frieda found me replacement clothing – the uniform of the animal house workers. Otherwise, I have nothing but good things to say about *Rps. viridis.*

The second half of my sabbatical was spent in Jim Barber's laboratory at Imperial College in London. It was exciting to be in the center of London in Spring and Summer and to see the Royal Albert Hall just outside the laboratory. Jim and Lyn Barber went out of their way to make the not-so-easy task of finding temporary housing in West London easier. Unfortunately though, I never got sufficiently settled to carry out a substantial project. I spent a lot of the six months traveling in England and France to give seminars. I visited Sir George Porter and his colleagues at the Royal Institution a few times, touting my thesis to an initially skeptical group, that proteins were involved in organizing photosynthetic pigments (I believe I made many converts); I wrote one review (Olson and Thornber 1979) while commuting by train to the lab from Sussex; Jim and I talked extensively about the content of a chapter for his forthcoming Volume 3 of Topics in Photosynthesis (Thornber and Barber 1979); and I met or re-met many contemporaries who would have been my immediate scientific colleagues had I ever returned to live in the United Kingdom.

Several interesting events occurred during the six months that led to long-lasting friendships and scientific interactions: John Bennett (University of Warwick) came to Imperial College to hear my seminar, and later on the way to the pub, dropped a bombshell – 'of course you know that your protein (LHCP) is phosphorylated'. John believed that the phosphorylation was connected with State 1–State 2 transitions. Jim thought that the rate at which phosphorylation occurred was too slow for this to be correct; however, he later changed his opinion. Richard Cogdell (University of Glasgow) came to Imperial College so that we could work for a week on the purple bacterial antenna complexes. I had known Richard from recent Gordon Conferences, and an International Photosynthetic Prokaryote meeting in Dundee in 1976. We got along famously, and his visit to Imperial College started our lifelong friendship. We have many things in common, such as a love of cricket and of cribbage. Richard is one of the few people I know who can watch sports on TV, carry out a scientific conversation and play cribbage all at the same time without missing a single beat in all three. In the lab I showed Richard how to make the *Chromatium* RC and its antenna complexes, and we also found time to do the same with *Rps. capsulata,* and *Rsp. rubrum,* using my recent experience at Brookhaven to good advantage. Richard did further work, particularly carotenoid analysis of the complexes, and we presented the data at a CIBA Foundation Symposium in 1978 (Cogdell and

Table 1. Characteristics of thylakoid pigment-protein complexes

	Apparent size (kDa):			% Total Chl	Alternative names
	Holocomplex	Apoprotein(s)	Chl *a/b*		
PS I	230	Multiple	6.0	38	CPI*
CC I	120	58 + others	*a* only	20	Complex 1, CPI, CHLa-P1, P700-CHLa-Protein, A1
LHC Ia	65[a]/25	24, 22	1.4	}18	LHCPIa
LHC Ib	65[a]/25	21.5, 21	2.3		LHCPIb
LHC Ic		17			psaF gene product
LHC Id		11			
CC II-RC	80	32(D2), 30(D1), 9, 4	*a* + pheophytin	1	
CC IIa	55	47	*a* only	}10	CPIII, CHLa-P2, CP47, A2
CC IIb	50	43	*a* only		CPIV, CHLa-P3, CP43
LHC IIa	35	31	2.25	4	CHLa/b-P1, CP29
LHC IIb	72[a]	28, 27, 25	1.33	40	Complex 2, CP II, CP2, LHCP, CHLa/b-P 2, LHCP1, 2 and 3, AB1, 2 and3
LHC IIc	30	29, 26.5	1.80	4	CP27
LHC IId	24	21	0.93	4	CP24
LHC IIe	12	11			

[a]Denotes the size when trimeric forms of the pigment-proteins have been isolated.
CPI* = CPI + LHCPI.

Thornber 1979; see also Cogdell and Thornber 1981). Richard made purple bacterial antenna complexes his major project over the next two decades. He has been remarkably successful. And then there was the famous England versus the Rest of the World Soccer game at the International Congress in Reading. Jim Barber (a cat-like goalkeeper par excellence) and I both wanted to be the English keeper, but loyalty prevailed and I had to keep goal for the Rest of the World. This proved to be a higher quality game than the organizers and certainly I expected. The Rest of the World lost, and at my recent retirement symposium at UCLA the major bone of contention was how many goals did I let England score. I thought it was two; others thought three or more. Fortunately, for our sanity, science is not all work and no play.

Most of the sabbatical leave was a very happy time. However, one unhappy event overrode all the good times, and that was the unexpected death of my father in June 1977.

1977–1982 – University of California, Los Angeles

On my return from sabbatical leave, the cast of the lab was much different from the one I had left. Sally Reinman and Richard E.B. Seftor were graduate students; John Markwell was a post-doc; and C. Donald Miles was on sabbatical leave from the University of Missouri, Columbia. John had been in June Lascelles' lab at UCLA for the prior year. He came to see me shortly after I returned from Britain, and in response to one of his first questions – 'what do you want me to work on?' he quotes me as saying, 'why don't you work on that bench over there, and see if you can get a paper out by the end of the year'. I'm not that quick-witted, but my response became part of lab folklore. Whatever the truth was, I wanted John to improve the SDS-PAGE system to reduce the amount of free pigment, for only then could we isolate some of those additional pigment-proteins that we knew must exist. Jan Brown and Randy Alberte (Brown et al. 1974) had obtained the following distribution of Chl in the P700-Chl *a*-protein (10–18%), LHCP (40–60%) and free pigment (20–50%). If we reduced the amount of free pigment, we expected to find at least the reaction center of PS II and a Chl *a*-protein of PS I. Working with Sally, John reduced the free pigment to 14–18% by using a Tris-glycine buffer and a low percentage of acrylamide/bisacrylamide, an empirically-arrived-at improvement of the system, which became known as MARS (Markwell and Reinman system) gels (Markwell et al. 1978). The banding pattern differed from the original Thornber system (often called the old Thornber system, behind my back of course). Four bands, in addition to free pigment, were present. They were given temporary names (see Table 1) until their function could be determined: A1, which had the same size as Chl-protein I (CPI) now accounted for 27–32% of the thylakoid Chl, AB1, AB2 and AB3 for 54%. AB1 had an apparent size of 80 kD, AB2 of 60 kD and AB3 of 46 kD. AB3 thus did not seem equivalent to CP II (30 kD) of the original system, and what their relationship was to the now well-known Chl *a/b*-protein trimer is not easily interpreted. It would have been all too easy to believe that they were trimer, dimer and monomer, but as far as I know, Roger Hiller's group has been the only group to report the existence of a Chl *a/b*-protein dimer (Hiller et al. 1974).

It was during this period that my personal life changed, and I began living with Elaine M. Tobin, also a faculty member in the Biology department at UCLA. We were very fortunate in the blending of our children, and my two girls and Elaine's two boys (David and Adam, seven and four years old, respectively) got along well from the very beginning, and are still very close to each other today.

John Bennett visited the lab for a few months in 1979 and showed (Bennett et al. 1981) that AB1, AB2, and AB3, after excision from a MARS gel, all yielded the LHCP component when rerun in the original system, and all contained the same two phosphorylatable ∼ 25 kD polypeptides. In addition, we were able to conclude that no matter what SDS treatment they received at room temperature, they were not dissociated into a colorless polypeptide and pigments. Interestingly, Jan Anderson used a slightly different system and also produced three Chl *a/b*-protein bands (LHCP1, LHCP2, and LHCP3). Table 1 shows the alternative names that have been used for essentially identical plant pigment-proteins.

Sally Reinman used MARS gels to study pigment organization in a cyanobacterium (Reinman and Thornber 1979). With the assistance of David Geffen, an undergraduate, she observed a putative PS II pigment-protein band termed A2. Sally also found that PS I in cyanobacteria appeared mainly in a 255 kD component (A), and a smaller portion, with an identical absorption spectrum, in a 118 kD component (A1) where the PS I component of most eukaryotes is located. We did not know what to make of this A band until about ten years later when papers appeared describing PS I as occurring as a trimer in cyanobacteria. Now it is even better understood as a result of work by a former graduate student of mine, Parag R. Chitnis (Chitnis and Chitnis 1993).

Don Miles used the new electrophoretic system to look at maize mutants. PS I-minus mutants lacked the A1 band further confirming it as the P700-Chl *a*-protein, and PS II-minus mutants lacked A2 further indicating its involvement with PS II activity (Miles et al. 1979). The biggest surprise was that in some mutants AB2 was absent which indicated that it did not fit into its being a simple oligomer of AB3, whereas AB1 was present and did. This different behavior of AB2 arose again three years later when John Markwell visited Jim Barber's Lab in London. Herbert Nakatani and John compared the MARS gel patterns of solubilized rapidly isolated, *intact* pea chloroplasts and solubilized thylakoids of hypotonically ruptured plastids. In the intact plastid extract, AB2 was absent and a larger band (L) at ∼ 100 kD spectrally similar to AB2 at ∼ 60 kD, was seen (Markwell et al. 1981b). This observation has not been explained, and I think it is worthy of reinvestigation now that all thylakoid polypeptides can be separated by SDS-PAGE and their functions are known. Such studies may tell us something of the quaternary structure of LHCP.

Don Miles and John were also involved in finding detergents that would not kill PS II activity so that we could look for active PS II component(s) in solubilized extracts. They found two such surfactants not previously used in biochemical research – Miranol and Deriphat D-160 (Markwell et al. 1979a). Kimiyuki Satoh came to the lab for a few months shortly thereafter and examined the effect of these and other surfactants on PS II activity and the subunit composition of PS II fractions.

How did we come up with these unusual surfactants? John Markwell read and thought deeply about surfactants and what was occurring during SDS-PAGE (e.g. Markwell et al. 1981a; Markwell and Thornber 1982). It was a lab custom for late afternoons on Fridays to be devoted to liquid refreshment, and it was at one of these 'lab meetings' that John proffered the notion that the ideal detergent would have a C_{12} tail and a quartenary N to which something with a negative charge weaker than that of dodecyl sulfate was bound. We hired Steve Manley to synthesize dodecyl-glutamate but before we were able to test its action, John found in a book (Rosen and Goldsmith 1972) that two such surfactants were marketed – Miranol and Deriphat D-160.

While these were useful for solubilizing and not killing PS II activity, it was their replacement of SDS in electrophoresis buffers that started the lab on what has been a fourteen-year in-depth characterization of the multiple pigmented components of the photosynthetic apparatus. Russell Boggs, an undergraduate lab dishwasher, and Merri Skrdla, a graduate student, took an early look at the altered electrophoresis conditions, and John Bennett, during his time in the lab, contributed much to our understanding of solubilization and electrophoresis. In 1979, a second paper was published on deriphat (Markwell et al. 1979b) describing the effect of substituting it for SDS in the electrophoresis buffers but not as the membrane solubilizing agent which remained SDS. Four green bands were obtained for higher plant thylakoids, and the free pigment zone (mainly carotenoid) now contained only 2–3% of the thylakoid pigment. We concluded that essentially all of the Chl in higher plant thylakoids occurred in conjugation with protein, and that it was the electrophoresis conditions *per se*, and not solubilization, which dissociated pigments from the apoproteins. The deriphat system proved to be also applicable to algal and bacterial photosynthetic membranes (e.g., Boczar et al. 1980; Ferguson et al. 1991).

Merri Skrdla in her dissertation research separated the band of intact PS I from deriphat gels into an antenna component and the P700-Chl*a*-protein complex. Mullet et al. (1980) had previously indicated the existence of a PS I antenna complex, and this fraction of Merri's represented a biochemical isolation and characterization of it; however, for whatever reason (maybe because my duties as chair of the department gave me less time to follow what was going on in the labs' research) we did not write up this work, and by the time that Merri submitted her thesis, her findings had been eclipsed by those of others (Jan Anderson, Jan Brown, Phil Haworth, Charlie Arntzen, Eric Lam and Richard Malkin) and even by others in my lab who were interested in PS I. Merri loved animals, birds and fishes so that during her time at UCLA, the lab was gradually filled with fish tanks and cages. This trait, at least as far as birds were concerned, was continued by Camille Peterson, who was my technician during my tenure as Chair of the 55-member Biology Department (1981–86). Even today, feathers and fur are still found in remote corners of the lab and office. Many people outside UCLA remember the menagerie either from visits or via the telephone ('sounds like you've got birds in your office'); I had. Those readers who have been in my office between 1980–90 will also have noticed large red footprints on its ceiling. No one has ever admitted putting them there, but I have always suspected that they originated from a Friday lab meeting that ran late into the night.

Richard Seftor studied the RC's of BChl *b*-containing photosynthetic bacteria for his dissertation research, particularly continuing my earlier work on *T. pfennigii* (Seftor and Thornber 1984). Most helpful to his training were extensive visits made to the lab by Richard Cogdell and Beverly Pierson over the period. The two Richards painstakingly determined the subunit composition and the BChl/BPh*b* ratio of *Rps viridis* RC's while Richard Cogdell found little specificity of the carotenoid molecule bound to the *Rps viridis* RC unit. It was with some trepidation that we subsequently awaited the three-dimensional structure of RC (Deisenhofer et al. 1985) because this would reveal unambiguously whether our determinations of molar ratios of pigments were indeed correct. To our considerable satisfaction they were.

We also did a lot of analysis between 1979–82 of the BChl spectral forms in the RC. We determined that the same spectral forms occur in vivo as in isolated RC's; i.e., detergents did not affect them. We proposed (Thornber et al. 1980; Seftor and Thornber 1984) that P960 had a higher energy transition at 835 nm, which has still to be confirmed, and that *T. pfennigii* was spectrally quite different from its *Rps. viridis* counterpart. *Rps. viridis* had obvious different spectral forms of BChl (the so-called voyeur BChl's) but not of BPh, while the reverse was true for *T. pfennigii*. We spent a great deal of time examining which chromophore(s) were reduced by light in sodium dithionite-reduced RC's. This was easier to do in the two organisms we studied than in other photosynthetic bacteria because the BChl *b*-containing RC's have very rapid rates for electron transfer from C553 to P$^+$960 which enables electrons photo-ejected from P960 to be trapped on components at the top of the photoact in reduced RCs. At room temperature only one BPh (I) was reduced, but at 77 °K both BPh and at least one BChl were reduced. We pointed out that this did not necessarily mean that all three were on a direct pathway between P960 and Q, and, in fact, we suggested that an alternative electron pathway existed (Thornber et al. 1981).

Beverly K. Pierson spent her 1981–82 sabbatical leave in my lab. She came with the express desire to isolate and examine the RC of a thermophilic green bacterium, *Chloroflexus aurantiacus*. She succeeded (see Pierson 1994). Its RC was more purple than green bacterial-like. It appeared to have a BChl/BPh ratio of 3/3 compared to 4/2 in purple bacteria, and it had greater thermal stability. The RC did not contain carotenoids (Pierson and Thornber 1983; Pierson et al. 1983). It was postulated that it represented a primitive evolutionary form of the purple bacterial RC. This was a very satisfying piece of work by Beverly, with some help from Rich Seftor.

In this same year, Neil R. Baker, whom I had known since his time in Warren Butler's lab in the mid-1970s, and Maxine Baker spent a sabbatical year in my lab. Neil and John Markwell with Maxine began a study of the relationship between phosphorylation of LHCP and the State 1–State 2 transitions. By then it was dogma that the redox state of PQ controlled the protein kinase activity. Neil and John obtained evidence that energy charge, as originally defined by Daniel Atkinson (UCLA), was also involved (Markwell et al. 1982); cyclic AMP but not AMP also had an effect (Baker et al. 1982). Later they showed that multiple kinases were involved in thylakoid protein phosphorylation (Markwell et al. 1983), and Baker et al. (1983) examined the control during chloroplast biogenesis. I always felt that our protein phosphorylation studies were neglected mainly because it was anti-dogma. I feel that the real story behind LHCP phosphorylation has yet to be delineated.

Improvement of PAGE systems so that more pigment was retained with their apoproteins during fractionation was undertaken by many other groups (see Thornber 1986 for review) in 1978–82 (Jan Anderson in Australia; Roderick Park at Berkeley; Otto Machold in Germany; Edith Camm and Beverly Green in Canada; and, Philippe Delepelaire and Nam-Hai Chua in New York). Each system successfully reduced the free pigment to 12–18% of the total Chl, and retained more pigment with PS I band(s) and/or with the core PS II (e.g. CP47 and CP43). In addition, the Camm-Green and Machold PAGE systems provided the first evidence for more than one biochemically distinct Chl *a/b*-protein in PS II (CP29 or Chl *a/b*-P1, see Table 1). This last observation was seminal in obtaining a more detailed picture of the pigment-protein composition of the photosynthetic apparatus. At least five more Chl *a/b*-proteins of PS I and PS II were identified over the next decade (Table 1). Furthermore, the Camm-Green

system was the first to use glycosidic surfactants to solubilize thylakoids; such surfactants were to prove the closest to 'ideal' solubilizing agents.

The large number of green bands on PAGE gels then being obtained was confusing to interested parties who did not work with higher plant pigment-proteins. The biochemical characteristics of supposedly equivalent pigment-proteins differed, often only slightly, between different PAGE systems, which made it extremely difficult for anyone to evaluate just how many pigment-proteins were present in the thylakoids. Furthermore, as was found later, different pigment-proteins often had very similar characteristics. It was clear that each green band in each system had to be even better characterized. The field was pressing us for a nomenclature for these complexes, particularly for the Chl *a/b*-proteins, but a delay in presenting one proved wise. It was only when the primary structures of all of the apoproteins (obtained from gene sequences) were obtained in the early 1990s that an unambiguous classification was possible (Table 1).

Elaine Tobin was a tremendous help in leading me towards a more molecular biological approach to the study of pigment-proteins during this 1977–82 period. Together we had tried to understand the subunit structure of LHCP (CP II), but because its apoproteins electrophoresed with many other polypeptides in the 20–32 kD region, we had difficulty making an unequivocal identification. We did not then know that other Chl *a/b*-proteins existed, and that their apoproteins also migrated in this region of the gel. We thought, however, that monoclonal antibodies might give us a good chance of identifying the LHCP apoproteins. These were made by Fiona Hunter at the University of Alabama in Birmingham, to where she had moved after graduating from UCLA. Their cross-reactivity was with two of what we now know to be three easily resolved LHCP polypeptides (see Thornber et al. 1986). We did this work during a three-month stay in the plant biology section of the Biological Sciences department at the University of Warwick, England where we worked with John Ellis and John Bennett, among others. Our visit coincided with the Falkland Islands War.

We spent the summer in Itzhak Ohad's laboratory at the Hebrew University of Jerusalem. Here our visit coincided with Israel's invasion of Lebanon, but surprisingly this did little to impact our visit. Elaine's boys (David and Adam Tobin, 10 and seven years old, respectively) and my girls (Karen and Emma, then 15 and 12 years old) were with us, and for six weeks, we explored Jerusalem and much of Israel in considerable depth. The time I spent there proved deeply emotional for me, resulting in my subsequent love for that country. Elaine had wondered how a non-Jew would take to Israel. In the three Thornbers she got a unanimous positive response. Perhaps the biggest impact of exploring Israel was that it removed forever my insular view that history did not start until A.D. 1066, a view that had been drilled into me by my English schooling.

Most of the research I did in Israel was crammed into the last two weeks of the visit. Itzhak and I tried to understand how the 77 K fluorescence of *C. reinhardtii* was correlated with the electrophoretically separated pigment-proteins of the wild type and *y-1* mutant of the alga. A few of the conclusions we arrived at are in Schuster et al. (1988); the remainder helped us with other projects of our own. Some of my fondest memories of working with Itzhak are of his boundless energy. He would drive me back to my apartment from the lab sometimes between midnight and 2 a.m., and then return to finish our experiment. For a short period there I was able to slow Itzhak's rapid-fire communications in English to a pace at which I could learn so much from him. I found him to be an incredibly warm and generous person.

1983 – the present

We returned to a 'peaceful' USA in Fall 1982 for me to resume my duties as chair of the department. Rich Seftor was finishing his dissertation research, Bill Dietrich spent a sabbatical year with us, and Camille Peterson held the lab together with her good humor and efficiency.

Some research done in Elaine's lab by Willem Stiekema while we were on sabbatical, and on-going research in the same lab by George Karlin-Neumann on the characterization of a genomic clone for *Lemna gibba* LHCP, was to greatly influence the focus of my research for the next few years. But in early 1983 some major events occurred in my life: I married Elaine immediately following a Gordon Conference on Photosynthesis in California, which enabled several friends who would otherwise not have been able to attend, to be present. Itzhak Ohad and Richard Cogdell played major roles as Best Man and Chief Bartender, respectively, and Alison Telfer (Imperial College, London) was there. Four weeks later I underwent emergency heart bypass surgery. After a brief recovery period, I resumed my duties as Chair using the sun-drenched deck of our house as an office. My administrative duties required considerable help and frequent visits from my ever-helpful Management Services Officer, John (Kip) Brott.

It was during this convalescence that Elaine brought home the LHCP amino acid sequence derived from the *L. gibba* clone (*cab* AB19). This gene, unlike those previously isolated for this protein from pea and petunia, contained an intron and had obvious differences in primary structure from the other two, even though the homology at the protein level was 85%. The *Lemna* gene coded for what became known as a Type II polypeptide, whereas those from pea and petunia were classified as Type I LHCP. I spent many happy hours in the sun joining and folding colored pipe cleaners, each color representing a different type of amino acid, into what I anticipated would be the protein's tertiary structure in the membrane. I used a standard listing of which amino acids had a preference to be in alpha-helices and which in beta-turns, etc., to accomplish my task. Obtaining the sequence and understanding how it was arranged in the lipid bilayer had been a major driving force for me ever since I saw the complex nearly 20 years previously. This I thought would be the culmination of my scientific dreams. It proved to be just a step along the way. I have often wondered what Lula Carr, our housekeeper for many years, must have thought when she saw me attempting this apparently childish wire-folding experiment. Lulu, as she is better known to many researchers who have stayed at our house, has been a savior to Elaine and myself over the years as we went about our two jobs and had as many as four children in the house on some days.

It took a visit from our neighbor Robert M. Sweet (then at UCLA in the protein structure group, and now at Brookhaven Laboratories) to put me on the right track for obtaining a more correct folding of the protein. He asked me if I realized that in David S. Eisenberg's laboratory at UCLA there was a computer program that would do what I was trying to do. This program identified for us *three* hydrophobic membrane spanning alpha-helices, and some other short amphiphilic helical segments, plus a small amount of beta-structure whereas by 'wire-folding' I had arrived at *four* such membrane-spanning regions. It was a triumph for computer over man, because the computer was ultimately proved to be correct, I placed three Chl *a*, three Chl *b* and one carotenoid molecule in the folded structure because those were the ratios we had measured in SDS-prepared material. We then had a model we could test.

This model was presented by Elaine (Tobin et al. 1984) in April 1983 at a Keystone conference on Biosynthesis of the Photosynthetic Apparatus, organized by L. Andrew Staehelin, Richard B. Hallick and myself. For obvious reasons, the bulk of arranging and running the meeting fell on the other two. The aim of the meeting was to bring together, essentially for the first time, many of the biophysicists and biochemists in photosynthesis with those working on the molecular biology of the chloroplast and photosynthetic bacteria, and those working on plants in industry. We hoped new collaborations would result, and in this I think we were reasonably successful.

In summer 1983 I described the model at the International Photosynthesis Congress in Brussels, and immediately prior to that meeting at a most timely symposium on pigment-proteins in photosynthetic organisms organized by Herbert Zuber, during which the temperature in Zürich reached a record high for this century. The problematic terminology for the plant Chl-proteins was discussed but not resolved. All participants, however, agreed that the photosynthetic apparatus in plants could be divided into four pigmented components each of which was thought likely to contain more than one pigment-protein. Thus, CC I and LHC I represented the two PS I components, and CC II plus LHC II those of PS II. A core complex (CC) represented the functional part of the photosystem plus its core antenna, whereas light-harvesting complex (LHC) represented the bulk of the antenna of a photosystem. This terminology has survived to the present day (see Table 1). It was at these meetings that I first met Rachel Nechushtai who was to have a major impact on my laboratory in the mid-1980's.

The arrangement of amino acids in our proposed folding of LHC II's major apoprotein appeared in Karlin-Neumann et al. (1985). This model was incorporated into several textbooks of that period, and it survived, without modification, for about five years; that is, until Werner Kühlbrandt's electron diffraction work provided more direct evidence of its structure. We had assumed that the alpha-helices would be perpendicular to the plane of the membrane; they were not. But, what had surprised us about our 1985 model was that almost one-half of the polypeptide was located outside the lipid bilayer and the three-dimensional structure confirmed this. We were also intrigued by the mechanism by which this polypeptide would have to be inserted into the membrane, and by the timing (before, during or after insertion) of pigment addition to the protein. These questions interested several

people in Elaine's and my laboratories, and we were fortunate to have people who were willing and eager to seek some answers. The two groups occupied much of the Plant Physiology building from 1978 onwards; I had moved there from the Botany building to occupy rooms vacated by Sam Wildman when he retired.

From 1984 onwards the two laboratories received an almost unbroken stream of post-docs and faculty from Israel. Each provided excellent examples to other occupants of the lab of hard work; furthermore, they did not seem to mind my practicing the 30 words of Hebrew I know with them. Rachel Nechushtai was the first to arrive, followed shortly thereafter by Eitan Harel, and later in 1986, Alexander Vainstein. Three graduate students joined the group during 1983–84 – Parag Chitnis, Saeid Nourizadeh and Gary Peter. For a short period there were more males than females in the lab – a rare situation during my career.

During 1984–85 Parag, Eitan and Bruce Kohorn, a post-doc in Elaine's group, examined uptake of precursor Chl *a/b*-proteins (pLHCP) into intact etiochloroplasts. They showed that after uptake, processing of pLHCP (Type I) to the mature protein occurs in developing plastids after the pLHCP is inserted into the LHC IIb complex in the thylakoids (Chitnis et al. 1986, 1988). Six pLHCP mutants with deletions of a small portion of the mature polypeptide were constructed by Bruce to test our putative model, but because they were not incorporated into the complex, the data neither proved nor disproved the structure (Kohorn et al. 1986). Parag and Rachel subsequently described a stromal protein factor that was needed for pLHCP to be inserted into isolated thylakoids (Chitnis et al. 1987). They and others have advanced this story subsequently in their own careers.

Gary Peter wanted to study the effects of plant growth conditions on the pigment content of the caroteno-Chl-proteins and on the pigment-protein composition of the photosynthetic unit for his dissertation. To do so he had to first fractionate the unit into individual pigment-proteins, describe the biochemistry of each one, and correlate his data with those of others working in the immediate area. Hardly surprisingly, this mammoth task occupied his entire time in the graduate program, and he did not get to his original aim. The students who followed Gary, however, have been grateful to him for laying the foundations for many of their dissertation projects.

Several other groups (e.g. Beverley Green's and Andrew Staehelin's) had success in the mid-1980's finding new LHC pigment-proteins (e.g., CP29 and

CP24) via electrophoresis of glycosidic detergent extracts of thylakoids. It was obvious that we should drop SDS solubilization and replace it by treatment with glycosidic detergent, of which there were several available. Because it was vital for Gary's original aim to retain all the pigment with each pigment-protein, we used John Markwell's Deriphat-PAGE system to resolve the detergent extracts. Gary soon found that the different glycosidic detergents yielded extracts that produced slightly different, sometimes substantially different, green band patterns on PAGE (see Peter and Thornber 1990, for details). By 1989 he had perfected a two-dimensional separation of all the thylakoid polypeptides – a non-denaturing resolution of pigmented bands in the first-dimension followed by fully denaturing SDS-urea-PAGE of their subunits in the second.

It was largely through Gary's own work and his collaborative efforts in the lab with Otto Machold (Peter et al. 1988) and with Rachel Nechushtai (Nechushtai et al. 1987) that we were able to fractionate and identify all the pigment-proteins that had been described by groups working in this area; furthermore, we were able to add some (LHC Ic, LHC Id and LHC IIe) to the list (Table 1) that had not been identified before. We described the subunit and pigment compositions of all of them that were resolved using Gary's system, concentrating on the LHC components (Peter and Thornber 1988). Thus, what had seemed an extremely complicated situation in the early 1980s (see above) was much clearer by 1990, and particularly importantly there was now general agreement among those working on these components. Details of exactly how we viewed the LHC's at that time, and of their occurrence in situ as homo- and hetero-oligomers as well as the existence of CC II as a dimer can be found in Peter and Thornber (1991a, b; see also Thornber et al. 1991, 1994).

Other students who joined my group after Gary (Shivanthi Anandan, Daryl Morishige, Beth Welty Dreyfuss and Susanne Preiss) found and characterized a few more pigment-proteins to yield our current view as set out in Table 1. These four students, to different extents, also isolated and sequenced some of the LHC I and II genes, and correlated the deduced amino acid sequences with a particular spot on Gary's two-dimensional PAGE system. Most recently, Beth, Angela I. Lee and Tracey Takeuchi have examined the biogenesis and turnover of the plant LHC's under normal growth conditions and under stress (Dreyfuss and Thornber 1994a, b; Takeuchi and Thornber 1994;

Lee and Thornber 1995). This has entailed a careful analysis of the carotenoid content of the LHC's (two lutein and either one neoxanthin or one violaxanthin molecule, depending on which LHC is considered, per polypeptide), and a study of the site of the photoprotective violaxanthin cycle among the LHC I and LHC II pigment-proteins. Lastly, I still maintain an interest in pigment organization in purple bacteria via research done by Cheryl Kerfeld and Richard Cogdell on *Chr. purpuratum*.

I have greatly condensed the lab's findings over the last six years because, although it has been a particularly enjoyable period of my research career, it is too soon to put the work in perspective. I hope that some of those who have been with me during this period will be asked to do so later in their own careers.

Final comments

I took advantage of the incentives offered by the University of California's early retirement program in July 1994. The research program had reached a most convenient point at which I could stop and let others continue it. A second heart operation in 1992 convinced me to seek a more relaxed life and to pursue other interests.

I suppose you never get every last answer to your major research interest, but I have come close. Stephen Gomez, my last graduate student, is tying up some of the 'loose ends'. I shall miss being among students, I shall regret the reduced contact with friends in photosynthesis, but I shall not miss the stress connected with obtaining grants, faculty meetings and other committee work. Those who know me, and hopefully those who read this perspective, will know or realize that I have never taken my science overly seriously; I always felt it was there to be enjoyed and to be balanced by other interests.

Why did I choose a career in plant biochemistry? Because my interest in the extensive and varied chemistry of plants and in a most unusual process, photosynthesis, was aroused as an undergraduate and has continued unabated. It is surely unusual and significant that five of the 23 undergraduates in the University of Cambridge Biochemistry (Part II) senior year class of 1958 have major positions in plant biochemistry (Geoffrey Hind, Derek Lamport and myself in the US; John Kirk in Australia; and Mike Johnson in the UK). We had many interesting lecturers but those covering plant biochemistry (Don Northcote, Robin Hill and Reggie

Trim) must have presented a particularly appealing picture. I wonder if we would work with plants if we were to start our careers again today. I know I would.

Acknowledgements

I wish to thank Carole Malkin and Bob Herbstman for reading and commenting on my text and showing how to make it livelier, and Elaine Tobin for encouragement, consultation, input, and continuously making me aware that space in this issue was not infinite.

References

Alberte RS and Thornber JP (1978) A rapid procedure for isolating the photosystem I reaction center in a highly enriched form. FEBS Lett 91: 126–130

Alberte RS, Thornber JP and Naylor AW (1972) Appearance of photosystem I and II in greening jack bean leaves. J Expt Bot 23: 1060–1069

Alberte RS, Thornber JP and Naylor AW (1973) Biosynthesis of the photosystem I chlorophyll-protein complex of greening leaves of higher plants. Proc Natl Acad Sci USA 70: 134–137

Alberte RS, McClure PR and Thornber JP (1976) Photosynthesis in trees: Organization of chlorophyll and photosynthetic unit size in isolated gymnosperm chloroplasts. Plant Physiol 58: 341–353

Baker NR, Markwell JP and Thornber JP (1982) Adenine nucleotide inhibition of phosphorylation of the light-harvesting chlorophyll *a/b*-protein complex. Photobiochem Photobiophys 4: 211–217

Baker NR, Markwell JP, Bradbury M, Baker MG and Thornber JP (1983) Thylakoid protein kinase activity and associated control of excitation energy distribution during chloroplast biogenesis in wheat. Planta 159: 151–158

Bennett J, Markwell JP, Skrdla MP and Thornber JP (1981) Higher plant chlorophyll *a/b*-protein complexes: Studies on the phosphorylated apoproteins. FEBS Lett 131: 325–330

Boardman NK and Anderson JM (1964) Isolation from spinach chloroplasts of particles containing different proportions of chlorophyll *a* and chlorophyll *b* and their possible role in the light reactions of photosynthesis. Nature 203: 166–167

Boczar B, Prezelin BB, Markwell JP and Thornber JP (1980) A chlorophyll *c*-containing pigment-protein complex from the marine dinoflagellate, *Glenodinium* sp. FEBS Lett 120: 243–247

Brown JS, Alberte RS and Thornber JP (1974) Comparative studies on the occurrence and spectral composition of chlorophyll-protein complexes in a wide variety of plant material. Proceedings of the Third International Congress on Photosynthesis, Rehovot, Israel, pp 1951–1962. Elsevier, Amsterdam

Butler WL and Kitajima M (1975) Energy transfer between photosystem II and Photosytem I in chloroplasts. Biochim Biophys Acta 396: 72–85

Case GD, Parson WW and Thornber JP (1970) Photooxidation of cytochromes in reaction center preparations from *Chromatium* and *Rps. viridis*. Biochim Biophys Acta 223: 122–128

Chance B, DeVault DC, Tasaki A and Thornber JP (1979) The effects of high hydrostatic pressure upon light-induced electron transfer and proton-binding in *Chromatium*. In: Chance B, DeVault DE, Frauenfelder H, Marcus RA, Schrieffer JA and Sutin N (eds) Tunneling in Biological Systems, pp 387–404. Academic Press, New York

Chitnis VP and Chitnis PR (1993) PsaL subunit is required for the formation of photosystem I trimers in cyanobacterium Synechocystis sp. PCC 6803. FEBS Lett 336: 330–334

Chitnis PR, Harel E, Kohorn BD, Tobin EM, and Thornber JP (1986) Assembly of the precursor and processed light-harvesting chlorophyll *a/b* protein of Lemna into the light-harvesting complex II of barley etiochloroplasts. J Cell Biol 102: 982–988

Chitnis PR, Nechushtai R and Thornber JP (1987) Insertion of the precursor of the light-harvesting chlorophyll-protein into the thylakoids requires the presence of a developmentally regulated stromal factor. Plant Mol Biol 10: 3–11

Chitnis PR, Morishige D, Nechushtai R and Thornber JP (1988) Assembly of the barley light-harvesting chlorophyll *a/b*-proteins in barley etiochloroplasts involves processing of the precursor on thylakoids. Plant Mol Biol 11: 95–107

Cogdell RJ and Thornber JP (1979) The preparation and characterization of different types of light-harvesting pigment-protein complexes from some purple bacteria. CIBA Symposium, New Series 61: 61–79

Cogdell RJ and Thornber JP (1981) Light-harvesting pigment-protein complexes of purple bacteria. FEBS Lett 122: 1–8

Criddle RS and Park L (1964) Isolation and properties of a structural protein from chloroplasts. Biochem Biophys Res Commun 17: 74–79

Davis BJ (1964) Disc electrophoresis – II. Methods and application to human serum proteins. Ann NY Acad Sci 121: 404–427

Davis MS, Forman A, Hanson LK, Thornber JP and Fajer J (1979) Anion and cation radicals of bacteriochlorophyll and bacteriopheophytin *b*. Their role in the primary charge separation in *Rhodopseudomonas viridis*. J Phys Chem 83: 3325–3332

Deisenhofer J, Epp O, Miki K, Huber R and Michel H (1985) Structure of the protein subunits in the photosynthetic reaction centre of *Rhodopseudomonas viridis* at 3 Å resolution. Nature 318: 618–624

Dietrich WE Jr and Thornber JP (1971) The P700-chlorophyll *a*-protein of a blue-green algae. Biochim Biophys Acta 245: 482–483

Dreyfuss BW and Thornber JP (1994a) Assembly of the major light-harvesting complex (LHC IIb) during biogenesis of the plastid. Plant Physiol 106: 829–839

Dreyfuss BW and Thornber JP (1994b) Organization of LHC I and its assembly during biogenesis of the plastid. Plant Physiol 106: 841–848

Dutton PL, Kihara T, McCray J and Thornber JP (1971) Cytochrome photooxidation at 77 °K in *Chromatium D* chromatophores and in a subchromatophore preparation. Biochem Biophys Acta 226: 81–87

Ferguson L, Halloran E, Hawthornthwaite AM, Cogdell RJ, Kerfeld C, Peter GF and Thornber JP (1991) The use of non-denaturing Deriphat-gel electrophoresis to fractionate pigment-protein complexes of purple bacteria. Photosynth Res 30: 139–143

Gingras G and Jolchine G (1969) Isolation of a P-870 enriched particle from *Rsp. rubrum*. In: Metzner H (ed) Progress in Photosynthesis Research, Vol 1, pp 206–216. International Union of Biological Sciences, Tübingen

Govindjee (1988) The discovery of chlorophyll protein complex by Emil L. Smith during 1937–1941. Photosynth Res 16: 285–289

Govindjee, Papageorgiou G and Rabinowitch E (1967) Chlorophyll fluorescence and photosynthesis. In: Gilbault GG (ed) Fluorescence: Theory Instrumentation and Practice, pp 511–564. Marcel Dekker, New York

Gregory RPF, Raps S, Thornber JP and Bertsch WF (1972) Chlorophyll-protein-detergent complexes compared with thylakoids by means of circular dichroism. In: Forti G, Avron M and Melandri A (eds) Proceedings Second International Congress of Photosynthesis, Stresa, pp 1503–1508. Dr. Junk, The Hague

Hayden DB and Hopkins WG (1977) Evidence for a second distinct chlorophyll *a*-protein complex in maize mesophyll chloroplasts. Can J Bot 55: 2525–2529

Highkin HR (1950) Chlorophyll studies on barley mutants. Plant Physiol 25: 294–306

Hiller RG, Genge S and Pilger D (1974) Evidence for a dimer of the light-harvesting chlorophyll-protein complex II. Plant Sci Lett 2: 239–242

Hiyama T and Ke B (1972) Difference spectra and extinction coefficients of P700. Biochim Biophys Acta 267: 160–171

Kan KS and Thornber JP (1976) The light-harvesting chlorophyll *a/b*-protein of *Chlamydomonas reinhardii*. Plant Physiol 57: 47–53

Karlin-Neumann GA, Kohorn BD, Thornber JP and Tobin EM (1985) A chlorophyll *a/b*-protein encoded by a gene containing an intron with characteristics of a transposable element. J Mol Appl Genetics 3: 41–60

Kohorn BD, Harel E, Chitnis PR, Thornber JP and Tobin EM (1986) Functional and mutational analysis of the light-harvesting chlorophyll *a/b* protein of thylakoid membranes. J Cell Biol 102: 972–981

Kühlbrandt W, Wang DN and Fujiyoshi Y (1994) Atomic model of plant light-harvesting complex. Nature (London) 367: 614–621

Kung SD and Thornber JP (1971) Photosystem I and II chlorophyll-protein complexes of higher plants. Biochim Biophys Acta 253: 285–289

Kung SD, Thornber JP and Wildman SG (1972) Nuclear DNA codes for the photosystem II chlorophyll-protein complex of chloroplast membranes. FEBS Lett 24: 185–188

Lee AI and Thornber JP (1995) Analysis of the pigment stoichiometyr of pigment-protein complexes from barley: The xanthophyll cycle intermediates occur mainly in the light-harvesting complexes of PS I and PS II. Plant Physiol, in press

Lin L and Thornber JP (1975) Isolation and partial characterization of the photochemical reaction center of a member of the Thiorhodaceae, *Chromatium vinosum* (strain D). Photochem Photobiol 22: 37–40

Malkin R, Bearden AJ, Hunter FA, Alberte RS and Thornber JP (1976) Properties of the low temperature Photosystem I primary reaction in the P700-chlorophyll *a*-protein. Biochim Biophys Acta 430: 389–394

Markwell JP and Thornber JP (1982) Treatment of the thylakoid membrane with detergent: Use of the chlorophyll *a* absorption spectrum to assess a surfactant's efficacy. Plant Physiol 70: 633–636

Markwell JP, Reinman S and Thornber JP (1978) Chlorophyll-protein complexes from higher plants: A procedure for improved stability and fractionation. Arch Biochem Biophys 190: 136–141

Markwell JP, Miles CD, Boggs RT and Thornber JP (1979a) Solubilization of Photosystem II from higher plant chloroplasts by two zwitterionic detergents. FEBS Lett 99: 11–14

Markwell JP, Thornber JP and Boggs RT (1979b) Evidence that in higher plant chloroplasts all the chlorophyll exists as chlorophyll-protein complexes. Proc Natl Acad Sci USA 76: 1233–1235

Markwell JP, Thornber JP, Reinman S, Satoh K, Bennett J, Skrdla MP and Miles CD (1981a) Supramolecular organization of the thylakoid membrane: Use of surfactants for ordered fractionation of membrane components. Proceedings Vth International Congress on Photosynthesis Research. Halkidiki, Greece 3: 317–325

Markwell JP, Nakatani HY, Barber J and Thornber JP (1981b) Chlorophyll-protein complexes fractionated from intact chloroplasts. FEBS Lett 122: 149–153

Markwell JP, Baker NR and Thornber JP (1982) Metabolic regulation of the thylakoid protein kinase. FEBS Lett 142: 171–174

Markwell JP, Baker NR and Thornber JP (1983) Evidence for multiple protein kinase activities in tobacco thylakoids. Photobiochem Photobiophys 5: 201–207

Miles CD, Markwell JP and Thornber JP (1979) Characterization of chlorophyll-protein complexes in maize photosynthetic mutants. Plant Physiol 64: 690–694

Mullet JE, Burke JJ and Arntzen CJ (1980) Chlorophyll-proteins of photosystem I. Plant Physiol 65: 814–822

Nechushtai R, Peterson CC, Peter GF and Thornber JP (1987) Purification and characterization of the light-harvesting chlorophyll *a/b*-protein of photosystem I (LHCI) of *Lemna gibba*. Eur J Biochem 164: 345–350

Olson JM (1994) Reminiscence about '*Chloropseudomonas ethylicum*' and the FMO-protein. Photosynth Res 41: 3–5

Olson JM and Romano CA (1962) A new chlorophyll from green bacteria. Biochim Biophys Acta 59: 726–728

Olson JM and Thornber JP (1979) Photosynthetic reaction centers. In: Capaldi RA (ed) Membrane Proteins in Energy Transduction, pp 279–340. Marcel Dekker, New York

Peter GF and Thornber JP (1988) The antenna components of Photosystem II with emphasis on the major pigment protein, LHC IIb. In: Scheer H and Schneider S (eds) Photosynthetic Light-Harvesting Systems, pp 175–186. W de Gruyter and Co, Berlin

Peter GF and Thornber JP (1990) Electrophoretic procedures for fractionation of Photosystem I and II pigment-proteins of higher plants and for determination of their subunit composition. In: Rogers LJ (ed) Methods in Plant Biochemistry, Volume 5: Amino Acids, Proteins and Nucleic Acids, pp 194–212. Academic Press, San Diego

Peter GF and Thornber JP (1991a) Biochemical composition and organization of higher plant Photosystem II light-harvesting pigment-proteins. J Biol Chem 266: 16745–16754

Peter GF and Thornber JP (1991b) Biochemical evidence that the Higher plant Photosystem II core complex is organized as a dimer. Plant Cell Physiol 32: 1237–1250

Peter GF, Machold O and Thornber JP (1988) Identification and isolation of Photosystem I and Photosystem II pigment-proteins from higher plants, In: Harwood J and Walton TJ (eds) Plant Membranes: Structure, Assembly and Function, pp 17–31. The Biochemical Society, London

Pierson BK (1994) Reflections on *Chloroflexus*. Photosynth Res 41: 7–15

Pierson BK and Thornber JP (1983) Isolation and spectral characterization of photochemical reaction centers from the thermophilic green bacterium *Chloroflexus aurantiacus* strain, J-10-Fl. Proc Natl Acad Sci USA 80: 80–84

Pierson BK, Thornber JP and Seftor REB (1983) Partial purification, subunit structure and thermal stability of the photochemical reaction center of the thermophilic green bacterium, *Chloroflexus aurantiacus*. Biochim Biophys Acta 723: 322–326

Prince RC, Tiede DM, Thornber JP, and Dutton PL (1977) Spectroscopic properties of the intermediary electron carrier in the reaction center of *Rhodopseudomonas viridis*: Evidence for its interaction with the primary acceptor. Biochim Biophys Acta 462: 467–490

Pucheu NL, Kerber NL and Garcia A (1976) Isolation and purification of reaction centers from *Rps. viridis* NHTC 133 by means of LDAO. Arch Microbiol 109: 301–305

Rabinowitch EI and Govindjee (1965) The role of chlorophyll in photosynthesis. Sci Am 213: 74–83

Reinman S and Thornber JP (1979) The electrophoretic isolation and partial characterization of three chlorophyll-protein complexes from blue-green algae. Biochim Biophys Acta 547: 188–197

Ridley SM, Thornber JP and Bailey JL (1967) A study of the water-soluble proteins of spinach beet chloroplasts with particular reference to Fraction I protein. Biochim Biophys Acta 140: 62–79

Rosen MJ and Goldsmith HA (1972) Systematic Analysis of Surface Active Agents. Wiley-Interscience, New York

Schuster G, Nechushtai R, Ferreira PCG, Thornber JP and Ohad I (1988) Structure and biogenesis of *Chlamydomonas Reinhardtii* photosystem I. Eur J Biochem 177: 411–416

Seely GR (1973) Energy transfer in a model of the photosynthetic unit of green plants. J Theor Biol 40: 189–199

Seftor REB and Thornber JP (1984) The photochemical reaction center of *Thiocapsa pfennigii*. Biochim Biophys Acta 764: 148–159

Shiozawa JA, Alberte RS and Thornber JP (1974) The P700-chlorophyll *a*-protein. Isolation and characteristics of the complex in higher plants. Arch Biochem Biophys 165: 388–397

Smith EL (1938) Solutions of chlorophyll protein compounds (phyllochlorins) extracted from spinach. Science 88: 170–171

Takeuchi TS and Thornber JP (1994) Biochemical effects of heat stress on the photosynthetic apparatus of a cereal plant (*Hordeum vulgare*). Austr J Plant Physiol 21: 759–770

Thornber JP (1969) Comparison of a chlorophyll *a*-protein isolated from a blue-green alga with chlorophyll-protein complexes isolated from green bacteria and higher plants. Biochim Biophys Acta 172: 230–241

Thornber JP (1970) The photochemical reactions of purple bacteria as revealed by studies of three spectrally different carotenobacteriochlorophyll-protein complexes isolated from *Chromatium*,, strain D. Biochemistry 9: 2688–2698

Thornber JP (1975) Chlorophyll-proteins: Light-harvesting and reaction center components of plants. Ann Rev Plant Physiol 26: 127–158

Thornber JP (1985) Citation Classic: Chlorophyll-proteins: light-harvesting and reaction center components of plants. Curr Contents/Agr Biol 16: 16

Thornber JP (1986) Biochemical characterization and structure of pigment-proteins of photosynthetic organisms. In: Staehelin LA and Arntzen CJ (eds) Encl Plant Physiol, New Series 19, pp 98–142. Springer-Verlag, Berlin

Thornber JP and Barber J (1979) Photosynthetic pigments and models for their organization in vivo. Top Photosynth 3: 27–70

Thornber JP and Highkin HR (1974) Composition of the photosynthetic apparatus of normal barley leaves and a mutant lacking chlorophyll *b*. Eur J Biochem 41: 109–116

Thornber JP and Olson JM (1968) The chemical composition of a crystalline bacteriochlorophyll-protein complex isolated from a green bacterium, *Chloropseudomonas ethylicum*. Biochemistry 7: 2242–2249

Thornber JP, Gregory RPF, Smith CA and Bailey JL (1967a) Studies on the nature of chloroplast lamella. 1. Preparation and some properties of two chlorophyll-protein complexes. Biochemistry 6: 391–396

Thornber JP, Stewart JC, Hatton MWC and Bailey JL (1967b) Studies on the nature of chloroplast lamella. 2. Chemical composition and further physical properties of two chlorophyll-protein complexes. Biochemistry 6: 2006–2014

Thornber JP, Olson JM, Williams DM and Clayton ML (1969) Isolation of the reaction center from *Rhodopseudomonas viridis*. Biochim Biophys Acta 172: 351–354

Thornber JP, Alberte RS, Hunter FA, Shiozawa JA and Kan KS (1977) The organization of chlorophyll in the plant photosynthetic unit. Brookhaven Symp Biol 28: 132–148

Thornber JP, Dutton PL, Fajer J, Forman A, Holten D, Olson JM, Parson WW, Prince RC, Tiede DM and Windsor MW (1978a) Isolated photochemical reaction centers from bacteriochlorophyll *b*-containing organisms. IVth International Congress on Photosynthesis Research. Reading, England, pp 55–70. The Biochemical Society, London

Thornber JP, Trosper TL and Strouse CE (1978b) Bacteriochlorophyll in vivo: Relation of spectral forms to specfic membrane components. In: Clayton RK and Sistrom WR (eds) The Photosynthetic Bacteria, pp 133–160. Plenum Press, New York

Thornber JP, Cogdell RJ, Seftor REB and Webster GD (1980) Further studies on the composition and spectral properties of the photochemical reaction centers of bacteriochlorophyll *b*-containing bacteria. Biochim Biophys Acta 593: 60–75

Thornber JP, Seftor REB and Cogdell RJ (1981) Intermediary electron carriers in the primary photosynthetic event of *Rhodopseudomonas viridis*. FEBS Lett 134: 235–239

Thornber JP, Peter GF, Nechushtai R, Chitnis PR, Hunter FA and Tobin EM (1986) Electrophoretic separation of chlorophyll-protein complexes and their apoproteins. In: Akoyunoglou G and Senger H (eds) Regulation of Chloroplast Differentiation, pp 249–258. AR Liss Inc, New York

Thornber JP, Morishige DT, Anandan S and Peter GF (1991) Chlorophyll-carotenoid-proteins of higher plant thylakoids. In: Scheer H (ed) The Chlorophylls, pp 549–585. CRC Press, Boca Raton

Thornber JP, Cogdell RJ, Chitnis P, Peter GF, Gomez S, Morishige DT, Anandan S, Preiss S, Welty BA, Lee A, Takeuchi TS and Kerfeld C (1994) Antenna pigment-protein complexes of higher plants and purple bacteria, In: Barber J (ed) Advances in Molecular and Cell Biology, Vol 10, pp 55–118. JAI Press Inc, Greenwich, CT

Tobin EM, Wimpee CF, Silverthorne J, Stiekema WJ, Neumann GA and Thornber JP (1984) Phytochrome regulation of the expression of two nuclear-coded chloroplast proteins. In: Thornber JP, Staehelin LA and Hallick R (eds) Biosynthesis of the Photosynthetic Apparatus: Molecular Biology, Development and Regulation, pp 325–334. AR Liss, Inc, New York

Trosper TL, Benson Dl, Thornber JP (1977) Isolation and spectral characteristics of the photochemical reaction center of *Rhodopseudomonas viridis*. Biochim Biophys Acta 460: 318–330

Wildman SG and Bonner J (1947) The proteins of green leaves. I. Isolation, enzymatic properties and auxin content of spinach cytoplasmic proteins. Arch Biochem 14: 381–413

Yamamoto HY and Vernon LP (1969) Characterization of a partially purified photosynthetic reaction center from spinach chloroplasts. Biochemistry 8: 4131–4137

Chapter 18

The Discovery of the Heat Shock Response in Plants

Pi-Fang Linda Chang[*] and Chu-Yung Lin
Department of Botany, National Taiwan University
Taipei 106

ABSTRACT

When soybean seedlings or cultured cells of soybean and tobacco are shifted from a normal growth temperature of 28°C to higher temperatures, the synthesis of a new set of proteins (heat shock proteins, or HSPs) is induced while the synthesis of most proteins made at 28°C is reduced, as first reported by J. L. Key, J. P. Mascarenhas and their co-workers in 1980. A number of other plants have since been shown to respond similarly to heat shock as soybeans do. This response in plant systems seems to have many properties in common with the heat shock response of *Droscphila* and other eukaryotic systems. The pattern of heat shock protein synthesis in soybean seedlings is closely paralleled by the accumulation and subsequent loss of heat shock-specific mRNAs. The acquisition of thermotolerance in plants correlates well with the presence of HSPs. Following the first reports of plant HSPs in the early 1980s, multifamilies of heat shock genes have been isolated and studied in different plant systems. The localization of heat shock proteins during stress and their roles in thermotolerance have also been studied. The heat shock elements of plant genes and their expression in transgenic plants have been described. Some HSPs in plants are also present during certain developmental stages, and some HSPs are induced by other stress agents. Recent studies of the heat shock response in plants have led to a better understanding of the heat shock system of plants at the molecular level, and of the possible physiological and biochemical significance of heat shock proteins in plants.

Introduction

The heat shock (HS) response is the universal induction of a small number of highly conserved proteins, the heat shock proteins (HSPs), when cultured cells, tissues, or whole organisms are exposed to elevated temperatures. The

---

[*] Current address: Department of Plant Pathology, National Chung Hsing University, Taichung 402

HS response is one of the most highly conserved biological systems known, and it has been observed in virtually every organism examined. It occurs in almost every cell and tissue type and in cultured cells of animals and plants. The HS response has great physiological importance since it appears to be critical in protecting cells from thermal damage by excessive temperature and possibly even from other stresses. HSPs (also referred to as stress proteins) have been applied for several uses in human medicine (e.g., diagnostics, drug development, and vaccines) (see Welch, 1993). Substantial progress has been made in understanding the HS response, and a number of reviews focusing on various aspects of the HS response have been published (Schlesinger *et al.*, 1982a; Neidhardt *et al.*, 1985; Nover *et al.*, 1984; Craig, 1985; Lindquist, 1986; Lindquist and Craig, 1988; Schlesinger, 1990; Nover, 1991; Parsell and Lindquist, 1993; Morimoto *et al.*, 1994). In addition, several reviews on HSPs and HS genes in plants (Key *et al.*, 1985a; Kimpel and Key, 1985; Nagao *et al.*, 1986; Nagao and Key, 1989; Nover *et al.*, 1989; Vierling, 1991; Yeh *et al.*, 1994) have been published. The properties and uses of HS promoters have been outlined (Pelham, 1987), and an extensive review of the expression of HS genes in homologous and heterologous systems has been compiled by Nover (1987). Recently, the evolution, structure and biochemical function of the small HSPs have been described by Waters *et al.* (1996). This chapter describes the discovery of HSPs, beginning with a brief history of *Drosophila* HSPs, then concentrating on studies of HSPs in plants, especially the plant low molecular mass (LMM) HSPs, through 1990. The similarities and differences among plant species, with selected reference to animal systems, are highlighted.

The soybean proteins and genes have been more extensively characterized than other plants, and this comparative analysis will emphasize these proteins and genes relative to those of other plants and animals.

The Puffs in *Drosophila*

Although the scientific roots of the HS response are much older (see Nover, 1991), the onset of the molecular biological approaches can be clearly traced back to the report of Ritossa (1962) on the selective induction of new sites of gene activity after heat treatment of *Drosophila* larvae. Ritossa (1962, 1963, 1964), a geneticist working in Naples, reported that when *Drosophila* (*D. busckii* and *D. melanogaster*) larvae, raised at 25°C, were exposed to 30–32°C for about 30 min, several new puffs appeared on the giant salivary gland chromosomes. Following a 30-minute HS, the new puffs disappeared after about 1 hour at 25°C. If the larvae were maintained at 30–32°C, the new puffs regressed after about 3 hours. Ritossa also reported that, in the case of prolonged HS treatment, the puffs active prior to the HS treatment regressed and even disappeared altogether in some experiments. These early observations demonstrated that the HS response involved the transient induction of the HS puffs and the inhibition of puffs, which were present at the lower temperature, after HS. In addition, Ritossa noted that the HS-induced puffing pattern in salivary gland chromosomes was also observed during different stages of larval development, and in the midgut and hindgut as well, indicating that it was not limited to salivary glands and was not tissue-specific. The HS puffs could be induced by treating excised salivary glands at the higher temperature. The puffing pattern seen after temperature elevation could also be induced by treating the isolated salivary glands with 2,4-dinitrophenol, sodium salicylate, sodium azide, dicoumarol, or by release from anoxia. Experiments with radioactive precursors showed that the heat-induced puffs were the sites of intense RNA transcription of active, induced genes. Thus, several important features of the HS response were already apparent from Ritossa's experiments (Morimoto *et al.*, 1990).

In the late 1960s, Berendes *et al.* (1965) using *D. hydei* and Ashburner (1970) using *D. melanogaster*, made a thorough study of the HS-induced puffs in these two strains. They tested a number of different inducers and found that the HS puff induction was very rapid, occurring within minutes after temperature increase. Would the presumed protein products, directed from mRNAs made at HS puff sites from the induced genes, be seen on the polyacrylamide gels? Mitchell and Tissières at Caltech wanted to resolve this question. Using SDS-slab gel electrophoresis to analyze radio-labeled proteins from heat-shocked *Drosophila* tissues, they found striking results in 1973: HS led to the appearance of approximately seven new protein species, whereas many proteins, made constitutively before the stress was applied, regressed and, in some experiments, disappeared altogether. This was reminiscent of the appearance of seven HS puffs on the polytene chromosomes and the parallel disappearance of most puffs active before HS. These results were obtained with all the different tissues examined, including salivary glands, brain, Malpighian tubes, and wing imaginal discs at several stages during development (Tissières *et al.*, 1974). Within a few months, these data were confirmed in Cambridge by Ashburner's group using *Drosophila* salivary glands as the experimental material (Lewis *et al.*, 1975).

From which HS puff does the mRNA for a particular protein band originate? The HS messages had to be characterized, and the answer would come from the newly devised technique of *in situ* hybridization of RNA to the polytene chromosomes (Gall and Pardue, 1971). Two reports appeared in 1975 making use of this technique and *Drosophila* tissue culture cells, which are convenient for RNA work and high specific-activity labeling. Lindquist-McKenzie *et al.* (1975) found that temperature elevation caused the rapid disappearance of preexisting polysomes, followed by the buildup of new larger polysomes. The newly made RNA in these polysomes consisted mainly of poly(A) RNA of a size sufficient to code for HSP70. This RNA hybridized *in situ* to chromosomal DNA at the sites of the two largest HS puffs, band sites 87A and 87C. Spradling *et al.* (1975) used a different strategy: they hybridized *in situ* to polytene chromosomes with poly(A) RNA extracted from control cells maintained at 25°C. The radio-labeled control cell RNA hybridized to about 50 chromosomal bands, whereas poly(A) RNA from heat-shocked cells hybridized to seven sites not seen with mRNA from control cells. These seven sites corresponded to those of the HS puffs. Further work showed that purified fractions of HS mRNA hybridized *in situ* to specific HS puffs. Moreover, when these purified HS mRNA fractions were added to *in vitro* systems for protein synthesis, they were translated into specific HSPs (Lindquist-McKenzie and Meselson, 1977; Spradling *et al.*, 1977; Mirault *et al.*, 1978). Taken together, these experiments clearly indicated that HS mRNAs, made at HS puff sites, were translated into HSPs (Morimoto *et al.*, 1990).

The Universality of the Heat Shock Response

For many years HS research was solely done with *Drosophila*, although there was always the question of how universal the phenomenon would turn out to be. At the end of the 1970s, however, a typical HS response had been reported for several other systems: in chicken embryonic fibroblasts (Kelley and Schlesinger, 1978), CHO cells (Bouche *et al.*, 1979), *Escherichia coli* (Lemeaux *et al.*, 1978; Yamamori *et al.*, 1978; Neidhardt and van Bogelen, 1981; Tilly *et al.*, 1983; Daniels *et al.*, 1984), *Naegleria* (Walsh, 1980), *Dictyostelium* (Loomis and Wheeler, 1980, 1982), *Tetrahymena* (Fink and Zeuthen, 1978; Glover *et al.*, 1981), mammals (Li and Werb, 1982; Wang *et al.*, 1981), and avian cells (Schlesinger *et al.*, 1982b). Similar observations were also made in yeast (Miller *et al.*, 1979; McAlister *et al.*, 1979; McAlister and Finkelstein, 1980), plants (Barnett *et al.*, 1980; Key *et al.*, 1981), and other organisms. Quite independently from the study of HS in *Drosophila*, biochemists studying the effects of trauma and other agents on vertebrate cells discovered that these agents could result in dramatic changes in the patterns of protein synthesis (Hightower and Smith, 1978; Hightower, 1980; White, 1980). The similarities between these responses and those of tissues to HS were soon recognized (Hightower and White, 1981). The HS response had been shown not only to be universal but also to occur under a wide variety of different stress conditions (Ashburner, 1982).

It is clear from studies on organisms including bacteria, fungi, insects, plants, mammals, and avian cells that the HS response is ubiquitous. Although specific structural or enzymatic functions for the HSPs have not yet been fully elucidated, the synthesis of HSPs has been strongly correlated with the development of thermotolerance. Thermotolerance is defined as the ability of organisms to tolerate normally lethal temperatures after an initial exposure to a sublethal, HSP-inducing temperature. Some HSPs are observed to associate with specific organelles during the heat treatment, and it is now widely suggested that this localization is important to HSP function. The question of how this set of proteins might mediate the development of thermotolerance in plants is the ultimate focus of much of the work in several laboratories (see Vierling, 1991).

Discovery of the Heat Shock Response in Plants

In 1980, using SDS-polyacrylamide gel electrophoresis (PAGE) to study the synthesis of proteins labeled with radioactive amino acids, Thomas Barnett and co-workers in Joseph P. Mascarenhas's laboratory reported that a new pattern of protein synthesis was observed in tobacco and soybean tissue culture cells exposed to a HS (Barnett *et al.*, 1980). Simultaneously, Joe L. Key, Chu-Yung Lin, and Yih-Ming Chen observed that protein synthesis changes rapidly and dramatically when the growth temperature of soybean seedlings is increased from 28°C to about 40°C, as analyzed on one-dimensional and two-dimensional PAGE (Key *et al.*, 1981). These two reports were the first evidence of HSPs in plants. A number of other plant species, including corn (Baszczynski *et al.*, 1982; Cooper and Ho, 1983; Key *et al.*, 1983a; Kelley and Freeling, 1982), tomato (e.g., Scharf and Nover, 1982; Nover and Scharf, 1984), tobacco (e.g., Barnett *et al.*, 1980; Meyer and Chartier, 1983), cotton, pea, millet, and sunflower (Key *et al.*, 1983a) have been shown to undergo the transition during a HS treatment from normal protein synthesis to predominantly HSP synthesis, usually maximal at about 10°C above the normal growth temperature. The pattern of HSP synthesis in soybean seedlings is closely paralleled by the accumulation at 40°C and subsequent loss at 28°C of HS-specific mRNAs, which are coupled with the rapid polyribosome-to-monoribosome transition that occurs at 40°C (Key *et al.*, 1981). These results imply a strongly regulated transcriptional and translational response to HS in soybean. The level of translational control of normal mRNAs during HS varies considerably among plants; and even in soybean (e.g., Key *et al.*, 1981), the level of translational control may not be as great as in *Drosophila*. While the HS response has been shown to occur in a wide range of plants by a number of laboratories, this review focuses on soybean, pea, and rice and primarily on the work done in the laboratories of Joe L. Key, Fritz Schöffl, Elizabeth Vierling, and ourselves.

When the incubation temperature of two-day-old etiolated soybean seedlings is increased from 30°C up to about 40°C (either abruptly or gradually), there is little change in the amount of externally supplied radioactive amino acids which become incorporated into proteins. Above this temperature, denoted 'breakpoint' in our studies, there is a precipitous drop in incorporation (e.g., Key *et al.*, 1982, 1983a). Other factors, such as amino acid uptake, translocation and/or activation in addition to decreased rates of protein synthesis may contribute to this abrupt drop in incorporation; however, there is clearly a dramatic change in the pattern of protein synthesis (Key *et al.*, 1981). At 35°C to 37.5°C, HSP synthesis appears against a background of normal proteins, and at 40°C or above, the synthesis of HSPs predominates with minimal synthesis of most normal proteins. The induction of HSP synthesis is rapid; the synthesis becomes predominant within the first 30 minutes at 40°C to 41°C. This pattern persists for 6 to 8 hours, at which time HSP synthesis declines dramatically, and the normal pattern of protein synthesis returns. The observed 'shut-down' of HSP synthesis with increasing time at HS temperatures is consistent with the experiments of Lindquist and co-workers (DiDomenico *et al.*, 1982) demonstrating autoregulation of HS protein synthesis in *Drosophila*. Their results indicate that accumulation of a specific level of functional HSPs represses further HSP production.

When soybean seedlings are returned to 28°C–30°C after a 2 to 4–hour incubation at 40°C, there is a rapid decay of HSP synthesis coupled with the appearance of normal protein labelling patterns. In 4-hours' time, at 30°C, HSP synthesis is essentially undetectable (Key *et al.*, 1981). However, those HSPs synthesized during the initial HS treatment are generally stable for at least 21 hours, regardless of whether the tissue is returned to 28°C or maintained at the HS temperature. The HSPs can accumulate to levels which permit their detection by Coomassie staining, in some cases becoming the major stainable proteins (see Key *et al.*, 1985a).

Although the molecular masses and isoelectric points of HSPs vary considerably from species to species (Key *et al.*, 1983a), all plants seem to synthesize a much more complex pattern of low molecular mass (LMM) HSPs than other organisms studied to date. In *Drosophila*, the four major LMM HSPs are 22, 23, 26 and 27 kD. In plants, there is a very complex group of 15 to 27 kD proteins, comprising of 20–30 different polypeptides. Interestingly, this large diversity of HSPs in plants compared to other organisms is restricted to the LMM polypeptides. The high molecular mass (HMM) HSPs (68–70, 84, 92 kD) are highly conserved between plants, animals and bacteria (Key *et al.*, 1985a). Due to the unique nature and complexity of the 15 to 18 kD HSPs of plants, many of the studies in soybean and rice have concentrated on these proteins, and the mRNAs and genes which encode them. In the following sections of this chapter, we will focus primarily on the LMM HSPs.

Plant Heat Shock Proteins

One striking feature of soybean HSPs, and plants generally, is the relatively complex constellation of LMM HSPs that is induced by elevated temperature. The complexity of the LMM HSPs has been reviewed for a number of plant species (Nagao *et al.*, 1986; Vierling, 1991). Differences in protein isolation procedures, tissue treatment, as well as separation techniques may account for differences in the literature regarding HS protein induction, for example, in the maize system (Baszczynski and Walden, 1983; Cooper and Ho, 1983; Cooper *et al.*, 1984). A comparative study using uniform methodology allowed estimates by two-dimensional PAGE analyses of the total number of LMM HSPs synthesized for various species of plants. Two classes of LMM HSPs were identified, namely polypeptides induced by HS from nondetectable levels and polypeptides that were present at control temperatures but which increased after HS (Mansfield and Key, 1987). For example, in soybean 27 HSPs in the 15 to 25 kD range were detected, with 6 being enhanced and 21 being induced by heat treatment. Other species examined included pea, sunflower, wheat, rice, maize, millet, and *Panicum miliaceum*. In each species examined, the LMM HSPs resolved into a diverse array of 12 to greater than 20 polypeptides upon electrophoresis. Since the electrophoresis system used resolved only acidic proteins, this number of HSPs detected after heat treatment would be a minimal value. Nover and Scharf (1984) identified 18 basic proteins induced by HS in tomato suspension cultures, of which nine were less than 30 kD. These observations taken together with previous *in vivo* labeling and *in vitro* translation analyses (Key *et al.*, 1983a, 1985b) illustrate the abundance of LMM HSPs in plants. This contrasts markedly with most other organisms where the LMM HSPs are a simple group consisting of usually one to five proteins (see Craig, 1985). The significance of these differences in complexity and abundance of the LMM HSPs between plants and other organisms is not known. Green plants have an additional organelle, the chloroplast, within which HSPs have been identified following heat treatment. These HSPs have been shown to be encoded by nuclear genes which are translated in the cytoplasm and then transported into the chloroplast (Kloppstech *et al.*, 1985; Vierling *et al.*, 1986).

As proposed by Elizabeth Vierling (1991), plant LMM HSPs can be further divided into four multigene families, according to the sequence similarities of the deduced amino acid sequences from HS genes isolated from several different plants: (1) the class I proteins are cytosolic; (2) the class II proteins are cytosolic with their deduced amino acid sequences less homologous to those of the class I proteins; (3) the class III proteins are associated with chloroplasts; (4) the class IV proteins are associated with endomembrane systems. The corresponding genes from different classes cross-hybridize only at very low stringency. Recently, at least one more class, the class V, of the LMM HSPs was suggested to be associated with mitochondria as proposed by Waters *et al.* (1996) based on the sequence analyses of soybean HS genes isolated by LaFayette *et al.* (1996). Lenne and Douce (1994) also reported that a nuclear-encoded 22 kD HSP was localized to pea mitochondria.

The patterns of HMM HSPs from different plant species show much less variation than the LMM HSPs when analyzed by two-dimensional PAGE. The distribution of HSPs of 68, 70, 83, and 92 kD are remarkably similar for soybean, pea, millet, corn, and cotton (Key *et al.*, 1983a). The abundance of these proteins is much less than the LMM HSPs, and for some plant species additional higher molecular mass proteins have been identified (see Nagao and Key, 1989).

Accumulation of Heat Shock mRNAs in Plants

Following HS at 40°C, a very abundant class of RNA appears in seedlings which is either absent or present at very low levels in control (30°C) tissue (Schöffl and Key, 1982; Dawson and Grantham, 1981). In the case of soybean, 20 or more poly(A)RNA sequences of about 800 nucleotides in length accumulate to as many as 20,000 copies per cell within two hours at 40°C based on cDNA/poly(A)RNA kinetic hybridization studies (Schöffl and Key, 1982). These mRNAs are undetectable in 30°C tissue as analyzed by Northern hybridization using cloned HS-specific cDNA probes (Schöffl and Key, 1982; Czarnecka *et al.*, 1984). The HS mRNAs are easily detected during the initial three to five minutes at HS temperatures, and they accumulate during the next one to four hours, depending on the exact temperature of the HS. At 40°C, the level of HS mRNAs remains approximately constant from two to six hours of HS, and then declines gradually over the next few hours. The decline in HS mRNAs during continuous 40°C treatment is much slower than when the tissue is returned to 30°C after a two to four hour-HS. Under the latter conditions, the HS mRNA levels decline with a half-life of about one hour. The relative contributions of the rate of HS mRNA synthesis and the rate of decay to these levels is not known. In both cases, however, the loss of HS mRNAs correlates with the loss of HSP synthesis under the same conditions, and these data are consistent with the view that autoregulation occurs at the level of transcription of HS mRNAs coupled with decreased translation of those mRNAs, as described in *Drosophila* (DiDomenico *et al.*, 1982). A large percentage of normal poly(A)RNAs persist during HS at 40°C, indicated by the minimal change in complexity of the mRNA populations during HS (Key *et al.*, 1985a). However, some mRNAs decline significantly during HS, for example, an auxin-regulated sequence, pJCW1 (Walker and Key, 1982) and ribulose bisphosphate carboxylase small subunit mRNA (Vierling and Key, 1985); while other sequences do not decline at all, for example, an actin mRNA (see Key *et al.*, 1985a) and a ribulose bisphosphate carboxylase large subunit mRNA (Vierling and Key, 1985).

Using the cDNA clones isolated by Tseng *et al.* (1993) as probes, our laboratory reported that, in etiolated rice seedlings, the mRNAs encoding the class I LMM HSPs accumulated during two hours of HS treatments, at the optimal temperature of 41°C. The mRNAs were first detected after five minutes and gradually accumulated over time for two hours. Prolonged incubation of rice seedlings at 41°C resulted in a decrease in the mRNA accumulation, which may be the result of autoregulation, as suggested above.

Development of Thermotolerance in Plants

Several lines of evidence support the view that in soybean, as in other organisms, an initial HS treatment at a tolerant temperature provides thermoprotection to otherwise lethal temperatures (Key *et al.*, 1982; Key *et al.*, 1983b; Lin *et al.*, 1984; Schlesinger *et al.*, 1982a). In soybean seedlings, several different pretreatments: (1) two hours at 40°C, (2) thirty minutes at 40°C followed by two to four hours at 30°C, (3) seven to ten minutes at 45°C followed by two hours at 30°C, all permit the soybean seedlings to tolerate a two-hour 45°C HS treatment, an otherwise lethal treatment (Lin *et al.*, 1984). All of these treatments, which lead to the development of thermotolerance, cause the accumulation of high levels of HSPs prior to the otherwise lethal treatment. Similar results of acquired thermotolerance were also reported recently by the authors' group (Jinn *et al.*, 1997). In addition, we found that the most heat-responsive region in soybean etiolated seedlings was in the root tip, where the HSPs synthesis was rapid and to a higher level than other regions. In different pretreatments followed by 45°C HS treatments, only the root meristematic regions were protected from the heat damage as indicated by the TTC (2,3,5-triphenyltetrazolium chloride) assay. These results are in good agreement with our previous report that the structure of plasmalemma, mitochondria, plastids and nuclei of root meristematic cells, observed under electron microscope, were protected from heat damage with a preincubation before HS (Chen *et al.*, 1988).

An enriched HSPs fraction, which contained HSPs of 15–18 kD, 68–70 kD, 24 kD and 22 kD, was prepared from the 70%–100% ammonium sulfate fraction of the postribosomal supernatant from heat-shocked soybean seedlings. These enriched HSPs were resistant to heat denaturation, as judged by their unpelletability after HS (Jinn *et al.*, 1989). This HSP-enriched fraction also protected soluble proteins of control seedlings from heat denaturation for at least one hour. The degree of protection was proportional to the HSP-enriched fraction added (Jinn *et al.*, 1989).

Our group has isolated two major proteins from the 15–18 kD class of soybean HSPs, obtained from an HSP-enriched ammonium sulfate fraction separated by two-dimensional PAGE, as antigens to prepare antibodies specific to individual proteins (Hsieh *et al.*, 1992). Each of these antibody preparations reacted with its antigen and cross-reacted with 12 other 15- to 18-kD HSPs. The accumulation of the 15–18 kD HSPs under various HS treatments was quantified. These proteins accumulated to a level of about 1% of the total protein and correlated well with the establishment of thermotolerance (Hsieh *et al.*, 1992). These antibodies, together with the antibody raised from a 16.9 kD rice HSP, cross-reacted with the same class I LMM HSPs in ten plant species tested (Jinn *et al.*, 1993). Additionally, the HSPs-enriched fraction was exchangeable among soybean, mung bean and rice for the thermostabilization assay, as reported by our group (Jinn *et al.*, 1993). When the HSP-enriched fraction mentioned above was separated by non-denaturing PAGE, the presence of an HMM complex (280 kD) was observed in soybean seedlings (Jinn *et al.*, 1995). This complex cross-reacted with antibodies raised against soybean class I LMM HSPs. Dissociation of the complex by denaturing PAGE revealed the complex to contain at least 15 proteins of 15–18 kD class I LMM HSPs detectable by staining, radiolabeling, and by Western blotting. A similar LMM HSP complex was observed in mung bean (295 kD), in pea (270 kD), and in rice (310 kD). The isolated HSP complex was able to protect up to 75% of soluble proteins from heat denaturation *in vitro* (Jinn *et al.*, 1995). A single recombinant 16.9 kD class I LMM HSP of rice also provided thermoprotection *in vitro* and formed a HMM complex (~310 kD) in a non-denaturing polyacrylamide gel, as reported by our group (Yeh *et al.*, 1995). Interestingly, this recombinant rice protein, when expressed in *E. coli* cells which generally do not synthesize this class of LMM HSPs, also provided thermotolerance to *E. coli* cells *in vivo* (Yeh *et al.*, 1997).

The research group in Georgia has found one other treatment, incubation in 50–75 µM arsenite for three hours, which also induces synthesis of nearly the full spectrum of HSPs. The treatment also confers the ability to tolerate a two-hour 45°C treatment. As described by Key *et al.* (1985a), arsenite is unique among many other stresses studied in its ability to closely mimic the HS response. A number of functions are clearly protected by these pretreatments, including the capacity to synthesize HSPs and HS mRNAs at these otherwise inhibitory conditions and the ability of the seedlings to survive and grow rather normally upon return to the normal (e.g., 30°C) growth conditions. Thus, all vital functions must be protected sufficiently during the otherwise lethal treatment to permit these functions to persist and recover to essentially a normal state within a few hours after returning to the normal temperature. The recovery time depends to some extent upon the temperature-time relationships of the HS treatment. Slower recovery is observed if the HS treatment exceeds the 'breakpoint', as compared to a HS treatment at or below the breakpoint temperature (Key *et al.*, 1985a).

Localization and Function of the Plant Heat Shock Proteins

The selective localization of HSPs with cellular structures during HS likely provides the basis of thermoprotection (Schlesinger *et al.*, 1982a; 1982b; Lin *et al.*, 1984). It is known in *Drosophila* that some HSPs localize specifically within the nucleus (Velazquez *et al.*, 1980; Arrigo *et al.*, 1980; Arrigo and Ahmad-Zadeh, 1981; Velazquez *et al.*, 1983). In addition, evidence suggests that HSPs also associate with other organelle fractions during HS, including the cytoskeleton in animal cells (Schlesinger *et al.*, 1982b), and mitochondria and ribosomes in soybean (Lin *et al.*, 1984). We do not know the mechanism of the association of HSPs with the plant organelle fractions, but the association is HS-dependent; HSPs made during arsenite treatment are found throughout the cytoplasm, but they become localized immediately after the temperature is shifted up. The HSPs become associated with organelles during HS, are dispersed from these fractions over a period of three to four hours after the tissue is returned to 30°C, and rapidly re-associate (within 15 minutes) during a subsequent HS in a temperature-dependent manner (Lin *et al.*, 1984).

In many organisms, the cytoplasmic HSPs can be detected in aggregates larger than one megadalton. These large LMM HSP aggregates, referred to as 'heat shock granules' (HSGs), form reversibly from the smaller particles, depending on the duration and severity of the imposed stress conditions (Arrigo, 1987; Arrigo and Welch, 1987; Arrigo *et al.*, 1988; Collier *et al.*, 1988; Nover *et al.*, 1989). Some of the association of soybean LMM HSPs with organelles, as described by Lin *et al.* (1984) and Nagao *et al.* (1986), most likely results from copurification of HSGs with different cell fractions (Mansfield and Key, 1988; Nover, 1991). For example, LMM HSP aggregates have now successfully been separated from ribosomes and prosomes (Nover *et al.*, 1989). Furthermore, electron microscopic studies of LMM HSPs in plants indicate HSGs accumulate in the perinuclear region rather than within the nucleus (Nover *et al.*, 1989; Nover, 1991). It is also clear that some reports of LMM HSP association with membranes resulted from the presence of the specific chloroplast and endomembrane LMM HSPs. It would appear that under most conditions class I and II LMM HSPs in plants are located in the cytoplasm (see Vierling, 1991).

In tomato, HSGs appear to contain LMM cytoplasmic HSPs, HSP70, other unidentified proteins, and an RNA component (Nover *et al.*, 1989). Nover *et al* (1989) reported that the RNA component was selectively composed of mRNAs present prior to heat stress and devoid of HSP mRNA. They suggest that the function of HSGs is to protect and store normal cellular mRNAs during stress. However, in chicken, *Drosophila*, and mammals, the particles are composed primarily of LMM HSPs, and the presence of other proteins or RNA has not been confirmed (Arrigo and Welch, 1987; Arrigo *et al.*, 1988; Collier *et al.*, 1988;). Recently, using the monospecific polyclonal antibodies (prepared by Hsieh *et al.*, 1992) to soybean class I LMM HSPs for immunolocalization analysis, Jinn *et al* (1997) in the authors' group have demonstrated that the class I LMM HSPs originally localized in the cytoplasm were translocated into the nucleus and nucleolus during HS. These soybean HSPs, found in the aggregated granular structure during HS, were randomly distributed in the cytoplasm and in the nucleus. However, when the HS treatment ended, the granular structure disappeared and the class I LMM HSPs became distributed homogeneously in the cytoplasm (Jinn *et al.*, 1997).

Two HSPs, 20 and 24 kD, appear to specifically localize in the mitochondrial fraction, and they do not disperse throughout the cell significantly during a subsequent 30°C incubation (Key *et al.*, 1982; Lin *et al.*, 1984). This may indicate specific transport of these proteins into the mitochondria. Using an oxygen electrode to monitor the O_2 consumption of isolated mitochondria, Chou *et al.* (1989) of our group have found that the association of soybean LMM HSPs with mitochondria provides protection of mitochondrial oxidative phosphorylation coupled to O_2 uptake at 42°C.

A striking feature of one subset of plant HS proteins is their localization into chloroplasts (Kloppstech *et al.*, 1985; Vierling *et al.*, 1986; Suss and Yordanov, 1986). These HS proteins are nuclear-encoded, synthesized in the cytoplasm, processed, and transported into the chloroplast. Localization of HSPs into chloroplasts might be expected, since thermoprotection of some chloroplast functions seems to occur (Kyle *et al.*, 1985). Based on observations that the pea chloroplast HSP is bound to thylakoid membranes, Kloppstech and colleagues (Kloppstech *et al.*, 1985; Glaczinski and Kloppstech, 1988) have proposed that the function of chloroplast HSP is to protect or repair photosystem II during stress. Photosystem II is one of the more heat-sensitive components of the chloroplast (Berry and Björkman, 1980). However, thylakoid localization of chloroplast HSP occurred only at temperatures above 38°C and at relatively high light intensities. Below 38°C the protein showed no strong association with membranes (Glaczinski and Kloppstech, 1988). Using a variety of stress conditions, including temperatures of 38°C or 40°C at moderate light intensities, Chen *et al.* (1990) did not find a significant proportion of the pea chloroplast HSP to be associated with thylakoids. Greater than 80% of the protein was present in the soluble chloroplast fraction. The chloroplast HSP was also produced at significant levels in root tissues, indicating it may function in all types of plastids, not only in photosynthetic organelles. The presence of the protein in nonphotosynthetic plastids and its homology to cytoplasmic and endomembrane LMM HSPs thus suggests that the chloroplast HSP performs other functions, in addition to protecting photosynthesis (Chen *et al.*, 1990).

Heat Shock Response in Field Grown Plants

The studies summarized above were accomplished using etiolated seedlings in shaking culture under laboratory conditions. Field conditions during the summer of 1983 in Georgia permitted an assessment of HS mRNA accumulation at ambient temperatures of 39°C to 40°C under both irrigated and non-irrigated conditions (Kimpel and Key, 1985). Under dry conditions where leaf temperatures approach or exceed ambient temperatures (Jung and Scott, 1980), HS mRNAs accumulated to very high levels between 10:00 a.m. and 5:00 p.m. Irrigated plants, where leaf temperatures probably remain significantly below ambient, accumulated low levels of HS mRNA at the peak ambient temperature of 39°C to 40°C, levels comparable to 35°C or so under laboratory conditions in etiolated seedlings. Silver-stained O'Farrell gels also showed the accumulation of HSPs in non-irrigated plants but not in irrigated plants (Mansfield, 1986). These results have been confirmed using growth chamber-grown green plants (Kimpel and Key,

1985). In other experiments, cotton grown under recurring water stress exhibited elevated leaf temperatures and accumulated proteins with molecular masses corresponding to those of HSPs (Burke *et al.*, 1985). Both of these studies, although limited to examining plants suffering from severe stress, established that HSP expression does occur in the natural environment.

Relationship of Heat Stress and Other Stresses in Soybean Seedlings

Based on the observations that a wide range of stress agents induce the 'HS response' in *Drosophila* (see Ashburner and Bonner, 1979), including recovery from anoxia, coupled with the dramatic effects of anaerobiosis on plant protein synthesis (Lin and Key, 1967; Sachs *et al.*, 1980; Sachs and Freeling, 1978), and general effects of water stress on the protein synthetic apparatus (Hsiao, 1970, 1973; Moriella *et al.*, 1973), a detailed analysis of the possible relationship of other physical stress agents to high temperature (HS) stress was performed by Key's group (Czarnecka *et al.* 1984; Edelman *et al.*, 1988). These studies generally utilized HS cDNA clone probes in Northern hybridization analyses of poly(A)RNAs isolated from 'stressed' soybean seedlings. In the case of arsenite, *in vivo* protein labeling was also achieved, but with most of these stress agents amino acid uptake and incorporation was so impaired that meaningful *in vivo* studies were not feasible. The 'stress' agents included: PEG (to impose up to -8 bars water stress), salt stress, anaerobiosis, high levels of various plant hormones, amino acid analogs, heavy metals, arsenite, and a range of respiratory and/or phosphorylation inhibitors (Czarnecka *et al.*, 1984). Generally, very low levels of 'HS mRNAs' were detected using various HS cDNA probes following treatment with this range of agents; the levels which were detectable above untreated controls were generally 100-fold (or more) lower than HS-induced mRNA levels. However, arsenite resulted in induction and accumulation of very significant levels of the HS mRNAs homologous to all of the available HS cDNA probes, although the induction was slower and often somewhat lower levels accumulated than with HS (Key *et al.*, 1985a).

HSP Expression During Plant Development

The transcription of HSP genes and HSP synthesis have been shown to occur in all vegetative tissues, in mid-maturation seeds (Altschuler and Mascarenhas, 1985; Atkinson and Walden, 1985; Chrispeels and Greenwood, 1987), in the aleurone of imbibed seeds (Brodl *et al.*, 1990; Sticher *et al.*, 1990), and in germinating embryos (Helm *et al.*, 1989; Helm and Abernathy, 1990;

Howarth, 1989). However, full activation of the HS response does not occur during very early embryo development (Pitto *et al.*, 1983; Zimmerman *et al.*, 1989) or pollen germination (Schrauwen *et al.*, 1986; van Herpen *et al.*, 1989; Xiao and Mascarenhas, 1985). Actually, early embryos contain HSP mRNAs that are not actively translated unless the embryo is heat stressed (Pitto *et al.*, 1983; Zimmerman *et al.*, 1989). Both germinating pollen (Xiao and Mascarenhas, 1985) and globular embryos (Zimmerman *et al.*, 1989) are very heat sensitive. Whether this is due to their inability to express a complete heat shock response is only a hypothesis at this time.

The expression of HSP mRNA and protein during development, in the absence of heat stress, has been studied in a number of eukaryotes (Bond and Schlesinger, 1987). LMM HSP mRNAs are found not only early in embryogenesis, but class I LMM HSP mRNAs are also found in fully developed seeds of pea and wheat (Helm *et al.*, 1989; Helm and Abernathy, 1990; Helm *et al.*, 1990; Vierling and Sun, 1989). In contrast to the situation in early embryo development, accumulation of LMM HSP mRNAs during late seed development is accompanied by accumulation of the corresponding proteins (Helm *et al.*, 1990; Helm and Abernathy, 1990). Additional studies on HSP expression in the natural environment may reveal that production of HSPs is an integral part of many phases of plant life history, even for individuals growing in optimal environments.

Low Molecular Mass Heat Shock Genes

Characterization of soybean HS cDNA and genomic clones has demonstrated that soybean LMM HS protein genes represent several multigene families with domains of homology to evolutionarily distant organisms including *Drosophila, Xenopus,* and *Caenorhabditis elegans* (see Nagao *et al.*, 1986). The area of highest conservation resides in the carboxyl portion of the proteins. Based on hybrid-select translation and DNA sequence analyses, the largest soybean family (class I) consists of 13 proteins of 15–18 kD, and DNA sequences of representative genomic clones of this family have been published (Schöffl *et al.*, 1984; Czarnecka *et al.*, 1985; Nagao *et al.*, 1985). Based on genes sequenced, the molecular mass range of this family is 17.3–18.5 kD (Key *et al.*, 1987). Comparative analysis of four class I soybean HS protein genes of 17.3–17.6 kD showed greater than 90% amino acid homology with approximately two-thirds of the nucleotide changes being silent substitutions (Nagao *et al.*, 1985). Comparison of an 18.5 kD HS protein sequence, Gmhsp18.5-V, with the four 17-kD sequences, showed approximately 75% amino acid identity among the five sequences but 96% identity when compared individually to each of the four 17-kD proteins

(Nagao and Key, 1989). Three cDNA clones (Tseng *et al.*, 1992; 1993; Lee *et al.*, 1995) and two genomic clones (Tzeng *et al.*, 1992, 1993) for class I LMM HSPs of rice have been isolated and characterized in our laboratories. These sequences encode proteins of 16.9–18.0 kD and show about 70% to 99.3% amino acid identity among them.

Based on hybrid-select translation and DNA sequence analyses, an additional family of soybean HSP genes encoding approximately 21–24 kD proteins is represented by cDNA clones pFS2033, pEV1, pEV2, and pEV6 and genomic clone Gmhsp22-K. Amino acid sequence alignment shows that homology variation between clones within this size class (e.g., Gmhsp22-K vs. pEV2, 42% identity in 193 amino acid overlap) may be as much as the variation between these and the 17-kD HS proteins (e.g., Gmhsp22-K vs. Gmhspl7.6-L, 47% identity in 151 amino acid overlaps (Nagao and Key, 1989). The region of maximum conservation is located toward the carboxyl terminus of the protein.

HS genes encoding 26–28 kD proteins can be divided into at least two classes. One class, represented by soybean cDNA clone pCE54, encodes a family of general stress proteins. This family of four to six genes is expressed constitutively at control temperatures, and RNA and protein synthesis are enhanced during HS and by numerous other stress agents including arsenite, heavy metals, high salt, anaerobiosis, water stress, and ABA treatment (Czarnecka *et al.*, 1984). The sequence of the gene corresponding to pCE54, Gmhsp26-A, containing a single intron of 388 bp located between codons 107 and 108 in an open reading frame of 225 codons, predicts a 26 kD protein. Processing of the intron was preferentially inhibited by treatment of soybean seedlings with $CdCl_2$ and $CuSO_4$, but it was not inhibited by elevated temperature (Czarnecka *et al.*, 1987).

Hydropathy analysis of the deduced amino acid sequence of pCE54, as compared to the smaller HSPs, indicate a high degree of relatedness within the carboxyl half of the protein. While clearly related to HSPs, the lower amino acid sequence identity suggests that this protein is highly diverged and may therefore be specialized for general stress adaptation in soybean (Czarnecka *et al.*, 1987).

Vierling *et al.* (1988) have isolated and sequenced cDNA clones from soybean and pea that specify nuclear-encoded HSPs which localize to chloroplasts. Nucleotide sequence comparison shows that the derived amino acid sequence of the mature pea and soybean proteins are 79% identical with all but two changes representing conservative replacements. The soybean cDNA encodes 181 amino acids or 20.5 kD of the 22-kD mature protein. Comparison of the pea chloroplast HSP sequence to the PIR protein data base from the National Biomedical Research Foundation (March 1987 version) identified significant homologies to soybean LMM cytoplasmic HSPs, *Drosophila* LMM HSPs, and α-crystallins from several eukaryotes. As noted previously in comparing soybean HSP genes of different families, the highest homology among these proteins occurs in the carboxyl-terminal half of the proteins (Nagao and Key, 1989).

Concluding Remarks

There are many striking and highly conserved features for the plant heat shock response. Much progress has been made since 1990, but much remains to be discovered. Additional information at the physiological, biochemical, molecular and genetic levels is needed for us to make significant progress in gaining an understanding of these phenomena, so that we might possibly manipulate the response in a beneficial manner. Hopefully, the enormous wealth of information accumulated in the preceding 20 years will increase our appreciation of the roots and help us continue to improve our understanding of the HS response.

Acknowledgment

The authors would like to thank Dr. Janice Kimpel for critical reading of this manuscript and Mr. Huai-Yi Chen for preparation of the manuscript. Part of the authors' research referred to in this review has been supported by grants from the National Science Council, Republic of China.

References

Altschuler, M. and Mascarenhas, J. P. (1985) Transcription and translation of heat shock and normal proteins in seedlings and developing seeds of soybean exposed to a gradual temperature increase. *Plant Mol. Biol.* **5:** 291-297.

Arrigo, A.-P. (1987) Cellular localization of HSP23 during *Drosophila* development and following subsequent heat shock. *Dev. Biol.* **22:** 39-48.

Arrigo, A.-P. and Ahmad-Zadeh, C. (1981) Immunofluorescence localization of the small heat shock proteins (hsp 23) in salivary gland cell of *Drosophila melanogaster*. *Mol. Gen. Genet.* **184:** 73-79.

Arrigo, A.-P. and Welch, W. J. (1987) Characterization and purification of the small 28,000-dalton mammalian heat shock protein. *J. Biol. Chem.* **262:** 15359-15369.

Arrigo, A.-P., Fakan, S. and Tissières, A. (1980) Localization of the heat shock-induced protein in *Drosophila melanogaster* tissue culture cells. *Dev. Biol.* **78:** 86-103.

Arrigo, A.-P., Suhan, J. and Welch, W. J. (1988) Dynamic changes in the structure and intracellular locale of the mammalian low-molecular-weight heat shock protein. *Mol. Cell. Biol.* **8:** 5059-5071.

Ashburner, M. (1970) Pattern of puffing activity in the salivary gland chromosomes of *Drosophila*. V. Response to environmental treatments. *Chromosoma* **31:** 356-376.

Ashburner, M. (1982) The effects of heat shock and other stress on gene activity: an introduction. In *Heat shock: From Bacteria to Man*, M. J. Schlesinger, M. Ashburner and Tissières A. eds., pp. 1-9. Cold Spring Harbor Laboratory Press, Cold Spring Harbor, New York.

Ashburner, M. and Bonner, J. J. (1979) The induction of gene activity in *Drosophila* by heat shock. *Cell* **17**: 241-254.

Atkinson, B. G. and Walden, D. B., eds. (1985) *Changes in Eukaryotic Gene Expression in Response to Environmental Streess*. Academic Press, Orlando, Florida, p. 379.

Barnett, T., Altschuler, M., McDaniel, C. N. and Mascarenhas, J. P. (1980) Heat shock induced proteins in plant cells. *Dev. Genet.* **1**: 331-340.

Baszczynski, C. L. and Walden, D. B. (1983) Regulation of gene expression in corn (*Zea mays* L.) by heat shock. II. *In vitro* analysis of RNAs from heat-shocked seedlings. *Can. J. Biochem. Cell Biol.* **61**: 395-403.

Baszczynski, C. L., Walden, D. B. and Atkinson, B. G. (1982) Regulations of gene expression in corn (*Zea mays* L) by heat shock. *Can. J. Biochem.* **60**: 569-579.

Berendes, H. D., van Breugel, F. M. A. and Holt, T. K. H. (1965) Experimental puffs in salivary gland chromosomes of *D. hydei*. *Chromosoma* **16**: 35-46.

Berry, J. A. and Björkman, O. (1980) Photosynthetic response and adaptation to temperatures in higher plants. *Annu. Rev. Plant. Physiol.* **31**: 491-543.

Bond, U. and Schlesinger, M. J. (1987) Heat-shock proteins and development. *Adv. Genet.* **24**: 1-28.

Bouche, G., Amalric, F. Caizergues-Ferres, M. and Zalta, J. P. (1979) Effects of heat shock on gene expression and subcellular protein distribution in Chinese hamster ovary cells. *Nucleic Acid Res.* **7**: 1739-1747.

Brodl, M. R., Belanger, F. C. and Ho, T.-H. D. (1990) Heat shock proteins are not required for the degradation of •-amylase mRNA and the delamellation of endoplasmic reticulum in heat-stressed barley aleurone cells. *Plant Physiol.* **92**: 1133-1141.

Burke, J. J., Hatfield, J. L., Klein, R. R. and Mullet, J. E. (1985) Accumulation of heat shock proteins in field-grown cotton. *Plant Physiol.* **78**: 394-398.

Chen, Q., Lauzon, L. M., DeRocher, A. E. and Vierling, E. (1990) Accumulation, stability, and localization of a major chloroplast heat-shock protein. *J. Cell Biol.* **110**: 1873-1883.

Chen, Y.-R., Chou, M., Ren, S.-S., Chen, Y.-M. and Lin, C.-Y. (1988) Observations of soybean root meristematic cells in response to heat shock. *Protoplasma* **144**: 1-9.

Chou, M., Chen, Y.-M. and Lin, C:-Y. (1989) Thermotolerance of isolated mitochondria associated with heat shock proteins. *Plant Physiol.* **89**: 617-621.

Chrispeels, M. J. and Greenwood, J. S. (1987) Heat stress enhances phytohemagglutinin synthesis but inhibits its transport out of the endoplasmic reticulum. *Plant Physiol.* **83**: 778-784.

Collier, N. C., Heuser, J., Levy, M. A. and Schlesinger, M. J. (1988) Ultrastructural and biochemical analysis of the stress granule in chicken embryo fibroblasts. *J. Cell Biol.* **106**: 1131-1139.

Cooper, P. and Ho, T.-H. D. (1983) Heat shock proteins in maize. *Plant Physiol.* **71**: 215-222.

Cooper, P., Ho, T.-H. D. and Hauptman, R. M. (1984) Tissue specificity of the heat shock response in maize. *Plant Physiol.* **75**: 431-440.

Craig, E. A. (1985) The heat shock response. *CRC Crit. Rev. Biochem.* **18**: 239-280.

Czarnecka, E., Edelman, L., Schöffl, F. and Key, J. L. (1984) Comparative analysis of physical stress responses in soybean seedlings using cloned heat shock cDNAs. *Plant. Mol. Biol.* **3**: 45-58.

Czarnecka, E., Gurley, W. B., Nagao, R. T., Mosquera, L. and Key, J. L. (1985) DNA sequence and transcript mapping of a soybean gene encoding a small heat shock protein. *Proc. Natl. Acad. Sci., USA.* **82**: 3726-3730.

Czarnecka, E., Nagao, R. T., Key, J. L. and Gurley, W. B. (1987) Characterization of Gmhsp26-A, a stress gene encoding a divergent heat shock protein of soybean; heavy metal induced inhibition of intron processing. *Mol. Cell. Biol.* **8:** 1113-1122.

Daniels, C. J., McKee, A. H. Z. and Doolittle, W. F. (1984) Archaebacterial heat-shock proteins. *EMBO J.* **3:** 745-749.

Dawson, W. O. and Grantham, G. L. (1981) Inhibition of stable RNA synthesis and production of novel RNA in heat stressed plants. *Biochem. Biophys. Res. Commun.* **100:** 23-30.

DiDomenico, B. J., Bugaisky, G. E. and Lindquist, S. (1982) The heat shock response is self-regulated at both the transcriptional and post-transcriptional level. *Cell* **31:** 593-603.

Edelman, L., Czarnecka, E. and Key, J. L. (1988) Induction and accumulation of heat shock-specific poly(A+)RNAs and proteins in soybean seedlings during arsenite and cadmium treatments. *Plant Physiol.* **86:** 1048-1056.

Fink, K. and Zeuthen, E. (1978) Heat shock proteins in *Tetrahymena*. *ICN-UCLA Sympos. Mol. Cell. Biol.* **12:** 103-115.

Gall, J. G. and Pardue, M. L. (1971) Nucleic acid hybridization in cytological preparations. *Methods Enzymol.* **21:** 470-480.

Glaczinski, H. and Kloppstech, K. (1988) Temperature-dependent binding to the thylakoid membranes of nuclear-coded chloroplast heat-shock proteins. *Eur. J. Biochem.* **173:** 579-583.

Glover, C. V. C., Vavra, K. J., Guttman, S. D. and Gorovsky, M. A. (1981) Heat shock and deciliation induce phosphorylation of histone H1 in *T. pyriformis*. *Cell* **23:** 73-77.

Helm, K. W. and Abernathy, R. H. (1990) Heat shock proteins and their mRNAs in dry and early imbibing embryos of wheat. *Plant Physiol.* **93:** 1626-1633.

Helm, K. W., DeRocher, A., Lauzon, L. and Vierling, E. (1990) Expression of low molecular weight heat shock proteins during seed development. *J. Cell. Biochem.* **14E:** 301 (Abstr.)

Helm, K. W., Petersen, N. S. and Abernathy, R. H. (1989) Heat shock response of germinating embryos of wheat. *Plant Physiol.* **90:** 598-605.

Hightower, L. E. (1980) Cultured animal cells exposed to amino acid analogues or puromycin rapidly synthesis several polypeptides. *J. Cell. Physiol.* **102:** 407-427 .

Hightower, L. E. and Smith, M. D. (1978) Effects of canavanine on protein metabolism in Newcastle disease-virus infected chicken embryo cells. In *Negative strand viruses and the host cell*, B. W. J. Mahey and R. D. Barry eds., pp. 395-405. Academic Press, London.

Hightower, L. E. and White, F. P. (1981) Cellular responses to stress: Comparison of a family of 71-73 kilodalton proteins rapidly synthesised in rat tissue slices and canavanine-treated cells in culture. *J. Cell. Physiol.* **108:** 261.

Howarth, C. (1989) Heat shock proteins in *Sorghum bicolor* and *Pennisetum americanum*. I. Genotypic and developmental variation during seed germination. *Plant Cell Environ.* **12:** 471-477.

Hsiao, T. C. (1970) Rapid changes in levels of polyribosomes in *Zea mays* in response to water stress. *Plant. Physiol.* **46:** 281-285.

Hsiao, T. C. (1973) Plant responses to water stress. *Annu. Rev. Plant Physiol.* **24:** 519-570.

Hsieh, M.-H., Chen, J.-T., Jinn, T.-L., Chen, Y.-M. and Lin, C.-Y. (1992) A class of soybean low molecular weight heat shock proteins: immunological study and quantitation. *Plant Physiol.* **99:** 1279-1284.

Jinn, T. L., Yeh, Y. C., Chen, Y. M. and Lin, C. Y. (1989) Stabilization of soluble proteins *in vitro* by heat shock proteins-enriched ammonium sulfate fraction from soybean seedlongs. *Plant Cell Physiol.* **30(4):** 463-469.

Jinn, T.-L., Chang, P.-F. L., Chen, Y.-M., Key, J. L. and Lin C.-Y. (1997) Tissue-type-specific heat-shock response and immunolocalization of class I low-molecular-weight heat-shock proteins in soybean. *Plant Physiol.* **114:** 429-438.

Jinn, T.-L., Chen, Y.-M. and Lin, C.-Y. (1995) Characterization and physiological function of class I low molecular weight heat shock protein complexes in soybean. *Plant Physiol.* **108:** 693-701.

Jinn, T.-L., Wu, S.-H., Yeh, C.-H., Hsieh, M.-H., Yeh, Y.-C., Chen, Y.-M. and Lin, C.-Y. (1993) Immunological kinship of class I low molecular weight heat shock proteins and thermostabilization of soluble proteins *in vitro* among plants. *Plant Cell Physiol.* **34(7):** 1055-1062.

Jung, P. K. and Scott, H. D. (1980) Leaf water potential, stomatal resistance, and temperature relations in field-grown soybeans. *Agronomy J.* **72:** 986-990.

Kelley, P. M. and Freeling, M. (1982) A preliminary comparison of maize anaerobic and heat shock proteins. In *Heat shock: From Bacteria to Man*, M. J. Schlesinger, M. Ashburner and A. Tissières, eds., pp. 315-319. Cold Spring Harbor Laboratory Press, Cold Spring Harbor, New York.

Kelley, P. M. and Schlesinger, M. J. (1978) The effect of amino acid analogues and heat shock on gene expression in chicken embryo fibroblasts. *Cell* **15:** 1277-1286.

Key, J. L., Czarnecka, E., Lin, C.-Y., Kimpel, J., Mothershed, C. and Schöffl, F. (1983a) A comparative analysis of the heat shock response in crop plants. In *Current Topics in Plant Biochemistry and Physiology*, D. D. Randall, D. G. Blevins, R. L. Larson and B. J. Rapp, eds., pp. 107-118. Univ. of Missouri Press, Columbia, Missouri.

Key, J. L., Kimpel, J. A., Lin, C. Y., Nagao, R. T., Vierling, E., Czarnecka, E., Gurley, W. B., Roberts, J. K., Mansfield, M. A. and Edelman, L. (1985a) The heat shock response in soybean. In *Cellular and Molecular Biology of Plant Stress*, J. L. Key and T. Kosuge, eds., pp. 161-179. Alan R. Liss, Inc., New York.

Key, J. L., Kimpel, J. A., Vierling, E., Lin, C.-Y., Nagao, R. T., Czarnecka, E. and Schöffl, F. (1985b) Physiological and molecular analyses of the heat shock response in plants. In *Changes in Eukaryotic Gene Expression in Response to Environmental Stress*, B. G. Atkinson and D. B. Walden, eds., pp. 327-348. Academic Press, Inc., New York.

Key, J. L., Lin, C. Y. and Chen, Y. M. (1981) Heat shock proteins of higher plants. *Proc. Natl. Acad. Sci., USA.* **78:** 3526-3530.

Key, J. L., Lin, C.-Y., Ceglarz, E. and Schöffl, F. (1983b) The heat shock response in soybean seedlings. In *Structure and Functions of Plant Genomes*, O. Ciferri, and L. Dure, eds, NATO ASI Series A, Life Sciences, Vol. 63, pp. 25-36. Plenum Press, New York.

Key, J. L., Lin, C.-Y., Ceglarz, E. and Schsöffl, F. (1982) The heat-shock response in plants: physiological considerations. In *Heat shock: From Bacteria to Man*, M. J. Schlesinger, M. Ashburner and A. Tissères, eds., pp. 329-336. Cold Spring Harbor Laboratory Press, Cold Spring Harbor, New York.

Key, J. L., Nagao, R. T., Czarnecka, E. and Gurley, W. B. (1987) Heat stress: Expression and structure of heat shock protein genes. *NATO Adv. Study Inst. Plant Mol. Biol.* **A140:** 385-397.

Kimpel, J. A. and Key, J. L. (1985) Heat shock in plants. *Trends Biochem. Sci.* **10:** 353-357.

Kloppstech, K., Meyer, G., Schuster, G. and Ohad, I. (1985) Synthesis, transport and localization of a nuclear coded 22-kd heat-shock protein in the chloroplast membranes of peas and *Chlamydomonas reinhardi. EMBO J.* **4:** 1901-1909.

Kyle, D. J., Ohad, I. and Arntzen, C. J. (1985) Molecular mechanisms of compensation to light stress in chloroplast membranes. In *Cellular and Molecular Biology of Plant Stress*, J. L. Key and T. Kosuge, eds., pp. 51-69. Alan R. Liss, Inc., New York.

LaFayette, P. R., Nagao, R. T., O'Grady, K., Vierling, E. and Key, J. L. (1996) Molecular characterization of cDNAs encoding low-molecular-weight heat shock proteins of soybean organelles. *Plant Mol. Biol.* **30:** 159-169.

Lee, Y.-L, Chang, P.-F. L., Yeh, K.-W., Jinn, T.-L., Kung, C.-C. S., Lin, W.-C., Chen, Y.-M. and Lin, C.-Y. (1995) Cloning and characterization of a cDNA encoding an 18.0-kDa class-I low-molecular-weight heat-shock protein from rice. *Gene* **165:** 223-227.

Lemeaux, P. G., Herendeen, S. L., Bloch, P. L. and Neidhardt, F. C. (1978) Transient rates of synthesis of individual polypeptides in *E. coli* following temperature shifts. *Cell* **13:** 427-434.

Lenne, C. and Douce, R. (1994) A low molecular mass heat-shock protein is localized to higher plant mitochondria. *Plant Physiol.* **105**: 1255-1261.

Lewis, M. J., Helmsing, P. and Ashburner, M. (1975) Parallel changes in puffing activity and patterns of protein synthesis in salivary glands of *Drosophila*. *Proc. Natl. Acad. Sci., USA*. **72**: 3604-3608.

Li, G. C. and Werb, Z. (1982) Correlation between synthesis of heat shock proteins and development of thermotolerance in Chinese hamster fibroblasts. *Proc. Natl. Acad. Sci., USA*. **79**: 3213-3222.

Lin, C.-Y. and Key, J. L. (1967) Dissociation and reassembly of polyribosomes in relation to protein synthesis in the soybean root. *J. Mol. Biol.* **26**: 237-247.

Lin, C.-Y., Robert, J. K. and Key, J. L. (1984) Acquisition of thermotolerance in soybean seedlings: synthesis and accumulation of heat shock proteins and their cellular localization. *Plant Physiol.* **74**: 152-160.

Lindquist, S. (1986) The heat-shock response. *Annu. Rev. Biochem.* **55**: 1151-1191

Lindquist, S. and Craig, E. A. (1988) The heat-shock proteins. *Annu. Rev. Genet.* **22**: 631-677.

Lindquist-McKenzie, S. L. and Meselson, M. (1977) Translation *in vitro* of *Drosophila* heat shock messages. *J. Mol. Biol.* **117**: 279-283.

Lindquist-McKenzie, S. L., Henikoff, S., and Meselson, M. (1975) Localization of RNA from heat-induced polysomes at puff sites in *Drosophila melanogaster*. *Proc. Natl. Acad. Sci., USA*. **72**: 1117-1121.

Loomis, W. F. and Wheeler, S. (1980) Heat Shock response of *Dictyostelium*. *Dev. Biol.* **79**: 399-408.

Loomis, W. F. and Wheeler, S. A. (1982) The physiological role of heat-shock proteins in *Dictyostelium*. In *Heat shock: From Bacteria to Man*, M. J. Schlesinger, M. Ashburner and A. Tissières, eds., pp. 353-359. Cold Spring Harbor Laboratory Press, Cold Spring Harbor, New York.

Mansfield, M. (1986) *The Heat Shock Response of Soybean: Synthesis, Accumulation, and Distribution of the Low Molecular Weight Heat Shock Proteins*, Ph.D. dissertation. University of Georgia, Athens, Georgia.

Mansfield, M. A. and Key, J. L. (1987) Synthesis of the low molecular weight heat shock proteins in plants. *Plant Physiol.* **84**: 1007-1017.

Mansfield, M. A. and Key, J. L. (1988) Cytoplasmic distribution of heat shock proteins in soybean. *Plant Physiol.* **86**: 1240-1246.

McAlister, L. and Finkelstein, D. B. (1980) Heat shock proteins and thermal tolerance in yeast. *Biochem. Biophys. Res. Commun.* **93**: 819-824.

McAlister, L., Strausberg, S., Kulaga, A. and Finkelstein, D. B. (1979) Altered patterns of synthesis induced by heat shock in yeast. *Curr. Genet.* **1**: 63-74.

Meyer, Y. and Chartier, Y. (1983) Long-lived and short-lived heat-shock proteins in tobacco mesophyll protoplasts. *Plant. Physiol.* **72**: 26-32.

Miller, M. J., Xuong, N.-H. and Geiduschek, E. P. (1979) A response of protein synthesis to temperature shift in the yeast *Saccharomyces cerevisiae*. *Proc. Natl. Acad. Sci., USA*. **76**: 5222-5225.

Mirault, M. E., Goldschmidt-Clermont, M., Moran, L., Arrigo, A.-P. and Tissières, A. (1978) The effect of heat shock on gene expression in *Drosophila melanogaster*. *Cold Spring Harbor Symp. Quant. Biol.* **42**: 819-827.

Morilla, C. A., Boyer, J. S. and Hageman, R. H. (1973) Nitrate reductase activity and polyribosomal content of corn (*Zea mays* L.) having low leaf water potentials. *Plant Physiol.* **51**: 817-824.

Morimoto, R. I., Tissières, A. and Georgopolous, C. (1990) The stress response, function of the proteins, and perspectives. In *Stress Proteins in Biology and Medicine*, R. I. Morimoto, A. Tissières and C. Georgopolous, eds., pp. 1-36. Cold Spring Harbor Laboratory Press, Cold Spring Harbor, New York.

Morimoto, R. I., Tissières, A. and Georgopolous, C., eds. (1994) *The biology of heat shock proteins and molecular chaperones.* Cold Spring Harbor Laboratory Press, Cold Spring Harbor, New York, p. 610.

Nagao, R. T. and Key, J. L. (1989) Heat shock protein genes of plants. In *Cell Culture and Somatic Cell Genetics of Plants,* I. Vasil and J. Schell, eds, Vol. 6, pp. 297-328. Academic Press, Orlando, Florida.

Nagao, R. T., Czarnecka, E., Gurley, W. B., Schöffl, F. and Key, J. L. (1985) Genes for low-molecular-weight heat shock proteins of soybeans: Sequence analysis of a multigene family. *Mol. Cell. Biol.* **5:** 3417-3428.

Nagao, R. T., Kimpel, J. A., Vierling, E. and Key, J. L. (1986) The heat shock response: A comparative analysis. In *Oxford Surveys of Plant Molecular and Cell Biology,* B. J. Miflin, ed., pp. 384-438. Oxford Univ. Press, Oxford, England.

Neidhardt, F. C. and van Bogelen, R. A. (1981) Positive regulatory gene for temperature-controlled proteins in *Escherichia coli. Biochem. Biophys. Res. Commun.* **100:** 894-900.

Neidhardt, F. C., van Bogelen, R. A. and Vaughn, V. (1985) The genetics and regulation of heat shock proteins. *Annu. Rev. Genet.* **18:** 295-329.

Nover, L. (1987) Expression of heat shock genes in homologous and heterologous systems. In *Enzyme and Microbial Technology,* S. W. May and R. E. Spier, eds., Vol. 9, pp. 129-192. Butterworth, London.

Nover, L. and Scharf, K.-D. (1984) Synthesis, modification and structural binding of heat-shock proteins in tomato cell cultures. *Eur. J. Biochem.* **139:** 303-313.

Nover, L., ed. (1991) *Heat shock response.* CRC Press, Boca Raton, Florida.

Nover, L., Hellmund, D., Neumann, D., Scharf, K.-D. and Serfling, E. (1984) The heat shock response of eukaryotic cells. *Biol. Zentralbl.* **103:** 357-435.

Nover, L., Scharf, K. D. and Neumann, D. (1989) Cytoplasmic heat shock granules are formed from precursor particles and are associated with a specific set of mRNAs. *Mol. Cell. Biol.* **9:** 1298-1308.

Parsell, D. A. and Lindquist, S. (1993) The function of heat-shock proteins in stress tolerance: degradation and reactivation of proteins. *Annu. Rev. Genet.* **27:** 437-496.

Pelham, H. (1987) Properties and uses of heat shock promoter. In *Genetic Engineering,* J. K. Setlow, ed., pp. 27-44. Plenum, New York.

Pitto, L., LoSchiavo, F., Giuliano, G. and Terzi, M. (1983) Analysis of heat-shock protein pattern during somatic embryogenesis of carrot. *Plant Mol. Biol.* **2:** 231-237.

Ritossa, F. M. (1962) A new puffing pattern induced by temperature shock and DNP in *Drosophila. Experientia* **18:** 571-573.

Ritossa, F. M. (1963) New puffs induced by temperature shock, DNP and salicylate in salivary chromosomes of *Drosophila melanogaster. Drosophila Inf. Service* **37:** 122-123.

Ritossa, F. M. (1964) Specific loci in polytene chromosomes of *Drosophila. Exp. Cell. Res.* **35:** 601-607.

Sachs, M. M. and Freeling, M. (1978) Selective synthesis of alcohol dehydrogenase during anaerobic treatment of maize. *Mol. Gen. Genet.* **161:** 111-115.

Sachs, M. M., Freeling, M. and Okimoto, R. (1980) The anaerobic proteins of maize. *Cell* **20:** 761-767.

Scharf, D.-K. and Nover, L. (1982) Heat-shock-induced alternations of ribosomal protein phosphorylation in plant cell cultures. *Cell* **30:** 427-437.

Schlesinger, M. J. (1990) Heat shock proteins. *J. Biol. Chem.* **265:** 12111-12114.

Schlesinger, M. J., Aliperti, G. and Kelley, P. M. (1982b) The response of cells to heat shock. *Trends in Biochem Sci.* **6:** 222-225.

Schlesinger, M. J., Ashburner, M., and Tissières, A., eds. (1982a) In *Heat shock: From Bacteria to Man.* Cold Spring Harbor Laboratory Press, Cold Spring Harbor, New York, p. 450.

Schöffl, F. and Key, J. L. (1982) An analysis of mRNAs for a group of heat shock proteins of soybean using cloned cDNAs. *J. Mol. Appl. Genet.* **1:** 301-314.

Schöffl, F., Raschke, E. and Nagao, R. T. (1984) The DNA sequence analysis of soybean heat-shock genes and idenitification of possible regulatory promoter elements. *EMBO J.* **3:** 2491-2497.

Schrauwen, J. A. M., Reijnen, W. H., DeLeeuw, H. C. G. M. and vanHerpen, M. M. A. (1986) Response of pollen to heat stress. *Acta Bot. Neerl.* **35:** 321-327.

Spradling, A., Pardue, M. L. and Penman, S. (1977) Messenger RNA in heat-shocked *Drosophila* cells. *J. Mol. Biol.* **109:** 559-587.

Spradling, A., Penman, S. and Pardue, M. L. (1975) Analysis of *Drosophila* mRNA by *in situ* hybridization: Sequences transcribed in normal and heat shock cultured cell. *Cell* **4:** 395-404.

Sticher, L., Biswas, A. K., Bush, D. S. and Jones, R. L. (1990) Heat shock inhibits •-amylase synthesis in barley aleurone without inhibiting the activity of endoplasmic reticulum marker enzymes. *Plant Physiol.* **92:** 506-513.

Suss, K.-H. and Yordanov, I. T. (1986) Biosynthetic cause of *in vivo* acquired thermotolerance of photosynthetic light reactions and metabolic responses of chloroplasts to heat stress. *Plant Physiol.* **81:** 192-199.

Tilly, K., McKittrick, N., Zylicz, M. and Georgopoulos, C. (1983) The dnaK protein modulates the heat-shock response of *Escherichia coli. Cell* **34:** 641-646.

Tissières, A., Mitchell, H. K. and Tracy, V. M. (1974) Protein synthesis in salivary glands of *Drosophila melanogaster:* Relation to chromosome puffs. *J. Mol. Biol.* **84:** 389-398.

Tseng, T. S., Tzeng, S. S., Yeh, K. W., Yeh, C. H., Chang, F. C., Chen, Y. M. and Lin, C. Y. (1993) The heat-shock response in rice seedlings: isolation and expression of cDNAs that encode class I low-molecular-weight heat-shock proteins. *Plant Cell Physiol.* **34 (1):** 165-168.

Tseng, T. S., Yeh, K. W., Yeh, C. H., Chang, F. C., Chen, Y. M. and Lin, C. Y. (1992) Two rice (*Oryza sativa*) full-length cDNA clones encoding low-molecular-weight heat-shock proteins. *Plant Mol. Biol.* **18:** 963-965.

Tzeng, S.-S., Chen, Y.-M. and Lin, C.-Y. (1993) Isolation and characterization of genes encoding 16.9 kD heat shock proteins in *Oryza sativa. Bot. Bull. Acade. Sin.* **34:** 133-142.

Tzeng, S.-S., Yeh, K.-W., Chen, Y.-M. and Lin, C.-Y. (1992) Two *Oryza sativa* genomic DNA clones encoding 16.9-kilodalton heat-shock proteins. *Plant Physiol.* **99:** 1723-1725.

van Herpen, M. M. A., Reijnen, W. H., Schrauwen, J. A. M., de Groot, P. F. M., Jager, J. W. H. and Wullems, G. J. (1989) Heat shock proteins and survival of germinating pollen of *Lilium longiflorum* and *Nicotiana tabacum. J. Plant Physiol.* **134:** 345-351.

Velazquez, J., DiDomenico, B. and Linquist, S. (1980) Intracellular localization of heat shock proteins in *Drosophila. Cell* **20:** 679-689.

Velazquez, J., Sonoda, S., Bugaisky, G. E. and Lindquist, S. (1983) Are HS proteins present in cells that have not been heat shocked. *J. Cell Biol.* **96:** 286-290.

Vierling, E. (1991) The roles of heat shock proteins in plants. *Annu. Rev. Plant Physiol. Plant Mol. Biol.* **42:** 579-620.

Vierling, E. and Key, J. L. (1985) Ribulose 1,5-bisphosphate carboxylase synthesis during heat shock. *Plant Physiol.* **78:** 155-162.

Vierling, E. and Sun, A. (1989) Developmental expression of heat shock proteins in higher plants. In *Environmental Stress in Plants,* J. Cherry, ed., pp. 343-354, Springer-Verlag, Berlin.

Vierling, E., Mishkind, M. L., Schmidt, G. W. and Key, J. L. (1986) Specific heat shock proteins are transported into chloroplasts. *Proc. Natl. Acad. Sci., USA.* **83:** 361-365.

Vierling, E., Nagao, R. T., DeRocher, A. E. and Harris, L. M. (1988) A heat shock protein localized to chloroplasts is a member of a eurkaryotic superfamily of heat shock proteins. *EMBO J.* **7:** 575-581.

Walker, J. C. and Key, J. L. (1982) Isolation of cloned cDNAs to auxin-responsive poly(A)+RNAs of elongating soybean hypocotyl. *Proc. Natl. Acad. Sci., USA.* **79:** 7185-7189.

Walsh, C. (1980) Appearance of heat shock proteins during the induction of multiple flagella in *Naegleria gruberi. J. Biol. Chem.* **225:** 2629-2632.

Wang, C. H., Gomer, R. H. and Lazarides, E. (1981) Heat shock proteins are methylated in avian and mammalian cells. *Proc. Natl. Acad. Sci., USA.* **78**: 3531-3535.

Waters, E. R., Lee, G. J. and Vierling, E. (1996) Evolution, structure and function of the small heat shock proteins in plants. *J. Exp. Bot.* **47**: 325-338.

Welch, W. J. (1993) How cells respond to stress. *Scientific American* **268**: 34-41.

White, F. P. (1980.) Differences in protein synthesized *in vivo* and *in vitro* by cells associated with the cerebral microvascalature. A protein synthesized in response to trauma? *Neuroscience* **5**: 1793-1799.

Xiao, C.-M. and Mascarenhas, J. P. (1985) High temperature-induced thermotolerance in pollen tubes of *Tradescantia* and heat-shock proteins. *Plant Physiol.* **78**: 887-890.

Yamamori, T., Ito, K., Nakamura, Y. and Yura, T. (1978) Transient regulation of protein synthesis in *Escherichia coli* upon shift-up of growth temperature. *J. Bacteriol.* **134**: 1133-1140.

Yeh, C.-H., Chang, P.-F. L., Yeh, K.-W., Lin, W.-C., Chen, Y.-M. and Lin, C.-Y. (1997) Expression of a gene encoding a 16.9-kDa heat-shock protein, Oshsp 16.9, in *Escherichia coli* enhances thermotolerance. *Proc. Natl. Acad. Sci., USA.* **94**: 10967-10972.

Yeh, C.-H., Yeh, K.-W., Wu, S.-H., Chang, P.-F. L., Chen, Y.-M. and Lin, C.-Y. (1995) A recombinant rice 16.9-kDa heat shock protein can provide thermoprotection *in vitro*. *Plant Cell Physiol.* **36(7)**: 1341-1348.

Yeh, K.-W., Jinn, T.-L., Yeh, C.-H., Chen, Y.-M. and Lin, C.-Y. (1994) Plant low-molecular-mass heat-shock proteins: their relationship to the acquisition of thermotolerance in plants. *Biotechnol. Appl. Biochem.* **19**: 41-49.

Zimmerman, J. L., Apuya, N., Darwish, K. and O'Carroll, C. (1989) Novel regulation of heat shock genes during carrot somatic embryo development. *Plant Cell* **1**: 1137-1146.

Chapter 19

Discovery of Photoregulated Gene Expression

John C. Watson
Department of Biology
Indiana University – Purdue University at Indianapolis
723 West Michigan Street, Indianapolis, IN 46404-5132, USA

ABSTRACT

Light exerts a profound influence on the growth and development of green plants. Many of the developmental responses of plants to light are initiated by activation of the photoreceptor phytochrome. In this chapter, emphasis is given to the pioneering studies establishing that light acting through phytochrome triggers differential expression of specific nuclear genes.

Introduction

Since most plants are sessile organisms, they must be able to sense and respond to changes in the environment to survive. Indeed, plants have evolved mechanisms that use environmental signals to regulate their growth and development. Light is perhaps the most important of the environmental factors that influence plant development. Apart from its role as the energy source for photosynthesis, light provides crucial environmental cues that control virtually every stage of the plant life cycle, from seed germination, through stem and leaf development, to flowering. Perhaps at no time are the effects of light on plant development more apparent then when angiosperm seedlings are grown in its absence. Etiolation, the response to growth in darkness, results in greatly elongated stems, a reduction or arrest in leaf development, and no chlorophyll accumulation. Complex morphological and biochemical changes occur when etiolated seedlings are exposed to light. One major consequence of de-etiolation (greening) is the assembly of the photosynthetic apparatus. The ability of an etiolated seedling to sense and respond to the light environment, and so trigger the transition to photosynthetic growth, is pivotal to its establishment and survival. This is particularly true for species whose seeds germinate deep within the soil, as with many important crop plants. Thus, de-etiolation has important ecological and agricultural consequences.

Plants are able to distinguish several important features of their light environment, such as wavelength, intensity, duration, and direction. To accomplish this, plants contain distinct photoreceptor families that allow light to cue plant responses during photomorphogenesis. These are the phytochromes (primarily for the red and far-red light), blue light photoreceptors, a UV-A photoreceptor, and a UV-B photoreceptor(s). Of the photoreceptors active in the visible portion of the spectrum, the phytochromes are the best understood. Phytochrome is a photointerconvertible pigment: it is synthesized in the dark as a red light absorbing form, called Pr, and upon illumination with red light is converted to a far-red light absorbing form, called Pfr. Far-red light drives Pfr back to the Pr form, thus providing the mechanism for the classic red/far-red reversibility of many phytochrome-mediated responses (see Fig. 1). Indeed, red/far-red reversiblilty is a powerful diagnostic tool to determine whether phytochrome controls a particular response that is often used with etiolated seedlings.

In his pioneering studies on phototropism, Charles Darwin (1881) siezed upon the utility of using seedlings as experimental material because they can obtained quickly. Moreover, he found that dark-grown seedlings are exquisitely sensitive to low levels of light. Still to this day, many plant biologists use seedlings in their investigations, and as can be seen below, elucidating how seedlings respond to light played an important part in the discovery of photoregulated gene expression. Following the discovery of phytochrome (Briggs, 1998), we have come to realize that the phytochrome sets in motion many of the events that we observe during de-etiolation in white light. It may seem ironic, however, that in many experiments on plant photoresponses, white light effects are characterized first, followed by an assessment of phytochrome involvement. We now realize that the profound changes in form and function that occur during de-etiolation are driven by major changes in the patterns of gene expression and protein synthesis. At least in my view, it might be inaccurate to describe the discovery of photoregulated gene expression in terms of one or two seminal experiments. Rather, the "discovery" begins in the 1950s, spans nearly three decades, and is closely linked with efforts to understand plant development and photoreceptors.

Light Affects Protein Levels and Enzyme Activities

Even a casual inspection of seedlings that have been transferred from darkness to white light tells the observer that chlorophyll accumulates and the growth of leaves is stimulated. In the 1950s, investigations began to ask whether other cellular constituents other than chlorophyll, such as proteins and enzymes, were affected. DeDeken-Grenson (1954) found that exposing etiolated chicory seedlings to white light caused a substantial accumulation of protein. As for enzyme activity, Hageman and Arnon (1955) found that the activity of glyceraldehyde-3-phosphate dehydrogenase increased during greening of pea seedlings. Subsequently, Marcus (1960) showed that illuminating kidney bean seedlings with a short pulse of red light triggered an increase in the activity of the enzyme, an effect that was reversed by far-red light. So we knew for the first time that phytochrome regulated enzyme activity. Over the ensuing twenty years, dozens of enzymes were reported to be regulated by light (see Schopfer, 1997; Lamb and Lawton, 1983), and in several cases the altered activity was associated with changes in the amount or synthesis of enzyme protein.

Light and the Gene Expression Machinery

With the 1960s came analyses of plant nucleic acids and the development of cell-free systems for protein and RNA synthesis. Since light can cause changes in the amount of protein, one can reasonably ask if it can also trigger changes in RNA and the ability to synthesize protein. With etiolated pea seedlings, Jaffe (1969) found that one day after a red light pulse, both the RNA content and ribosome fraction increased within the apical buds. Phytochrome was involved since the effect of red was reversible with far-red. With regard to protein synthesis, Williams and Novelli (1964) found that ribosomes isolated from etiolated maize, bean, and soybean seedlings illuminated with white light had an increased ability to incorporate amino acids into protein *in vitro* compared to their dark counterparts. Subsequently, they went on to show that brief treatments with red light were considerably more effective than blue or green light, and nearly as effective as white light (Williams and Novelli, 1968). Moreover, the stimulation was still observed when exogenous poly(U) was added to direct the incorporation of phenyalanine. At least as judged by the cell-free incorporation system, the ribosomes themselves were activated by treating the seedlings with light. Using the red/far-red reversibility criterion, Travis *et al.* (1974) convincingly showed that phytochrome mediates the stimulation of cytoplasmic 80S ribosome activity. Thus, the elevation in protein levels elicited by light might be facilitated by increasing ribosome activity.

Willams and Novelli (1968) also found that the polysome content increased after treating etiolated seedlings with light. Since changes in polysomes were likely to involve mRNA, another question emerged: does the increase in the polysome to monoribosome ratio reflect an increase in the amount of mRNA, or an increase in the availability of stored mRNA? In the early 1970s, it was shown that the majority of eukaryotic mRNAs are polyadenylated at their 3'-ends. Thus, it was possible to isolate this fraction from total cellular RNA by affinity chromatography on oligo(dT)-cellulose, and to measure the amount of poly(A) easily by hybridization with radioactive poly(U). Using these methods, Grierson and Covey (1975) found that exposing etiolated mung bean seedlings to white light stimulated the accumulation of poly(A)$^+$ RNA in the primary leaves. This suggested that the supply of mRNA template increased during greening. Definitive evidence that the phytochrome-mediated increase in the polysome/monosome ratio in mustard seedlings resulted from rapid increases in the amount of mRNA was provided by Mösinger and Schopfer (1983).

Light and Differential Gene Expression

In the mid-1960s, a model was proposed by Mohr's laboratory (Hock and Mohr, 1964; Mohr, 1966) that most likely was both startling and controversial at the time. The basic hypothesis was that the mode of action of phytochrome in regulating seedling development was through differential gene expression. The model was based largely on Mohr *et al.*'s analyses of photomorphogenesis in seedlings of white mustard (*Sinapis alba* L.; see Mohr, 1966). As in other dicotyledonous seedlings, exposing etiolated mustard seedlings to light, inhibited hypocotyl elongation and stimulated cotyledon expansion in a phytochrome-dependent manner. In contrast to some species, phytochrome also regulated anthocyanin accumulation in mustard seedlings. Mohr's group found that phytochrome mediated an increase in both protein and RNA contents that was accompanied by an increase in the incorporation of radioactive precursors. Importantly, the RNA synthesis inhibitor — actinomycin D prevented the phytochrome-dependent expansion of cotyledons and anthocyanin accumulation, and the protein synthesis inhibitor — puromycin blocked the rise in anthocyanin. These observations were consistent with the notion that gene expression must be involved.

The model, as summarized in Fig. 1, began with the quite reasonable supposition that the 'pattern of "primary" differentiation' of an etiolated seedling would be determined by those genes that were actively expressed at that stage of development. The genes in the genome were thought of as being of four types: (1) active genes, whose expression did not change during seedling development; (2) inactive genes that were not expressed at any time in the seedling; (3) repressible genes — genes that were active in darkness but whose expression could be repressed by phytochrome; and (4) potentially active genes — genes that were inactive during etiolation but whose expression could be induced by phytochrome. The induction or repression of enzyme synthesis by phytochrome was proposed to result from the switching on or off of the appropriate genes. This in turn would account for the developmental changes triggered by phytochrome (the 'pattern of "secondary" differentiation'). The model also included the idea that the formation of Pfr would initiate a signal chain leading to the nucleus, where the effect on the potentially active genes and repressible genes would occur.

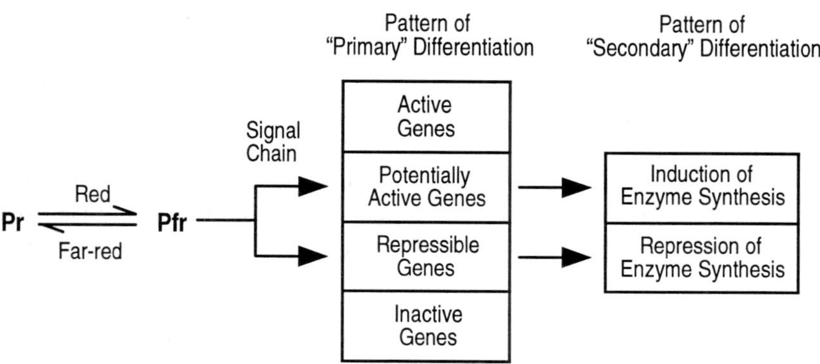

Fig. 1. Mohr's model on differential gene expression triggered by phytochrome. The details of the model are described in the text. Redrawn from Mohr (1966 & 1972).

Put into today's parlance, the model predicts that the phytochrome signal transduction chain causes transcriptional induction or repression of specific nuclear genes. The transcriptional events lead to changes in mRNA levels, which in turn lead to alteration in protein levels, which plays a causal role in the developmental transition initiated by phytochrome. The excellent experiments described below show that the major points of this remarkable model are indeed accurate.

Light Induces the Synthesis of Certain RNAs

One way to test Mohr's model would be to compare directly the amount of the mRNA encoding a particular protein from etiolated seedlings with that from light treated seedlings. However, in the late 1960s and early 1970s, the experimental tools did not exist to examine the abundance of a specific mRNA species. Nevertheless, DNA:RNA hybridization methods existed to compare RNA populations. Thompson and Cleland (1972) provided us with a seminal finding by using competition hybridization to compare the RNAs from dark-grown and greening pea seedlings. In this type of competition hybridization, the amount of identity between an unlabeled competitor RNA and a radiolabled reference RNA is judged by the extent to which the competitor reduces hybridization of the reference RNA to filter-bound DNA. If the unlabeled competitor is from the same source as the labeled RNA (or is identical in sequence content), then "complete" competition occurs and only very little hybridization occurs. On the other hand, if the competitor contains RNA sequences that are absent from the reference RNA, then the competition curve reaches a plateau above that obtained with unlabeled reference RNA. In the Thompson and Cleland experiments, denatured pea DNA was bound to a filter, and then hybridized with RNA from 6 to 7 day-old seedlings exposed to white light for 30 hours. During illumination, the RNA was labeled by feeding the seedlings [^{32}P]-phosphate so the extent of hybridization could be measured easily. Unlabeled RNA from etiolated seedlings was added to the standard hybridization reactions in increasing amounts as the competitor, and compared with competitor RNA from greening seedlings. The results showed that RNA from dark-grown seedlings failed to compete fully with the RNA from greening seedlings. Clearly then, we knew for the first time that the illumination of seedlings led to the synthesis of RNA species that were absent in the dark.

Photoregulation of Translatable mRNA

In the early 1970s, convenient cell-free translation systems were developed that were dependent on exogenous mRNA to drive protein synthesis. These *in vitro* translation systems consist of an extract, usually from either wheat germ or rabbit reticulocytes, containing ribosomes and the other macromolecular components needed for translation. Protein is synthesized when the extract is supplemented with an ATP generating system, amino acids, and an mRNA template.   Usually, the mRNA is supplied as a polyadenylated RNA [poly(A)$^+$] fraction, and one of the amino acids is radioactively labeled (most often [^{35}S]-methionine is used).   Separating the polypeptides labeled *in vitro* by gel electrophoresis, followed by fluorography of the gel, allows the detection of those polypeptides encoded by relatively abundant mRNAs.   These methods were employed by Tobin and Klein (1975) to ask whether there were differences in the ability of poly(A)$^+$ RNA from dark adapted *Lemna gibba* and plants exposed to white light for one day after dark adaptation to direct protein synthesis *in vitro*. Of the labeled polypeptides produced *in vitro*, a few were more intensely labeled using RNA from light treated plants as template.   Moreover, a few polypeptides were more intensely labeled with the dark treated RNA sample.   The implication of these results was important: light exerted a differential effect on the abundance (or translatability) of particular mRNAs.

It was, of course, desirable to know the identity of proteins encoded by mRNAs exhibiting a differential light effect.   This was approached by combining immunological methods with *in vitro* translation.   If an antibody to a protein of interest is available, then one can immunoprecipitate the labeled protein from an *in vitro* translation reaction, and so measure the specific protein synthesized from its translatable mRNA.   Studies on four proteins led the way to understanding differential expression during photomorphogenesis.   These four proteins shared three significant features: (1) they were sufficiently abundant to purify biochemically (a necessary prerequisite for generating antibodies), (2) they had important cellular functions, and (3) the levels of these proteins were known to change in a phytochrome-dependent   manner   when   dark-grown   plants   were illuminated.   The   small   subunit   of   ribulose-1,5-bisphosphate carboxylase/oxygenase (RbcS) and the light harvesting chlorophyll *a/b*-binding apoprotein (Cab) photosystem II accumulated during de-etiolation.   NADPH-protochlorophyllide   oxidoreductase   (Por)   and phytochrome declined in amount.   Although all four are encoded by nuclear genes, the RbcS, Cab, and Por proteins are localized in chloroplasts while phytochrome is not.

With antibodies in hand, it was possible to measure the levels of specific translatable mRNAs.   In turn, two fundamental questions could be answered.   Did illumination with white light trigger an alteration in translatable mRNA?   If so, could phytochrome action be demonstrated by showing red/far-red reversibility?   Not surprisingly, publications on the up-regulated proteins appeared before those dealing with the down-regulated ones. Thus, Apel and Kloppstech (1978) found that exposing etiolated barley seedlings to white light strongly promoted the accumulation of translatable *Cab* mRNA. Likewise, Tobin (1978) reported that white light triggered substantial accumulation of translatable *RbcS* mRNA in *Lemna*. To test for phytochrome control, Apel (1979) found that a brief pulse of red light given to etiolated barley seedlings induced the accumulation of translatable *Cab* mRNA. Because a far-red light pulse immediately following the red light pulse effectively reversed the effect of red alone, the response was clearly mediated by phytochrome. Subsequently, similar results for translatable *RbcS* mRNA in *Lemna* were reported by Tobin (1981). As for the proteins that showed negative regulation by light, Apel (1981) studied translatable *Por* mRNA in barley seedlings and found that white triggered a decline, and the response exhibited red/far-red reversibility. Translatable phytochrome mRNA in oat seedlings was shown to decline in response to white light by Gottmann and Schäfer (1982). This result was confirmed and extended by Colbert *et al.* (1983), who also showed red/far-red reversibility. These experiments with phytochrome mRNA were particularly exciting, because they demonstrate autoregulation of phytochrome expression. Not only can phytochrome trigger dramatic shifts in the translatable mRNAs encoding other genes, it can do so for its own gene!

Light Alters Specific Transcript Abundance

It seemed extremely likely that the differences in translatable mRNA levels just described reflected alterations in amount of these mRNAs within the total mRNA population. Although much less likely, an alternative explanation was formally possible: that light might in some way change the efficiency with which these mRNAs could direct protein synthesis *in vitro*. An excellent way to measure mRNA abundance directly would be to employ nucleic acid hybridization. Purifying specific mRNAs by biochemical means was, at best, impractical. However, with the advent of complementary DNA (cDNA) cloning techniques, it became possible to "purify" mRNAs by molecular cloning. Since expression vectors had yet to be constructed, identifying cDNA clones encoding a protein of interest was tremendously laborious. Generally, a method called "hybrid selection" was used. For this, a series of individual clones were hybridized to an mRNA preparation, the mRNA "selected" by each clone was recovered, translated *in vitro*, and the protein products tested by immunoprecipitation using an antibody directed against the protein of interest. Once an appropriate clone was characterized, its cDNA insert could be used as a specific hybridization probe to detect its corresponding mRNA. The DNA was [^{32}P]-labeled to high specific activity *in vitro*, denatured, and then hybridized in probe excess to filters containing either total cellular RNA or poly(A)$^+$ RNA. With careful controls for how much RNA was loaded onto the filters, one could measure the specific transcript as a percentage of total RNA in the sample.

Armed with cDNA clones, several laboratories began asking whether light changed the abundance of specific transcripts. Smith and Ellis (1981), employing the cDNA:RNA hybridization approach just described, reported that illumination of etiolated pea seedlings with white light caused *RbcS* transcripts to accumulate. Subsequently, similar white light effects were reported for *Cab* transcripts in barley (Gollmer and Apel, 1983), and both *RbcS* and *Cab* transcripts in peas (Thompson *et al.*, 1983; Jenkins *et al.*, 1983), mung bean (Thompson *et al.*, 1983), and *Lemna* (Stiekema *et al.*, 1983). Moreover, all these authors found that phytochrome controlled transcript accumulation by showing red/far-red reversibility. As for *Por* transcripts, Batschauer and Apel (1984) showed that white light causes them to decline and showed photoreversibility mediated by phytochrome. Likewise, Colbert *et al.* (1985) found similar responses for phytochrome transcripts in oat seedlings. Taken together, the elegant work just described convincingly showed that phytochrome mediates increased expression of some genes at the mRNA level, but decreased expression of others. After almost 20 years of stepwise progress, we had obtained solid experimental confirmation of the main facet of Mohr's model (see Fig. 1). Phytochrome does indeed trigger differential gene expression.

Transcriptional or Post-transcriptional Regulation?

Knowing that specific mRNA levels change during a developmental process begs the question of the mechanism underlying the change. Does the rate of gene transcription change? If not, does some post-transcriptional event dictate the change in mRNA abundance? Are both transcriptional and post-transcriptional controls at work? Although not without vagaries, one way to examine the synthesis of pre-mRNAs is through run-on transcription in isolated nuclei. This method allows engaged RNA polymerases to continue RNA synthesis after the nuclei are isolated. Nuclei are pulse labeled with a radioactive RNA percursor, the labeled nuclear RNA is then hybridized to filters containing cloned DNA of the genes of interest. Using run-on transcription, Gallagher and Ellis (1982) were the first to show that light increased the transcription of specific genes. They showed that nuclei from light-grown pea seedlings exhibited considerably higher transcription of the *RbcS* and *Cab* genes than nuclei from etiolated seedlings. With *Lemna*, Silverthorne and Tobin (1984) demonstrated phytochrome control of *RbcS* and *Cab* transcription using the run-on assay. Subsequently, phytochrome was shown to mediate a decline in transcription of the *Por* (Mösinger *et al.*, 1985) and phytochrome genes (Colbert *et al.*, 1985). Although these experiments did not rule out the role of post-transcriptional events for photoregulated changes in transcript abundance, they showed that, at least for the *RbcS*, *Cab*, *Por*, and phytochrome gene, transcriptional regulation plays a major role. Moreover, phytochrome exerts an influence on nuclear events (i.e., transcription). Once again, this harkens back to Mohr's model, and the prediction that a signal emanating from phytochrome would enter the nucleus and regulate gene expression.

Conclusions

The elegant studies described above laid a solid foundation for the explosion of work that has occurred in recent years. Continued studies of the four genes described above, and the many photoregulated genes that were isolated and characterized since, showed that an intriguing diversity of response patterns exists. We learned that blue and UV light photoreceptors, in addition to phytochrome, regulated the expression of certain genes. Transcriptional control came under intense scrutiny, so that we have an understanding of the promoter elements that confer light responsiveness and the transcription factors that recognize those elements. We have seen an ever increasing awareness of the importance of post-transcriptional events in photoregulated mRNA accumulation. Lastly, at least some components of the signal transduction have been identified. We have indeed come a long way since DeDeken-Grenson reported in 1954 that treating etiolated seedlings with light caused protein to accumulate.

Acknowledgements

I thank Gene R. Williams, for it was in his laboratory that I first became interested in de-etiolation, and Winslow R. Briggs, Gerald F. Deitzer, Lon S. Kaufman, Stefan J. Surzycki, and William F. Thompson for so many "illuminating" discussions over the years.

References

Apel, K. (1979) Phytochrome-induced appearance of mRNA activity for the apoprotein of the light-harvesting chlorophyll *a/b* protein of barley (*Hordeum vulgare*). *Eur. J. Biochem.* **97:** 183-188.

Apel, K. (1981) The protochlorophyllide holochrome of barley (*Hordeum vulgare* L.). Phytochrome-induced decrease of translatable mRNA coding for the NADPH:prochlorophyllide oxidoreductase. *Eur. J. Biochem.* **120:** 89-93.

Apel, K. and Kloppstech, K. (1978) The plastid membranes of barley (*Hordeum vulgare*). Light-induced appearance of mRNA coding for the apoprotein of the light-harvesting chlorophyll *a/b* protein. *Eur. J. Biochem.* **85:** 581-588.

Batschauer, A. and Apel, K. (1984) An inverse control by phytochrome of the expression of two nuclear genes in barley (*Hordeum vulgare* L.). *Eur. J. Biochem.* **143:** 593-597.

Briggs, W. (1998) Discovery of phytochrome. In *Discoveries in Plant Biology, Vol. 2*, Kung, S. D. and Yang, S. F., eds., pp. 115-135, World Scientific Publishing, Pte., Singapore.

Colbert, J. T., Hershey, H. P. and Quail, P. H. (1983) Autoregulatory control of translatable phytochrome mRNA levels. *Proc. Natl. Acad. Sci., USA.* **80:** 2248-2252.

Colbert, J. T., Hershey, H. P. and Quail, P. H. (1985) Phytochrome regulation of phytochrome mRNA abundance. *Plant Mol. Biol.* **5:** 91-101.

Darwin, C. (1881) *Power of Movement in Plants.* D. Appleton, New York.

DeDeken-Grenson, M. (1954) Grana formation and synthesis of chloroplastic proteins induced by light in portions of etiolated leaves. *Biochim. Biophys. Acta* **14:** 203-211.

Gallagher, T. F. and Ellis, R. J. (1982) Light-stimulated transcription of genes for two chloroplast polypeptides in isolated pea leaf nuclei. *EMBO J.* **1:** 1493-1498.

Gollmer, I. and Apel, K. (1983) The phytochrome-controlled accumulation of mRNA sequences encoding the light-harvesting chlorophyll *a/b* protein of barley (*Hordeum vulgare* L.). *Eur. J. Biochem.* **133:** 309-313.

Gottmann, K. and Schäfer, E. (1982) *In vitro* synthesis of phytochrome apoprotein directed by mRNA from light and dark grown *Avena* seedlings. *Photochem. Photobiol.* **35:** 521-525.

Grierson, D. and Covey, S. (1975) Changes in the amount of ribosomal RNA and poly(A)-containing RNA during leaf development. *Planta* **127:** 77-86.

Hageman, R. H. and Arnon, D. I. (1955) Changes in glyceraldehyde phosphate dehydrogenase during the life cycle of a green plant. *Arch. Biochem. Biophys.* **57:** 421-436.

Hock, B. and Mohr, H. (1964) Die regulation der O_2-aufnehme von senfkeinlingen (*Sinapis alba* L.) durch licht. *Planta* **61:** 209-228.

Jaffe, M. J. (1969) Phytochrome controlled RNA changes in terminal buds of dark grown peas (*Pisum sativum* cv. Alaska). *Plant Physiol.* **22:** 1033-1037.

Jenkins, G. I., Hartley, M. R. and Bennett, J. (1983) Photoregulation of chloroplast development: transcriptional, translational and post-translational controls? *Phil. Trans. R. Soc. Lond. B* **303:** 419-431.

Lamb, C. J. and Lawton, M. A. (1983) Photocontrol of gene expression. In *Encylopedia of Plant Physiology, New Series*, Vol. 16A, Shropshire, W. and Mohr, H., eds., pp. 213-287, Springer-Verlag, Berlin.

Lissemore, J. L. and Quail, P. H. (1988) Rapid transcriptional regulation by phytochrome of the genes for phytochrome and chlorophyll *a/b*-binding protein in *Avena sativa*. *Mol. Cell. Biol.* **8:** 4840-4850.

Marcus, A. (1960) Photocontrol of formation of red kidney bean leaf triosphosphopyridine nucleotide linked triosephosphate dehydrogenase. *Plant Physiol.* **35:** 126-128.

Mohr, H. (1966) Differential gene activation as a mode of action of phytochrome 730. *Photochem. Photobiol.* **5:** 469-483.

Mohr, H. (1972) *Lectures on Photomorphogenesis.* Springer-Verlag, New York.

Mösinger, E. and Schopfer, P. (1983) Rolysome assembly and RNA synthesis during phytochrome-mediated photomorphogenesis in mustard cotyledons. *Planta* **158**: 501-511.

Mösinger, E., Batschauer, A., Schäfer, E. and Apel, K. (1985) Phytochrome control of *in vitro* transcription of specfic genes in isolated nuclei from barley (*Hordeum vulgare*). *Eur. J. Biochem.* **147**: 137-142.

Schopfer, P. (1977) Phytochrome control of enzymes. *Ann. Rev. Plant Physiol.* **28**: 223-252.

Silverthorne, J. and Tobin, E. M. (1984) Demonstration of transcriptional regulation of specific genes by phytochrome action. *Proc. Natl. Acad. Sci., USA.* **81**: 1112-1116.

Smith, S. M. and Ellis, R. J. (1981) Light-stimulation accumulation of transcripts of nuclear and chloroplast genes for ribulosebisphosphate carboxylase. *J. Mol. Appl. Genet.* **1**: 127-137.

Stiekema, W. J., Wimpee, C. F., Silverthorne, J. and Tobin, E. M. (1983) Phytochrome control of the expression of two nuclear gene encoding chloroplast proteins in *Lemna gibba* G-3. *Plant Physiol.* **72**: 717-724.

Thompson, W. F. and Cleland, R. (1972) Effect of light and gibberellin on ribonucleid acid species of pea stem tissues as studied by deoxyribonucleic acid-ribonucleic acid hybridization. *Plant Physiol.* **50**: 289-292.

Thompson, W. F., Everett, M. L., Polans, N. O., Jorgensen, R. A. and Palmer, J. D. (1983) Phytochrome control of RNA levels in developing pea and mung-bean leaves. *Planta* **158**: 487-500.

Tobin, E. M. (1978) Light regulation of specific mRNA species in *Lemna gibba* L. G-3. *Proc. Natl. Acad. Sci., USA.* **75**: 4749-4753.

Tobin, E. M. (1981) Phytochrome-mediated regulation of messenger RNAs for the small subunit of ribulose 1,5-bisphosphate carboxylase and the light-harvestign chlorophyll *a/b*-protein in *Lemna gibba. Plant Mol. Biol.* **1**: 35-51.

Tobin, E. M. and Klein, A. O. (1975) Isolation and translation of plant messenger RNA. *Plant Physiol.* **56**: 88-92.

Travis, R. L., Key, J. L. and Ross, C. L. (1974) Activation of 80S maize ribosomes by red light treatment of dark-grown seedlings. *Plant Physiol.* **53**: 28-31.

Williams, G. R. and Novelli, G. D. (1964) Stimulation of an in vitro amino acid incorporating system by illumination of dark-grown plants. *Biochem. Biophys. Res. Comm.* **17**: 23-27.

Williams, G. R. and Novelli, G. D. (1968) Ribosome changes following illumination of dark-grown plants. *Biochim. Biophys. Acta* **155**: 183-192.

Chapter 20

Organization and Regulation of Nitrogen Fixation Genes: 1974-1995

San Chiun Shen
*Laboratory of Molecular Genetics, Shanghai Institute of Plant Physiology
Academia Sinica, Shanghai 200032*

ABSTRACT

Studies of the nitrogen fixation (*nif*) genes began in the Laboratory of Molecular Genetics in 1974. I was privileged to work with a group of young scientists from different areas of biology and chemistry, on what was regarded a fancy subject in those days. Results of fine-structure mapping and complementation tests of *nif* genes demonstrated that the *nif* genes of *Klebsiella pneumoniae* reside as a single cluster near the histidine operon (*his*) on the chromosome. Intragenic complementation occurred between two *nifJ* mutants, indicating the *nifJ* product as a dimeric protein of identical subunits.

For study of the regulation of *nif* genes, a test system for the role of *nifA* product NifA on *nif* gene expression was created. NifA was proven to be the key protein in the regulation of *nif* genes. There was evidence that the repressive action of the *nifL* gene product, NifL, is in its interaction with NifA. A two-level regulation hypothesis for NifA production has been proposed. Under oxygen, the transcription of *nifLA* operon is repressed, and NifA produced will be inactivated by its interaction with NifL. This hypothesis is now substantiated by experiments.

In *Rhizobium meliloti*, *nod/nif* genes were shown to express sequentially during the development of the rhizobia. NifA functions both for nodulation and nitrogen fixation.

Back to My Primary Interest, Genetics

In the beginning of this century, most of the investigations on biological nitrogen fixation were focused on the enzymology of the process. Attempts to study the genetic basis of nitrogen fixation met great difficulties. It is because nitrogen fixation by many nitrogen-fixing microbes, especially the symbiotic bacteria *Rhizobium*, involves complex interactions between rhizobia and host plants, thus the prospect for research was not good.

In 1971, S. Streicher, E. Gurney, and R. Valentine published their work on the nitrogen fixation (*nif*) genes in *Klebsiella pneumoniae* (Streicher *et al.*, 1971). It was exciting to demonstrate for the first time the *nif* genes. Also, this area of research was facilitated by the correct choice of nitrogen-fixing bacteria to work with — *K. pneumoniae* (close relative to *Escherichia coli*). With its similarity of genetic background, including the genetic map of *E. coli*, *K. pneumoniae* has been proven as a good model for the study of nitrogen fixation genetics. R. A. Dixon and J. R. Postgate of University of Sussex in England, successfully transferred the cluster of *nif* genes from *K. pneumoniae* into *E. coli* to yields of *E. coli* capable of fixing nitrogen (Dixon *et al.*, 1972). Furthermore, the elaborate genetic manipulation developed in *E. coli* can be readily applied to *K. pneumoniae*. So finally, a cluster of *nif* genes was mapped near the histidine operon on the *K. pneumoniae* chromosome (MacNeil *et al.*, 1978). Since then *K. pneumoniae* became a model organism for studying the structure, regulation and conservation of *nif* genes. In pursuing the knowledge of the *nif* genome of agriculturally important species, studies on *nif* genes of *Rhizobium* were initiated.

In 1974, three years after reading the publication on *nif* genes by Streicher *et al.*, I was unexpectedly able to go back to what had become my primary interest – Genetics. That was the time when the unprecedented "Cultural Revolution" was being launched in this country. Anyone who carried out research should first consider if it was politically reasonable. I chose the project on genetics of biological nitrogen fixation, which was not exactly my real interest. Nitrogen fixation genetics occurred to me as a good title that would attract support from the authorities, as it followed the dogma "theory must be connected with practice", as well as the fact that its final goal would be making crops capable of fixing nitrogen. Furthermore, I thought that the *nif* gene research started only a few years ago, and we can match up with our effort. Overall, our main purpose was to do our utmost to make up an important area of biological research, genetics, which has been deserted for more than thirty years. In spite of inadequate equipment and technology, we were all enthusiastically collecting the bacterial strains and picking up the necessary methods for transduction, conjugation, and biochemical tests. Through the efforts of my young colleagues, our first paper on genetic analysis of nitrogen fixation system in *K. pneumoniae* was published in *Scientia Sinica* 1977 (Xue *et al.*, 1977). I remember hurrying the editor of that journal to publish it as soon as possible. The response to our published work was warm and encouraging, especially from Professor Winston J. Brill of University of Wisconsin.

The year 1977 is memorable not only because we published our first paper on *nif* genes, but also because it marked the beginning of genetic research after the so called "Cultural Revolution" in China.

The following year, I was invited to attend the 50th Anniversary Symposium for Building of Biology Division of California Institute of Technology by T. H. Morgan.   There, I gave a talk about genetics of bacterial nitrogen fixation (Shen, 1978).

Transduction and Three-factor Reciprocal Cross for Fine Structure Mapping of *nif* Mutations

The nitrogen-fixing bacteria, *Klebsiella pneumoniae M5a1* which is bacteriophage *P1*-sensitive was used by Valentine group at University of California, Berkeley for the study of *nif* genes. They treated the wild type *K. pneumoniae* with mutagen, nitrosoguanidine, and screened for the Nif⁻ mutants which were unable to use N_2 as the sole source of nitrogen for growth, and thus deficient of nitrogenase activity. The dithionite assay was used to monitor nitrogenase activity and select the Nif⁻ mutants. The Nif⁻ mutants were converted back to Nif⁺ by transduction with phage *P1* that had been grown on wild type bacteria. The transductional analyses have revealed a cluster of *nif* genes in a segment proximal to the histidine operon (*his*) on the chromosome. This finding was previously substantiated by experiments that showed the conjugational transfer of a functionally complete set of *nif* genes from *K. pneumoniae* to *E. coli*. Since the hybrid *E. coli* had acquired the activity of fixing N_2, it was inferred that the structural and regulatory *nif* genes were all located near the *his* operon. The data also showed that the *nif* genes linked to the *his* operon were separated into two clusters by a physical distance of about 9 Kb in length which was then called the "silent region".

Some of my colleagues had much knowledge of bacteriophage and some were experts in biochemistry, but none of them were familiar with genetics. Therefore, I provided assistance in elementary genetics. I also shared with them at leisure, my knowledge on the personal and intellectual life of the great scientists at CalTech, such as George Beadle and Max Delbruck, with the hope of inspiring them. Simultaneously, I was also inspired. We had several strains of *K. pneumoniae* obtained from abroad through the help of my friend, Dr. M. C. Niu of Temple University. Most of the *nif⁻* mutants were prepared by ourselves. We even prepared the common biochemicals, for instance, L-histinal, the precursor of L-histidine used for detecting the *his⁻* mutants. In order to study the fine-structure mapping of the *nif* mutations, we re-examined the results previously published by Kennedy and Dixion (Kennedy *et al.*, 1977), and adopted a more reliable method for measuring the co-transduction. We used a given *his*D mutant CH80 as the common recipient and the various *nif* mutants to be mapped as donors in *P1*-transduction. We consistently found that the maximal distance of co-transduction percentage is only 6%. From the data of co-transduction, the order of *nif* mutations was roughly estimated, final decision of the precise order of *nif* mutations was obtained from the test of three-point reciprocal transductional crosses. According to the equation, f = (1-d/L)^3, where f = co-transductional frequency, d = distance between the two alleles co-transducted and L = length of transducing DNA (here taken to be 80 Kb), the physical distance between the two most distant mutations is about 1-2 Kb. Therefore, no "silent region" exists within the *nif* cluster located near the *his* operon. Our results were eventually confirmed elsewhere.

Also, we discovered some *his*D-unlinked Nif⁻ strains (Xue *et al.*, 1980). According to their biochemical phenotypes, they are glutamine auxotrophs and they regulate the overall expression of *nif* operons. Now, we know that the products of the nitrogen assimilation genes, *ntr*C or *ntr*A are necessary for *nif* gene expression.

We were all happy with the results obtained and rejoiced over the good beginning.

Complementation Test and Delineation of *nif* Genes

The complementation test was conducted according to the procedure of Dixon *et al.* The *E. coli rec*⁻ strain was used for the complementation test in order to distinguish it from recombination. *K. pneumoniae nif* mutant was converted to a *his*⁻*nif*⁻ double mutant by mutagenesis, which was then allowed to cross with the *E. coli recA* harbouring the *nif* PRD1 plasmid. The resultant conjugants (*his*⁻*nif*⁻/*his*⁺*nif*⁺Km^{r}carb^{r}Tc^r) were crossed with *K. pneumoniae* deletion mutant (Δ*his-nif*) again. The recombinant plasmids PRD1 *his*⁻*nif*⁺ in *Kp* were selected and transferred back to *E. coli recA* ready for complementation test, versus the given chromosomal *nif* mutant. We also carried out biochemical tests to investigate the function of the cistrons determined by complementation. Though the work was heavy and tedious, we were all happy with the success of the search for *nif* cistrons among numerous mutants and always eager to find a new one.

Our results confirmed W. J. Brill's description on the 17 *nif* genes in *K. pneumoniae* (MacNeil *et al.*, 1978). However, we found that a mutant C1005 isolated in this laboratory located between *nif*H and *nif*J gene behaved differently. A complementation test was run using *nif* mutant C1005 either in plasmid as donor or in chromosome as recipient versus the known *nif* strains reciprocally.   It showed that this mutant was capable of complementing all known *nif* genes.   Therefore, it allowed us to speculate that a new *nif* gene, we call *nifC*, would be present.   According to the P1-transduction and three-factor reciprocal crosses, this assumed gene was mapped between *nif*H and *nifJ* (Jin *et al.*, 1980). The following year after the publication of our results, we cooperated with Brill's group in Wisconsin to repeat the *nifC* work.   At that time, my colleague Jiabi Zhu, was a visiting scientist in Brill's laboratory.   She conducted genetic analysis while Gary Stacey of Brill's group conducted the biochemical test.   They confirmed our previous observations: (1) *nifC1005* mutant complemented all other known *nif* genes, including *nifJ*; (2) Fe-Mo-Co and Component I stimulated the nitrogenase activity (Stacey *et al.*, 1982). However, by taking advantage of the large number of *nifJ* mutants from Brill's laboratory, the result of fine structure mapping indicated that *nifC1005* was located within *nifJ* gene. It was bounded by mutation on both sides, exhibiting the normal *nifJ⁻* phenotype and showing no complementation with other *nifJ* mutations. Biochemical studies illustrated that *nifJ⁻* product was determined as the dimeric proteins, its molecular weight is ca 257, 000, a dimer of identical subunits.   Some *nifC⁻/nifJ* or *nifJ⁻/nifJ⁻* merodiploids produced active but unstable *nifJ* proteins.   Summarizing these results indicated that *nifC1005* did not define a separate gene from *nifJ*. The data was consistent with the occurrence of intragenic complementation between two defective *nifJ* polypeptides.

Zhu learned a lot from Brill's group, but in turn also made contributions.   Her intelligence, modesty and skills won her a good reputation there.   Since then, a friendly bond between the two laboratories is established, particularly between Winston and I.

Conservation of *nif* Genes Between *K. pneumoniae* and Other Diazotrophs

As the cloning of *nif* DNA was developed, the cloned DNA fragments were used to obtain a detailed restriction map of *nif* gene.   To correlate the restriction map and the genetic map, the precise location of *K. pneumoniae* *nif* genes was determined by Cannon and Ausubel (Ausubel *et al.*, 1981). The cloned *nif* DNA fragments from *K pneumoniae* were used as a hybridization probe to identify the homologous sequence in other diazotrophs.   We found that in the nitrogen-fixing bacteria, *Enterobacter cloacae* has similar organization and regulatory property of *nif* genes (Zhu *et al.*, 1986).

In *Azotobacter, Rhizobium,* some nitrogen-fixing fungi (actinomycetes) and blue green algae, the homologous *nif* genes were successively demonstrated.

My colleague, Runzhi Jin recently has turned his interest to the *nif/nod* genes of *R. astragali*. It is a symbiotic rhizobium of clover, which serves as the main green manure in the southern part of China. He has successfully shown the homologous *nif/nod* genes in the rhizobium (Jin *et al.*, 1993).

Regulation of *nif* Genes

K. pneumoniae has been for some time the model organism for studies on the regulation of *nif* genes. The *K. pneumoniae nif* genes are subject to two-tier cascade control. The first is *nif*-specific and is mediated by the products, NifL and NifA of *nif*LA operon. NifA activates transcription of all *nif* genes, so it is the positive regulator, whereas NifL acts to inhibit NifA activity in response to oxygen or fixed nitrogen, thus NifL behaves as the negative regulator of *nif* genes. Transcription of *nif*LA operon is in turn regulated by components of the general "nitrogen control" (*ntr*) system, mediated by the products of *ntr*A and *ntr*C genes (Ow *et al.*, 1985). So this constitutes the second tier of indirect control of *nif* genes.

Zhu and Brill first demonstrated that NifA undergoes a reversible inactivation at high temperature, which prevents the activation of transcription of *nif* operons by NifA (Zhu *et al.*, 1981). Repression by high temperature does not require the NifL. Later, Zhu *et al.* found in our laboratory, a strain of nitrogen-fixing bacteria, *Enterobacter cloacae*, capable of growth with N_2 as the sole nitrogen source at 37°C. They demonstrated that *E. cloacae* NifA was less sensitive than NifA from *K. pneumoniae* (Zhu *et al.*, 1986). These results indicated that temperature regulation of *nif* genes is controlled by the temperature-sensitive NifA.

We were probably the first to create a test system to test the role of NifA for *nif* genes, in which the cloned constitutive *nif*A was coexisted with the *nif*H-*lacZ* fusion (Kong *et al.*, 1982). We found that NifA enhanced the transcription of nitrogenase genes *nif*HDK; and the oxygen and ammonium repression of *nif* genes can be relieved by an abundance of NifA (Zhu *et al.*, 1983). We also provided the evidence that the repressive action of NifL lies in its inactivation of NifA.

Q.T. Kong is highly skilled in laboratory work. His associates and him demonstrated in this laboratory, the oxygen sensitivity of *nif*LA promoter and pointed out that NifL is responsible for *nif* repression by oxygen (Kong *et al.*, 1986). They thus speculated that oxygen regulation of *nif* genes is mediated by NifA at two different levels. First, the oxygen sensitivity of the *nif*LA promoter ensures that all promoters, including the *nif*LA promoter, are repressed by oxygen; second, NifL acts as a repressor in the presence of oxygen, modulating the transcription of other *nif* genes.

The evidence for the two-level hypothesis has been substantiated by our subsequent investigations (Deng *et al.*, 1995) and the observations reported in other laboratories. Recently, we illustrated the structure of *nif*LA promoter with the characteristic anaerobic box and the transfactor for the sequence in the box. We also showed the possibility of NifL-NifA interaction with the result of inactivation of NifA under oxygen.

We also did some work on the regulatory property of the promoter of nitrogenase genes. In 1984, David W. Ow, after completing his doctoral degree from F. M. Ausubel's laboratory of Harvard University, joined us as a visiting scientist. In view of the similarity of *nif*A and *ntr*C genes, he proposed to carry out the mutagenesis experiment on *K. pneumoniae* promoter to see whether the NifA dependent *nif*H promoter can be converted to the NtrC dependent (Ow *et al.*, 1985). He devised a *nif*H promoter-lacZ fusion plasmid with the parental promoterless plasmid, heteroduplex formation method for mutagenesis. Curiously, a class of promoter mutants which could be activated either by NifA or NtrC was observed. Sequencing data showed that the consensus heptamer CCCT$^-$ ^{14}GCA of *nif*A promoter region seemed to play a regulatory role. Any C to T transition at -12 to -13 rendered the *nif*H promoter capable of being activated by NifA or NtrC (Shen *et al.*, 1983). The reason was unclear until the NifA binding site upstream of *nif*H promoter was illustrated by Dixon *et al.*

nif Genes of Symbiotic Nitrogen-Fixing Rhizobia

The establishment of symbiotic nitrogen fixation involves a series of differentiated gene activation through the interactions between the rhizobium and its partner, the legume. Early symbiosis involves stages of the infection process, whereby the invading rhizobia begin the initiation of root nodule formation of the host plant. The ability of rhizobia to nodulate the host plant is determined by several sets of nodulation (*nod*) genes. At the late stage of symbiosis, the rhizobia within the nodules have developed into bacteriods, in which the nitrogen fixation genes, named as *nif* and *fix* genes are activated. The genes *nif*HDK and *fix*ABC encode the nitrogenase polypeptides and other functions known to be required for nitrogen fixation. As we know, in the symbiotic stage, bacteriods are mainly working for nitrogen fixation. The fixed nitrogen NH_4^+ is assimilated as L-glutamate and transported into the cells of the host plant. Energy and building staffs for the protoplasts of rhizobia are evidently derived from the host plant cells. As far as the biological economy is concerned, it seems that the *nif/fix* genes are active only in bacteriods. After discussing with my colleagues, G. Q. Yu and J. B. Zhu, an attempt was made to construct the rhizobia to harbour plasmids carrying *nif/fix-lacZ* or *nod-lacZ* fusion and test the *nif/fix* and *nod* gene activity in the bacteriods of nodules by the simple histochemical method. Our observation confirmed our prediction: the *nod* genes are only active in the free-living rhizobia and in the rhizobia before differentiation into bacteriods. As soon as the rhizobia develop into bacteriods in the nodules, the *nod* genes become repressed. At the same time, the regulatory gene *nif*A is induced and the *nif*A-mediated *nif/fix* genes are thus switched on. This indicates the fact that *nod* and *nif/fix* genes are sequentially expressed in *Rhizobium* during development (Shen *et al.*, 1989).

Pleiotropic Function of *Rhizobium nif*A

Mutation of *Rhizobium nif*A resulted in the inactivity of *nif/fix* genes (Wang *et al.*, 1991) and the abnormal development of root nodules. Zhu designed an experiment in which the multicopied *nif/fix* promoter was introduced into the wild type rhizobia, which then infected the host legume. She found that the phenotype of such wild type rhizobia containing the multicopied *nif/fix* promoter was identical to that of *nif*A mutant. They were all defective in nitrogen fixation and produced many small-sized defective root nodules. Clearly, the *nif*A product, NifA was not available in the bacteriods, as it had been thoroughly titrated off by multiple *nif/fix* promoters (Wu *et al.*, 1995).

The results indicated that NifA functions for both nodulation and nitrogen fixation. However, does NifA signal the host plant to express normal level of nodulation? G. Q. Yu, a former physicochemist is currently studying this subject. He solved the complex problem of *Alfalfa-Rhizobia nod*D3 allele of *nod*D gene (Yu *et al.*, 1993, 1994).

References

Ausubel, F. M. and Cannon, F. C. (1981) Molecular Genetic Analysis of *Klebsiella pneumoniae* Nitrogen Fixation (*nif*) Genes. Cold Spring Harb. Symp. Quant. *Biol.* **45**: 487-499.

Deng, X. and Shen, S. C. (1995) Structure and Oxygen Sensitivity of *nif*LA Promoter of *Enterobacter cloacae. Science in China (Series B)* **38(1)**: 60-66.

Dixon, R. A. and Postgate, J. R. (1972) Genetic Transfer of Nitrogen Fixation from *Klebsiella pneumoniae* to *E. coli. Nature* **237**: 102-103.

Jin, R. Z., Huang, Y. C., Shen, M. C. and Shen, S. C. (1980) Complementation Analysis and Characterization of the Nitrogen Genes *nif*H, *nif*C and *nif*J in *Klebsiella pneumoniae*. *Scientia Sinica* **23(1)**: 108-118.

Jin, R. Z., Zhu, J. S., Jiang, Q. Y., Shen, S. S. and Shen, S. C. (1993) Evidence of *nod* and *nif* Genes on Megaplasmid in *Rhizobium astragali* 159. *Acta Microbiologica Sinica* **33(3)**: 170-176.

Kennedy, C. and Dixon, R. (1977) The nitrogen fixation cistrons of *Klebsiella pneumoniae*, presentation at the Conference on Genetic Engineering for Nitrogen Fixation, March 13-17, Brookhaven , New York, USA.

Kong, Q. T., Wu, Q. L., Ma, Z. F. and Shen, S. C. (1986) Oxygen Sensitivity of *nif*A Promoter of *Klebsiella pneumoniae*. *J. Bacteriol* **166**: 353-356.

Kong, Q. T., Wu, Q. L., Syvaneu, M., Lin, E. C. C. and Shen, S. C. (1982) Effect of *nif*A Gene Product on Expression of *lacZ* under *nif*H promoter in *E. coli*. *Scientia Sinica* **25(10)**: 1061-1070.

MacNeil,T., MacNeil, D., Roberts, G. P., Supiano, M. A. and Brill, W. J. (1978) Fine-structure mapping and complementation analysis of *nif* (nitrogen fixation) genes in *Klebsiella pneumoniae*. *J. Bacteriol* **136**: 253-266.

Ow, D. W., Guo, Q., Xiong, Y. and Shen, S. C. (1985) Mutational Analysis of the *Klebsiella pneumoniae* Nitrogenase Promoter: Sequences Essential for Positive Control by *nif*A and *ntr*C (*glnG*) Products. *J. Bacteriol* **161**: 868-874.

Ow. D. W., Guo, Q., Xiong, Y., Zhu, J. B. and Shen, S. C. (1985) Regulation of *Klebsiella pneumoniae* Nitrogen Fixation Gene Promoters by Regulatory Proteins *ntr*C, *nif*A and *nif*L. Nitrogen Fixation Research Progress, pp. 461-467.

Shen, S. C. (1978) Genetics of Bacterial Nitrogen Fixation from «Genes, Cells and Behavior», 50th Anniversary Symposium on the Founding of the Division of Biology, Norman Horowitz and Edward Hutchings, Jr., eds., CalTech, pp. 93-95.

Shen, S. C., Wang, S. P., Yu, G. Q. and Zhu, J. B. (1989) Expression of the Nodulation and Nitrogen Fixation Genes in *Rhizobium meliloti* during Development. *Genome* **31**: 354-360.

Shen, S. C., Xue, Z. T., Kong, Q. T. and Wu, Q. L. (1983) An Open Reading Frame Upstream from the *nif*H Gene of *Klebsiella pneumoniae*. *Nucleic Acid Research* **11**: 4241-4250.

Stacey, G., Zhu, J. B., Shah, V. K., Shen, S. C. and Brill, W. J. (1982) Intragenic Complementation by the *nif*J-coded Protein of *Klebsiella pneumoniae*. *J. Bacteriology* **150**: 293-297.

Streicher, S., Gurney, E. and Valentine, R. C. (1971) Transduction of the Nitrogen Fixation Genes in *Klebsiella pneumoniae*. *Proc. Nat. Acad. Sci. Vol. 68*, 6: 1174-1177.

Wang, S. P., Zhu, J. B., Yu, G. Q., Wu, Y. F. and Shen, S. C. (1991) Studies on the Heterogous Expression of *Rhizobium meliloti nif*A Gene and Oxygen Sensitivity of its Product. *Science in China (Series B)* **34**: 71-77.

Wu, T., Zhu, J. B., Yu, G. Q. and Shen, S. C. (1995) Inhibition of Nodule Development by Multicopy Promoters of *Rhizobium meliloti nif/fix* Genes. *Science in China (Series B)* **38(9)**: 1108-1116.

Xue, C. T., Jin, R. Z., Yu, Y. Y., Chen, H. Z., Li, W. J., Shen, M. J., Jiang, Q. Y. and Shen, S. C. (1977) Genetic Analysis of the Nitrogen Fixation System in *Klebsiella pneumoniae*. *Scientia Sinica* **20(6)**: 807-817.

Xue, Z. T., Jiang, Q. Y. and Shen, S. C. (1980) Mapping and Characterization of the *his*D-Unlinked *nif* Mutants in *Klebsiella pneumoniae*. *Scientia Sinica* **23**: 261-267.

Yu, G. Q., Zhu, J. B., Gao, Y. F. and Shen, S. C. (1994) Further Studies on Structure of *nod*D3 Gene in *Rhizobium meliloti*. *Science in China (Series B), Vol. 37*. 8: 975-983.

Yu, G. Q., Zhu, J. B., Gu, J., Deng, X. B. and Shen, S. C. (1993) Evidence that the Nodulation Regulatory Gene *nod*D3 of *Rhizobium meliloti* is Transcribed from Two Separate Promoters. *Science in China (Series B), Vol. 36*. 2: 225-236.

Zhu, J. B. and Brill, W. J. (1981) Temperature Sensitivity of the Regulation of Nitrogenase Synthesis by *Klebsiella pneumonia*. *J. Bacteriol, Vol. 145*. 2: 1116-1118.

Zhu, J. B., Li, Z. G., Wang, L. W., Shen, S. S. and Shen, S. C. (1986) Temperature Sensitivity of *nif*A-like Gene in *Enterobacter cloacae*. *J. Bacteriol* **166:** 357-359.

Zhu, J. B., Yu, G. Q., Wang, L. W., Shen, S. S. and Shen, S. C. (1983) Effect of *nif*A Product on Derepression of the *nif* genes in *Klebsiella pneumoniae*. *Scientia Sinica* **26(12):** 1258-1268.

Chapter 21

The Ti-plasmid and Plant Molecular Biology

Jeff Schell and Csaba Koncz
Max-Planck-Institut für Züchtungsforschung,
Carl-von-Linné-Weg 10, D-50829 Köln, Germany

ABSTRACT

In 1907, Smith and Townsend identified *Agrobacterium* as the causative agent of crown gall, the most common form of neoplasia in plants. Armin Braun, elaborating on Smith's idea about infectious plant cancer, predicted during the early fifties that *Agrobacterium* transfers a Tumor Inducing Principle (TIP) into plants that incites the proliferation of crown gall tumors by triggering the autonomous synthesis of plant growth hormones auxin and cytokinin. Also in 1970, Morel's group in France suggested that TIP is an *Agrobacterium*-derived DNA that specifies the production of unique compounds in crown galls termed opines that are characteristic for and catabolized by the tumor-inciting bacteria. That the TIP is indeed carried by a large tumor-inducing (Ti) plasmid of *Agrobacterium tumefaciens* was demonstrated in 1973 by my research group in Gent, Belgium. We and others later showed that TIP corresponds to a segment of the Ti plasmid which is transferred by agrobacteria and stably integrated into the nuclear genome of plant cells. The transferred DNA (T-DNA) was found to encode eukaryotic genes required for the production of opines and plant hormones in crown galls. Inactivation of the T-DNA encoded oncogenes involved in hormone synthesis allowed us in 1981 to obtain fertile transgenic plants showing stable Mendelian inheritance of a T-DNA encoded opine synthase gene. These results opened the way to broad range exploitation of T-DNA in plant transformation, physiology and genetics.

Agrobacterium and the Tumor-Inducing Principle

After obtaining an undergraduate degree in zoology (1957), I (J. Schell) studied microbiology and taxonomy in Prof. De Ley's laboratory in Gent before proceeding on to Prof. K. C. Winkler's laboratory at the State University of Utrecht to finish both my PhD thesis and my formal training in microbiology. This was the blossoming period of bacterial phage genetics. Thus, I decided to learn more about phages λ, P1, T2, T3, host restriction-modification systems and circular DNA species in bacteria as a postdoctoral fellow in the laboratories of Prof. W. Hayes (Hammersmith Hospital, London), A. Weissbach and J. Hurwitz (NIH and Albert Einstein College of Medicine), and L. Siminovitch (University of Toronto). The emergence of molecular biology and the advent of recombinant DNA techniques found me again in Belgium, where I became a professor and director of the Laboratory of General Genetics at the University of Gent in 1967. Based on my basic zoology education and experience in bacterial and phage genetics, I started to work with the plant pathogenic *Agrobacterium* that was already well-known as the causative agent of plant tumors. Together with my colleague and friend, Marc Van Montagu, and a small group of enthusiastic students, including initially M. Holsters, M. Zabeau, N. Van Larebeke, I. Zaenen, and G. Engler, we jointly surveyed and verified the data available in 1969 on *Agrobacterium*, plant tumors and transformation. At that time, crown gall was still considered as a potential model for cancer research as the etiology of animal and plant tumors appeared to be similar. A supporting argument for this was the experiment of the Danish oncologist, Carl O. Jensen in 1910 who successfully transplanted bacterium-free red mangel crown gall tumor tissues onto white sugarbeet stems. In particular, the American plant pathologist, Erwin F. Smith devoted his life work to verify a relationship between crown gall and cancer. His ideas were followed by Armin Braun, who between 1941 and 1978 published many seminal observations concerning the possible mechanism of *Agrobacterium*-induced tumor formation in plants.

Braun wished to resolve the controversy raised by Riker and Berge in 1935 who stated that primary and secondary crown gall tumor formation results from a continuous stimulation of plant cells by agrobacteria, rather than from stable neoplastic transformation of normal cells to tumor cells. In his experiments, Braun used an observation of Riker (1926) who recognized that crown gall tumor formation occurs only in a narrow window of temperature between 26 °C and 28 °C. Braun demonstrated that the incitement of crown galls by *Agrobacterium* requires a wounding of plants and subsequent wound healing for at least 24 to 48 h. Furthermore, he found in 1947 that the exposure of plants to 32 °C at any time within four to five days after *Agrobacterium* infection inhibits tumor formation. Based on these observations, Braun (1950) proposed that *Agrobacterium* can cause a permanent change in the growth of plant cells by temperature-sensitive transmission of a Tumor Inducing Principle (TIP) which is responsible for the neoplastic transformation. Braun (1959) also repeated Jensen's experiments with clonal, bacterium-free crown gall tissues demonstrating that plant tumors are indeed transplantable, as well as that neoplastic transformation is due to a stable genetic alteration of transformed plant cells.   He noticed that different bacterial strains incite tumors with a characteristic morphology, e.g., that strain B2 incites compact tumors in contrast to strain T37 which leads to the formation of shooting teratomas (Braun, 1948).   On the eve of the discovery of cell division promoting cytokinins, Braun and Naf (1954) and Braun (1956) reported that crown gall tumors overproduce auxin (the old leptohormone of Haberlandt) as well as cytokinesins (i.e., cytokinins) which were proposed to be N^6-substituted hypoxanthine derivatives. Braun thus demonstrated that crown gall tissues synthesize auxin and cytokinin, and that they can be maintained indefinitely in hormone-, and bacterium-free axenic cultures *in vitro*.   The TIP was therefore defined more precisely by Braun as a hypothetical principle which is required for the maintenance of cell division activity of crown galls via the production of plant growth factors.

Further suggestions about the nature of TIP were derived from the work of French chemists, Lioret (1956) and Morel (Menagé and Morel, 1964; Goldman *et al.*, 1969), who found that crown galls produce unique arginine derivatives termed opines.   Petit *et al.* (1970) demonstrated that the chemical nature of these opines, such as octopine and nopaline, strictly depended on which type of *Agrobacterium* strain was used for tumor induction.   They also showed that the opine synthesized by a crown gall is specifically catabolized by the bacterial strain which incited the tumor.   Together, these data suggested that Braun's TIP might be bacterial DNA which, in addition to carrying genetic information required for oncogenic plant hormone production, would also be responsible for directing the synthesis of specific opines in crown gall tumors.

An additional line of evidence indicating that TIP could be a transmissible bacterial DNA, was provided by the Australian microbiologist, A. Kerr.   His experiments were based on the early observations of Hendrickson *et al.* (1934) and Locke *et al.* (1939) who isolated an avirulent derivative (A66) of *Agrobacterium* strain A6 and found that co-inoculation of the avirulent and virulent strains in different positions along the stem allowed the avirulent A66 strain to incite tumors.   In 1953, Klein and Klein also reported on the transfer of tumor-inducing ability to avirulent agrobacteria.   However, Kerr (1969, 1971) was the first to use properly marked bacterial strains for co-infection of plant wound sites and thereby unequivocally demonstrated the transfer of the virulence trait between pathogenic and non-pathogenic isolates of *Agrobacterium*.   Later, together with ours (Genetello *et al.*, 1977) and J. Tempé's group in France, Kerr *et al.* (1977) also demonstrated that the transfer of virulence traits between agrobacteria is induced by the opines present in crown galls incited by virulent agrobacteria.

The Discovery of Tumor-Inducing (Ti) Plasmids

In the early 1960s, little attention was paid to the crown gall problem in plant research. Between 1965 and 1974, L. Ledoux reported on the direct transformation of many animal and plant organs with exogenous DNA. Thus, the resolution of the crown gall problem seemed to be very simple to many researchers who accepted the view that nucleic acids could be freely taken up and expressed by plant cells. In fact, Braun's TIP was soon identified as bacterial chromosomal and phage DNA, as well as infective RNA of various size and origin. Parsons and Bearsley reported in 1968 that temperate *Agrobacterium* phages, such as PS8 present in axenic crown gall cultures, may play a role in tumor induction. Because of our experience in phage genetics, we examined this hypothesis. We identified numerous inducible lysogenic phages closely related to phage Ω of Bearsley (1955), as well as defective non-plaque-forming prophages, which could only be observed under the electronmicroscope, in various *Agrobacterium* strains. The phage isolates were classified using virulent and avirulent *Agrobacterium* hosts as well as by detecting DNA sequence homology by filter hybridization. Although certain results obtained in collaboration with others suggested that some phage DNAs hybridized with crown gall DNA (Schilperoort *et al.*, 1974), we were unable to find a clear correlation between the presence of common phages in virulent *Agrobacterium* strains and their absence in non-pathogenic avirulent isolates (Schell, 1975). Despite these negative results, our studies of prophage DNAs led to the recognition that *Agrobacterium* contained one or more large circular plasmids that could be separated from the chromosomal and phage DNAs on alkaline sucrose gradients and purified by ethidium bromide-cesium chloride dye-buoyant density centrifugation. In addition to these separation methods, we characterized the plasmids and defective prophages by electron microscopy (EM), assayed for Ω-type prophage content and pathogenicity of many different *Agrobacterium* species, including virulent strains of *A. tumefaciens* (B_6, B_2A, 11158, TT-111, A_6, 396, 3/1 and 925), *A. rubi* (TR-2), *A. rhizogenes* (Kerr 38) and *A. species* (0362), as well as avirulent strains of *A. radiobacter* (S1005, TR1, 4718, 8149, 417, $M_{2/1}$) and *A. species* (0363). Ω-phage was found only in *A. tumefaciens* B_6 and B_2A, but not in other virulent strains, whereas defective prophages were detected by EM e.g., in the avirulent *A. radiobacter* strains TR1 and 8149, but not in some virulent strains. Thus, these data showed no correlation between pathogenicity and the presence of phages in *Agrobacterium*. On the contrary, our analysis revealed that all virulent strains contained a large plasmid which was absent from avirulent agrobacteria.

To demonstrate that a large plasmid carried the TIP defined by Braun, we obtained isogenic avirulent derivatives of virulent *Agrobacterium* strains. First, we examined the *A. tumefaciens* strains IIB and IIBNV6 provided by Braun and found that the crown gall-inducing IIB strain carried a large plasmid in contrast to the isogenic avirulent IIBNV6 isolate. At about this time, Hamilton and Fall (1971) reported that *A. tumefaciens* strains C58 and Ach-5, in which we also detected large plasmids, lost their virulence when grown at 37 °C. We grew these strains for five days at 37 °C and, after plating, tested 150 independent colonies for tumor-inducing ability: all of them were avirulent! Twelve of these avirulent isolates were tested for plasmids with the above described methods and we found that all had lost their large plasmids. Similar observations made with strain C58 supported our conclusion that the crown gall-inducing ability was indeed carried by a large plasmid which was therefore named tumor-inducing (Ti) plasmid. Following these experiments, we also assigned other genetic markers to the Ti plasmids. Kerr and Htay (1974) found that avirulent *A. radiobacter* strains 84 and S1005, which carried no plasmid, produced a bacteriocin termed agrocin 84 which efficiently killed the crown gall-inducing *A. tumefaciens* strains, but not their avirulent derivatives. As we knew that the loss of virulence was due to the loss of Ti plasmids, we isolated and tested an agrocin 84-resistant derivative (B6S3) of virulent *A. tumefaciens* strain B6. The B6S3 strain proved to be non-oncogenic and contained no Ti plasmid. During our phage studies, we had isolated a particular phage, AP1, which formed plaques only on a Ti plasmid cured avirulent strain C58C9, but not on its virulent Ti plasmid containing derivative C58. This observation showed that genes responsible for the AP1 phage exclusion were also encoded by the Ti plasmid. In collaboration with R. A. Schilperoort (University of Leiden), we tested the model proposed by Petit (1970) in the group of G. Morel and J. Tempé, which predicted that the TIP (i.e., the Ti plasmid) was responsible for the production of specific opines in crown galls, as well as for the degradation of opines by the tumor-inducing bacteria. This group in France showed that strain B6 induces octopine synthesizing compact tumors, whereas C58 incites shooting teratomas producing nopaline. Indeed, it was found that B6, but not its Ti plasmid cured derivative B6S3, catabolized octopine but not nopaline, whereas strain C58 grew on nopaline but not on octopine, whereas its avirulent derivative C58C9 could grow on neither. The Ti plasmids in these strains thus had to encode specific genes for both opine biosynthesis in crown galls and opine degradation in bacteria, and had a different specificity in B6 and C58. Therefore, the agrobacterium strains could be simply classified based on the type of opine synthesis and degradation genes carried by their Ti plasmids.

To unequivocally demonstrate that all these traits were indeed encoded by the Ti plasmid, we exploited Kerr's technique demonstrating the transfer of crown gall-inducing ability between properly marked, but otherwise isogenic, virulent and avirulent derivatives of *A. tumefaciens* strains B6 and C58. Upon mixed infection of tumors, the transfer of virulence from strains B6 and C58 correlated with the transfer of Ti plasmids to the cured B6S3 and C58C9 strains, which also gained the properties of pathogenicity, AP1 phage exclusion, agrocin 84 sensitivity and opine catabolism characteristic for B6 and C58.

In 1973, I reported these studies at a NATO Symposium on "Genetic Manipulation with Plant Materials" in Liège and the data were subsequently published by Schell (1975), Van Larebeke *et al.* (1974, 1975), Zaenen *et al.* (1974), and Engler *et al.* (1975), as well as together with R. A. Schilperoort by Bomhoff *et al.* (1976), and independently confirmed by Watson *et al.* (1975). The significance of these reports might have remained unnoticed by the plant research community which was busily testing Ledoux's DNA transformation data. However, in 1974 at a congress in Szeged (Hungary), G. P. Rédei (1976) provided the ultimate genetic evidence that transformation with bacterial DNA in Ledoux's experiments did not lead to complementation of the *Arabidopsis* thiamine mutants, as Ledoux had reported (1974) and F. P. Lurquin (1976) demonstrated that Ledoux's transformations did not resultd in the integration of exogeneous DNA into either cytoplasmic or nuclear genomes of plants. Thus, the problem of plant transformation was placed in the focus of general scientific attention and the Ti plasmid evolved to be a chief candidate as a potential transformation vehicle, in addition to some plant DNA and RNA viruses (for review see Howell, 1982).

Identification of the Transferred DNA (T-DNA)

The years of discoveries, 1971–1975, provided us with basic molecular biology tools, including DNA reassociation kinetic analysis, restriction endonucleases, DNA fragment separation by gel electrophoresis, labeling of nucleic acids with ^{32}P, detection of nucleotide sequence homologies by Southern blotting and hybridization, and cloning in plasmids and DNA sequencing (Nathans and Smith, 1975; Southern, 1975; Sinsheimer, 1977). Since all earlier DNA hybridization data were proven to be artificial (Schilperoort *et al.*, 1974; Chilton *et al.*, 1974a; Eden *et al.*, 1974) and the Ti plasmids were found to carry the TIP, recurrent efforts were devoted to detect the presence of *Agrobacterium*-derived DNA in crown galls. However, using the whole Ti plasmid as a probe in reassociation kinetic analysis,

Chilton *et al.* (1974b) and Merlo and Kemp (1976) were unable to detect the integration of Ti plasmid in crown gall DNA. A more detailed analysis using individually isolated, and later cloned, restriction endonuclease fragments of the Ti plasmid however demonstrated that DNA from a well-defined and conserved region of octopine- and nopaline-type Ti plasmids was present in different crown gall tumors (Chilton *et al.*, 1977, 1978a; Depicker *et al.*, 1978). To determine the boundaries of transferred DNA (termed thereafter T-DNA), the Ti plasmids were mapped by restriction endonucleases using different cloned fragments as probes in Southern hybridizations with crown gall DNAs (Chilton *et al.*, 1978b; Depicker *et al.*, 1980; Lemmers *et al.* 1980; Thomashow *et al.*, 1980; De Vos *et al.*, 1981). Cellular localization and reisolation of T-DNA fragments from crown gall tumors subsequently demonstrated that the T-DNA was covalently linked to plant nuclear DNA and flanked within the Ti plasmids by two imperfect 25-bp repeats (Chilton *et al.*, 1980; Willmitzer *et al.*, 1980; Yadav *et al.*, 1980; Zambryski *et al.*, 1980). Hybridizations of crown gall RNAs with T-DNA fragments also revealed that the T-DNA carries genes that are transcribed to poly A$^+$ mRNA by RNA polymerase II in plant tumors (Drummond *et al.*, 1977; Gurley *et al.*, 1979; Willmitzer *et al.*, 1981).

To precisely map and determine the function of T-DNA encoded genes, the transcript mapping and DNA sequencing methods were coupled with suitable insertional mutagenesis techniques allowing the modification of these genes in *Agrobacterium*. To perform mutagenesis using site-specific recombination, it was necessary to identify the mutant Ti plasmids after conjugation into a cured recipient strain. As Ti plasmid transfer could be previously achieved only in crown gall tissues obtained by mixed infection with donor and recipient strains, bacterial genetic techniques were developed to establish an efficient conjugation transfer of Ti plasmids *in vitro*. The observation of Kerr *et al.* (1977) and Genetello *et al.* (1977), which shows that Ti plasmid conjugation could be specifically stimulated by opines (e.g., by octopine for B6, Ach-5 and nopaline for C58), facilitated the isolation of transfer constitutive (Tra^c) Ti plasmid mutants. These studies also demonstrated that conjugation and opine catabolism are coordinately regulated in *Agrobacterium* (Klapwijk *et al.*, 1978). Agrobacteria "colonize" plants by inducing tumors to synthesize opines, which in turn provide a selective stimulation for bacteria living in the tumor to utilize the opines as sole carbon and nitrogen source, as well as to maintain the virulence by inducing the conjugation of Ti plasmids to avirulent bacteria. This mechanism was formulated as the concept of "genetic colonization" (Schell *et al.*, 1979) and as the "opine" concept by J. Tempé (Tempé and Schell, 1977; Tempé *et al.*, 1979).

The application of transposon insertion and site-specific mutagenesis techniques in conjunction with transcript mapping and sequencing studies led us and others to precise mapping, mutagenesis and functional analysis of T-DNA encoded genes.   In one of the pioneering experiments, my colleague and former student, J.-P. Hernalsteens, introduced a *Tn7* transposon into a gene located close to the right boundary of the T-region of a nopaline Ti plasmid. Tumor tissues induced with the *Agrobacterium* strain carrying this construct failed to synthesize nopaline, confirming that the insertion was located in the nopaline synthase gene (*nos*) carried by the T-DNA. On the other hand, mapping of the T-DNA in transformed plants confirmed the presence of this large *Tn7* transposon insert within the T-DNA transferred to crown gall cells demonstrating that the T-DNA of Ti plasmid can be used as a versatile tool to transform plant cells with foreign DNA (Hernalsteens *et al.*, 1980). This was the real beginning for us of the era of genetic engineering of plants.

The Use of T-DNA as Plant Transformation Vector

In 1978, I (J. Schell) was invited to be the director of the Max-Planck Institute for Plant Breeding in Cologne, Germany and encouraged to develop a new project based on the use of T-DNA in plant transformation and genetic engineering. As in Gent, I also found enthusiastic collaborators in Cologne, including the biochemists L. Willmitzer, J. Schröder and L. Otten; cell biologists H.-H. Steinbiß, O. Schieder, and newly arriving post-docs, C. Shaw, C. Koncz, D. Llewellyn, B. Baker and later many others. In addition, I received support not only from the Max-Planck Society, but also from the Genetics Institute of the University of Cologne, in particular from P. Starlinger, a well-known expert in plant transposon genetics. I was also able to maintain a strong bond and collaboration with the group in Gent until the late 1980s. Thus, we set up a very productive team to develop T-DNA as a tool for the study of molecular aspects of plant biology.

Functional analysis of the T-DNA encoded oncogenes and their protein products provided us with the possibility of studying the molecular action of plant hormones auxin and cytokinin, as it allowed us to show that the T-DNA encodes the genes for enzymes involved in the synthesis of these hormones. Inactivation of the *IaaH* and *IaaM* genes required for auxin synthesis was observed to decrease the auxin to cytokinin ratio leading to teratoma shoot formation. In contrast, abolishing the function of the *ipt* gene resulted in high auxin to cytokinin ratio promoting root formation. In addition to the functional analysis of T-DNA genes, which has been extensively reviewed (see e.g., Nester and Kosuge, 1981; Bevan and Chilton,

1982; Nester *et al.*, 1984; Morris, 1986; Binns and Thomashow, 1988; Zambryski *et al.*, 1989), we could devise various approaches to study cell differentiation and organ development, and use these tools to attempt the regeneration of fertile transformed plants. Braun and Wood (1976) claimed that teratoma shoots regenerate to fertile plants by suppression of the neoplastic state, and these results were supported by data of Sacristan and Melchers (1977) who achieved regeneration of plants from *Agrobacterium*-transformed single cell clones. However, repetition of these experiments showed that plants regenerated from teratoma shoots contained no T-DNA (see e.g., Yang *et al.*, 1980). These negative results led to the conclusion that "T-DNA cannot pass meiosis." Using a "brute force" approach, we assayed a large number of shoots regenerated from a teratoma which was induced with a T-DNA construct carrying inactivated *IaaM* and *IaaH* genes. We found, to our surprise, fertile plants which produced octopine and inherited this character in a Mendelian fashion (Otten *et al.*, 1981). The molecular analysis of these plants revealed that except for the octopine synthase gene, all other T-DNA encoded genes were deleted in these first transgenic plants. Thus, we learnt the lesson that by removal or inactivation of the T-DNA encoded oncogenes, T-DNA transformed plants can easily be obtained (De Greve *et al.*, 1982). Based on the observation that the Ti plasmid T37 incites shooting teratomas, showing very low activity of *IaaM* and *IaaH* genes, Barton *et al.* (1983) inserted a yeast alcohol dehydrogenase gene in the *ipt* gene inactivating cytokinin production. Their effort resulted in a transformed plant carrying a full-length T-DNA with the *ADH* gene, confirming the conclusion that T-DNA can successfully be used as a plant transformation vector if its oncogenes are removed and the T-DNAs are thus "disarmed."

To facilitate the introduction of foreign DNA sequences into the T-DNA, we built a commonly used cloning vector, pBR322, into the nopaline synthase gene. Foreign genes cloned in pBR322, were introduced into *Agrobacterium* by pBR322 mobilization and, because this vector was unable to replicate in *Agrobacterium*, could be stabilized by homologous recombination with pBR322 sequences within the T-DNA (Koncz *et al.*, 1984). To construct a disarmed vector, pGV3850, all DNA sequences between the T-DNA border repeats were replaced by pBR322 (Zambryski *et al.*, 1983). To select for T-DNA transformed cells, however, suitable plant selectable marker genes had to be constructed. Our experiments showed that bacterial, yeast and animal genes were either not expressed, or not properly expressed, in plants. L. Herrera-Estrella (a Mexican student) and A. Depicker used the promoter and polyadenylation signal sequences of the nopaline synthase gene and linked them to bacterial antibiotic resistance genes to construct chimeric selectable marker genes, and introduced them into tobacco using the pGV3850 vector. Following a simple selection for antibiotic resistance, they could regenerate fully normal and fertile transgenic plants from transformed tobacco calli (Herrera-Estrella *et al.*, 1983a, b). These experiments enabled wide-range application of T-DNA as a plant transformation vector.

The Birth and Growth of Plant Molecular Biology

In the early 1980s, many of us working in tough competition on *Agrobacterium*, the Ti plasmid and T-DNA realized that the real potential for further development lays in the application of these tools to explore the function and regulation of plant genes involved in metabolism, differentiation and development. Our experience with *Agrobacterium* could be immediately applied to the study of mechanisms by which plants recognize and mount a defense against pathogens. The construction of chimeric genes and the study of their expression in plants provided a tool to examine *cis*-regulatory DNA sequences of promoters controlling the expression of plant genes in response to developmental, hormonal and environmental stimuli, as well as the nature of regulatory factors binding to these sequences. It was also clear that these approaches would never reach their goals without agricultural application, and industrial and political interest. The potential of using transgenic plants as a suitable replacement for herbicides, insecticides, fungicides, and the use of them for environmentally friendly raw material production was recognized by plant breeders and industry. This was reflected in the publication of an overview on *Agrobacterium*-mediated plant transformation and its further applications in the 1987 *Annual Review of Plant Physiology* by a leading industrial laboratory (Klee *et al.*, 1987). Concerning the recent fate of *Agrobacterium* and Ti plasmid research, we wish to emphasize only a few developments which we think merit highlighting.

Among those who devoted their research activity to clarifying the mechanisms by which *Agrobacterium* recognizes plant cells and transfers its T-DNA into the plant cell nucleus, E. W. Nester and P. Zambryski and their students and collaborators have made outstanding contributions throughout the last decade. The biology of *Agrobacterium*-mediated gene transfer has been worked out in fine detail, with the exception of the T-DNA integration mechanism which still awaits the identification of plant nuclear factors mediating this process (for review see Zambryski 1988, 1989, 1992; Lynn and Chang, 1990; Citovsky and Zambryski, 1993; Lanka and Wilkins, 1995). T-DNA vectors and *Agrobacterium* helper strains are available for plant laboratories all over the world. Simplification of the use of T-DNA vectors goes back to the recognition that the T-DNA does not carry any gene necessary for its transfer to the plant nucleus. For DNA transfer, it is sufficient to insert genes between the two border repeats of the T-DNA which can be maintained in a plasmid replicon separated from the Ti plasmid providing the virulence (*vir*) gene functions in *Agrobacterium*. (Hoekema *et al.*, 1983). There is a wealth of different T-DNA vectors which can be simply used for cloning in such a binary vector system in order to study the activity of plant promoters with different reporter genes, or to construct chimeric genes for engineering new traits in plants, or to use antisense and overexpression approaches for inhibition or suppression of a particular gene activity by transformation. (for review see e.g., Weising *et al.*, 1988; Mazur and Falco, 1989; Klee and Estelle, 1991; Schell, 1992; Walden *et al.*, 1997).

The Use of T-DNA in Plant Genetics

Since 1983, plant molecular biology has gradually overtaken various fields of plant biology. Today, the analysis of plant gene functions, physiological and developmental processes, and many other aspects of plant biology cannot be meaningfully performed without the use of transgenic plants. Although several other transformation technologies, such as PEG-mediated direct DNA uptake into protoplasts and biolistic methods, are also available, an overwhelming majority of transformed plants are generated with T-DNA as gene vector (Birch, 1997).

In addition to the major impact of T-DNA transformation in the analysis of regulation of plant gene expression and engineering of plants with agriculturally useful properties, the use of T-DNA has contributed to substantial achievements in plant genetics. It was earlier realized that T-DNA integration into plant chromosomal DNA could occur within genes and thus cause gene mutations. To test the efficiency of T-DNA as an insertional mutagen, we used a gene fusion technology by linking promoterless reporter genes to the T-DNA end, such that T-DNA insertions in plant genes could easily be identified either by selecting or screening for the activation of reporter genes. Measurement of the frequency of T-DNA-induced gene fusions in various plants showed that 20% to 30% of all T-DNA inserts were located in genes (Koncz *et al.*, 1989)! T-DNAs carrying known DNA sequences, plant marker genes and, optionally, bacterial plasmid replicons could thus be used as efficient insertional mutagens for gene tagging, allowing simple isolation and characterization of mutant plant genes. As first examples of the application of this genetic technique, we reported on the T-DNA tagging of the *CH42* gene (Koncz *et al.*, 1990), whereas others found T-DNA inserts in the *GL1* and *AG1* loci of *Arabidopsis* (Marks and Feldmann, 1989; Yanofski *et al.*, 1990). In our laboratory, B. Baker was the first to use T-DNA to deliver the maize *Ac* and *Ds* transposons into tobacco and to develop this method to a highly efficient insertional mutagenesis system, later together with others (Baker *et al.*, 1986, 1987).

Based on the pioneering work of G. P. Rédei (1970, 1974), *Arabidospsis*, a species with the smallest known genome among angiosperms and a short generation time, became the model system in plant biology during the late 1980s (Koncz *et al.*, 1992a; Meyerowitz and Somerville, 1994). Today, thousands of T-DNA- and transposon-tagged genes are available in *Arabidopsis* and soon T-DNA tags will be isolated in each gene of this species (Koncz *et al.*, 1992b; Forsthoefel *et al.*, 1992; Bechtold *et al.*, 1993; Feldmann *et al.*, 1994). As shown recently, genes can be identified by sequencing the junctions of T-DNA inserts in the plant DNA (Mathur *et al.*, 1998). Alternatively, insertions in sequenced genes can be found by screening approaches based on the polymerase chain reaction. After sequencing over 30,000 *Arabidopsis* cDNAs, genomic research is expected to yield the complete sequence of the *Arabidopsis* genome within a few years. We who were amazed by the speed of research developments following the "green revolution," are presently faced with a new metamorphosis in plant biology termed "functional genomics," which allows sequence-based identification of plant gene mutations and subsequent analysis of plant gene functions. It is now possible for plant biologists to gain immediate information about a favorite gene just by turning on their computers; they can begin deciphering in a systematic way the protein interactions controlling such basic cellular functions as signaling pathways and cell division. Will this type of research leave space for intuition, fantasy, and creativity leading to basic discoveries as during earlier times? In our view, certainly. Nonetheless, we agree with G. P. Rédei, who taught his students an old Indian proverb concerning history: "who does not look back, can easily lose his path."

References

Baker, B., Coupland, G., Fedoroff, N., Starlinger, P. and Schell, J. (1987) Phenotypic assay for excision of the maize controlling element *Ac* in tobacco. *EMBO J.* **6**: 1547-1554.

Baker, B., Schell, J., Lörz, H. and Fedoroff, N. (1986) Transposition of the maize controlling element *activator* in tobacco. *Proc. Natl. Acad. Sci. USA* **83**: 4844-4848.

Barton, K. A., Binns, A. N., Matzke, A. J. M. and Chilton, M.-D. (1983) Regeneration of intact tobacco plants containing full length copies of genetically engineered T-DNA and transmission of T-DNA to R1 progeny. *Cell* **32**: 1033-1043.

Bearsley, R. E. (1955) Phage production by crown gall bacteria and the formation of plant tumors. *American. Naturalist.* **89:** 175.

Bechtold, N., Ellis, J. and Pelletier, G. (1993) In *planta Agrobacterium*-mediated gene transfer by infiltration of adults *Arabidopsis thaliana* plants. *C.R. Acad. Sci. [III]* **316:** 1194-1199.

Bevan, M. W. and Chilton, M.-D. (1982) T-DNA of the *Agrobacterium* Ti and Ri plasmids. *Ann. Rev. Genet.* **16:** 357-384.

Binns, A. N. and Thomashow, M. F. (1988) Cell biology of *Agrobacterium* infection and transformation of plants. *Ann. Rev. Microbiol.* **42:** 575-606.

Birch, R. G. (1997) Plant transformation: Problem and strategies for practical application. *Annu. Rev. Plant Physiol. Plant Mol. Biol.* **48:** 297-326.

Bomhoff, G. H., Klapwijk, P. M., Kester, H. C. M., Schilperoort, R. A., Hernalsteens, J.-P. and Schell, J. (1976) Octopine and nopaline synthesis and breakdown genetically controlled by a plasmid of *Agrobacterium tumefaciens. Mol. Gen. Genet.* **145:** 177-181.

Braun A. C. (1947) Thermal studies on the factors responsible for tumor initiation in crown gall. *Am. J. Bot.* **34:** 674-677.

Braun, A. C. (1948) Studies in the origin and development of plant teratomas incited by the crown gall bacterium. *Am. J. Bot.* **25:** 5111-519.

Braun., A. C. (1950) Thermal inactivation studies on the tumor-inducing principle in crown gall. *Phytopath.* **40:** 3.

Braun, A. C. (1956) The activation of two growth-substance systems accompanying the conversion of normal to tumor cells in crown gall. *Cancer Res.* **16:** 53-56.

Braun, A. C. (1959) A demonstration of the recovery of the crown gall tumor cell with the use of complex tumors of single cell origin. *Proc. Natl. Acad. Sci. USA* **45:** 932-938.

Braun, A. C. and Naf, U. (1954) A non-auxinic growth promoting factor present in crow-gall tissue. *Proc. Soc. Exp. Biol. Med.* **86:** 212-214.

Braun, A. C. and Wood, H. N. (1976) Suppression of the neoplastic state with the acquisition of specialized function in cells, tissues, and organs of crow gall teratomas of tobacco. *Proc. Natl. Acad. Sci. USA* **73:** 496-500.

Chilton, M.-D., Currier, T. C., Farrand, S. K., Bendich, A. J., Gordon, M. P. and Nester, E. W. (1974a) *Agrobacterium tumefaciens* and PS8 bacteriophage DNA not detected in crown gall tumors. *Proc. Natl. Acad. Sci. USA* **71:** 3672-3676.

Chilton, M.-D., Drummond, M. H., Gordon, M. P., Merlo, D. J., Montoya, A. L., Sciaky, D., Nutter, R. and Nester, E. W. (1977) Stable incorporation of plasmid DNA into higher plant cells: The molecular basis of tumorogenesis. *Cell* **11:** 263-271.

Chilton, M.-D., Drummond, M. H., Merlo, D. J. and Sciaky, D. (1978a) Highly conserved DNA of Ti plasmids overlaps T-DNA maintained in plant tumors. *Nature* **275:** 147-149.

Chilton, M.-D., Farrand, S. K., Eden, F. C., Currier, T. C., Bendich, A. J., Gordon, M. P. and Nester, E. F. (1974b) Is there foreign DNA in crown gall tumor DNA? *In Modification of the Information Content of Plant Cells*, R. Markham, D.R. Davies, D. Hopwood, and R.W. Horne, eds., Elsevier, New York, pp. 297.

Chilton, M.-D., Montoya, A. L., Merlo, D. J., Drummond, M. H., Nutter, R. C., Gordon, M. P. and Nester, E. W. (1978b) Restriction endonuclease mapping of a plasmid that confers oncogenicity upon *Agrobacterium tumefaciens* strain B6-806. *Plasmid* **1:** 147-149.

Chilton, M.-D., Saiki, R. K., Yadav, N., Gordon, M. P. and Quétier, F. (1980) T-DNA from *Agrobacterium* Ti plasmid is in the nuclear DNA fraction of crown gall tumor cells. *Proc. Natl. Acad. Sci. USA* **77:** 4060-4064.

Citovsky, V. and Zambryski, P. (1993) Transport of nucleic acids through membrane channels: snaking through small holes. *Annu. Rev. Microbiol.* **47:** 167-197.

De Greve, H., Leemans, J., Hernalsteens, J.-P., Thia-Toong, L., De Beuckeleer, M., Willmitzer, L., Otten, L., Van Montagu, M. and Schell, J. (1982) Regeneration of normal and fertile plants that express octopine synthase, from tobacco crown galls after deletion of tumour-controlling functions. *Nature* **300:** 752-755.

De Vos, G., De Beuckeleer, M., Van Montau, M., and Schell, J. (1981) Restriction endonuclease mapping of the octopine tumor inducing pTiAch5 plasmid of *Agrobacterium tumefaciens*, *Plasmid* **6**: 249-253.

Depicker, A., DeWilde, M., De Vos, G., De Vos, R., Van Montagu, M. and Schell, J. (1980) Molecular cloning of the nopaline Ti plasmid pTiC58 and its use for restriction endonuclease mapping. *Plasmid* **3**: 193-211.

Depicker, A., Van Montagu, M. and Schell, J. (1978) Homologous sequences in different Ti plasmids are essential for oncogenicity. *Nature* **275**: 150-152.

Drummond, M. H., Gordon, M. P., Nester, E. W., and Chilton, M.-D. (1977) Foreign DNA of bacterial plasmid origin is transcribed in crown gall tumors. *Nature* **269**: 535-536.

Eden, F. C., Farrand, S. K., Powell, J., Benedich, A. J., Chilton, M.-D., Nester, E. W. and Gordon, M. P. (1974) Attempts to detect deoxyribonucleic acid from *Agrobacterium tumefaciens* and bacteriophage PS8 in crown gall tumors by complementary ribonucleic acid/deoxyribonucleic acid-filter hybridization. *J. Bacteriol.* **119**: 547-553.

Engler, G., Holsters, M., Van Montagu, M., Schell, J., Hernalsteens, J.-P. and Schilperoort, R. A. (1975) Agrocin 84 sensitivity: A plasmid determined property in *Agrobacterium tumefaciens*. *Mol. Gen. Genet.* **138**: 345-349.

Feldmann, K. A., Malmberg, M. and Dean, C. (1994) Mutagenesis in *Arabidopsis*. In *Arabidopsis*,. M. Meyerowitz and C. R. Somerville, eds., Cold Spring Harbor Laboratory Press, Cold Spring Harbor, pp. 137-172.

Forsthoefel, N. R., Wu, Y., Schultz, B., Bennett, M. J. and Feldmann, K. A. (1992) T-DNA insertion mutagenesis in *Arabidopsis*: Prospects and perspectives. *Aust. J. Plant Physiol.* **19**: 353-366.

Genetello, C., Van Larebeke, N, Holsters, M., Depicker, A., Van Montagu, M. and Schell, J. (1977) Ti plasmids of *Agrobacterium* as conjugative plasmids. *Nature* **265**: 561-562.

Goldman, A., Thomas, D. W. and Morel, G. (1969) Sur la structure de la nopaline, métabolite anormale de certaines tumeurs de crown gall. *C. R. Acad. Sci Paris* **268**: 852-854.

Gurley, W. B., Kemp, J. D., Albert, M. J., Sutton, D. W. and Callis, J. (1979) Transcription of Ti plasmid-derived sequences in three octopine-type crown gall tumor lines. *Proc. Natl. Acad. Sci. USA* **76**: 2828-2832.

Hamilton, R. H. and Fall, M. Z. (1971) The loss of tumor-initiating ability in *A, tumefaciens* by incubation at high temperature. *Experimentia* **27**: 229-230.

Hendrickson, A. A., Baldwin, I. L. and Riker, A. J. (1934) Studies on certain physiological characters of *Phytomonas tumefaciens, Phytomonas rhizogenes* and *Bacillus radiobacter*. Part II. *J. Bacteriol.* **28**: 597-618.

Hernalsteens, J.-P., van Vliet, F., De Beuckeleer, M., Depicker, A., Engler, G., Lemmers, M., Holsters, M., Van Montagu, M. and Schell, J (1980) The *Agrobacterium tumefaciens* Ti plasmid as a host vector system for introducing foreign DNA in plant cells. *Nature* **287**: 654-656.

Herrera-Estrella, L., Depicker, A., Van Montagu, M. and Schell, J. (1983a) Expression of chimeric genes transferred into plant cells using a Ti plasmid-derived vector. *Nature* **303**: 209-213.

Herrera-Estrella, L., DeBlock, M., Messens, E., Hernalsteens, J.-P., Van Montagu, M. and Schell, J. (1983b) Chimeric genes as dominant selectable markers in plant cells. *EMBO J.* **2**: 987-995.

Hoekema, A., Hirsch, P. R., Hooykaas, P. J. and Schilperoort, R. A. (1983) A binary plant vector strategy based on separation of *vir* and T-region of the *Agrobacterium*. *Nature* **303**: 179-181.

Howell, S. H. (1982) Plant molecular vechicles: Potential vectors for introducing foreign DNA into plants. *Ann. Rev. Plant Physiol.* **33**: 609-650.

Kerr, A. (1969) Transfer of virulence between isolates of *Agrobacterium*. *Nature* **223**: 1175-1176.

Kerr, A. (1971) Acquisition of virulence by non-pathogenic isolates of *Agrobacterium radiobacter*. *Physiol. Plant Pathol.* **1**: 214-246.

Kerr, A. and Htay, K. (1974) Biological control of crown gall through bacteriocine production. *Physiol. Plant Pathol.* **4:** 37-42.

Kerr, A., Manigaults, P. and Tempé, J. (1977) Transfer of virulence *in vivo* and *in vitro* in *Agrobacterium. Nature* **265:** 560-561

Klapwijk, P. M., Scheulderman, T. and Schilperoort, R. A. (1978) Co-ordinated regulation of octopine degradation and conjugative transfer of Ti plasmids in *Agrobacterium tumefaciens.* Evidence for a common regulatory gene and separate operons. *J. Bacteriol.* **136:** 775-785.

Klee, H. and Estelle, M. (1991) Molecular genetic approaches to plant hormone biology. *Annu. Rev. Plant Physiol. Plant Mol. Biol.* **42:** 529-551.

Klee, H., Horsch R. and Rogers, S. (1987) *Agrobacterium*-mediated plant transformation and its further applications to plant biology. *Ann. Rev. Plant Physiol.* **38:** 467-480.

Klein, D. T. and Klein, R. M. (1953) Transmittance of tumor-inducing ability to avirulent crown gall and related bacteria. *J. Bacteriol.* **66:** 220-228.

Koncz C, Kreutzaler F, Kálmán Z. and Schell J (1984) A simple method to transfer, integrate and study expression of foreign genes, such as chicken ovalbumin and alpha-actin in plant tumours. *EMBO J.* **3:** 1929-1937.

Koncz, C., Chua, N.-H. and Schell, J. (1992a) Methods in *Arabidopsis* Research. World Scientific, Singapore.

Koncz, C., Martini, N., Mayerhofer, R., Koncz-Kálmán, Z., Körber, H., Rédei, G. P. and Schell, J. (1989) High-frequency T-DNA-mediated gene tagging in plants. *Proc. Natl. Acad.Sci. USA* **86:** 8467-8471.

Koncz, C., Meyerhofer, R., Koncz-Kálmán, Z., Nawrath, C., Reiss, B., Rédei, G. P. and Schell, J. (1990) Isolation of a gene encoding a novel chloroplast protein by T-DNA tagging in *Arabidopsis thaliana. EMO J.* **9:** 1337-1346.

Koncz, C., Németh, K., Rédei, G. P. and Schell, J. (1992) T-DNA insertional mutagenesis in *Arabidopsis. Plant Mol. Biol.* **20:** 963-976.

Lanka, E. and Wilkins, B. M. (1995) DNA processing reactions in bacterial conjugation. *Annu. Rev. Biochem.* **64:** 141-169.

Ledoux, L., Huart, R. and Jacobs, M. (1974) DNA-mediated correction of thiaminless *Arabidopsis thaliana. Nature* **249:** 17-21.

Lemmers, M., De Beuckeleer, M., Holsters, M., Zambryski, P., Depicker, A., Hernalsteens, J.-P., van Montagu, M. and Schell, J. (1980) Internal organization, boundaries and integration of Ti plasmid DNA in nopaline crown gall tumours. *J. Mol. Biol.* **144:** 353-376.

Lioret, C. (1956) Sur la mise en évidence d'un acide aminé non idenfié particulier aux tissus de crown gall. *Bull. Soc. Fr. Physiol. Veg.* **2:** 76.

Locke, S. B., Riker , A. J. and Duggar, B. M. (1939) The nature of growth substance originating in crown gall tissue. *J. Agric. Res.* **300:** 535-539.

Lurquin, F. P. (1976) Integration of exogeneous DNA in plants: A hypothesis awaiting clear-cut demonstration. In *Cell Genetics of Higher Plants*, D. Dudits, G. L. Farkas and P. Maliga, eds., Akadémiai Kiadó, Budapest, pp. 77-90.

Lynn, D. G. and Chang, M. (1990) Phenolic signals in cohabitation: Implications for plant development. *Annu. Rev. Plant Physiol. Plant Mol. Biol.* **41:** 497-526.

Marks, M. D. and Feldmann, K. A. (1989) Trichome development in *Arabidopsis thaliana.* I. T-DNA tagging of the *Glabrous 1* gene. *Plant Cell* **1:** 1043-1050.

Mathur, J., Szabados, L., Schaefer, S., Grunenberg, B., Lossow, A., Jonas-Straube, E., Schell, J., Koncz, C. and Koncz-Kálmán, Z. (1998) Gene identification with sequenced T-DNA tags generated by transformation of *Agrobacterium* cell suspension. *Plant J.* **13:** 707-716.

Mazur, B. J. and Falco, S. C. (1989) The development of herbicide resistant crops. *Annu. Rev. Plant Physiol. Plant Mol. Biol.* **40:** 441-470.

Menagé, A. and Morel, G. (1964) Sur la présence d'octopine dans les tissu de crown gall. *C. R. Acad. Sci. Paris.* **259:** 4795-4796.

Merlo, D. J. and Kemp, J. D. (1976) Attempts to detect *Agrobacterium tumefaciens* DNA in rown gall tumor tissue. *Plant Physiol.* **58:** 100-106.

Meyerowitz, E. M. and Somerville, C. R. (1994) *Arabidopsis,* Cold Spring Harbor Laboratory Press, Cold Spring Harbour, N. Y.

Morris, R. O. (1986) Genes specifying auxin and cytokinin biosynthesis in phytopathogens. *Ann. Rev. Plant Physiol.* **37:** 509-538.

Nathans, D. and Smith, H. O. (1975) Restriction endonucleases in the analysis and restructuring of DNA molecules. *Ann. Rev. Biochem.* **44:** 273-293.

Nester, E. W. and Kosuge, T. (1981) Plasmids specifying plant hyperplasias. *Ann. Rev. Microbiol.* **35:** 531-565.

Nester, E. W., Gordon, M. P., Amasino, R. M. and Yanofsky, M. F. (1984) Crown gall: A molecular and physiological analysis. *Ann. Rev. Plant Physiol.* **35:** 387-413.

Otten, L., De Greve, H., Hernalsteens, J.-P., Van Montagu, M., Schieder, O., Straub, J. and Schell, J. (1981) Mendelian transmission of genes introduced into plants by the Ti plasmid of *Agrobacterium tumefaciens. Mol. Gen. Genet.* **183:** 209-213.

Parsons, C. L. and Beardsley, R. E. (1968) Bacteriophage activity of crown gall tissue. *J. Virol.* **2:** 651.

Petit, A., Delhaye, S., Tempé, J. and Morel, G. (1970) Reserches sur les guanidines des tissus de crown gall. Mise en évidence d'une relation biochemique spécifique entre les souches d'*Agrobacterium tumefaciens* et les tumeurs qu'elles induisent. *Physiol. Veg.* **8:** 205-213.

Rédei, G. P. (1970) *Arabidopsis thaliana* (L.) Heynh. A review of genetics and biology. *Bibliogr. Genet.* **20:** 1-150.

Rédei, G. P. (1974) *Arabidopsis* as genetoc tool. *Annu. Rev. Genet.* **9:** 111-127.

Rédei, G. P., Acedo, G., Weingarten, H. and Kiehr, L. D. (1976) Has DNA corrected genetically thiaminless mutants of *Arabidopsis*? In *Cell Genetics of Higher Plants,* D. Dudits, G. L. Farkas and P. Maliga, eds., Akadémiai Kiadó, Budapest, pp. 91-94.

Riker, A. J. (1926) Studies on the influence of some environmental factors on the development of crown gall. *J. Agric. Res.* **26:** 425-435.

Sacristan, M.-D. and Melchers, G. (1977) Regeneration of plants from "habituated" and "*Agrobacterium*-transformed" single-cell clones of tobacco. *Mol. Gen. Genet.* **152:** 111-117.

Schell, J. (1975) The role of plasmids in crown gall formation by *A. tumefaciens.* In *Genetic manupulations with Plant Materials,* L. Ledoux, ed., Plenum Press, New York, pp. 163-181.

Schell, J. (1992) Plant biotechnology: A powerful tool to use plant resources and to improve the environmental impact of agriculture. In *Natural Resources and Human Health,* S. Baba, ed., Elsevier, New York, pp. 49-60.

Schell, J., Van Montagu, M., De Beuckeleer, M., De Block, M., Depicker, A., De Wilde, A., Engler, G., Genetello, C., Hernalsteens, J.-P., Holsters, M., Seurinck, J., Silva, A., Van Vliet, F. and Villarroel, R. (1979) Interactions and DNA transfer between *Agrobacterium tumefaciens,* the Ti plasmid and the plant host. *Proc. R. Soc. Lond.* **B204:** 251-266.

Schilperoort, R. A., Dons, J. J. M. and Ras, H. (1974) Characterization of the complex formed between PS8 cRNA and DNA isolated from A6-induced sterile crown gall tissues. In *Modification of the Information Content of Plant Cells,* R. Markham, D. R. Davies, D. Hopwood and R. Horne, eds., Elsevier, New York, pp.253-268.

Sinsheimer, R. L. (1977) Recombinant DNA. *Ann. Rev. Biochem.* **46:** 415-438.

Smith, E. F. and Townsend, C. O. (1907) A plant tumor of bacterial origin. *Science* **25:** 671-673.

Southern, E. (1975) Detection of specific sequences among DNA fragments separated by gel electrophoresis. *J. Mol. Biol.* **98:** 503-517.

Tempé, J. and Schell, J. (1977) Is crown gall a natural instance of gene transfer? In *Translation of Natural and Synthetic Polynucleotides,* A. B. Legocki, ed., University of Agriculture, Poznan, pp. 416-420.

Tempé, J., Guyon, P., Tepfer, D. and Petit, A. (1979) The role of opines in the ecology of the Ti plasmids of *Agrobacterium*. In *Plasmids of Medical, Environmental, and Commercial Importance*, K. Timmis and A. Pühler, eds., Elsevier, Amsterdam, pp. 353-363.

Thomashow, M. F., Nutter, R., Montoya, A., Gordon, M. P. and Nester, E. W. (1980) Integration and organization of Ti plasmid sequences in crown gall tumors. *Cell* **19**: 729-739.

Van Larebeke, N., Engler, G., Holsters, M., Van den Elsacker, S. Zaenen, I., Schilperoort, R. A., and Schell, J. (1974) Large plasmid in *Agrobacterium tumefaciens* essential for crown gall-inducing ability. *Nature* **252**: 169-170.

Van Larebeke, N., Genetello, C., Schell, J., Schilperoort, R. A., Hermans, A. K., Hernalsteens, J.-P. and Van Monagu, M. (1975) Acquisition of tumor-inducing ability by non-oncogenic agrobacteria as a result of plasmid transfer. *Nature* **255**: 742-743.

Walden, R., Reiss, B., Koncz, C. and Schell, J. (1997) The impact of Ti plasmid-derived gene vectors on the study of the mechanism of action of phytohormones. *Annu. Rev. Phytophatol.* **35**: 45-66.

Watson, B., Currier, T. C., Gordon, M. P., Chilton, M.-D. and Nester, E. W. (1975) Plasmid required for virulence of *Agrobacterium tumefaciens*. *J. Bacteriol.* **123**: 255-264.

Weising, K., Schell, J. and Kahl, G. (1988) Foreign genes in plants: Transfer, structure, expression, and applications. *Annu. Rev. Genet.* **22**: 412-477.

Willmitzer, L., De Beuckeleer, M., Lemmers, M., Van Montagu, M. and Schell, J. (1980) The Ti plasmid derived T-DNA is present in the nucleus and absent from plastids of plant crown gall cells. *Nature* **287**: 359-361.

Willmitzer, L., Otten, L., Simons, G., Schmalenbach, W., Schröder, J., Schröder, G., Van Montagu, M. and Schell, J. (1981) Nuclear and polysomal transcripts of T-DNA in octopine crown gall suspension and callus cultures. *Mol. Gen. Genet.* **182**: 255-262.

Yadav, N. S., Postle, K., Saiki, R. K., Thomashow, M. F. and Chilton, M.-D. (1980) T-DNA of a crown gall teratoma is covalently joined to host plant DNA. *Nature* **287**: 458-461.

Yang, F.-M., Montoya, A. L., Merlo, D. J., Drummond, M. H., Chilton, M.-D., Nester, E. W. and Gordon, M. P. (1980) Foreign DNA sequences in crown gall teratomas and their fate during the loss of the tumor traits. *Mol. Gen. Genet.* **177**: 704-714.

Yanosfski, M. F., Ma, H., Bowman, J. L., Drews, G. N., Feldmann, K. A. and Meyerowitz, E. M. (1990) The protein encoded by the *Arabidopsis* homeotic gene *agamous* resembles transcription factors. *Nature* **346**: 35-38.

Zaenen, I., Van Larebeke, N., Teuchy, H., Van Montagu, M. and Schell, J. (1974) Supercoiled circular DNA in crown gall inducing *Agrobacterium* strains. *J. Mol. Biol.* **86**: 109-127.

Zambryski, P. (1988) Basic process underlying *Agrobacterium*-mediated DNA transfer to plant cells. *Annu. Rev. Genet.* **22**: 1-30.

Zambryski, P. C. (1992) Chronicles from the *Agrobacterium*-plant cell DNA transfer story. *Annu. Rev. Plant Physiol.* **43**: 465-490.

Zambryski, P., Holsters, M., Kruger, K., Depicker, A., Schell, J., Van Montagu, M. and Goodman, H. M. (1980) Tumor DNA structure in plant cells transformed by *A. tumefaciens*. *Science* **209**: 1385-1391.

Zambryski, P., Joos, H., Genetello, C., Leemans, J., Van Montagu, M. and Schell, J. (1983) Ti plasmid vector for the introduction of DNA into plant cells without alteration of their normal regeneration capacity. *EMBO J.* **2**: 2143-2150.

Zambryski, P., Tempé, J. and Schell, J. (1989) Transfer and function of T-DNA genes from *Agrobacterium* Ti and Ri plasmids in plants. *Cell* **56**: 193-201.

Chapter 22

Active Ion Transport in Plants

Arnold J. Bloom
Department of Vegetable Crops, University of California
Davis, California 95616, USA

Alison R. Taylor
Marine Biological Association
Citadel Hill, Plymouth PL1 2PB, UK

ABSTRACT

The major principles of active transport were first elucidated in plants. These include the principles (a) that ions may enter an organism against their concentration gradients and independently of water movement, (b) that the processes of diffusion and osmosis are inadequate to sustain the observed ion gradients, (c) that ion movement depend upon the generation of metabolic energy, (d) that ion movement is carrier-mediated, (e) that electrochemical gradients across biological membranes drive ion transport, and (f) that at least two types of mechanisms are involved in the transport of virtually all ions. Plant materials such as giant algal cells remain model systems for ion transport studies.

Foundations

Active ion transport is the process through which organisms expend metabolic energy to translocate a particular ion between compartments. Active transport includes primary and secondary active transport. In primary active transport, energy is expended to alter the conformation of a transporter that moves an ion (e.g., a proton pump driven by the hydrolysis of ATP). In secondary active transport, energy is expended to produce a local deformation of the electrochemical gradient between compartments, and this deformation drives the movement of an ion through a co-transporter or channel (e.g., proton-nitrate cotransport). Active ion transport, be it primary or secondary, is distinct from strictly physical processes such as diffusion, mass flow, adsorption, and capillary action because in the physical processes, the movement of ions is independent of metabolism and lacks specificity for a given ion or a given class of ions.

The physiological mechanisms for active ion transport have proved to be highly conserved among organisms. As a consequence, current investigations on plant transport systems rely heavily upon information derived from the more extensively studied microbial and animal systems. Microbes and animals, however, obtain a large portion of their nutrition from ingestion of organic materials. These sources provide microbes and animals with highly concentrated supplies of nutrients. By contrast, plants must scavenge nutrition from dilute inorganic sources in their environment. Thus, plants usually absorb ions against their concentration gradients. Perhaps for this reason, nearly all of the modern concepts about active ion transport originated from experiments on plants.

The traditional view of nutrient acquisition, which persisted through the eighteenth century, was that ions enter an organism indiscriminately with the mass flow of water (Duggar, 1911). With experiments on the nutrient requirements of plants (Epstein, Chapter 1), evidence began to accumulate that did not support this theory. As early as 1804, de Saussure (1804) documented that plants absorb only a subset of the elements present in the soil solution. Subsequent studies established that plants, when subjected to appropriate nutrient solutions, could absorb water and dissolved substances at substantially different rates (Knop, 1859; Wolff, 1864); for example, root uptake of phosphate could exceed that of water by nearly three orders of magnitude (Sachs, 1882). The discovery of cellular plasmolysis (Nägeli and Cramer, 1855) led to experiments with plant cell protoplasts and to the conclusions that cell membranes are selectively permeable and that cellular water and ion exchange could occur independently of one another (Hofmeister, 1867; Pfeffer, 1900). As these and other inconsistencies with the mass flow theory of ion transport became apparent, a new theory — one that involved diffusion and osmosis — gained momentum.

Diffusion-Osmosis

Du Trochet (1837) defined cellular diffusion as the movement of substances through a porous protoplasmic membrane that continues until the concentrations in the cell and the surrounding medium reach an equilibrium. To account for the disparity between the ion concentrations within plant tissues and their surroundings, Mülder in 1851 (cited in Pfeffer, 1900; and in Baker and Hall, 1975) proposed that specific ions absorb to cell materials, thus shifting the diffusion equilibrium and permitting these ions to accumulate. Pfeffer (1900) recognized that adsorption alone could not support continuous ion influx. He stated, "A continuous disturbance of equilibrium may be maintained, if the absorbed substances at once undergo a more or less marked chemical change or alteration into soluble or insoluble compounds of different character... Moreover, the nature of the plasma is such as to render it possible that a substance may combine chemically with the plasmatic elements, thus being transmitted internally, and then set free again." These surprisingly modern definitions of facilitated diffusion and carrier-mediated transport failed to gain broad acceptance (Baker and Hall, 1975), perhaps, because they were buried in a chapter of ninety pages filled with less modern ideas about the "diosmotic" properties of cells and descriptions of artificial semi-permeable membranes (Pfeffer, 1900).

Studies on such artificial semi-permeable membranes became fashionable at the turn of the century. Donnan (1911, 1914) conducted experiments on the movement of Congo-red dye through membranes of vegetable parchment and the movement of potassium ferrocyanide through copper ferrocyanide membranes (Fig. 1). From these data, he developed a simple thermodynamic framework to explain how, in compartments separated by a semi-permeable membrane, the unequal distribution of an impermeable substance could drive the movement of a permeable ion against its concentration gradient (Donnan, 1911). This "Donnan equilibrium" became the foundation of the diffusion-osmosis theory that dominated thinking about ion transport for several decades. Although contradictory evidence continued to mount, remnants of this theory persisted until the 1960s.

Algae as Model Systems

Agriculture in the USA through the beginning of the twentieth century depended heavily upon potash supplies from Europe. When World War I disrupted this source of potassium, the search began for alternatives. As part of this effort, Hoagland (1944) at the University of California in Berkeley was enlisted to study the potassium-rich giant kelps found along the Pacific Coast of the USA. He observed that the levels of different ions in the kelps diverge radically from those of the seawater in which they are immersed. In particular, potassium in the kelps and seawater are about 200 mM and 10 mM, respectively. Sodium is skewed in the opposite direction with the levels in the kelps and seawater being about 50 mM and 450 mM, respectively. This puzzled Hoagland because potassium and

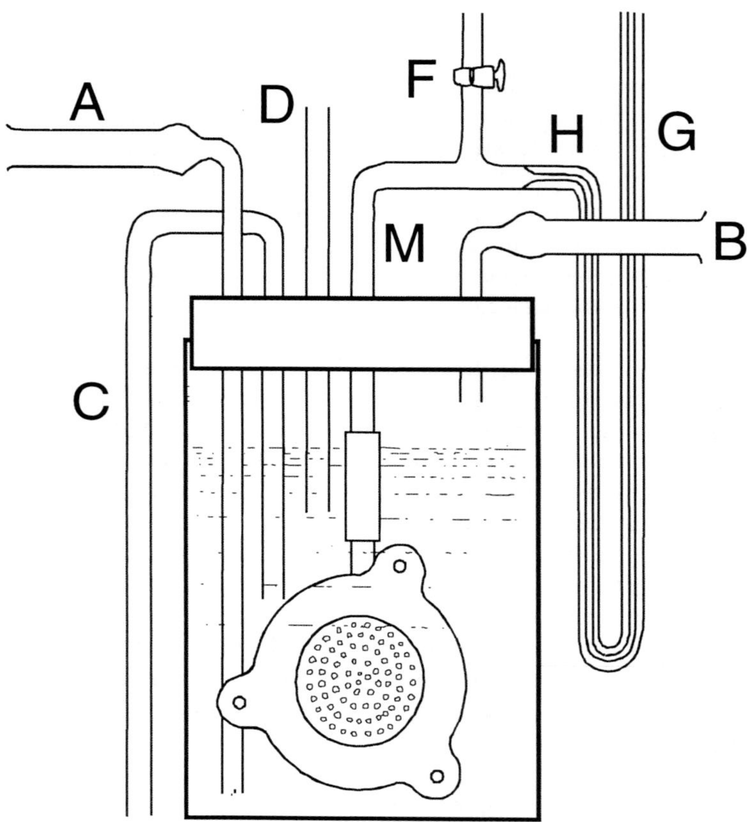

Fig. 1. The artificial semi-permeable membrane osmometer developed by Donnan and Harris (1911). The osmometer was suspended in a glass jar of about 1300 ml that was closed by a cork. Through the cork passed tube M to a mercury manometer, two soda-lime tubes (A and B) that allowed CO_2-free air to pass, and two tubes (C and D) for changing the liquid external to the osmometer. Parchment paper was sealed in the perforated silver plate. (Reproduced from Donnan and Harris, 1911).

sodium are similar monovalent cations that are neither adsorbed nor precipitated to any significant extent within the kelps, and he could not reconcile the accumulation of potassium versus the exclusion of sodium with a diffusion-osmosis theory.

The diffusion-osmosis theory was also assaulted with mounting evidence that ion transport required cellular energy. Hoagland (1923, 1929) found that the movement of ions, particularly bromide, into the large cells of the marine alga *Valonia* and the giant internodal cells of the fresh water alga *Nitella* increase in the light and change with temperature in a manner analogous to cellular respiration. Experiments in the early 1930s with barley (Hoagland and Broyer, 1936) and wheat (Lundegårdh, 1946) supported the conclusion that ion movement into roots depends on respiration. Diffusion-osmosis should not directly involve respiration.

During this period, there was a growing awareness that ion distributions across cell membranes had an electrical component. Nernst (1893) outlined the thermodynamic theory of electrochemical potentials in a monograph *Theoretische Chemie vom Standpunkte der Avogadro'schen Regel und der Thermodynamik* that underwent ten editions, the last of which appeared in 1921. Animal physiologists pursuing research on nerve conduction became attracted to the marine alga *Valonia* and the fresh water alga *Nitella* as model systems because these organisms had giant cells that were electrically excitable and amenable to the physiological techniques available at the time. The first recording of membrane potentials were obtained when Blinks (1930a; 1930b) placed crude intracellular electrodes into *Valonia* and *Nitella* cells. These electrical properties indicated that membranes have two lipid layers and prompted the development of the membrane bilayer model of Danielli and Davson (1935).

Rediscovery of the squid giant axon (Young, 1936) furnished Cole and Curtis (1938b) at Woods Hole, Massachusetts and Hodkin and Huxley (1939) at Plymouth, England with an animal alternative for the study of nerve conduction. These research groups, however, continued to use giant algal cells when squid were not available. Indeed, it was in *Nitella* that Cole and Curtis (1938a) first demonstrated how electrical changes across the plasma membrane derive from changes in ion permeability. They inserted fine capillary tubes filled with potassium chloride down the central axis of a segment of giant algal cell or squid axon and recorded the electrical potential between the inside of the cell and the external bath under various external potassium concentrations (Fig. 2). At high potassium concentrations, these measurements were consistent with the log-linear relationship between potassium concentration and membrane potential

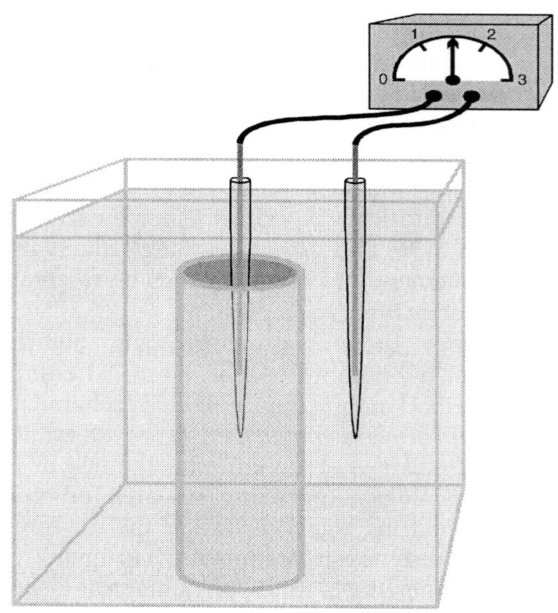

Fig. 2. A diagram of the system used to measure the membrane potential of giant algal cells. An excised internodal cell is placed in a bath. One glass capillary electrode is inserted into the cell, and another is left in the bath. These electrodes contain an Ag-AgCl reference wire immersed in a KCl solution. A voltmeter measures the electrical potential difference between the two electrodes.

predicted by the Nernst equation. This relationship broke down, however, at low potassium concentrations. To fit the data at low concentrations, Goldman (1943) derived an equation based on the permeabilities and concentrations of several ions. This equation has since become the standard.

Active Transport

Radioactive tracers became available after World War II, providing a powerful tool for ion transport studies in animals and plants. Ussing (1949) developed a frog skin preparation that permitted simultaneous measurements of the electrical potential and ionic composition on either side of an epithelial layer. He demonstrated that iodide movement against an electrochemical gradient resulted from active ion transport, not from diffusion-osmosis. Epstein (1952), based on measurements of potassium and rubidium absorption by barley roots as a function of external concentration, proposed that the absorption process involves the interaction of an ion with a membrane-bound carrier and approximates the kinetics of enzyme reactions (Fig. 3).

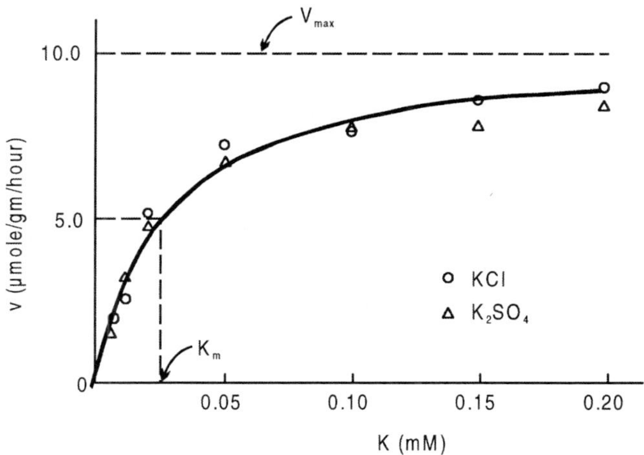

Fig. 3.   Rate of ^{42}K absorption (v) by excised barley roots as a function of potassium concentration when it was supplied as either KCl or K$_2$SO$_4$. The line is a plot of the Michaelis-Menten equation that best fits the data. From Epstein, E. (1972): Mineral Nutrition of Plants: Principles and Perspectives. ©John Wiley & Sons, Inc. Reprinted by permission of John Wiley & Sons, Inc.

Back at the marine stations, Cole (1968) exploited improvements in electronics and control theory to develop the membrane voltage clamp. Hodgkin and Huxley (1952b, 1952a, 1952) employed this technique to establish that changes in sodium and potassium permeabilities triggers action potentials in the squid axon. Caldwell and Keynes (1957) perfused squid axon preparations to control the ionic composition of both sides of a single cell membrane. They showed that a sustained extrusion of Na$^+$ from the axon against a concentration gradient requires ATP inside the cell. Similar results in crab nerve cells (Skou, 1957) and reconstituted erythrocyte ghosts (Whittam, 1958) led to the model that active transport of sodium and potassium are tightly linked in animals and utilizes the energy transferred during dephosphorylation of ATP (Caldwell, 1960).

Plant Physiology

The development of microelectrodes permitted membrane potential recordings on a variety of algal and higher plant cells (Hope, 1951; Walker, 1955; Kishimoto, 1957). Many of the cells of higher plants and fungi were found to have resting membrane potentials far more negative than the diffusion potential for the most permeable ions (Etherton and Higinbotham, 1960; Slayman, 1965). This forced plant physiologists to consider not only the chemical, but also the electrical gradients driving ion transport. Etherton and Higinbotham (1960) determined that the chief electrical barrier in cells of higher plants is the plasmalemma and that metabolic inhibitors cause a rapid depolarization of membrane potential across the plasmalemma, which is consistent with electrogenic ion transport. The first voltage clamp recordings in a plant cell were reported, once again, in giant algal cells (Findlay, 1961; Hope, 1961).

The nature of transport into roots was being well-characterized. Epstein (1963), continuing his work with potassium and rubidium absorption by barley roots, recognized two types of transport systems. One has a high affinity for potassium, high specificity for potassium, and low capacity (Mechanism 1 or HATS). The other has a low affinity, low specificity, and high capacity (Mechanism 2 or LATS). Such duality has now been demonstrated for most ions. In a current context, the Mechanism 1 corresponds to an ion carrier or pump whereas Mechanism 2 corresponds to an ion channel. Poole (1966), applying the same criterion as Ussing, concluded that potassium is actively transported inward at the plasmalemma of beet roots. Researchers working on a number of different plant systems showed sodium efflux and chloride influx to be active processes (Higinbotham, 1973). The last vestiges of the diffusion-osmosis model were abandoned.

The chemiosmotic hypothesis of Mitchell (1961), first forwarded to explain phosphorylation in mitochondria, was soon found applicable to plant solute transport. The sensitivity of solute fluxes and membrane potential to inhibitors of either respiration or photosynthesis pointed to the existence of an ATP-driven membrane pump (Etherton & Higinbotham, 1960; MacRobbie, 1965). The pH dependence of the membrane potential in giant algal cells (Kitasato, 1968; Saito and Senda, 1974) indicated that this pump is an electrogenic H^+-ATPase similar to one first identified in yeast cells (Slayman, 1970). By the mid 1970s, the importance of electrogenic H^+ pumps in generating pH and electrical gradients across the plasmalemma of plants was firmly established (Higinbotham and Anderson, 1974).

During the last twenty-five years, the similarities in transport systems among all organisms have become apparent. Molecular and structural analyses provide insight on how carriers, pumps, and channels function (Doyle *et al.*, 1998; Poole, 1998). Active transport of a wide range of organic molecules and ions is coupled to proton gradients generated by H^+-translocating ATPases (Leonard and Hodges, Chapter 15). The present challenge is to unravel the complex regulatory mechanisms among transporters through which plants adjust to both slow and rapid environmental changes.

References

Baker, D. A. and Hall, J. L. (1975) Ion transport — introduction and general principles. In *Ion Transport in Plant Cells and Tissues*. D. A. Baker & J. L. Hall, eds., pp. 1-37. North-Holland Publishing, Amsterdam.

Blinks, L. R. (1930a) The direct current resistance of *Nitella*. *J. Gen. Physiol.* **13**: 495-508.

Blinks, L. R. (1930b) The direct current resistance of *Valonia*. *J. Gen. Physiol.* **13**: 361-378.

Caldwell, P. C. (1960) The phosphorus metabolism of squid axons and its relationship to the active transport of sodium. *J. Physiol.* **152**: 545-560.

Caldwell, P. C. and Keynes, R. D. (1957) The utilization of phosphate bond energy for sodium extrusion from giant axons. *J. Physiol.* **137**: 12P-13P.

Cole, K. and Curtis, H. (1938a) Electrical impedance of *Nitella* during activity. *J. Gen. Physiol.* **22**: 37-64.

Cole, K. and Curtis, H. (1938b) Electrical impedance of the squid giant axon during activity. *J. Gen. Physiol.* **22**: 649-670.

Cole, K. S. (1968) *Membranes, ions and impulses; A chapter of classical biophysics*. University of California Press, Berkeley, pp.569.

Danielli, J. F. and Davson, H. (1935) A contribution to the theory of permeability of thin films. *Journal of Cellular and Comparative Physiology* **5**: 495-508.

de Saussure, T. (1804) *Recherches Chimiques sur la Végétation*. V. Nyon, Paris, pp. 327.

Donnan, F. G. (1911) Theorie der Membrangleichgewichte und Membranpotentiale bei Vorhandensein von nicht dialysierenden Elektrolyten. *Z. Elektrochem.* **17**: 572-581.

Donnan, F. G. and Allmand, A. J. (1914) Ionic equilibria across semi-permeable membranes. *J. Chem. Soc.* **105**: 1941-1963.

Donnan, F. G. and Harris, A. B. (1911) The osmotic pressure and conductivity of aqueous solutions of Congo-red and reversible membrane equilibria. *J. Chem. Soc.* **99**: 1554-1577.

Doyle, D. A., Cabral, J. M., Pfuetzner, R. A., Kuo, A. L., Gulbis, J. M., Cohen, S. L., Chait, B. T. and MacKinnon, R. (1998) The structure of the potassium channel: Molecular basis of K^+ conduction and selectivity. *Science* **280**: 69-77.

Du Trochet, H. M. (1837) *Memoires pour servir a l'histoire anatomique et physiologique des vegetaux et des animaux*. Chez J.-B. Bailliere, Paris, pp. 36.

Duggar, B. M. (1911) *Plant Physiology, with Secial Reference to Plant Production*. Macmillan Co., New York, pp. 516.

Epstein, E. (1952) A kinetic study of the absorption of alkali cations by barley roots. *Plant Physiol.* **27**: 457-474.

Epstein, E. (1973) Mechanisms of ion transport through plant cell membranes. *International Review of Cytology* **34**: 123-168.

Epstein, E., Rains, D. W. and Elzam, O. E. (1963) Resolution of dual mechanisms of potassium absorption by barley roots. *Proc. Natl. Acad. Sci. USA.* **49**: 684-692.

Etherton, B. and Higinbotham, N. (1960) Transmembrane potential measurements of cells of higher plants as related to salt uptake. *Science* **131**: 409-410.

Findlay, G. P. (1961) Voltage-clamp experiments with *Nitella. Nature* **191**: 812-814.

Goldman, D. E. (1943) Potential, impedance and rectification in membranes. *J. Gen. Physiol.* **27**: 37-60.

Higinbotham, N. (1973) The mineral absorption process in plants. *Bot. Rev.* **39**: 16-69.

Higinbotham, N. and Anderson, W. P. (1974) Electrogenic pumps in higher plant cells. *Canadian Journal of Botany* **52**: 1011-1021.

Hoagland, D. R. (1944) *Lectures on the Inorganic Nutrition of Plants.* F. Verdoorn, ed. Vol XIV. A New Series of Plant Science Books. Chronica Botica Co., Waltham, MA, pp. 226.

Hoagland, D. R. and Broyer, T. C. (1936) General nature of the process of salt accumulation by roots with description of experimental methods. *Plant Physiol.* **11**: 471-507.

Hoagland, D. R. and Davis, A. R. (1923) Further experiments on the absorption of ions by plants, including observations on the effect of light. *J. Gen. Physiol.* **6**: 47-62.

Hoagland, D. R. and Davis, A. R. (1929) The intake and accumulation of electrolytes by plant cells. *Protoplasma* **6**: 610-626.

Hodgkin, A. L. and Huxley, A. F. (1939) Action potentials recorded from inside a nerve fibre. *Nature* **144**: 710-711.

Hodgkin, A. L. and Huxley, A. F. (1952a) The components of membrane conductance in the giant axon of Loligo. *J. Physiol.* **116**: 473-496.

Hodgkin, A. L. and Huxley, A. F. (1952b) Currents carried by sodium and potassium ions through the membrane of the giant axon of Loligo. *J. Physiol.* **116**: 449-472.

Hodgkin, A. L., Huxley, A. F. and Katz, B. (1952) Measurement of current-voltage relations in the membrane of the giant axon of Loligo. *J. Physiol.* **116**: 424-448.

Hofmeister, W. F. B. (1867) *Die Lehre von der Pflanzenzelle.* Engelmann, Leipzig, xii, pp. 404.

Hope, A. (1961) The action potential in cells of *Chara. Nature* **191**: 811-812.

Hope, A. B. (1951) Membrane potential differences in bean roots. *Aust. J. Sci. Res. Series B* **4**: 265-274.

Kishimoto, U. (1957) Studies on the electrical properties of a single plant cell. Internodal cell of *Nitella flexilis. J. Gen. Physiol.* **42**: 1167-1187.

Kitasato, H. (1968) The influence of H+ on the membrane potential and ion fluxes of *Nitella. J. Gen. Physiol.* **52**: 60-87.

Knop, W. (1859) *Landwirtschaftlich Versuchsstation* **1**: 194.

Lundegårdh, H. (1946) Transport of water and salts through plant tissues. *Nature* **157**: 575-577.

MacRobbie, E. A. C. (1965) The nature of coupling between light energy and active ion transport in *Nitella translucens. Biochim. Biophys. Acta* **94**: 64-73.

Mitchell, P. (1961) Coupling of phosphorylation to electron and hydrogen transfer by a chemi-osmotic type of mechanism. *Nature* **191**: 144-148.

Nägeli, C. and Cramer, C. (1855) *Pflanzenphysiologische Untersuchungen.* Vol 1. F. Schulthess, Zurich.

Nernst, W. (1893) *Theoretische chemie vom standpunkte der Avogadro'schen regel und der thermodynamik.* F. Enke, Stuttgart, xiv, pp. 589.

Pfeffer, W. (1900) *The Physiology of Plants: A Treatise upon the Metabolism and Sources of Energy in Plants.* Clarendon Press, Oxford, pp. 632.

Poole, R. J. (1966) The influence of the intracellular potential on potassium uptake by beetroot tissue. *J. Gen. Physiol.* **49**: 551-563.

Poole, R. J. (1998) Solute Transport. In *Plant Physiology, Second Edition*, L. Taiz and E. Zeiger, eds., pp. 125-152. Sinauer Associates, Sunderland, Massachusetts.

Sachs, J. (1882) *Vorlesungen uber Pflanzen-Physiologie*. W. Engelmann, Leipzig, pp. 991.

Saito, K. and Senda, M. (1974) The electrogenic ion pump revealed by the external pH effect on the membrane potential of *Nitella*. Influences of external ions and electric current on the pH effect. *Plant Cell Physiol.* **15**: 1007-1016.

Skou, J. C. (1957) The influence of some cations on a adenosinetriphosphatase from perimpheral nerves. *Biochim. Bicphys. Acta* **23**: 394-401.

Slayman, C. L. (1965) Electrical properties of *Neurospora crassa*. Effects of external cations on the intracellular potential. *J. Gen. Physiol.* **49**: 69-72.

Slayman, C. L. (1970) Movement of ions and electrogenesis in microorganisms. *American Zoologist* **10**: 377-392.

Ussing, H. (1949) The distinction by means of tracers between active transport and diffusion. *Acta. Physiol. Scand.* **19**: 43-56.

Walker, N. A. (1955) Microelectrode experiments on *Nitella*. *Aust. J. Biol. Sci.* **8**: 476-489.

Whittam, R. (1958) Potassium movements and ATP in human red cells. *J. Physiol.* **140**: 479-497.

Wolff, W. (1864) *Landwirtschaftlich Versuchsstation* **6**: 203.

Young, J. Z. (1936) Structure of nerve fibres and synapses in some invertebrates. *Cold Spring Harbor Symp. Quant. Biol.* **4**: 1.

Chapter 23

Discovery of Chilling Injury

Mikal E. Saltveit
Mann Laboratory, Department of Vegetable Crops
University of California, Davis, CA 95616-8631, USA

ABSTRACT

It must have been obvious to farmers for millennia that plants differ in their sensitivity to low temperatures. Freezing damages most growing plants, but some plants are also injured by exposure to low, non-freezing temperatures. These chilling-sensitive plants are among the most important agricultural crops grown today and are major contributors to world food and fiber production. A partial list includes bananas, beans, citrus, cotton, maize, melons, millet, peppers, rice, sorghum, soybeans, squash, sugar cane, and tomatoes. The development of cultural practices from antiquity to the pre-scientific age that alleviated the severity of chilling injury symptoms indicates that the temperature limitations of many chilling-sensitive crops were well known. These practical methods to mitigate chilling injury far predate the scientific studies that were undertaken only during the past century to identify the physiological cause of chilling. The dominant paradigm for the past 25 years involves a causative progression from temperature-induced membrane phase transitions, to altered metabolism, to visual chilling injury symptoms. Much has been discovered about the practical effects of chilling, but the underlying molecular and physiological causes remain elusive and still awaits discovery.

Introduction

Chilling is the exposure of tissue to non-freezing temperatures below a certain threshold temperature. The word chilling has two dissimilar meanings in plant science. Chilling has been used in the horticulturally related sciences for many years to describe the beneficial effect of low temperatures in breaking dormancy and promoting the germination of seeds of annual plants and the growth and flowering of biannual or perennial plants. For example, a certain number of hours of chilling are required to promote the breaking of dormancy and the induction of uniform flowering of temperate fruit trees. In contrast, chilling can also describe the physiological disorders that develop in some plants after exposure to low, non-freezing temperatures. The latter use of chilling is the focus of this chapter.

Molisch (1897) is often credited with having suggested the term chilling injury (Erkältung) as opposed to freezing injury (Erfrieren). Most early literature dealing with chilling referred to it as *low temperature injury*. In their article covering a wide range of chilling-sensitive crops and treatments to reduce the symptoms of chilling injury, Sellschop and Salmon (1928) were among the first to use the term chilling injury in its present context. Confusion about the terminology used in describing research on chilling injury prompted Raison and Lyons (1986) to issue a plea for uniform terminology. The first research article on *low temperature injury* in the *Proceedings of the American Society for Horticultural Science* that used the term *chilling injury* was in 1938 (Morris and Platenius, 1938). The *Journal of the American Society for Plant Physiology* published their first article on *low temperature injury* in 1942 (Jones, 1942). Chilling injury was discussed in an early review in the *Annual Review of Plant Physiology* (Pentzer and Heinze, 1954). The first use of the term *chilling injury* in the indexes to Biological Abstracts occurred in 1959. A graphic representation of the literature cited in the comprehensive book on chilling injury edited by Wang (1990) arranged by year of publication shows that the level of interest, the focus of effort and the volume of work on the physiological basis of chilling injury have all increased dramatically since 1950 (Fig. 1).

Symptoms of chilling injury include reduced growth and vigor, inhibition of photosynthesis, accelerated senescence, increased water loss (with associated wilting and surface pitting), discoloration, and abnormal ripening (Lyons, 1973; Saltveit and Morris, 1990). The severity of these injuries depends on many factors including temperature, duration of exposure, stage of development, tissue, prior conditions, intensity of light, atmospheric composition (i.e., oxygen, carbon dioxide, and humidity), and subsequent handling. A dichotomy has developed between researchers studying whole, growing plants chilled in the field or greenhouse, and those studying harvested crops chilled during cold storage in the dark. Chilling in the field is usually caused by uncontrolled climactic factors, while chilling after harvest is usually part of an attempt to maintain quality and prolong the storage life by storing crops at low temperatures. The former is studied by scientists concerned with crop production and environmental stresses, while the latter is studied by postharvest physiologists and food scientists.

The major difference between these two areas of study is that chilling injury in the field or greenhouse is exacerbated by concomitant exposure to sunlight and by the time of day when chilling is started. These two factors can confound interpretation of experimental results. In contrast, by its very nature postharvest chilling of susceptible fruit, vegetables and ornamentals does not involve exposure to intense light. The contribution of diurnal variability to postharvest chilling sensitivity, although responsible for significant variation in chilling sensitivity of whole tomato plants (King *et al.*, 1988) and harvested tomato fruit (Saltveit and Cabrera, 1987), has received very limited attention. Changes in carbohydrate status appeared to be the determining factor in whole plants, while changes in membrane fluidity appeared to be the determining factor in harvested fruit.

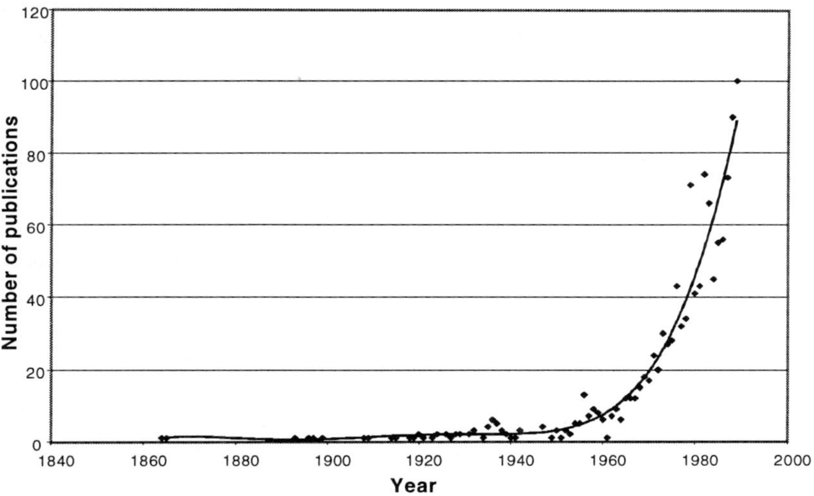

Fig. 1. The literature cited in the comprehensive book on chilling injury edited by Wang (1990) arranged by year of publication.

Whether in the field or in storage, plants adapted to tropical and subtropical growing conditions are most chilling-sensitive, while plants indigenous to temperate areas are for the most part chilling-tolerant (Table 1) (Saltveit and Morris, 1990). For example, chilling sensitivity occurs in many important agricultural crops of tropical or subtropical origin (e.g., avocados, bananas, corn, cotton, melons, rice, sorghum, sugarcane, and tomatoes). Other, non-chilling sensitive plants, appear to exhibit various levels of tolerance to chilling since a sufficiently severe chilling stress (i.e., a combination of temperature, duration, and contributing factors) will induce physiological disorders in most plants. For example, some plants indigenous to temperate areas (e.g., apples and peaches) are sensitive to chilling after a few months storage at temperatures below 5°C. Asparagus is another temperature crop that is injured at chilling temperatures. But it stores better for a few weeks at 5°C than at 0°C where chilling injury is more rapid, or at 10°C where it rapidly deteriorates. In most cases, however, the period required to induce an observable chilling symptom in temperate plants often significantly exceeds the duration the plant will remain free of senescent changes or microbial infections that terminate its life. Under modern methods of postharvest storage this duration can be exceeded. The breakdown of the flesh of some apple and peach cultivars during long-term storage at 0°C is thought to reflect metabolic dysfunction at such low temperatures.

Table 1. Some examples of commercial crops that are susceptible to chilling injury. Susceptibility depends on the plant part and cultivar tested.

FIELD CROPS	**FRUIT**	**VEGETABLES**
Corn**	Apples*	Asparagus
Peanut	Avocado	Basil
Rice	Banana	Beans, French
Sorghum	Citrus	Beans, Mung
Soybean	Mango	Cassava
Sugar cane	Papaya	Cucumber
	Peaches*	Eggplant
	Pineapple	Melons***
ORNAMENTALS	Plums*	Okra
African violet		Pepper
Coleus		Pumpkin
Orchids	**OTHER**	Squash
Poinsettia	Coffee	Sweet potato
Tradescantia	Ginger	Taro (and other aroids)
		Tomato

*   Only fruit of certain cultivars are chilling sensitive.
**   Only the plant is sensitive.
***   Probably all types of melons are sensitive.

It appears that most, if not all plants are intrinsically chilling-sensitive, and vary in their sensitivity to chilling because they have acquired various mechanisms to avoid or tolerate chilling during their evolutionary expansion into more temperate climates.   Despite an ever increasing research effort over the past quarter century, the molecular and physiological cause of chilling injury still awaits discovery.   For contemporary information and concepts on chilling, see reviews by Graham and Patterson (1982), Lyons (1973), Lyons and Breidenbach (1987), Markhart (1986), Minorsky (1985), Nishida and Murata (1996), Saltveit and Morris (1990), and Wang (1990).

Origin of Chilling Injury

Seed plants originated about 350 million years ago, while angiosperms (i.e., flowering plants), the division in which almost all agricultural crops are found, evolved from ancestral seed plants about 125 million years. Most of the extant angiosperm families are found in the tropics and it is in this environment that they presumably first arose. As in the tropics today, the environment was warm, humid and moderate, with the temperature probably fluctuating within narrow limits throughout the year.   As flowering plants continued to evolve, they radiated into the subtropics, and later into temperate areas. Continental drift and climactic changes during the ensuing millions of years altered local environments in which the plants grew and evolved, so that flowering plants presently grow in climates ranging from steamy rain forests to frozen tundra.

Many species adapted to the rigorous environmental conditions in temperate areas by adopting a variety of reproductive and growth strategies and a number of adaptive physiological mechanisms.   Some plants avoided low temperatures by coordinating their cycle of reproduction and growth with favorable local conditions of temperature and water availability, while others became physiologically tolerant of low temperatures and drought. Plants currently indigenous to the temperate areas of the world tolerate exposure to low, even freezing temperatures during parts of their life cycle. Important agricultural plants in this group include alfalfa, apple, beets, cabbage, peas, rye, strawberries and wheat. Those plants that remained in the tropics do not share this tolerance to cold temperatures.

Many species were apparently unable to adapt to extremes in temperature and remained confined to tropical areas of the world, or to the warm tropical environments during the summer months within temperate latitudes. In temperate areas, these chilling-sensitive crops are grown as "warm-season" crops, while the chilling-tolerant crops are grown as "cool-season" crops. A large number of important agricultural plants were only recently (in an evolutionary sense) moved to temperate growing areas from their tropical or subtropical centers of origin. These warm-season crops include beans, corn, cucumber, cotton, pepper, rice, sugar cane, sunflower, and tomato. Kotowski (1926) showed that the seed germination of several warm season crops failed at temperatures below 11°C, the threshold for chilling injury of many crops. In temperate areas, the requirement that their growing season fall within the warm tropical portion of the year severely limits the latitude and elevation at which these crops can be grown.

Pre-scientific Period

From the very beginning of agriculture and the dawn of civilization, the cultivation of crops has been an arduous task that has eclipsed all other human activities. With very few exceptions, the lives of men, women and children were short, brutish and filled with drudgery that produced meager harvests. Until recently, (and even now not an uncommon occurrence in many parts of the Third World), famine and pestilence were common features of everyday life. Famine occurred when there was insufficient food and was caused by poor harvests or by the appropriation of food by marauding armies or by callous overlords. However, the most common and reoccurring cause of famine resulted from the farmer's inability to store enough food to last from one harvest to another. The summer was often not an idyllic time of plenty, but a time when many went hungry. That period of time in late summer has been termed the "hunger gap" (Sweeney, 1995). It is not surprising then that the harvest festivals of agrarian societies are such joyous times for they herald the end of the yearly famine and a time of plenty.

In this environment, where securing enough food to survive was the paramount activity of almost every waking hour, the selection, cultivation, harvesting and storage of crops was not of passing consequence. Subsistence farmers, by necessity of their precarious existence, were very conservative and worked along traditional lines that were only slowly modified over the centuries as new ideas of practice significance occurred to the more observant farmers and spread by example. But change there was, pedestrian though it may have been. Only much later, when the wealth of the Roman and later empires supported the production of "luxuries" was there interest in growing exotic plants and crops, and even then their production was so limited as to make no impression on the diet of the average person.

A practical understanding of the effects of chilling on the growth of sensitive crops must have been repeatedly observed over millennia by attentive farmers. It is hard to imagine that the same perceptive farmers in ancient Greece and Rome who realized their fields could be fertilized by growing legumes (e.g., lupine, field bean, and vetch) (Fussell, 1972) were unaware that some plants were injured by exposure to low temperatures. Lucerne (alfalfa) reached Greece from Persia in the 5th century BC, followed by other legumes from Babylon and Egypt. Greek and Roman farmers learned empirically which crops flourished and which succumbed to the vagaries of climate. During the early centuries of the Roman Empire, many exotic plants were brought to Rome for display and study. Special environments were constructed to protect some of these plants from adverse climatic conditions implying a tacit awareness if not a knowledge of chilling sensitivity. Even the biblical admonishment in Ecclesiastes that "For everything there is a season", though metaphorical in its intent, implies a deep-seated understanding that there are climactic requirements and limitations that prescribe the successful agricultural environment for growing crops. For example, starting with the hegemony of Rome, herbs that were indigenous to the Mediterranean area were grown in carefully tended gardens throughout all of Europe. A few of the common herbs, e.g., basil, are very chilling-sensitive and their culture again implies a practical understanding of the effects of chilling on plants. It must have been obvious to early farmers not only in Europe, but also in Asia, India, Africa, and Mesoamerica that some crops were sensitive to low temperatures and had to be protected by enclosures or judicious planting dates to avoid exposure to potentially detrimental low temperatures. The successful cultivation of chilling-sensitive crops such as rice, sorghum, and corn in these areas testifies to the farmer's practical knowledge of the crop's climactic limitations.

My cursory reading of the tomes written to promote agriculture by the Greeks Mago the Carthaginian, Hesiod, and Xenophon around 500 BC, and by the Romans Varro, Cato the Censor, and Columella writing from the 2nd century BC to the 1st century AD, failed to identify any specific reference to or knowledge of chilling injury. Arabian agricultural scholars, such as ibn al Awam, were well acquainted with tropical and sub-tropical farming and their writings promoted the introduction of many chilling sensitive crops (e.g., cotton, melons, oranges, rice, saffron and sugar cane) into Europe after their conquest of Spain in the 5th century AD (Fussell, 1972). Most of the later encyclopedic writings on agriculture, even well into the

Renaissance, contained a scholastic recycling of these and other seminal authors and contained very little newly discovered empirical knowledge (Comet, 1997; Sweeney, 1995). We must proceed well into the 18th century before there are recognizable references to chilling injury in the extant literature.

Scientific Study of Chilling

The first published scientific work dealing with chilling injury was written in 1778 by Bierkander (Molisch, 1896), who listed several species of plants that were killed when exposed to low, non-freezing temperatures. Similar results were reported by Göeppert (1830). These reports obviously did not engender much interest in the plant scientists of their day because it took many years for the next paper to appear. In 1844, Hardy reported that 25 of 56 tropical plants tested were killed by exposure to temperatures in the range of 1 to 5°C (Molisch, 1896). Later, Sachs (1865) reported that exposure to chilling temperatures for one to a few days resulted in the death of plants of tropical and subtropical origin. These mainly descriptive reports reflected the expanding interest of European scientists in plant science in general, and their fascination with plants being discovered in the New and Old World tropics.

The 1850's and 60's was a time of extensive exploration in tropical and subtropical areas around the world, and many new plants were being brought back to Europe for study and display. The Swedish botanist Carolus Linnaeus had developed the modern system of taxonomic nomenclature only a century before and the study and naming of plants from exotic areas of the world was one of the major scientific endeavors of the day. Darwin fueled interest in the natural sciences with the publication of *The Origin of Species* in 1859. Commodore M. C. Perry visited Japan in 1853 and rekindled an interest in the Orient among Westerners. Tropical Africa and South America were just starting to be extensively explored during this period. In 1858, Lake Victoria was identified as the headwaters of the Nile and the interior of Africa was yielding some of its secrets. Rubber trees in the Amazon basin were being tapped to supply rubber to Europe in the late 1860's and the search for more trees fueled extensive exploration of the South American tropics by many naturalists.

Chilling injury was also of economic importance, and expanding trade and storage of fresh fruits and vegetables made the effects of chilling on agricultural crops more obvious. Chilling injury in fresh market tomatoes appears to have been first reported by Benson (1893) in Australia. He reported that storage at 5°C was harmful to tomato fruit, and that sensitivity to chilling was significantly influenced by the maturity of the fruit. Adams (1923) reported that mature-green tomato fruit stored at 0°C failed to ripen but that ¾ ripe fruit stored at 2°C for up to three weeks ripened to an acceptable level of quality. Diehl (1924) also reported that turning fruit was less chilling-sensitive than mature-green fruit. Rosa (1927) noted that both the retardation of ripening and tissue breakdown increased as the period of cold storage was extended and the temperature lowered for all maturities. However, Baker (1928) reported that both green and fully ripe fruit were injured by chilling. Wright *et al.* (1931) found that although firm, fully ripe fruits kept well at 0°C for 8–10 days, they were still chilling-sensitive because they soon broke down when transferred to warmer temperatures.

Belehradek (1935) listed a large number of changes that were associated with chilling injury, and pointed out that since all these changes were complex, none was likely to be the primary cause of the phenomenon of chilling injury. However, his review was almost exclusively concerned with injury that occurred after relatively long exposures. Luyet and Gehenio (1940) included cases of rapid injury to chilling and suggested that the mechanism causing injury soon after exposure was probably different from the mechanism responsible for injury occurring after prolonged exposure. Gehenio and Luyet (1939) had previously proposed that a reversible phase transition of the protoplasm from sol to gel occurs at the chilling temperature. After sufficient time, the gel would set and cause irreversible injury. Lewis (1956a), studying the effects of chilling on protoplasmic streaming in tomato trichome cells, found that the protoplast formed into aggregated clumps after 2 days at 0°C. Smith (1947) objected to any all-embracing theory for explaining chilling injury because of the fact that there are at least several types of injury that can be symptomatically distinguished.

Although these reports show that a few plant scientists were interested in the effects of chilling, a knowledge of the importance of chilling was lacking among most agricultural scientists. Many early recommendations for the storage of chilling-sensitive perishable fruit and vegetables were often at temperatures and durations now known to cause chilling injury. For example, snap beans (*Phaseolus vulgaris* L.) should not be stored longer than 2 weeks at chilling temperatures, but recommendations for their storage include 3 weeks at 0 to 3.5°C (Krause, 1927), 4 weeks at 0.5°C (Williams, 1936), or 1 to 3°C (Smith, 1935). Other researchers (e.g., Plank and Schneider, 1928) reported injury under these conditions (e.g., 0.5°C for 10 days), but it is doubtful that they were aware of the chilling injury phenomenon. For example, Platenius *et al.* (1934) reported that snap beans developed an injury known as "weathering" within one day at 21°C, after storage at 0°C for 8 days, 4.4°C for 12 days, or 10°C for 18 days.

The increase in the number of publications dealing with chilling after 1930 appears to have been interrupted by the Economic Depression and World War II (Fig. 1). Concomitant with the economic expansion following these social upheavals, the scientific investigation into the chilling injury of plants expanded greatly after 1950, and continues to increase as we enter the 21st century.

Theories of Chilling Injury

Many theories have been proposed to explain the physiological causes of chilling injury. They include cessation of protoplasmic streaming, accumulation of toxic compounds, membrane phase changes, altered metabolism, and general biochemical changes. Surprisingly most of these theories have persisted to this day in one form or another, and most of their major aspects are incorporated into currently accepted theories.

Protoplasmic Streaming

Sachs (1864) studied protoplasmic streaming in petiole hairs of pumpkin and tomato. He cited observations by earlier workers that streaming in several lower plants ceased near 0°C. It appeared that the minimum temperature for protoplasmic streaming in certain chilling susceptible plants approximated that of the upper limit of their chilling range. This suggested the possible involvement of cessation of streaming in the series of events resulting in chilling injury. This idea was later revived by Lewis (1956b), who observed that streaming in trichomes of chilling-sensitive honeydew melon, sweet potato, tobacco, tomato, and watermelon ceased almost immediately at 10°C or below. For example, protoplasmic streaming in petiole trichomes of tomato ceased after 1 to 2 minutes at 10°C. In contrast, streaming proceeded even at 0°C in chilling-tolerant radish and carrot. Prolonging the chilling exposure of tomato trichomes at 0°C delayed resumption of streaming when transferred to 20°C, and if chilling exceeded 24 hours, streaming failed to resume upon warming. Lewis suggested that chilling injury and cessation of protoplasmic streaming were separate symptoms of some fundamental disorder that is induced in chilling-sensitive plants at temperatures below 10°C. Alternatively, he suggested that the loss of streaming might induce altered respiration leading to chilling injury.

More recently, Das *et al.* (1966), Patterson and Graham (1977), and Woods *et al.* (1984) extended the studies of protoplasmic streaming to other plants and characterized the effects of various rates of cooling on streaming. A massive influx of calcium at the chilling temperature is currently proposed to be the event which depolymerizes elements of the cytoskeleton and stops streaming. Streaming is therefore thought to be a secondary effect of changes in membrane permeability. However, the kinetics of depolymerization occurs separately from the calcium rise, and whereas calcium levels return to normal levels in the cold, microtubules remain depolymerized. The depolymerization of cytoskeletal elements may be a primary cause of chilling.

Accumulation of Toxic Compounds

One of the earliest theories related to the cause of chilling injury proposed that it was the result of the accumulation of toxic levels of metabolites caused by temperature-induced imbalances in metabolism. Molisch (1896) thought that the development of severe brown spotting on tropical plants after a short exposure to 1 to 4°C was a type of cellular breakdown resulting from the accumulation of toxic substances in the cells due to incomplete oxidation. Nelson (1926) suggested that surface pitting was associated with the products of abnormal respiration. He thought that a fragment of a hydrolyzed glucoside was the toxic material and that chilling prevented its detoxification. The marked effect of low temperatures on the rate of respiration in papaya was interpreted by Jones (1942) as indicating that chilling injury is associated with an abnormal course of respiration. In fact, accumulation of toxic compounds as the result of abnormal metabolism was the only theory discussed at length in an early review by Pentzer and Heinze (1954).

Certain treatments, such as intermittent warming, appear to reduce the development of chilling injury because they facilitate the metabolism of accumulated toxic compounds. Intermittent warming is used to break up a period of chilling exposure that would cause injury into shorter segments, each one of which will not alone cause permanent injury. Enhanced respiration and ethylene production during the warming period is considered to be evidence of the metabolism and detoxification of accumulated toxic compounds. Kidd and West (1935) reported that apple scald (a chilling symptom) could be controlled in susceptible varieties by briefly warming the fruit during cold storage. They postulated that a toxic substance accumulated at cold storage temperatures and was metabolized at warm temperatures. Smith (1947) provided support for the accumulation of toxic material by showing that severe chilling injury could be prevented by a short period of warming. Morris *et al.* (1947) also found that the severity of injury to cucumbers and summer squash could be reduced by interrupting storage at 0°C with a short period at 21°C.

Plank (1938, 1941) mathematically analyzed chilling injury data obtained from plums and other fruit in South Africa and concluded that two main metabolic reactions are involved in chilling injury. One reaction produces a toxic substance while the other reaction consumes it. By adjusting the temperature coefficient for the two reactions he showed that the toxic substance would accumulate below a critical temperature and produce chilling injury. In support of this idea, Jones (1942) reported that the decline in respiration with temperature showed a marked deviation at chilling temperatures in papaya. Morris *et al.* (1947) found a lower respiratory quotient in summer squash at chilling versus non-chilling temperatures, and interpreted this as evidence of incomplete oxidation at chilling temperatures. However, Platenius (1942) had previously found that the respiratory quotients of ten chilling-sensitive fruits and vegetables did not show deviations over a range of temperatures from 0.5 to 24°C that would suggest an abnormal rate or course of respiration at chilling temperatures.

Many compositional changes occur in sugars and organic acids after chilling, but chemical analyses have generally failed to find significant correlations between these changes and changes in chilling sensitivity. For example, Miller and Heilman (1952) proposed that the destruction of ascorbic acid was the initial step in the development of chilling injury in pineapple. However, it was shown that decreases in ascorbic acid level of tomato fruits occurred only after the development of visible symptoms of injury and is certainly not one of the first phases in the development of chilling injury (Craft and Heinze, 1954; Lewis, 1956a).

If a toxic substance is produced, it is not very mobile. Eaks and Morris (1957) showed that when one end of a cucumber was chilled while the other was kept warm, injury occurred only in the chilled half. The demarcation separating injured from healthy tissue was sharp, showing that the cause of chilling injury was not systemic. However, although there is abundant circumstantial evidence for the toxic compound, no compound has been isolated from chilled tissue that fulfills the criteria of the toxic material. The failure of chilled tissue to produce an essential metabolite could just as well be proposed as a cause of chilling injury. This idea has received scant examination.

Membrane Phase Changes

Guignard (1909) may have been the first to suggest an involvement of cellular membranes in chilling injury. He observed the release of hydrocyanic acid from sorghum when exposed to chilling temperatures or chloroform vapors, and proposed that both caused increases in the permeability of certain cell membranes. Pantanelli (1919) reported changes in the cellular permeability of mandarin endocarp tissue held for 12 hours at temperatures near freezing, while Seible (1939) found changes in the permeability of chilled coleus cells to calcium nitrate.

A major component of cellular membranes are the phospholipids which form a bilayer that regulates the passive permeability of the membrane. These molecules are in turn composed of fatty acids which have different physical properties, e.g., melting points. Changes in the fatty acids making up a membrane can, therefore, significantly alter the temperature at which it undergoes a phase transition from solid to gel. Since the implied phase transition occurs at different temperatures in chilling-sensitive and tolerant species, the fatty acid composition of chilling-sensitive and tolerant plants was of interest to early researchers. As early as 1927, analyses of the fatty acid composition of membranes from plants of tropical origin showed that they generally possessed more highly saturated lipids than plants from temperate regions (Pearson and Raper, 1927). Similar analyses and results have been reported by many researchers over the past 70 years for many species subjected to a variety of pre- and post-chilling treatments. Although indicative of some relationship between membrane composition and chilling sensitivity, the purely correlative nature of the data and the inherent dissimilarity among the species being compared prevented this line of inquiry from reaching a definite conclusion.

A model describing the physiological cause of chilling that was based on the phase transition of membranes was presented in 1973 by Lyons (1973). It incorporated previous proposals by a generation of scientists including Lyons and Raison (1970), Raison *et al.* (1971b), Singer (1971), Levitt (1972), and Singer and Nicolson (1972), and brought together the ideas of a phase change of the membrane phospholipids and the resulting alterations in the metabolism of chilled cells. This model proposed the existence of a primary event (i.e., membrane phase transition) that resulted in a series of secondary events which in turn produced the symptoms of chilling injury (Raison, 1973) (Fig. 2).

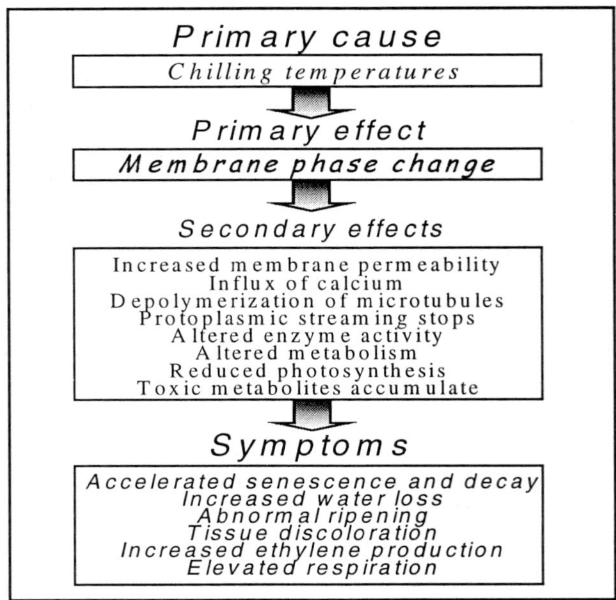

Fig. 2. Relationships among the factors involved in chilling injury according to the paradigm proposed by Lyons (1973).

A great deal of research has been done in the ensuing years to show that a membrane phase transition actually does occur at the chilling temperature and that it is confined to chilling-sensitive plants. However, much of this work involved the extraction and analysis of lipids from whole tissue or from isolated membranes. It has frequently been forgotten that work predating the proposal of the model in 1973 indicated that only a small portion of the phospholipids were involved in the critical phase transition (Raison *et al.*, 1971a). It appears that only certain discrete domains of the membrane phospholipids actually undergo the phase transition which results in alteration of the activation energy of membrane bound enzymes. If these domains constitute a small portion of the total lipids in the cell, then an analysis of differences in bulk lipids or lipids extracted from all the cellular membranes would probably not show the critical differences between tolerant and sensitive species or between conditioning treatments that modified chilling sensitivity. Recent studies with genetically transformed plants may provide a conclusive answer to the contribution that membrane composition has to chilling sensitivity.

Chilling-sensitive and tolerant plants have been genetically transformed to have altered levels of saturated and unsaturated fatty acids in their membrane lipids, and these modifications alter the plant's chilling sensitivity in the light (Nishida and Murata, 1996). For example, decreasing the component of unsaturated fatty acids from 64% to 24% increased chilling-induced inhibition of photosynthesis from 25% to 88% in tobacco leaf discs, while increasing unsaturated fatty acids from 66% to 72% decreased photosynthetic inhibition from 25% to 7%. These and other similar results clearly show that increasing the proportion of unsaturated fatty acids in the chloroplast increases the plant's resistance to chilling in the light. However, the corollary of increasing the chilling sensitivity of chilling tolerant plants by increasing the saturation of their fatty acids did not occur (Wu and Browse, 1995). Obviously, a simple phase transition model based on the degree of saturation of fatty acids in the membrane cannot totally account for chilling injury.

If a phase transition did occur at the chilling temperature and if that change altered membrane permeability, then there should be an immediate and measurable change in permeability as the tissue transits the critical chilling temperature. In fact, a rapid increase in membrane permeability was proposed to accompany the phase change in the lipids of chilling-sensitive tissues at the critical temperature (Lyons, 1973). Yet this immediate increase in permeability has never been observed. Kinetic studies (e.g., Saltveit, 1989) showed that enhanced permeability develops slowly during chilling and could require days or weeks to become significantly greater than the controls. It appears likely that some aspect of chilling injury produces damage to cellular membranes which then results in increased leakage.

A membrane phase change at the chilling temperature could also alter the activity of associated enzymes. An Arrhenius plot of enzyme activity versus temperatures should therefore show a discontinuity at the critical temperature for chilling-sensitive plants, whereas no such discontinuity should be observed for chilling-tolerant plants. Lyons and Raison (1970) showed that such discontinuities were present in chilling-sensitive sweet potato roots and, tomato and cucumber fruit, but not present in chilling-tolerant beet roots, potato tubers and cauliflower buds. Contemporary scientists quickly pointed out that the data could as easily show a smooth transition across the critical temperature as an abrupt discontinuity. Later, researchers suggested that the method of analysis could significantly affect both the appearance of the discontinuity and the apparent critical temperature (Iwaya-Inoue *et al.*, 1989; Wolfe and Bagnall, 1980). However, despite these reservations in the validity of the analysis, the preponderance of research articles have identified a discontinuity in Arrhenius plots at approximately the critical temperature for chilling. Whether this discontinuity is causally related to a membrane phase change or to some other temperature-induced change is still debated.

Although membrane lipid phase separation is associated with chilling of tomato fruit, the phase separation has been ascribed to an indirect effect of low temperatures (Sharom *et al.*, 1994). The phase transition was not reversible when the tissue was warmed above the chilling temperature. In fact, there was an increase in gel phase lipids upon returning the tissue to 25°C that coincided with development of visual chilling symptoms. They concluded that active phospholipid catabolism and accumulation of free fatty acids in the membrane followed chilling and preceded formation of gel phase lipids.

The membrane phase transition model has remained compatible with much of the empirical observations and experimental data collected over the years. Although this model is currently considered to have many critical shortcomings, it has been the predominant hypothesis over the past 25 years and has stimulated a great deal of interest and research in the area of chilling injury. However, the involvement of a membrane phase transition in chilling injury is apparently still unresolved.

Altered Metabolism

Blackman and Parija (1928) noted that the respiration of apples increased when they were transferred from a chilling to a non-chilling temperature. This increase was later observed for cucumbers (Eaks, 1952), tomatoes (Lewis, 1956a), asparagus (Morris and Watada, 1959), and citrus (Eaks, 1960). Blackman and Parija (1928) suggested three possible causes for this stimulation in efflux of carbon dioxide: 1) a change in the solubility of carbon dioxide, 2) a change in the equilibrium between starch and sugar, and 3) a change in the concentration of intermediate respiratory compounds. The first source was conclusively ruled out by Morris and Watada (1959), who showed that the change in gas solubility was insufficient to account for the carbon dioxide produced, and that there was a concomitant rise in oxygen uptake by the tissue. The second source is a possibility, but there is no correlation between total and reducing sugars and the respiration of sweet potatoes (Hasselbring and Hawkins, 1915; Appleman and Smith, 1936), or in potatoes, beets, turnips and dahlia roots (Appleman and Smith, 1936). The third possibility, often interpreted as the accumulation of toxic intermediates, is intriguing, but no one has yet isolated such a compound.

A pattern of increased respiration during early exposure to chilling temperatures has been reported for cucumbers (Eaks and Morris, 1956), tomatoes (Lewis, 1956a), green string beans (Tewfik and Scott, 1954), and sweet potatoes (Lewis and Morris, 1956; Wheaton, 1963), which are all sensitive to chilling. However, it has also been reported for potatoes (Hopkins, 1924; Platenius, 1942), peas (Tewfik and Scott, 1954), and *Vicia faba* seedlings (Palladine, 1899), which are all insensitive to chilling.

The respiration rate of several crops were plotted according to the Arrhenius equation to identify the temperature at which changes occurred in the rate of respiration. Chilling-sensitive crops like tomatoes (Lewis, 1956a), cucumber and summer squash (Eaks, 1952), and asparagus (Lipton, 1957) showed a marked change at 19, 16, and 8.5°C, respectively. In contrast, chilling-insensitive crops like lettuce (Pratt *et al.*, 1954), artichoke (Rappaport and Watada, 1958) and Brussels sprouts (Lyons and Rappaport, 1959) showed essentially no change. If the activation energies of a few key enzymes have large temperature coefficients near the chilling temperature and their enzyme activities change at different rates, then at some critical temperature there could be a crossover of control, and metabolism could shift to the unregulated production of inappropriate compounds that could accumulate to toxic levels. Such changes could also occur during temperature-induced phase changes in critical membranes if the activities of the enzymes were tightly coupled with the affected membranes.

Biochemical Changes

Shichi and Uritani (1956) were perhaps the first to use a biochemical approach to study the cause of chilling injury. They measured the ATP content of chilled disks of sweet potato roots and they concluded that ATP formation was retarded after 8–10 days at 0°C. However, the dinitrophenol test they used would only indicate a loosely coupled system and provided no direct evidence that less ATP was produced. They concluded that the primary effect of chilling was a coagulation of the lipid components of the cellular particulates, which leads to destruction of cellular structure and phosphorylation capacity.

Lieberman *et al.* (1958) compared the oxidative and phosphorylative activity of mitochondria isolated from control and chilled sweet potato roots. Mitochondrial activity began to decline by the 5th week in storage and was completely lost by the 10th week. There was a parallel between the increased concentration of chlorogenic acid (a phenolic compound) and the loss of mitochondrial activity.   Phenolic compounds can inhibit mitochondrial oxidative activity (Lieberman and Biale, 1956).   Since the mitochondria were exposed to cellular phenolic compounds during their extraction and isolation, the loss of activity could have been the result of this exposure to the increasing concentrations of chlorogenic acid and polyphenolic compounds rather than to chilling. Using the results of later studies, Lewis (1961) suggested that chilling caused a decline in oxidative phosphorylation which led to a dearth of energy needed to maintain cellular organization, and eventually to symptoms associated with prolonged chilling.

Other Causes

Many symptoms of chilling injury involve an accelerated rate of water loss from the damaged tissue.   Elevated relative humidity surrounding the commodity during or after chilling will reduce the rate of water loss and reduce the development of these chilling injury symptoms. Molisch (1896) studied the effect of a water-saturated atmosphere during the chilling of susceptible leaves and plants. He found that preventing water loss did not prevent death from chilling. His studies followed those of Sachs (1865), who proposed that the chilling-induced leaf wilting observed in many chilling-sensitive plants was the result of an inability of the roots to absorb and transport sufficient water from the cold soil to the leaves. Death was proposed to result from a transpirational deficit and dehydration. While reduced water uptake and transport has been reported for about 10 species of chilling-sensitive plants to date, water stress probably does not account for the rapid death of many chilling-sensitive species. Tissues within a larger organ still exhibit chilling sensitivity even though they are in a water-saturated environment and lose no water. Moreover, chilling injury occurs in harvested crops held at 100% relative humidity. Water loss is therefore not so much a cause as a symptom of chilling injury. Morris and Platenius (1938) emphasized that physical injury of the tissue and relative humidity of the environment had a significant influence on the development of chilling injury symptoms such as wilting and pitting, but not on the level of chilling sensitivity.

Alleviation of Chilling Injury

Kidd *et al.* (1927) appear to have been the first to use controlled atmospheres to reduce chilling injury.   Low oxygen inhibited the development of internal browning in a chilling-sensitive apple cultivar, while high carbon dioxide levels increased its incidence and severity. Avocados (Spalding and Reeder, 1972) and stone fruit (O'Reilly, 1947) usually benefit from controlled atmosphere, but certain treatments produce off-flavors in some cultivars. The response of chilled citrus to controlled atmospheres is variable (Miller, 1946), whereas it leads to additional injury in many crops. Mature-green and breaker tomato fruit showed no reduction in chilling injury when held 10 days at 0°C in 3 to 21% oxygen (Morris and Kader, 1975). In addition, 5 to 20% carbon dioxide injured the fruit and the level of injury was additive to that caused by chilling alone. Cucumber fruit reacted in a similar fashion (Eaks, 1956). Chilled asparagus, bell peppers and okra also did not benefit from altering the atmosphere (Forney and Lipton, 1990).

Wartenburg (1929) found that repeated exposure of young chilling-sensitive plants (e.g., *Phaseolus vulgaris*) to a few hours of chilling at 0°C had a cumulative effect on the subsequent development of chilling injury symptoms. The poorer shelf-life of autumn grown "warm season" crops has been attributed to partial chilling in the field. Although the crops recover and show no chilling symptoms, they do become more sensitive to subsequent stresses like chilling.

In contrast, a more severe exposure near the limits necessary to induce chilling injury itself, will induce resistance to subsequent chilling. Spranger (1941) observed that ornamental plants were less chilling-sensitive when grown for a few months at temperatures just above chilling compared to plants grown at room temperature. Wheaton and Morris (1967) found that a 48 hour exposure of tomato seedlings to 12.5°C produced the maximum protection against chilling injury. This conditioning effect has been commercially used to increase the chilling resistance of some harvested crops (Lyons and Breidenbach, 1979).

Conditioning can also be done at higher temperatures. One effect of warm temperatures is to enhance wound healing and curing after harvest. Since many chilling injury symptoms are exacerbated by prior physical injury, this could certainly be a part of the conditioning effect. Interestingly, holding tomato pericarp tissue for 6h at ever increasing temperatures from 10 to 32°C resulted in an increase, not a decrease, in subsequent chilling sensitivity (Saltveit, 1991). However, if the conditioning temperature was raised just a few more degrees to 34°C, chilling sensitivity abruptly and dramatically decreased. This induced chilling tolerance has been subsequently confirmed by others (Lurie and Klein, 1991) and been found to correlate with the appearance of heat shock proteins (Lafuente *et al.*, 1991; Collins *et al.*, 1995; Sabehat *et al.*, 1996). The mode of action by which heat shock induces chilling tolerance awaits elucidation.

Is There a Primary Cause of Chilling?

The assumption is commonly made that a symptom developing soon after chilling is more closely related to the primary effect of chilling than is a symptom that develops later. Similarly, a symptom that develops after brief chilling is considered to be more closely related to the primary event than is a symptom arising only after prolonged chilling. Using these criteria, many of the changes that occur after chilling can temporally be eliminated as the cause of chilling. For example, the repeated observation that membrane permeability does not increase immediately upon chilling, and the recent report that the phase change in membranes of chilling-sensitive plants does not occur until after chilling, both call into question the primacy of a membrane phase transition in causing chilling injury.

The sensitivity to chilling ranges over a wide spectrum of temperatures and exposure times, from tropical plants that are injured by a few hours at 18°C to temperate plants that are injured only after many weeks at 0°C. It seems unreasonable to ascribe similar physiological causes of chilling injury to plants that exhibit these extremes in temperature sensitivity. It is commonly thought that a large number of economically important plants share a similar threshold temperature for chilling injury, and that this similarity implies that a common temperature dependent physical event is involved. However, the common threshold temperature is an illusion that disappears when one includes such factors as duration of exposure, tissue type, and prior and subsequent exposure to other stresses in the factors affecting chilling sensitivity.

There is often a wide variation in chilling sensitivity within a family or genus, with some species being very chilling-sensitive while others display varying levels of chilling tolerance. The appearance of chilling tolerance in diverse species that are unrelated evolutionarily indicates that chilling tolerance arose independently a number of times and probably by a number of different mechanisms. As plant species radiated out from the tropics into the harsher temperate climate, they probably evolved a number of mechanisms to avoid injury at low temperatures. It is unlikely that an abrupt mutation occurred conferring chilling resistance to chilling-sensitive plants. Rather, it is more likely that a number of small changes accumulated over time, each conferring a small level of chilling tolerance. These changes could include modifications to cytoskeletal elements, membrane components (e.g., saturation of fatty acids, ion pumps), isoenzymes, metabolic pathways, chloroplasts, mitochondria, etc.   The multiplicity of possible changes is reflected in the independent appearance of chilling tolerance in evolutionarily divergent lines, and with the occurrence of chilling-sensitive and tolerant species within the same family. For example, within the genus *Lycopersicon* there is a wide variation in chilling sensitivity (Vallejos *et al.*, 1983).   The commercial tomato, *L. esculentum* is much more sensitive than the *L. hirsutium*.   If tolerance to chilling did in fact accumulate over time by successive small changes, then the current emphasis on finding a single cause for chilling injury would be unproductive, since there would be no single cause to discover.   Rather a number of factors each conferring a small level of tolerance would need to be sequentially identified and characterized.   The profusion of chilling injury symptoms and the vast differences in temperature thresholds and durations of exposure that induce chilling injury argue for a multiplicity of causes.

A shift in the basic paradigm which guides research in a specific area of science often precedes rapid advances in that field of study (Kuhn, 1970). For a quarter of a century the basic paradigm in the study of chilling injury has been that a phase change in some cellular membrane is the transducing factor leading from low, chilling temperatures to physiological consequences (Lyons, 1973).   Recent studies using molecular biology and genetic engineering to change the saturation of fatty acid in membranes have failed to definitively show that the phase transition of a membrane is the primary cause of chilling injury.   Although decreasing the saturation of fatty acids in chloroplast membranes reduced the sensitivity of chilling-sensitive tobacco (Nishida and Murata, 1996), the corollary of increasing the saturation of fatty acids in membranes did not increase the sensitivity of chilling-tolerant *Arabidopsis* plants (Wu and Browse, 1995).

The study of chilling would be different if we start from the premise that all seed plants were initially chilling-sensitive and that during their evolutionary expansion into temperate climates they acquired various levels of tolerance by various biochemical and physiological modifications. The changes which conferred tolerance within a specie in one family were likely different from those conferring tolerance within a specie in another family.   Of the many temperature-induced changes that do occur in the physiology of chilling-sensitive plants, the most important change that causes chilling injury in one specie is probably different from the change that is most important in causing injury in another.   Employing these criteria, it would be more reasonable to search for the many mechanisms that cause injury and the many mechanisms that have evolved to confer tolerance than to search for the single mechanism that supposedly underlies the basis for all chilling sensitivity (Fig. 3). A new paradigm based on a multiplicity of causes may now be warranted.

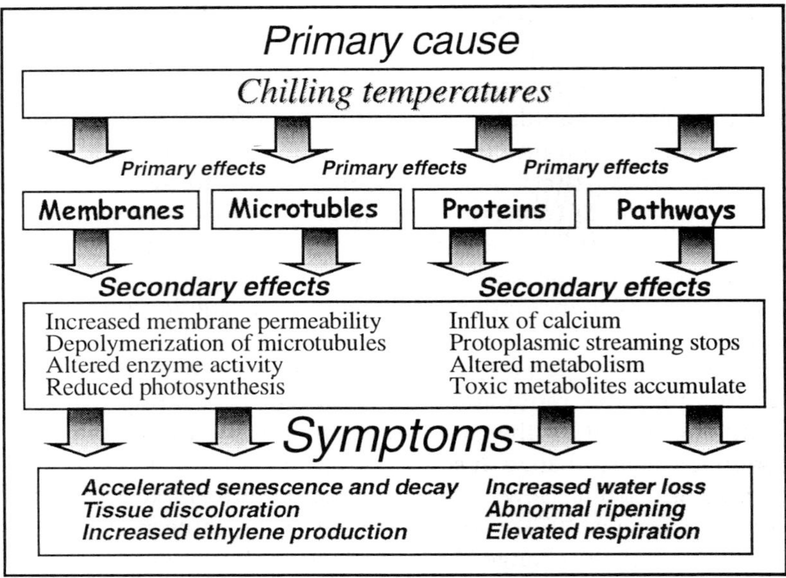

Fig. 3. Possible interrelationships among the primary and secondary factors that produce the symptoms of chilling injury.

References

Adams, D. B. (1923) Cool storage of tomatoes. Experiments in keeping qualities. *J. Dept. Agric. Victoria* **21**: 621-622.

Appleman C. O. and Smith, C. L. (1936) Effect of previous cold storage on the respiration of vegetables at higher temperatures. *J. Agr. Research* **53**: 557-580.

Baker, J. (1928) Cold storage trials with tomatoes. *D.S.I.R. Rept. Food Invest. Bull.* **1927**: 43-44.

Belehradek, J. (1935) Temperature and living matter. *Protoplasm Monographien* **8**, Berlin, pp. 277.

Benson, A. H. (1893) Cold storage of fruit. *Agric. Gaz. N.S.W.* **4**: 870-877.

Blackman, F. F. and Parija, P. (1928) Analytic studies in plant respiration. I. The respiration of a population of senescent ripening apples. *Proc. Roy. Soc (London), B.* **103**: 412-445.

Collins, G. G., Nie, X. L. and Saltveit, M. E. (1995) Heat shock proteins and chilling injury of mung bean hypocotyls. *J. Expt. Bot.* **46**: 795-802.

Comet, G. (1997) Technology and Agricultural Expansion in the Middle Ages: The example of France north of the Loire. In *Medieval Farming and Technology*, G. Astill and J. Langdon, eds., Vol. 1, Koninklijke Brill, Leiden, The Netherlands, pp. 321. ISBN 9-0041-0582-4.

Craft, C. C. and Heinze, P. H. (1954) Physiological studies of mature-green tomatoes in storage. *Proc. Amer. Soc. Hort. Sci.* **64**: 343-350.

Das, T. M., Hilderbrandt, A. T. and Riker, A. J. (1966) Cinephotomicrography of low temperature effects on cytoplasmic streaming, nucleolar activity and mitosis in single tobacco cells in microculture. *Am. J. Bot.* **53**: 253-259.

Diehl, H. C. (1924) The chilling of tomatoes. *U.S. Dept. Agric. Circ.* **315**: 5.

Eaks, I. L. (1952) The post-harvest physiology of cucumber (Cucumis sativus L.) and summer squash (Cucurbita pepo L.) at chilling and non-chilling temperatures. Ph.D. dissertation. Univ. Calif., Davis, pp. 137.

Eaks, I. L. (1956) Effect of modified atmospheres on cucumbers at chilling and non-chilling temperatures. *Proc. Amer. Soc. Hort. Sci.* **67**: 473-478.

Eaks, I. L. (1960) Physiological studies of chilling injury in citrus fruits. *Plant Physiol.* **35**: 632-636.

Eaks, I. L. and Morris, L. L. (1956) Respiration of cucumber fruits associated with physiological injury at chilling temperatures. *Plant Physiol.* **31**: 308-314.

Eaks, I. L. and Morris, L. L. (1957) Deterioration of cucumbers at chilling and non-chilling temperatures. *Proc. Amer. Soc. Hort. Sci.* **69**: 388-399.

Forney, C. F. and Lipton W. J. (1990) Influence of controlled atmosphere and packaging on chilling sensitivity. In *Chilling Injury of Horticultural Crops*, Wang C. Y., ed., pp. 257-267.

Fussell, G. E. (1972) *The Classical Tradition in West European Farming*. David & Charles Limited, South Devon House, Newton Abbot, Devon, England, pp. 237, ISBN 0-7153-5564-3.

Gehenio, P.M. and Luyet, B. J. (1939) A study of the mechanism of death by cold in the myxomycetes. *Biodynamic* **2**: 1-22.

Göeppert, H. R. (1830) *Uber die Wärmeentwicklung in den Pflanzen, deren Gefrieren und die Schutzmittel gegen dasselbe*. Max and Comp., Berlin.

Graham, D. and Patterson, B. D. (1982) Responses of plants to low, non-freezing temperatures: proteins, metabolism and acclimation. *Annu. Review Plant Physiol.* **33**: 347-372.

Guignard, L. (1909) Influence de l'anesthesie et du gel sur le dedoublement de certaine glucosides chez les plantes. *Compt. rend. de l'Acad. Sci.* **149**: 91-93.

Hasselbring, H. and Hawkins, L. A. (1915) Respiration experiments with sweet potatoes. *J. Agric. Res.* **5**: 509-517.

Hopkins, E. F. (1924) Relation of low temperatures to respiration and carbohydrate changes in potato tubers. *Bot. Gaz.* **78**: 311-326.

Iwaya-Inoue, M., Sakaguchi, K. and Kaku, S. (1989) Statistical studies using AIC method to decide the question of 'break' or 'straight' in Arrhenius plots of water proton NMR relaxation times in chilling-sensitive Vigna and insensitive Pisum seedlings. *Plant & Cell Physiol.* **30**: 309-316.

Jones, W. W. (1942) Respiration and chemical changes of the papaya fruit in relation to temperature. *Plant Physiol.* **17**: 481-486.

Kidd, F. and West, C. (1935) The cause and control of superficial scald of apples. *D.S.I.R. Rept. Food Invest. Bd.* **1934**: 111-117.

Kidd, F., West, C. and Kidd, M. N. (1927) Gas storage of fruit. Special Rep. No. 30, Sci. Industrial Research Dept., London 1927, 87.

King, A.I., Joyce, D.C. and Reid, M. S. (1988) Role of carbohydrates in diurnal chilling sensitivity of tomato seedlings. Plant Physiol. **86**: 764-768.

Kotowski, F. (1926) Temperature relations to germination of vegetable seed. *Proc. Amer. Soc. Hort. Sci.* **23**: 176-184.

Krause, M. (1927) Versuche uber die kaltlagerung von obst und gemuse. From abstract in *Institut International DuFreie*. **6**: 1089-1092.

Kuhn, T. S. (1970) *The Structure of Scientific Revolutions*. 2nd ed., Univ. Chicago Press, Chicago, pp. 210.

Lafuente, M. T., Belver, A., Guye, M. G. and Saltveit, M. E. (1991) Effect of temperature conditioning on chilling injury of cucumber cotyledons. *Plant Physiol.* **95**: 443-449.

Levitt, J. (1972) *Responses of plants to environmental stress.* Academic Press, New York, pp. 697.

Lewis, D. A. (1956a) Physiological studies of tomato fruit injured by holding at chilling temperatures. Ph.D. dissertation, Univ. Calif., Davis, p. 153.

Lewis, D. A. (1956b) Protoplasmic streaming in plants sensitive and insensitive to chilling temperatures. *Science* **124**: 75-76.

Lewis, D. A. and Morris, L. L. (1956) Effect of chilling storage on respiration and deterioration of several sweet potato varieties. *Proc. Amer. Soc. Hort. Sci.* **68**: 421-428.

Lewis, T. L. (1961) Physiological studies of chilling injury in tomato fruits. Ph.D. dissertation, Purdue Univ., p. 85.

Lieberman, M. and Biale, J. B. (1956) Oxidative phosphorylation by sweet potato mitochondria and its inhibition by polyphenols. *Plant Physiol.* **31**: 420-424.

Lieberman, M., Craft, C. C., Audia, W. V. and Wilcox, M. S. (1958) Biochemical studies of chilling injury in sweetpotatos. *Plant Physiol.* **33**: 307-311.

Lipton, W. J. (1957) Physiological changes in harvested asparagus (*Asparagus officinalis*) as related to temperature of holding. Ph.D. dissertation, Univ. Calif., Davis. p. 116.

Lurie, S. and Klein, J. D. (1991) Acquisition of low-temperature tolerance in tomatoes by exposure to high-temperature stress. *J. Amer. Soc. Hort. Sci.* **116**: 1007-1012.

Luyet, B. J. and Gehenio, P. M. (1940) The mechanism of injury and death by low temperature. A review. *Biodynamic* **3**: 33-103.

Lyons, J. M. (1973) Chilling injury in plants. *Annu. Rev. Plant. Physiol.* **24**: 445-466.

Lyons, J. M. and Breidenbach, R. W. (1979) Strategies for altering chilling sensitivity as a limiting factor in crop production. In *Stress Physiology in Crop Plants*, Mussell, H. and Staples, R. C., eds., John Wiley & Sins, New York, 179-196.

Lyons, J. M. and Breidenbach, R. W. (1987) Chilling injury. In *Postharvest Physiology of Vegetables*, Weichmann, J., ed., Marcel Dekker, New York, pp. 305-326.

Lyons, J. M. and Raison, J. K. (1970) Oxidative activity of mitochondria isolated from plant tissue sensitive and resistant to chilling injury. *Plant Physiol.* **45**: 386-389.

Lyons, J. M. and Rappaport, L. (1959) Effect of temperature on respiration and quality of Brussels sprouts during storage. *Proc. Amer. Soc. Hort. Sci.* **73**: 361-366.

Markhart, A. H. (1986) Chilling injury: A review of possible causes. *Hort. Science* **21**: 1329-1333.

Miller, E. V. (1946) Physiology of citrus in storage. *Bot. Rev.* **12**: 393-423.

Miller, E. V. and Heilman, A. S. (1952) Ascorbic acid and physiological breakdown in the fruits of the pineapple (*Ananas comosus* L. Merr.). *Science* **116**: 505-506.

Minorsky, P. V. (1985) An heuristic hypothesis of chilling injury in plants: a role for calcium as the primary physiological transducer of injury. *Plant Cell and Environment* **8**: 75-94.

Molisch, H. (1896) Das Erfriern von Pflanzen bei Temperaturen uber dem Eispunkt. *Sitzungsber. d. Kaiserlichen Akademie der Wissenschaften zu Wien. Bd.*, Abth. I, 105.

Molisch, H. (1897) *Untersuchungen uber das Erfieren der Pflanzen.* Fischer, Jena, pp. 1-73.

Morris, L. L. and Kader, A. A. (1975) Postharvest physiology of tomato fruits. *Univ. Calif., Davis, Veg. Crops Ser.* **171**: 36.

Morris, L. L. and Platenius, H. (1938) Low temperature injury to certain vegetables after harvest. *Proc. Amer. Soc. Hort. Sci.* **36**: 609-613.

Morris, L. L. and Watada, A. E. (1959) Stimulation of the respiration rates of cucumber and asparagus subsequent to removal from chilling conditions. *Amer. Soc. Hort. Sci.* **56th** Annual Meeting, Abst. 218.

Morris, L. L., Mann, L. K. and Lorenz, O. A. (1947) Chilling injury to certain vegetables. *Annu. Rept. Veg. Crops Dept., Univ. Calif., Davis.* Proj. 1175, Pt. 3: 418-447.

Nelson, R. (1926) Storage and transportational diseases of vegetables due to suboxidation. *Mich. Agric. Exp. Sta. Tech. Bull.* No. **81**.

Nishida, I. and Murata, N. (1996) Chilling sensitivity in plants and cyanobacteria: The crucial contribution of membrane lipids. *Annu. Rev. Plant Physiol. Plant Mol. Biol.* **47:** 541-568.

O'Reilly, H. J. (1947) Peach storage in modified atmosphere. *Proc. Amer. Soc. Hort. Sci.* **49:** 99-106

Palladine, W. (1899) L'influence des changements de temperatur sur la respiration des plantes. *Rev. Gen. Bot.* **110:** 241-257.

Pantanelli, E. (1919) Alterazioni del ricambio e della permeabilita cellulare a temperature prossime al congelamento. (Changes in metabolism and cellular permeability at temperatures near the freezing point.) *Accad. Dei Lincei, Rome. Scienze Fisiche Atti, Ser. 5.* **28:** 205-209.

Patterson, B. D. and Graham, D. (1977) Effect of chilling temperatures on the protoplasmic streaming of plants from different climates. *J. Exp. Bot.* **28:** 736-743.

Pearson, L. K. and Raper, H. S. (1927) The influence of temperature on the nature of the fat formed by living organisms. *Biochem. J.* **21:** 875-879.

Pentzer, W. T. and Heinze, P. H. (1954) Postharvest physiology of fruits and vegetables. *Annu. Rev. Plant Physiol.* **5:** 205-224.

Plank, R. (1938) Contribution to the theory of cold injury to fruit. *Food Res.* **3:** 175-187.

Plank, R. (1941) Zur Theorie der Kaltlagerkrankheiten von Fruchten. *Planta* **32:** 364-390.

Plank, R. and Schneider, E. (1928) *Versuche uber die kaltlagerung von obst und gemuse.* Zweiter Bericht. Beihefte Zur Zeitschrift fur die gesamte Kalte-Industrie. Gesellschaft Fur Klatewesen m.b.H Berlin, 37-40.

Platenius, H. (1942) Effect of temperature on the respiration rate and the respiratory quotient of some vegetables *Plant Physiol.* **17:** 179-197.

Platenius, H., Jamison, F. S. and Thompson, H. C. (1934) Studies on cold storage of vegetables. *New York (Cornell) Agr. Expt. Sta. Bull.* 602, p. 24.

Pratt, H. K., Morris, L. L. and Tucker, C. L. (1954) Temperature and lettuce losses. *Calif. Agric.* **8:** 14-15.

Raison J. K. (1973) The influence of temperature induced phase changes on the kinetics of respiratory and other membrane associated enzyme systems. *Bioenergetics* **4:** 285-309.

Raison, J. K. and Lyons, J. M. (1986) Chilling injury: a plea for uniform terminology. *Plant Cell & Environment* **9:** 685-686.

Raison, J. K., Lyons, J. M. and Thomson, W. W. (1971b) The influence of membranes on the temperature-induced changes in the kinetics of some respiratory enzymes of mitochondria. *Arch Biochem Biophys* **142:** 83-90.

Raison, J. K., Lyons, J. M., Mehlhorn, R. J. and Keith, A. D. (1971a) Temperature-induced phase changes in mitochondrial membranes detected by spin labeling. *J. Biol. Chem.* **246:** 4036-4040.

Rappaport, L. and Watada, A. E. (1958) Effect of temperature on artichoke quality. *Proc. Conf. Transporation of Perishables, Davis Calif., Calif. Terminal Railroads.*, pp. 142-146.

Rosa, J. T. (1927) Ripening and storage of tomatoes. *Proc. Amer. Soc. Hort. Sci.* **23:** 232-242.

Sabehat, A., Weiss, D. and Lurie, S. (1996) The correlation between heat-shock protein accumulation and persistence and chilling tolerance in tomato fruit. *Plant Physiol.* **110:** 536-541.

Sachs, J. (1864) Ueber die obere Tempertatur-Grenze der Vegetation. *Flora (Jena)* **47:** 5-12.

Sachs, J. (1865) Hardbuch der Experimental-Physiologie der Planzen. In *Handbuch der Physiologischen Botanik, Vol. 4*, W. Hofmeister, Leipzig, p. 514.

Saltveit, M. E. (1989) A kinetic examination of ion leakage from chilled tomato pericarp disks. *Acta Horticulturae* **258:** 617-622.

Saltveit, M. E. and Cabrera, R. M. (1987) Tomato fruit temperature before chilling influences ripening after chilling. *Hort. Science* **22:** 452-454.

Saltveit, M. E. and Morris, L. L. (1990) Overview of chilling injury of horticultural crops. In *Chilling Injury of Horticultural Crops.*, Chien Yi Wang, ed., CRC Press, pp. 3-15.

Saltveit, M.E. (1991) Prior temperature exposure affects subsequent chilling sensitivity. *Physiol. Plant.* **82:** 529-536.

Seible, D. (1939) Ein Betrag zur Frage der Kalteschaden an Pflanzen bei Temperaturen uber dem Gefrierpunkt. *Beitrage zur Biologie der Pflanzen* **26:** 289-330.

Sellschop, J.P.F. and Salmon, S. C. (1928) The influence of chilling, above the freezing point, on certain crop plants. *J. Agric. Res.* **37:** 315-338.

Sharom, M., Willemot, C. and Thompson, J. E. (1994) Chilling injury induced lipid phase changes in membranes of tomato fruit. *Plant Physiol.* **105:** 305-308.

Shichi, H. and Uritani, I. (1956) Alterations of metabolism in plants at various temperatures. Part I. Mechanisms of cold damage of sweet potato. *Bull. Agr. Chem. Soc. Japan.* **20:** 284-288.

Singer, S. J. (1971) Structure and function of biological membranes. L. I. Rothfield, ed., pp. 145-222. Academic Press, New York, p. 486.

Singer, S. J. and Nicolson, O. L. (1972) The fluid mosaic model of the structure of cell membranes. *Science* **175:** 720-731.

Smith, F. E. V. (1935) Low temperature investigation. *Annu. Rept. Dept. Agr. Jamaica for 1935.*

Smith, W. H. (1947) Extending the life of Victoria plum. *J. Pomol. Hort. Sci.* **23:** 92-98.

Spalding, D. H. and Reeder, W. R. (1972) Quality of 'Booth 8' and 'Lula' avocados stored in a controlled atmosphere or air. *Proc. Fla. State Hortic. Soc.* **85:** 337-340.

Spranger, E. (1941) Das erfrieren der pflanzen uber o" mit besonderer berucksichtigung der warmhauspflanzen. Gartenbausiss. **16:** 90-128.

Sweeney, D. (ed.) (1995) *Agriculture in the Middle Ages.* Univ. Penn. Press, Philadelphia, p. 371. ISBN 0-8122-1551-7.

Tewfik, S. and Scott, L. E. (1954) Vegetable storage. Respiration of vegetables as affected by post-harvest treatment. *J. Agric. Food Chem.* **2:** 415-417.

Vallejos, C. E., Lyons, J. M., Breidenbach, R. W. and Miller, M. F. (1983) Characterization of a differential low-temperature growth response in two species of Lycopersicon: the plastochron as a tool. *Planta* **159:** 487-496.

Wang, C. Y. (1990) *Chilling injury of horticultural crops.* CRC Press, Boca Raton, Florida, U.S.A. p. 313.

Wartenburg, H. (1929) Uber primare und sekundare Kalteryesistenz bei Bohnensippen. Eine Vorstudie zur Genetik der Kalteempfindlichkeit. *Planta Archiv fur Wissenschatliche Botanik* **7:** 347-381.

Wheaton, T. A. (1963) Physiological comparison of plants sensitive and insensitive to chilling temperatures. Ph.D. dissertation, Univ. Calif., Davis, p. 93.

Wheaton, T. A. and Morris, L. L. (1967) Modification of chilling sensitivity by temperature conditioning. *Proc. Amer. Soc. Hort. Sci.* **91:** 529-533.

Williams, W. J. (1936) The cold storage of vegetables. *Ice and Refrig.* XCI: **5.**

Wolfe, J. and Bagnall, D. J. (1980) Arrhenius plots-curves of straight lines? *Ann. Bot. London.* **45:** 485-488.

Woods, C. M., Reid, M. L. and Patterson, B. D. (1984) Response to chilling stress in plant cells. I. Changes in cyclosis and cytoplasmic structure. *Protoplasma* **121:** 8-16.

Wright, R. C., Pentzer, W. T., Whiteman, T. M. and Rose, D. H. (1931) Effect of various temperatures on storage and ripening of tomatoes. *U.S. Dept. Agric., Tech. Bull.* **268,** 34.

Wu, J. and Browse, J. (1995) Elevated levels of high-melting-point phosphatidylglycerols do not induce chilling sensitivity in an Arabidopsis mutant. *Plant Cell* **7:** 17-21.

Chapter 24

In Vitro Induced Haploids in Plant Genetics and Breeding

Hu Han and Guo Xiangrong
*State Key Laboratory of Plant Cell and Chromosome Engineering
Institute of Genetics, Chinese Academy of Sciences, Beijing 100101*

Introduction

Haploids have attracted the interest of plant physiologists, embryologists, geneticists and breeders since the first discovery of haploid plants in *Datura stramonium* as early as (Blakeslee *et al.*, 1922). In the beginning, haploid plants were regarded as a special biological phenomenon. Up to 1960s, the occurrence of haploids were reported in 71 species of angiosperms belonging to 39 genera and 14 families (Kimber & Riley, 1963). However, the low frequency of spontaneously arising haploid plants severely limited the utilization of haploids for crop improvement and genetic studies. In the last three decades, many efficient and simple techniques, especially *in vitro* culture have been developed to produce haploid plants in larger numbers. For instance, by using anther culture, haploids were induced in 247 species of angiosperms including some hybrids, which belong to 88 genera of 34 families (Maheshwari *et al.*, 1983). Meanwhile, chromosome elimination (the bulbosum technique), and *in vitro* culture of unfertilized ovaries and ovules have been proved to be an efficient means of haploid induction in some species. These advancements opened up the way for studying and utilizing the haploids in higher plants.

This paper mainly discusses these advancements of haploids in higher plants in terms of its origin, genetics and application in breeding.

1. Origin of Haploids

There are two different generations in the life cycle of higher plants, namely the spore-producing asexual generation and the gamete-produc sexual generation producing gametes. The sporophyte in the asexual generation is diploid with two chromosome sets derived from both parents. Before the formation of the spores through meiosis, the zygotic (diploid) chromosome number is reduced to the gametic (haploid) number: the characteristic for the gamete or the haploid phase of the life cycle. This means that haploids of higher plants are sporophyte with gametic chromosomes.

The haploids of higher plants are naturally produced via abnormal fertilization, therefore they are rarely seen, and the frequency of haploids arising is very low. Since the first report of haploid plants in the early 1920s, many efforts have been made to induce haploids of higher plants, which could be roughly classified into two categories: *in vivo* induction of haploids by various physical, chemical or biological stimulants, and *in vitro* culture. The latter is far more efficient than the former.

From the life cycle point of view, there are three pathways to induce haploid sporophyte of higher plants (Hu Han, 1997) :

I.   Haploid sporophyte originating from meiotic spore by means of pollen (anther) or/and, ovary and ovule culture.
II. Haploid sporophyte originating from zygote due to chromosome elimination.
III. Haploid sporophyte originating from male and female gametes, their precursor cells and gametophytes by the isolation and manipulation of reproductive cells and protoplasts. (Yang and Zhou, 1992).

Since it is most difficult to obtain haploids using the third pathway, this technique has not been applied to cereal crops up to now. This technique of the experimental plant reproductive biology shows the great potential in providing new means for biotechnology and the way eventually to permit direct reproductive cell engineering. Meanwhile, it serves to deepen our understanding in the control of reproductive processes.

2.   Genetics of Haploids

Pollen-derived haploids are an ideal material for investigating the important genetic questions concerning heredity, variation, recombination and expression, which possess all the basis of the mechanism of formation of genetic characteristics as haploids possess only one set of chromosomes with special genetic characters. The main genetic characteristics of *in vitro* induced haploid plants are:

2.1. Genetic (chromosomal) stability and variability

From pollen-derived plants, stable, homozygous diploid strains can be obtained via chromosome doubling of haploids. For many years, we have been studying karyotypic analysis of somatic cells and pollen mother cells

(PMCs) derived from pollen plants. By the investigation of chromosome configuration of root tip cell, PMCs and genetic analysis of pollen plants, we found that genetic (chromosomal) stability and variability co-existed in anther culture during the same process (Hu, 1983). It is an important genetic feature of pollen-derived plants.

The genetic analysis and cytological observations were carried out in unselected populations of pollen plants. The results obtained from such populations in wheat (Hu *et al.*, 1978, 1980; Yang *et al.*, 1978), rice (Chen and Li, 1978), maize (Gu *et al.*, 1986) and tobacco (Xu *et al.*, 1980) over a period of several years indicated that about 90% of the diploid lines were genetically uniform. These results suggested that, although diploids, heteroploids as well as haploids occur through anther culture, it might be considered that mainly homozygous recombinant lines (90%) could be produced in anther culture. The co-existing variants make up only about 10% of pollen plants.

2.2. Microspore clonal variation

Variability of the chromosome number and structure in the plant cells regenerated *in vitro* is a common phenomenon (D'Amato, 1978). The same phenomenon has also been observed in pollen plants. The technique to induce genetic variation using cell culture has been termed somaclonal and gametoclonal variation (Sharp *et al.*, 1984). But pollen plants derived from the microspore clones are not gametes, and should be termed microsporeclonal variation. The recovery of aneuploid plants using tissue culture may appear as an unwanted aspect of instability during *in vitro* culture, nevertheless, there are many analytical and practical applications in aneuploids; pollen clonal variation is easily produced during anther culture *in vitro*. This kind of variation possesses universality (Hu *et al.*, 1978; Hu *et al.*, 1980). Types of variation are not only in number but also in structure of chromosome (Hu *et al.*, 1989), the main mechanism of chromosome structure variation is in breakage and reunion. These principles paved new ways for producing translocation lines and for gene-mapping (Hu *et al.*, 1999).

2.3. Genetic genotypes fully expressed at homozygous plant level

Since pollen grains from F_1 hybrids are heterozygous, different gene combinations of both parents of a cross occur in every grain. As no sexual processes are involved in *in vitro* anther culture, the competition of gametic types in fertilization is avoided. Different gametic types, including recombinants and variants, may be fully kept which could be in one generation fully expressed and rapidly stabled at pollen plant level.

Meanwhile, increasing the efficiency of selection can greatly shorten the cycle of investigation and breeding (Hu, 1996).

3.   Application of *In Vitro*-Induced Haploids

3.1. The possibility of using DH plants in crop improvement

The biometrical studies with DH lines (Snape, 1989) indicated that DH systems have the unique genetic property of allowing completely homozygous lines to be developed from heterozygous parents in a single generation. In self-pollinating species, such as wheat, barley and rice, this property can be used to increase the efficiency of cultivar production. Time can be saved in getting selected material ready for commercialization. There is also an increase in selection efficiency relative to conventional practices because of an increase in additive genetic variation, an absence of dominance variation and within-family segregation, as well as a decrease in environmental variation effects through greater replication possibilities.

3.2. Developing and releasing new varieties

In 1979, Ciba-Geigy seeds licensed a barley cultivar, Mingo, the first licensed cultivar of *in vitro*-induced haploid based on the chromosome elimination, bulbosum method (Kasha *et al.*, 1980). Since the anther culture technique was applied to breeding programs in China, great progress has been achieved. Hu (1985) described the achievement of pollen haploid breeding. Using anther culture, combined with the conventional breeding method, Chinese scientists have succeeded in developing and releasing a number of cultivars in wheat, rice, maize, pepper and fruits etc., which had good agronomic characters of high yield, fine adaptation, disease and drought resistance etc.   Meanwhile, many Chinese breeding units have so far established the procedure and system of pollen haploid breeding as a routine method of crop improvement.

3.3. Chromosome engineering in Triticeae using pollen-derived plants

Hu Han and his group combined chromosome engineering with anther culture and modified identification methods to transfer the desirable chromosomes (genes) into cultivars and thus to create new strains of wheat. As chromosome substitution, addition and translocation lines [especially the non-Robertsonian translocation lines (E. M. Wang *et al.*, 1998)] have been obtained, this method has significant potential as a tool in germplasm enhancement (Hu *et al.*, 1992).

3.4. Genetic manipulation

Microspore culture makes selection on the single cell level possible, and offers new prospects for genetic manipulation, e.g., mutagenesis and transformation.   Direct gene transfer by microinjection of isolated multicellular pollen embryoids offer the possibility of transgenic plant formation in all cereals (Potrykus *et al.*, 1985; Potrykus, 1988).   It can be expected that new and interesting information on the genetic manipulation using microspore culture of cereals will be obtained in the near future.

3.5. Usage of DH populations

A new field in the usage of haploids is molecular genome identification, particularly for QTL analysis.   DH populations are an important tool to obtain reproducible DNA polymorphisms in barley (Heun *et al.*, 1991) and rice (McMouch *et al.*, 1992, Xu *et al.*, 1994) etc.   One method for the RFLP map of these cereals was based on DH populations.

Conclusion

In relation to the life cycle, there are three pathways to induce haploid sporophytes of higher plants:

I.   From meiotic spores by anther (pollen) culture and unpollinated ovary (ovule) culture;
II.   From reduced zygotes by chromosome elimination; and
III. From male and female gametes by isolation and manipulation of reproductive cells and protoplast.

Among them, approaches I and II have been successfully developed and utilized in production.   This progress has paved the way for fundamental studies and practical utilization of DHs.

Haploids possessing only one set of chromosomes have special genetic characteristics: I. Genetic (chromosomal) stability and variability; II. Microspore clonal variation; and III. Genetic genotypes fully expressed at homozygous plant level.   It is an ideal material for investigating the important genetic fundamental questions concerning heredity, variation, recombination and expression.

In practice, DHs combined with different breeding methods, such as conventional breeding, chromosome engineering, mutagenesis and wide hybridization, will efficiently recover and screen a lot of recombinants and variants. Meanwhile, it can also create new types which are usually difficult to obtain by conventional methods for wheat improvement.

References

Blakeslee, A. F., Belling, J., Farnham, M. E. and Begner, A. D. (1922) A haploid mutant in the jimson weed Datura stramonium. *Science* **55**: 1433.

Chen, Y. and Li, L. (1978) Investigation and utilization of pollen derived haploid plants in rice and wheat. In *Proc. of Sino-Australian Symposium on Plant Tissue Culture*, pp. 199-212, Science Press, Beijing.

D'Amato, F. (1978) Chromosome number variation in cultured cells and regenerated plants. In *Frontiers of Plant Tissue Culture*, Thorpe, T. A., ed., pp. 287-295, Univ. Calgary Press, Calgary.

Gu, M. G. (1986) Cytogenetic stability and variability of calli and cell clones originated from maize pollen and their regenerated plants. *In vitro Haploids in Higher Plants*, Hu, H. and Yang, H. Y., eds., China Academic Publishers, Beijing and Springer Verlag, Berlin, Heidelberg, New York, Tokyo, pp. 79-90.

Han, H. (1997) *In vitro* induced haploids in wheat. *In vitro* Haploid Production in Higher Plants, Jain, S. M. , Sopory, S. K. and Veilleux, R. E., eds., Vol. 4, 73-97.

Heun, M., Kennedy, A. E., Anderson, J. A., Lapitan, N. L. V., Sorrelis, M. E. and Tanksley, S. D. (1991) Construction of a restriction fragment length polymorphism map for barley. *Genome* **34**: 437-447.

Hu, H. (1983) Genetic Stability and Variability of Pollen-Derived Plants. In *Plant Cell Culture in Crop Improvement*, Sen, S. K. and Giles, K. L., eds., Pleunum Press, New York and London, pp. 145-157.

Hu, H. (1985) Use of Haploids in Crop Improvement. In *Biotechnology in International Agricultural Research*, International Rice Research Institute, Manila, Philippines, pp. 75-84.

Hu, H. (1992) Germplasm enhancement by anther culture in Triticeae. *Hereditas* **116**: 151-154.

Hu, H. (1996) Chromosome engineering in the Triticeae using pollen-derived plants (CETPP). *In vitro* Haploid Production in Higher Plants, Jain, S. M., Sopory, S. K. and Veilleux, R. E., eds. Vol. 2, pp. 203-223.

Hu, H., His, T. and Chia, S. (1978) Chromosome variation of somatic cell of pollen calli and plants in wheat (*Triticum aestivum* L.). *Scientia Sinica* **23**: 905-912.

Hu, H., His, T. Y. and Ouyang, J. W. (1980) Chromosome variation of pollen mother cells of pollen-derived plants in wheat (*Triticum aestivum* L.). *Scientia Sinica* **23**: 905-912.

Hu, H., Xi, Z. Y., Jia, S. E. (1978) Chromosome variation of somatic cells of pollen calli and plants in wheat (Triticum aestivum L.). *Acta Genetica Sinica* **5**: 23-30.

Hu, H., Xi, Z. Y., Ouyang, J. W. *et al.* (1980) Chromosome variation of pollen mother cell of pollen derived plants in wheat (Triticum aestivum L.). *Sci. Sin. Ser B.* **23**: 905-912.

Hu, H., Ziying, X. and Yuezhi, T. (1989) Gametoclonal variation in Triticeae. In *Plant Chromosome Research 1987, Proc 1st Sino-Jpn Symp Plant Chromosomes*, pp. 231-234.

Hu, H., Zhang, X. Q., Zhang, W. *et al.* (1999) Chromosome engineering of pollen wheat. *Chinese Science Bulletin*, Vol. 44. **1**: 6-11.

Kasha, K. J. and Reinbergs, E. (1980) Achievements with Haploids in Barley Research and Breeding. In *Plant Genome*, Davies and Hopwood, eds. The John Innes Charity, pp. 215-230.

Kimber, G. and Riley, R. (1963) Haploid angiosperms. *Bot. Rev.* **29(4):** 480-531.

Maheshwari, S. C., Raishid, A. and Tyagi, A. K. (1983) Anther pollen culture for production of haploids and their utility. *IAPTC Newsletter* **41:** 2-7.

McMouch, S. R. and Tanksley, S. D. (1992) Development and use of RFLP in rice breeding and genetics. *Rice Biotechnology* 109-133.

Potrykus, I. (1988) Direct gene transfer to plants. In *Application of Plant Cell Tissue Culture (Ciba Foundation Symposium 137)* John Wiley Sons, New York, pp. 144-162.

Potrykus, I., Paszkowski, J., Saul, M., Kruger-Lebus, S., Muller, T., Schocher, R., Negrutiu, I., Kunzler, P. and Shillito, R. (1985) Direct gene transfer to protoplast: an efficient and generally applicable method for stable alteration of plant genomes. In *Plant Genetics,* Feeling, M., ed., Liss, New York, pp. 181-199.

Sharp, W. R., Evans, D. A. and Ammirato, P. V. (1984) Plant genetic engineering: Designing crops to meet food industry specifications. *Food Technology,* pp. 112-119.

Snape, J. W. (1989) Doubled haploid breeding: theoretical basis and practical applications. In *Review of Advances in Plant Biotechnology,* Mujeeb-Kazi, A. and Sitch, L. A., eds., 1985-1989, Rice Research Institute, Manila, pp. 19-30.

Wang, E. M, Xing, H. Y, Wen, Y. X. *et al.* (1998) Molecular and biochemical characterization of a non-Robertsonian wheat-rye chromosome translocation line. *Crop Sci.* **38:** 1076-1080.

Xu, H., Ai, S., Chen, Z. and Jia, X. (1980) Report on heredity and viability of progenies of pollen-derived tobacco. *Zhongguo Yancao* **1:** 8-10.

Xu, J. C., Zhu, L. H., Chen, Y., Lu, C. F. and Cai, H. W. (1994) Construction of a rice molecular linkage map using a doubled haploid population. *Acta Genet Sin* **21:** 205-214.

Yang, D. and Gao, J. (1979) Observation of genetic stability and variability of the progenies of pollen-derived haploid plants in wheat. *Ann. Rep. Inst. Agric. Sci. Shandong, Chang Wei* **1978:** 23-32.

Yang, H. Y. and Zhou, C. (1992) Experimental plant reproductive biology and reproductive cell manipulation in higher plants: Now and the future. *Amer. J. Bot.* **79(3):** 354-363.

INDEX

A

A. radiobacter 396
A. rhizogenes 396
A. tumefaciens 396
Abe, S. 248
Absorption kinetics 416
Acetate/mevalonate pathway 141,
153
Acetoacetyl-CoA 144
Acetobacter 186
Acetobacter xylinum 200, 204, 205,
210
Acetylated cellobiose 181
Acetyl-CoA 144, 153
Acidic subunit 242, 244
Active transport 411
Addition line 452
Additive genetic variation 452
Adenosine diphosphate D-glucose
(ADPG) 68, 69
Adenosine diphosphoglucose 223
Agrobacterium *394*
Agrobacterium-mediated gene
transfer 402
Agrocin 84 397
Agros, P. 245
Akazawa, Takashi 223
Akhtar 114, 116
Albersheim 53
Albumin 240, 241
2S albumin 246
2S albumins 255-260, 263
Alcohol-scluble protein 240, 241
Aleuron 240
grain 240

tissue 247
Algae 146, 150
Algal cellulose 177, 178, 183
Allosteric mutants 228
Alt, Juliane 282
Aluminum 10
Aminoacyl-tRNA synthetase 97
1-aminocyclopropane-1-carboxylic
acid (ACC) 17, 37
acid synthase (ACC synthase) 39,
40
Aminoethoxyvinylglycine (AVG)
17, 23
2-aminoisobutyric acid 31, 34-37
5-aminolevulinate biosynthesis 95
bacteria 96, 97, 103
glutamate-1-semialdehyde 102,
103
glutamate-1-semialdehyde
aminotransferase 103
glutamyl-tRNA^{Glu} reductase 100
mammals 77
plants 83, 84, 88
tRNA^{Glu} 98
5-aminolevulinate synthase 100
5-aminolevulinic acid dehydratase
110
Aminotransferases 50
Amorphous cellulose 185
Amylases 218
Amylopectin 217, 222
synthetic pathway 230, 231
Amylose 217, 219, 222
synthetic pathway 230, 231
Amylose-extender mutant 232
Aneuploid 451
Angiosperm 449

D

G

H

N

Q

T